# MEDICAL MOLECULAR GENETICS

# Medical Molecular Genetics

## Patricia A. Hoffee, Ph.D.

*Professor of Molecular Genetics and Biochemistry*

*Department of Molecular Genetics and Biochemistry*

*University of Pittsburgh School of Medicine*

*Pittsburgh, Pennsylvania*

**Fence Creek Publishing**

**Madison, Connecticut**

Typesetter: Pagesetters, Brattleboro, VT
Printer: Port City Press, Baltimore, MD
Illustrations by Visible Productions, Fort Collins, CO
Distributors:

**United States and Canada**

Blackwell Science, Inc.
Commerce Place
350 Main Street
Malden, MA 02148
Telephone orders: 800-215-1000 or 781-388-8250
Fax orders: 781-388-8270

**Australia**

Blackwell Science, PTY LTD.
54 University Street
Carlton, Victoria 3053
Telephone orders: 61-39-347-0300
Fax orders: 61-39-347-5001

**Outside North America and Australia**

Blackwell Science, LTD.
c/o Marston Book Service, LTD.
P.O. Box 269
Abingdon Oxon, OX 14 4XN England
Telephone orders: 44-1-235-465500
Fax orders: 44-1-235-465555

The authors of this book would like it to be known that it is the decision of the publishers to include an apostrophe "s" in syndromic names, despite the fact that this practice is often no longer used.

2 3 4 5 6 7 8 9 10

# TABLE OF CONTENTS

# CONTRIBUTORS

**Kenneth L. Garver, M.D., Ph.D.**
Adjunct Professor of Human Genetics
Department of Human Genetics
Emeritus Clinical Professor of Pediatrics
University of Pittsburgh School of Medicine
Consultant, Department of Human Genetics
Allegheny General Hospital
Pittsburgh, Pennsylvania

**W. Allen Hogge, M.D.**
Professor of Obstetrics, Gynecology, Repro-
    ductive Sciences, and Human Genetics
Department of Obstetrics, Gynecology, and
    Reproductive Sciences
University of Pittsburgh School of Medicine
Medical Director, Department of Genetics
Magee-Womens Hospital
Pittsburgh, Pennsylvania

**Mary L. Marazita, Ph.D.**
Professor of Human Genetics and Oral and
    Maxillofacial Surgery
Department of Oral and Maxillofacial Surgery
Department of Human Genetics
Director, Cleft Palate-Craniofacial Center
University of Pittsburgh School of Medicine
Pittsburgh, Pennsylvania

**Patricia A. Mowery-Rushton, Ph.D.**
Assistant Professor of Pediatrics
Department of Pediatrics
University of Pittsburgh School of Medicine
Laboratory Director of Cytogenetics
Division of Medical Genetics
Children's Hospital of Pittsburgh
Pittsburgh, Pennsylvania

**Urvashi Surti, Ph.D.**
Associate Professor of Pathology and Human Genetics
Department of Pathology
Department of Human Genetics
University of Pittsburgh School of Medicine
Director, Cytogenetic Laboratory
Magee-Womens Hospital
Pittsburgh, Pennsylvania

# PREFACE

Traditionally, genetics has been taught to medical students piecemeal—partly in microbiology, partly in biochemistry, sometimes as part of a pathology course, or as a course in medicine. This separation of genetics into basic science and clinical science resulted from the limited knowledge of human genetics and the increasing knowledge in the field of molecular biology of microbial cells. Six years ago, the University of Pittsburgh School of Medicine made a dramatic change in their curriculum, creating an integrated rather than a departmental teaching program. One of the courses resulting from this change covered molecular and human genetics, the goal of which was the integration of all of the genetics taught in various courses in the first 2 years of medical school into a single 3-week course! The first attempt to organize this course resulted in essentially an end-to-end ligation of molecular genetics and human genetics, each comprising about half the allotted time. The course was unsuccessful and needed to be reorganized. The genetics professors decided that in today's world genetics *is* molecular, whether one is dealing with bacteria, mice, or humans.

Our objective in this text is to relay that message to medical students taking a course in medical genetics, to house officers who suddenly find themselves at a loss with the new genetics they encounter, and to physicians who want and need to apply current methodologies in genetics to the diagnosis, screening, and treatment of various human diseases. The message in this text and the materials presented all point to the dramatic changes occurring in the field of human genetics. Understanding the molecular basis of human genetics has become essential in diagnosing and treating human disease. Our goal is to present to the reader the basic principles needed to use the discoveries of this remarkable field. Each day a new inherited disease gene is cloned, and by the year 2005, it is predicted that the entire sequence of the human genome of 3 billion base pairs will be known. Can cures by gene therapy be far behind?

*Medical Molecular Genetics* covers the basic structure, properties, and functions of nucleic acids; mutations and their consequence on the function of proteins; DNA repair processes and their relationship to human disease; chromosome structure and the basis of inheritance; modern methods for detecting defective genes; and the basis of positional cloning and how it is currently being used to map disease genes. Other topics covered include regulation of gene expression, chromosomal abnormalities and how they lead to disease, how genes are inherited, and how genes are mapped. The recent information on the role of unstable repeats of trinucleotides in diseases such as fragile X syndrome and Huntington's chorea are covered, as are the various mitochondrial disorders. The clinical aspects of genetics and discussions of syndromes and dysmorphology, prenatal diagnosis, and genetic counseling are also presented. The most up-to-date information on cancer genetics includes the role of oncogenes and tumor suppressor genes in tumorigenesis and the use of molecular tests to diagnose cancer genes. In a final chapter, the current state of gene therapy and the potential of this therapy for treating human disease are presented.

Each chapter contains a section with practice questions, and several chapters also present problem sets to help the reader deal with actual experimental data and their interpretation. Wherever possible, a clinical case pertinent to the material in the text is included. Tables and illustrations are used to clarify and enhance the written material to present the reader with a sufficient, but not overly detailed, view of the information.

Our hope is that this text will provide readers with a level of understanding of human genetics that will enable them not only to use the current knowledge of the field in the practice of medicine today but also will enable them to understand future findings and their impact on the treatment of human disease.

Patricia A. Hoffee

# ACKNOWLEDGMENTS

I would like to thank:

All of the contributors who brought their expertise to this text: Dr. Allen Hogge, Dr. Ken Garver, Dr. Mary Marazita, Dr. Urvashi Surti, and Dr. Patricia Mowery-Rushton.

Dr. Eric Hoffman, Dr. Salem Khan, and Dr. Urvashi Surti for reviewing several chapters and for their helpful and constructive comments.

Dr. Stephen Phillips for his generous contribution of some of the figures.

All of my colleagues for their passion for science. I hope their enthusiasm for human genetics and its potential for understanding and treating human disease is felt by all who read this textbook.

The people of Fence Creek Publishing Company, particularly Matt Harris, who worked so hard to make this project a reality. A special thanks to the editorial staff for their outstanding job, especially Jane Edwards for her patience, persistence, and professionalism in orchestrating the editing and completion of this text.

*I would like to dedicate this book to four special women whom I admire and who have shaped my life and its commitment to science and genetics. My mother, Alberta H. Hoffee, taught me dignity and graciousness. I never really appreciated her until she was gone. My former colleague and friend, Dr. Ora Mendelson Rosen, was a brilliant teacher and scientist who died too young. She taught me the patience to ask the right questions and instilled in me the persistence to find the correct answers. My friend, Nancy Block, is a talented artist and painter who sees the beauty in the world and taught me to appreciate so much. My friend, Dr. Patricia J. Jarrett, is a respected physician and psychiatrist who constantly strives for perfection. She gave me strength when I needed it the most. Thank you all for being there for me.*

# INTRODUCTION

*Medical Molecular Genetics* is one of ten titles in the *Integrated Medical Sciences (IMS) Series* from Fence Creek Publishing. These books have been designed as course supplements and aids for board review for first- and second-year medical students. Rather than focusing on the individual basic science disciplines, the books in the *IMS Series* have been designed to highlight the points of integration between the sciences, including clinical correlations where appropriate. Each chapter begins with a clinical case, the resolution of which requires the application of basic science concepts to clinical problems. Extensive use of margin notes, figures, tables, and questions illuminates core biomedical concepts with which medical students often have difficulty.

Each book in the *IMS Series* shares common features and formats. Attempts have been made to present difficult concepts in a brief and focused format and to provide a pedagogical aid that facilitates both knowledge acquisition and also review.

Given the long gestation period necessary to publish a book, it is often impossible for publishers to keep pace with the changes and advances that occur so rapidly. However, the authors and the publisher recognize the need to have access to the most current information and are committed to keeping *Medical Molecular Genetics* as up to date as possible between editions. As the field of genetics evolves, updates to this text may be posted on our web site periodically at http://www.fencecreek.com.

We hope that the student finds the format and the text material relevant, interesting, and challenging. The authors, as well as the Fence Creek staff, welcome your comments and suggestions for use in future editions.

# INTRODUCTION: THE PATHWAY TO MEDICAL MOLECULAR GENETICS

## PATHWAY TO MEDICAL MOLECULAR GENETICS

Genetics is the study of the mechanism of heredity, which is defined as the transmission of characteristics from parents to their offspring. Genetics also is the study of the variation of an organism's characteristics from generation to generation, a variation that forms the basis of evolution. Scientists have different opinions on exactly when the science of genetics began; some believe it was in 1839 with the first description of a cell by Matthias Schleiden and Theodor Schwann. Most scientists, however, agree that the studies of Johann Gregor Mendel on the garden pea in 1865 launched the field of genetics. During the past 150 years, many amazing discoveries in genetics have occurred. A chronologic list of some of the most important discoveries is presented in Table 1-1. Although far from inclusive, Table 1-1 highlights the advances in genetics that are considered to be the most important in the field.

The history of genetics can be viewed much like the painting career of an accomplished artist. First came a classical period, with its fine attention to detail and its description of scientific phenomena. Then, the middle or molecular period dealt with atoms and molecules as the building blocks needed to create the final picture. Finally, the current period has produced the discovery of recombinant DNA technology and the beginning of molecular human genetics. For the first time, scientists are beginning to have a complete blueprint of the human genome, which hopefully will lead to dramatic advances in the understanding and treatment of inherited human disorders.

### Classical Period

Our basic understanding of heredity is attributed to the work of Mendel, who was born in 1822 and entered an Augustinian monastery as a boy. By the age of 22 years, he had become a monk and begun his studies at Vienna University on natural science, a discipline he later taught when he returned to the monastery at Brno. In 1854, Mendel initiated his work on inheritance in the garden pea, a study that was published in 1865. It is widely known that Mendel's work was neither understood nor appreciated by other geneticists of the day. The significance of his findings was rediscovered in 1900,

**TABLE 1-1** ▶
*Important Advances and Discoveries in Genetics*

| Year | Scientist | Description of the Findings |
|------|-----------|----------------------------|
| **Classical Period** | | |
| 1839 | Schleiden and Schwann | Formulated the "cell theory" |
| 1860 | Weismann | Identified germ cells |
| 1865 | Mendel | Described inheritance patterns and introduced the concept of a gene |
| 1871 | Miescher | Described organic acids from nuclei (nucleic acids) |
| 1877 | Fleming | Described chromosomes (colored bodies) in mitosis |
| 1897 | Buchner | Described enzymes |
| 1903 | Sutton | Localized genes to chromosomes |
| 1908 | Hardy-Weinberg | Established laws to determine how genes behave in populations |
| 1909 | Garrod | Described the first "inborn errors of metabolism" |
| 1909 | Johannsen | Made the distinction between genotype and phenotype |
| 1911 | Wilson | Associated color blindness with the X chromosome |
| 1915 | Morgan | Formulated the concept that genes are inherited on a chromosome and established that map distance is a reflection of recombination frequency |
| 1927 | Muller | Induced mutations in flies by x-rays |
| 1941 | Beadle and Tatum | Formulated the one gene–one enzyme hypothesis |
| **Modern Molecular Period** | | |
| 1943 | Luria and Delbruck | Established the spontaneous nature of mutation in bacteria |
| 1944 | Avery, MacLeod, and McCarty | Identified DNA as the transforming (genetic) material |
| 1946 | Lederberg and Tatum | Described recombination in bacteria (conjugation) |
| 1947 | Monod | Performed the initial work on enzyme induction |
| 1949 | Pauling | Associated a defect in hemoglobin with sickle cell anemia |
| 1949 | Barr | Identified Barr bodies as sex chromatin |
| 1950 | Chargraff | Showed that DNA is made up of adenine, guanine, cytosine, thymine and deoxyribose |
| 1950 | Lwoff | Established that bacterial viruses can exist in cells in a latent form |
| 1951 | Hershey and Chase | Identified DNA as the genetic material of phage |
| 1952 | Zinder and Lederberg | Showed that gene transfer in bacteria can be mediated by virus vectors (the process of transduction) |
| 1953 | Hayes | Discovered plasmids (F factor) in bacteria |
| 1953 | Watson and Crick | Established the structure of DNA as a double helix |
| 1953 | Howard and Pelc | Described the stages of the cell cycle in mammals |
| 1953 | McClintock | Discovered transposable elements |
| 1955 | Wollman and Jacob | Utilized Hfr mating in *Escherichia coli* for mapping genes |
| 1956 | Sanger | Determined the amino acid sequence of insulin |
| 1956 | Tjio and Levan | Established that humans have 46 chromosomes |
| 1956 | Kornberg | Identified DNA polymerase activity in vitro |
| 1956 | Ingram | Determined the amino acid sequence of hemoglobin and showed that the sickle cell mutation was a result of a single amino acid change |
| 1958 | Puck | Established procedures to clone single mammalian cells |
| 1957 | Benzer | Introduced the concept of a cistron, a muton and a recon |
| 1958 | Pontecorvo | Helped establish the field of somatic cell genetics |
| 1958 | Zamecnik | Identified ribosomes |
| 1958 | Messelson and Stahl | Established that DNA is replicated semi-conservatively |
| 1959 | LeJeune | Described trisomy 21 in Down's syndrome as the first chromosomal abnormality |
| 1960 | Doty and Marmur | Developed methods to study nucleic acid hybridization |

| Year | Scientist | Description of the Findings |
|------|-----------|----------------------------|
| 1960 | Jacob and Monod | Formulated the operon theory |
| 1960 | Nowell | Identified the Philadelphia chromosome |
| 1960 | Ames | Described genetic control of histidine biosynthesis in *E. coli* |
| 1961 | Yanofsky | Established that single base changes in the DNA lead to single amino acid changes in the protein |
| 1961 | Hurwitz and Ochoa | Identified RNA polymerase activity in vitro |
| 1961 | Spiegelman and Brenner | Discovered messenger RNA |
| 1961 | Nirenberg | Elucidated the genetic code |
| 1961 | Barski and Euphressi | Described the formation of somatic cell hybrids |
| 1961 | Lyon | Formulated the X inactivation hypothesis |
| 1962 | Okada | Showed that Sendai virus increases cell–cell fusion |
| 1964 | Littlefield | Adapted the HAT medium for selecting somatic cell hybrids |
| 1965 | Englesberg | Formulated the concept of regulation by positive control |
| 1965 | Sutherland | Established that cyclic AMP is the second messenger |
| 1965 | Hayflick | Described senescence in cultured cells and formulated a theory for aging |
| 1965 | Khorana | Established the structure of transfer RNA |
| 1966 | Breg and Steele | Made the first prenatal diagnosis using chromosome analysis |
| 1967 | Weiss and Green | Developed the somatic cell hybrid procedure for mapping human genes |
| 1968 | Cleaver | Showed that xeroderma pigmentosum is a defect in DNA repair |
| 1969 | Harris | Described the suppression of malignancy in a fusion between a normal cell and a malignant cell |

### Recombinant DNA Era and Human Molecular Genetics

| Year | Scientist | Description of the Findings |
|------|-----------|----------------------------|
| 1969 | Gall and Pardue | Described the procedure of in situ hybridization |
| 1970 | Nathans and Smith | Established that restriction endonucleases create sticky DNA ends that can be recombined to form chimeric molecules |
| 1970 | Caspersson | Discovered chromosome banding techniques |
| 1971 | Knudson | Formulated the two-hit hypothesis to explain retinoblastoma inheritance |
| 1972 | Temin and Baltimore | Discovered reverse transcriptase (RNA to DNA) |
| 1972 | Kerr | Described programmed cell death (apoptosis) |
| 1973 | Boyer, Cohen, and Berg | Formed recombinant DNA molecules in vitro |
| 1974 | Kornberg | Described the structure of the nucleosome |
| 1975 | Southern | Developed the Southern blotting technique |
| 1976 | Leder | Cloned the mouse β-globin gene |
| 1977 | Sanger, Maxam, and Gilbert | Developed methods for sequencing DNA |
| 1977 | Chambon, Jeffreys, Flavell, Tilghman, and Leder | Discovered introns and exons and RNA splicing in mammalian cells |
| 1978 | Bishop and Varmus | Discovered oncogenes |
| 1979 | Lane, Crawford, and Levine | Discovered p53 |
| 1979 | Weinberg | Showed that oncogenes can be transferred by DNA into cells |
| 1981 | Anderson | Sequenced human mitochondrial DNA |
| 1981 | Ruddle, Palmiter, and Brinster | Created the first transgenic mice and germline transmission of a transgene |
| 1982 | Weinberg | Identified tumor suppressor genes |
| 1985 | Mullis | Developed the polymerase chain reaction (PCR) |
| 1986 | Friend | Cloned the retinoblastoma (Rb) gene |
| 1987 | Kunkel | Cloned the Duchenne muscular dystrophy gene using positional cloning |
| 1988 | Blackburn | Isolated the first telomerase |
| 1989 | Rommens and Riorden | Cloned the cystic fibrosis gene |
| 1990 | Nurse, Pines, and Hunter | Established that the cell cycle is controlled by cyclins and cyclin-dependent kinases |

| Year | Scientist | Description of the Findings |
|------|-----------|----------------------------|
| 1990 | Blaese and Anderson | Carried out the first gene therapy trial using a retrovirus vector carrying the adenosine deaminase gene to treat two children with severe combined immunodeficiency disease |
| 1992 | Multiple contributors | Developed the beginnings of the complete human gene map |
| 1993 | Vogelstein, Modrich, and Kolodner | Showed that colon cancer is caused by defects in DNA mismatch repair |
| 1995 | Multiple contributors | Continued isolation of an increasing number of human genes by the positional candidate gene approach |
| 1996 | International effort | Established the first expressed sequence tag (EST) map and the first sequence tagged site (STS) map of the human genome |

when Mendel's results were verified by the work of Hugo deVries in Holland, Erich Tschermak in the United States, and Karl Correns in Germany.

Three basic principles arose from Mendel's work on the garden pea. His first principle clearly indicated that *characteristics were inherited as units* and not blended. Although characteristics might not be expressed in the offspring of the first generation, they could appear unchanged in the second generation. The second principle—*the principle of segregation*—presented the concept of two alleles of a gene, with each allele segregating separately in the gametes, as illustrated in Figure 1-1. Mendel's third principle—*the law of independent assortment*—stated that different unit pairs assorted randomly and independently of one another in the gametes. Mendel's understanding of his findings is nicely illustrated by the following quotation from a letter that he wrote to Carl Nageli in 1867 [1]:

> The course of development consists simply of this: That in each generation the two parental characteristics appear, separated and unchanged, and there is nothing to indicate that one of them has either inherited or taken over anything from the other.
>
> My experiments with single characters all lead to the same result: That from the seeds of the hybrids, plants are obtained, half of which in turn carry the hybrid character (Aa), the other half, however receive the parental characters A and a in equal amounts. Thus, on the average among four plants, two have the hybrid character Aa,

**FIGURE 1-1** ▶

*Mendel's Principle of Segregation. Each gene has two alleles. One parent has two copies of the dominant allele A and is denoted AA, whereas the other parent has two copies of the recessive allele a and is denoted aa. One copy of each allele is passed to the gametes at meiosis; at fertilization, one allele from each parent is combined to form the diploid F1 offspring. When one parent is homozygous for A and one parent is homozygous for a, all the F1 offspring will be heterozygous, or Aa. At the next generation, mating between two F1 offspring yields the F2 offspring, who will show a 3:1 ratio of the dominant phenotype to the recessive phenotype and the genotypes AA, 2 Aa, and aa. The dominant phenotype is indicated by the shaded squares, and the recessive phenotype is denoted by the unshaded squares.*

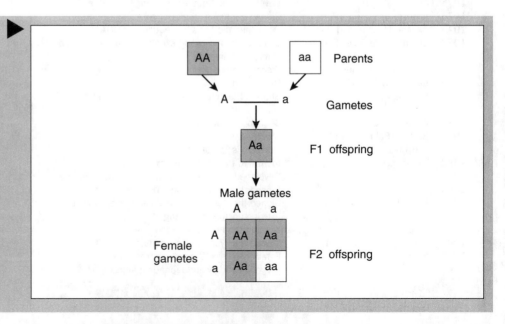

one the parental character *A*, and the other the parental character *a*. Therefore, 2 *Aa* + *A* + *a* or *A* + 2 *Aa* + *a* is the empirical simple, developmental series for two differentiating characters. (Reprinted with permission from Genetics Society of America.)

Within 2 years of the rediscovery of Mendel's laws, William Bateson and Reginald Punnett in England and Lucien Cuenot in France extended Mendel's studies to animals and found the same principles applied. These findings were made even more significant when William Sutton described the behavior of chromosomes in the germ cells, providing a potential mechanism that could account for the units that were described years earlier by Mendel. A simple explanation for the physical basis of independent assortment was that the different pairs of hereditary units were located on different pairs of chromosomes. These units of inheritance were given the name "genes" in 1909 by Wilhelm Johannsen.

The first understanding of the relationship between mendelian inheritance and human disease is attributed to the physician–scientist Archibald Garrod. As early as 1897, he noticed abnormal discoloration in the urine of some of his patients and first described the disease alkaptonuria. He realized that alkaptonuria was present early in infancy and continued throughout the life of the patient, suggesting the defect must be congenital. Later work led to his finding that children with alkaptonuria who had healthy parents were the result of consanguineous marriages. Consultations with Bateson led to the conclusion that alkaptonuria could be regarded as one of the rare recessive characters described by Mendel 35 years earlier. The following quotation taken from Garrod's *Inborn Errors of Metabolism* and presented as the Croonian Lectures in 1908 indicates Garrod's understanding of the relationship of Mendel's laws of inheritance to human disease [2]:

> It was pointed out by Bateson, and has recently been emphasized by Punnett, that the mode of incidence of alkaptonuria finds a ready explanation if the anomaly in question be regarded as a rare recessive character in the Mendelian sense. Mendel's law asserts that as regards two mutually exclusive characters, one of which tends to be dominant and the other recessive, cross-bred individuals will tend to manifest the dominant character, but when they interbreed the offspring of the hybrids will exhibit one or other of the characters and will consist of dominants and recessives in definite proportions. Mendel's theory explains this by the supposition that the germinal cells or gametes of each generation are pure as regards the qualities in question, and accounts for the numerical results observed by the production of dominant and recessive gametes in equal numbers. Of the offspring of two hybrids, one quarter will result from the union of two dominant gametes and will produce such gametes only; another quarter will result from the union of recessive gametes and will produce only recessive gametes. The remaining half will themselves manifest the dominant character, but will be hybrids like their parents and will produce gametes of both varieties. Only when two recessive gametes meet in fertilization will the resulting individual show the recessive character. If the recessive character be a rare one, many generations may elapse before the union of two such gametes occurs, for the families in which they are produced will be few in number and the chance that in any given marriage both parents will contribute such gametes will be very small. When, however, intermarriage occurs between two members of such a family, the chance will be much greater and of the offspring of such marriage several are likely to exhibit the peculiarity. The rarer the anomaly the more conspicuous should be the influence of consanguinity. (Reprinted with permission from Oxford University Press.)

Garrod's second important contribution to human genetics was his understanding of biochemical reactions and their relationship to human disease. His studies on several diseases, which he described as inborn errors of metabolism, included alkaptonuria, albinism, pentosuria, and cystinuria. Garrod's conclusion from his studies suggested for the first time that each of these conditions could be explained by a single block in metabolism and that this block in the normal metabolism of a cell was caused by the inherited deficiency of a specific enzyme. Thus, Garrod was the first to suggest the relationship between a gene and a specific enzyme in the cell.

The next major contribution to the current understanding of genetics came from the work of Thomas Hunt Morgan, who was the first to describe linkage analysis and recombination in the fruit fly, *Drosophila melanogaster*. His work was instrumental in establishing the concept that genes are linked in a linear order on each chromosome and that recombination can occur between these genes during meiosis and chiasma formation. From Morgan's studies on recombination came the important concept that the map distance on a chromosome is a reflection of the recombination frequency between two genes; that is, the closer together two genes are on a chromosome, the lower the recombination frequency between them. Thus, distance on a map can be expressed as a map unit or a centimorgan (cM), which is equivalent to 1% recombination frequency.

As geneticists began to develop a picture of genes as physical units localized in a linear array on each chromosome, they began to question the mechanism by which alterations or mutations occurred in genes and gave rise to phenotypic changes in the offspring. A major contribution to our understanding of the basis of mutation came from the work of Hermann Joseph Muller in 1927. Working with *Drosophila*, Muller discovered that he could induce the appearance of mutations by x-rays and that these x-ray-induced mutations were in all respects identical to the mutations that occurred naturally.

Although the early studies on the fruit fly provided the field of genetics with many important basic concepts, and even today remains a favorite species for study by many geneticists, the understanding of the biochemical basis of genetics became possible with the initiation of studies on single-cell organisms. The first such studies began in 1941 with the work of George Beadle and Edward Tatum using the fungus *Neurospora*. The advantage that *Neurospora* offered over more complex organisms was its ability to grow on a simple, well-defined medium. This characteristic provided an opportunity to select for a single, specific change in the growth requirements of the organism, as illustrated by its inability to grow on the defined medium without a growth supplement. Beadle, in his description of this work, was quick to point out that Garrod, in his studies on alkaptonuria, clearly had in mind the concept of a gene-enzyme-chemical reaction interrelationship, the fundamental underlying hypothesis for their work.

In the same year, Tatum received the Nobel prize, and in his Nobel lecture, he elaborated on the findings he and Beadle had made. The following quotation is from Tatum's Nobel Prize Lecture [3].

Now for a brief and necessarily somewhat superficial mention of some of the problems and areas of biology to which these relatively simple experiments with *Neurospora* have led and contributed. First, however, let us review the basic concepts involved in this work. Essentially, these are (1) that all biochemical processes in all organisms are under genic control; (2) that these overall biochemical processes are resolvable into a series of individual stepwise reactions; (3) that each single reaction is controlled in a primary fashion by a single gene, or in other terms, in every case a 1:1 correspondence of gene and biochemical reaction exists, such that (4) mutation of a single gene results only in an alteration in the ability of the cell to carry out a single primary chemical reaction. As has repeatedly been stated, the underlying hypothesis, which in a number of cases has been supported by direct experimental evidence, is that each gene controls the production, function, and specificity of a particular enzyme. Important experimental implications of these relations are that each and every biochemical reaction in a cell of any organism, from a bacterium to man, is theoretically alterable by gene mutation, and that each such mutant cell strain differs in only one primary way from the non-mutant parental strain. (Reprinted with permission from the Noble Foundation.)

Thus, by 1941 the stage was set for the beginning of the molecular period of genetics. The concepts of biochemistry and genetics had been firmly merged, and the scientific world was ready to embark on the detailed understanding of the physical nature of genes and proteins and how the genetic material and the information it contains is passed from generation to generation.

## Molecular Period

The achievements of the molecular period came quickly and provided most of the basic concepts in molecular genetics that exist today. First and foremost, the 1940s began with compelling evidence that a molecule known as deoxyribonucleic acid (DNA)—composed of the bases adenine, thymine, guanine, and cytosine; the sugar deoxyribose; and the compound phosphate—was the genetic material. This concept was first evident from the work of Oswald Avery, Colin MacLeod, and Maclyn McCarty, which was published in 1944 [4]. In this classic paper, the authors demonstrated that the chemical agent responsible for transforming the bacterium *Pneumococcus* was the sodium salt of DNA. The following quotation is from the discussion of this paper, which was published in the *Journal of Experimental Medicine* [4]:

> It is, of course, possible that the biological activity of the substance described is not an inherent property of the nucleic acid but is due to minute amounts of some other substance adsorbed to it or so intimately associated with it as to escape detection. If, however, the biologically active substance isolated in highly purified form as the sodium salt of deoxyribonucleic acid actually proves to be the transforming principle, as the available evidence strongly suggests, then nucleic acids of this type must be regarded not merely as structurally important but as functionally active in determining the biochemical activities and specific characteristics of pneumococcal cells. Assuming that the sodium deoxyribonucleic acid and the active principle are one and the same substance, then the transformation described represents a change that is chemically induced and specifically directed by a known chemical compound. If the results of the present study on the chemical nature of the transforming principle are confirmed, then nucleic acids must be regarded as possessing biological specificity the chemical basis of which is as yet undetermined. (Reprinted with permission from the Rockefeller University Press.)

Much of the scientific world remained unconvinced that a simple molecule like DNA could serve such a specific role in genetic inheritance. The chemical composition of DNA, consisting of only four simple bases, did not seem to have the properties that most scientists considered necessary for the complex process of inheritance. It was thought that proteins, with 21 different amino acids, were much more likely candidates. However, additional experiments were published showing that the transfer of genetic material from one bacterial cell to another involved DNA, and they provided further support for the work of Avery and his coworkers.

In 1946, Joshua Lederberg and Tatum, working with the bacterium *Escherichia coli*, first described the process of *conjugation*, in which the DNA of one bacterial cell, the donor cell, was transferred by cell-to-cell contact to a recipient bacterial cell, resulting in a permanent alteration of the DNA of the recipient cell. Although the initial conclusion from this work that the cells underwent cell-to-cell fusion was incorrect, the results clearly implicated DNA as the genetic material. The support for DNA as the genetic material continued to build, with the finding by Alfred Hershey and Martha Chase in 1951 that the genetic material in a bacterial virus was almost certainly DNA. In 1952, Norton Zinder and Lederberg showed that DNA also could be transferred from one bacterial cell to another bacterial cell by way of a virus vector—a gene transfer process termed *transduction*. Thus, by 1952, a significant body of evidence had accumulated from the studies on transformation, transduction, and conjugation in bacteria, implicating DNA as the genetic material. These three methods of gene transfer in bacteria are illustrated in Figures 1-2–1-4.

During the next 30 years, additional studies on gene transfer in bacteria contributed an enormous volume of information to the field of molecular genetics. The study of conjugation and transduction provided information on genetic mapping, linkage analysis, virus-host interactions, mutation analysis, and gene regulation and control. Without the initial work on *E. coli*, with its simple growth requirements, its haploid chromosome number, and its ability to be genetically manipulated, the science of genetics would probably not be where it is today. Even now, important functions and pathways that were

**FIGURE 1-2** ▶

*Gene Transfer in Bacteria by Transformation.* In the process of transformation, linear double-stranded molecules of DNA are bound by recipient cells. One strand of the double-stranded DNA molecule enters the recipient cell, with the simultaneous degradation of the complementary strand. Once inside the cell, the incoming single-stranded DNA molecule replaces the homologous sequence of DNA on the DNA of the recipient cell, with the loss of the recipient allele. The replacement of the recipient allele by the donor allele results in a permanent alteration of the recipient's genotype.

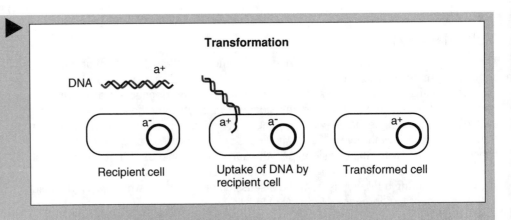

**Transformation**

**FIGURE 1-3** ▶

*Gene Transfer in Bacteria by Transduction.* Transduction is the transfer of genetic material from one cell (the donor cell) to another cell (the recipient cell) via a virus vector. DNA-containing bacterial viruses, known as temperate viruses, infect the donor cell. Inside the donor cell, the virus undergoes a lytic cycle, with subsequent degradation of the DNA of the donor cell. During the process of maturation of the virus, at which time viral DNA is enclosed in a protein virus head, a mistake is made. At a frequency of approximately 1 in 1 million, the virus mistakenly incorporates in the protein head a piece of bacterial DNA instead of viral DNA. Upon subsequent infection of another bacterial cell—the recipient cell—the piece of bacterial DNA enclosed in the viral head is injected into the recipient cell. The incoming linear double-stranded DNA molecule replaces the homologous DNA sequence in the recipient cell, with the loss of the recipient allele. As in transformation, replacement of the recipient allele by the donor allele results in a permanent alteration of the recipient's genotype.

**Transduction**

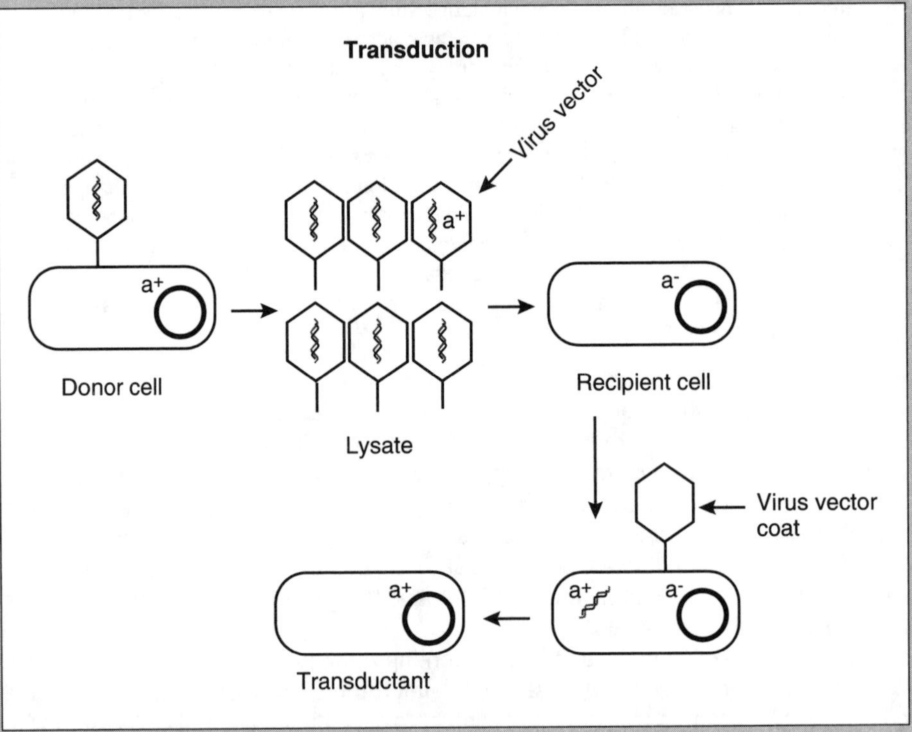

initially elucidated in *E. coli* are being rediscovered in human cells. An important case in point is the recent discovery of the mismatch repair system and its association with colon cancer (see Chapter 4).

In 1953, even the most serious doubters began to believe the critical role of DNA in inheritance when James Watson and Francis Crick published their classic paper on DNA structure, which showed two complementary DNA strands held together by the hydrogen bonding of adenine (A) to thymine (T) and guanine (G) to cytosine (C) [5]. The proposed structure provided a basis for understanding the preciseness of replication and how a simple molecule composed of four bases could serve as the genetic material.

Once the structure of DNA was known and DNA was firmly established as the genetic material, the next important question to be addressed was how a DNA molecule transmitted its information within the cell to provide the necessary proteins needed to carry out normal growth, metabolism, and reproduction. Only 3 years later, in 1956, Jacques Monod combined his talents with those of Francois Jacob to investigate this very question.

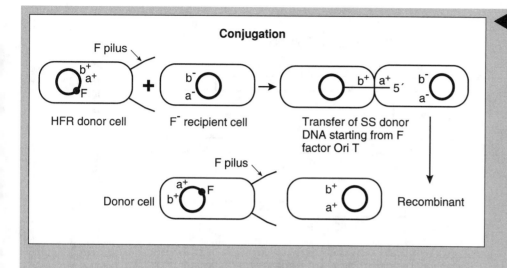

Conjugation

HFR donor cell    F⁻ recipient cell    Transfer of SS donor DNA starting from F factor Ori T

Donor cell    Recombinant

**FIGURE 1-4**

**Gene Transfer in Bacteria by Conjugation.** *In the process of conjugation, two viable bacterial cells come in contact by way of an organelle, the F pilus, which is present on male donor cells. Male bacterial cells contain a plasmid, the fertility (F) factor, which confers on the cells the ability to transfer DNA to a recipient cell. Recipient or female cells do not contain the F factor. In the process of conjugation, the two cells come in contact and form a cell-to-cell conjugation bridge. A single strand (SS) of the DNA from the donor cell is transferred starting with the 5' end through the conjugation bridge to the recipient cell. The incoming single-stranded DNA is replicated inside the recipient cell and replaces the homologous DNA sequences on the recipient DNA. As in transformation and transduction, the replacement of the recipient allele by the donor allele results in the permanent alteration of the recipient's genotype. HFR donor = high frequency recombinant donor; Ori T = the site within the F factor DNA that determines the origin of transfer of DNA.*

Meanwhile, geneticists working with mammalian cells also were making significant discoveries. In 1956, Frederick Sanger reported the complete amino acid sequence of the protein insulin, and Joe-Hin Tijo and Albert Levan finally established the normal chromosome number in humans as 46. Vernon Ingram, with his studies on hemoglobin, identified a single amino acid change in the hemoglobin S molecule synthesized by patients with sickle cell anemia. This finding formed the basis for the understanding that a single base change in the DNA can result in a single amino acid change in a protein. By 1958, Theodore Puck and G. Pontecorvo had begun the field of somatic cell genetics, in which single cells from mammals were cultured much like bacterial cells, allowing them to be studied in vitro. In 1959, Jerome LeJeune identified the first chromosome abnormality associated with an inherited disease by showing that Down's syndrome patients had three copies of chromosome 21.

During 1960 and 1961, geneticists began to determine how information is transferred from DNA to protein. Many key findings on information transfer were presented at the 1961 Symposium on Cellular Regulatory Mechanisms, which took place at Cold Spring Harbor, New York [6]. Although that symposium is best remembered for the presentation by Jacob and Monod of the operon model for gene regulation, a number of other key discoveries were also introduced to the scientific world at that time. Among these were the findings of Charles Yanofsky, who showed that single base changes in the DNA affect the amino acid sequence of the protein and who provided the first evidence for the concept of *colinearity* of a gene with its protein.

Colinearity, diagrammed in Figure 1-5, remained the dogma for the next 25 years as the only way to describe the relationship between a base sequence in the DNA and the amino acid sequence of a protein. Not until 1977, when intervening sequences were found in mammalian genes, was another relationship even considered possible. The remarkable part about Yanofsky's results is that, at the time, the genetic code was still unknown and would not be elucidated until the studies of Marshall Nirenberg 1 year later. When the genetic code was elucidated, Yanofsky's work was completely verified using the newly identified amino acid codons.

Another basic finding presented at this symposium was the identification of a rapidly produced and unstable RNA molecule, which is known today as messenger RNA (mRNA). One of the early candidates thought to be responsible for the transfer of information from DNA to the ribosomes was the RNA associated with ribosomes known as ribosomal RNA (rRNA). However, there were a number of problems associated with this

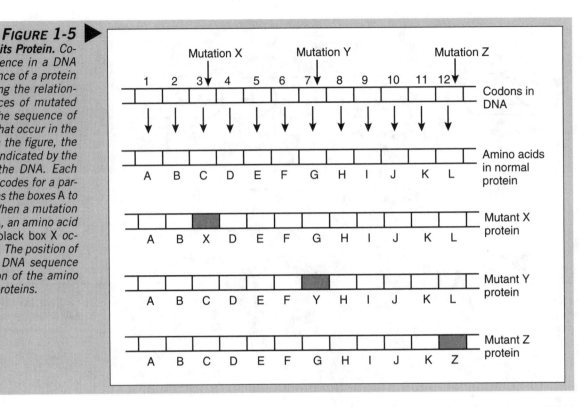

FIGURE 1-5 ▶

**FIGURE 1-5**

**Colinearity of a Gene and its Protein.** Colinearity of the base sequence in a DNA and the amino acid sequence of a protein is demonstrated by showing the relationship between the sequences of mutated sites within a gene and the sequence of amino acid substitutions that occur in the corresponding proteins. In the figure, the mutations X, Y, and Z are indicated by the arrows to the codons in the DNA. Each codon, or triplet of bases, codes for a particular amino acid, noted as the boxes A to L in the normal protein. When a mutation (e.g., X) occurs in the DNA, an amino acid change indicated by the black box X occurs in the mutant protein. The position of the mutated sites in the DNA sequence coincides with the position of the amino acid substitutions in the proteins.

hypothesis. The base composition of rRNA was relatively constant, whereas the AT/GC content of DNA was quite divergent. The rRNA molecule was very stable, yet information on the induction of the protein β-galactosidase indicated that the "message" could appear in minutes and just as quickly decay. Thus, a search was made for a molecule that would have the proper base ratios to DNA and would have a high rate of turnover, a molecule that would act like a messenger. Several investigators, including Saul Spiegelmen and Sidney Brenner, simultaneously found such a molecule and called it mRNA. Shortly afterward, reports from the laboratories of Severo Ochoa and Jerry Hurwitz demonstrated in vitro the activity of an enzyme known as RNA polymerase, which could synthesize an RNA molecule using DNA as a template.

One might wonder why mRNA was so difficult to find. In bacteria, mRNA is turned over within minutes. Additionally, mRNA is not free, in the sense that as the mRNA is transcribed from a DNA template, the ribosomes immediately attach and begin translation, followed by the rapid decay of the mRNA. With the techniques available at the time and given the fact that mRNA makes up only a small component of the total RNA in the cell, it is amazing that it could be isolated at all. Very important at the time was a new technique developed by Paul Doty and Julius Marmur known as *nucleic acid hybridization*. This technique, which was instrumental in the final isolation of mRNA, became fundamental to future studies in molecular genetics.

In this myriad of incredibly important discoveries came the formulation of the *operon model* by Jacob and Monod. The operon model as it is known today is presented in Figure 1-6. Although the key features of the operon model remain the same as first presented by Jacob and Monod, certain parts of the model were modified in subsequent years to arrive at the one presented in Figure 1-6 [7]. Jacob and Monod originated the idea that a cytoplasmic regulator could bind to a specific site on the DNA, (i.e., the operator site) and inhibit the transcription of the structural genes that were part of an operon. This new idea presented the concept of *cis*-acting controlling sequences that were recognized by *trans*-acting cytoplasmic regulators, which would eventually explain regulation at the transcriptional level in all organisms, including humans.

The original model of the operon proposed that the cytoplasmic regulator of the lactose operon was a RNA molecule that could recognize by base complementarity a sequence of bases on the DNA. In the presence of a small molecule or inducer, the regulator would be altered in a way that it could no longer bind to the DNA, and RNA polymerase would be free to initiate transcription of the structural genes. The initial idea

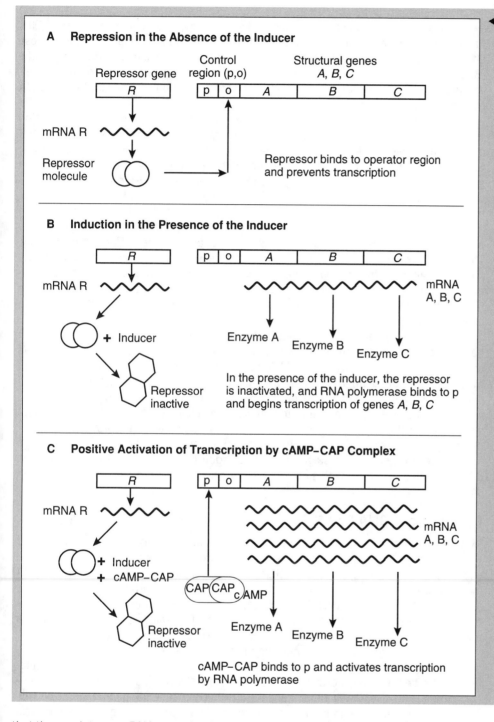

**A    Repression in the Absence of the Inducer**

Repressor binds to operator region and prevents transcription

**B    Induction in the Presence of the Inducer**

In the presence of the inducer, the repressor is inactivated, and RNA polymerase binds to p and begins transcription of genes A, B, C

**C    Positive Activation of Transcription by cAMP–CAP Complex**

cAMP–CAP binds to p and activates transcription by RNA polymerase

**FIGURE 1-6**

***Positive and Negative Control in a Bacterial Operon.*** *Inducible bacterial operons contain structural genes denoted as genes A, B, and C, along with their closely linked controlling sites, the operator site (o) and the promoter site (p). The expression of the structural genes is regulated by the production of a protein repressor molecule coded for by the repressor (R) gene. (A) In the absence of an inducer, the repressor protein binds to the operator site on the DNA and prevents transcription by RNA polymerase from the promoter site. (B) When the cells are exposed to an inducer, often the substrate of the enzymes coded for by the structural genes, the inducer binds to the repressor protein, altering its structure and preventing its binding to the operator site. In the absence of the binding of the repressor protein at the o site, RNA polymerase binds to the p site and initiates transcription of the structural genes in the operon. The mRNA produced is a polycistronic mRNA containing the information for the translation of all of the enzymes coded for by the operon. (C) A second level of control, positive control, is necessary for maximum rates of transcription of the structural genes. When the levels of cyclic adenosine monophosphate (cAMP) are high in the bacterial cell, a transcription factor known as the cAMP activator protein (CAP) binds cAMP and is activated to bind to the p site on the DNA. When the cAMP–CAP complex is bound to the p site, RNA polymerase activates transcription, resulting in maximum expression of the structural genes of the operon.*

that the regulator was RNA proved to be wrong. Genetic and biochemical studies carried out over the next 4 years clearly showed that the repressor molecule was a DNA-binding protein.

The operon model also presented for the first time the concept of a polycistronic mRNA molecule, which is produced directly from a DNA template and contains the information for the synthesis of all of the proteins in the operon. The production of a single polycistronic mRNA by RNA polymerase binding at a promoter site allowed for the rapid and efficient turning on and turning off of all the genes in a operon in response to environmental changes. Control of an operon according to the operon model of Jacob and Monod was strictly negative control. That is, the only event necessary to turn on the operon and begin transcription was the removal of the repressor, which negatively controlled expression of the structural genes. As one might predict, this concept was appealing because of its simplicity, but it was incomplete because it was *too* simple.

A major modification of operon control by negative control came from work of Ellis Englesberg on the arabinose fermentation pathway and the investigation of the glucose effect; that is, the ability of glucose to repress the transcription of many different operons [8, 9]. From these studies, the concept that operons are subject to control by positive-acting factors, proteins that are essential for the activation of transcription by RNA polymerase, was developed.

The theory of a more complex positive control system to explain enzyme induction was not readily accepted by a scientific community that was completely convinced and persuaded by the simplicity of Jacob and Monod's operon model. For nearly a decade, Englesberg continued to challenge the simplicity of negative control with his elegant genetic experiments supporting positive control of the L-arabinose operon. The discovery that glucose caused a dramatic decline in levels of cyclic adenosine monophosphate (cAMP) in *E. coli* eventually led to the identification of the cAMP activator protein (CAP), which binds cAMP and is essential for activating transcription of operons in a positive manner. These findings resulted in the final acceptance of Englesberg's theory. Thus, the operon model as it is known today has new dimensions and added flexibility for controlling gene expression at the transcriptional level. Both positive- and negative-acting cytoplasmic products can interact with *cis*-acting binding sites on DNA to modulate the transcription of genes and give a cell maximum flexibility to survive in a changing environment.

As bacterial geneticists continued to elucidate genetic control and gene mapping in bacterial cells, mammalian geneticists were beginning their studies on cultured human somatic cells. The discovery by Mary Weiss and Howard Green in 1965 that somatic cell hybrids formed by fusing cells from different species specifically lost chromosomes of one species opened up the field of somatic cell genetics. This provided a new and innovative way to map human genes to specific chromosomes. The use of mouse–human somatic cell hybrids, in which only human chromosomes are lost, provided a way of mapping human genes without relying on pedigree and linkage analysis of transmitted human disease genes. The stage was set for additional molecular techniques to be discovered that would allow this new method of gene mapping to become crucial in the study of human genetics.

## Recombinant DNA Era

In 1970, the recombinant DNA era began with the discovery by Dan Nathans and Hamilton Smith of the properties of restriction endonucleases and the ability of these enzymes to create sticky DNA ends that could be recombined to form chimeric DNA molecules. When this property of restriction endonucleases to form sticky DNA ends was applied in vitro by Herb Boyer, Stan Cohen, and Paul Berg to form recombinant DNA molecules made up of DNA from two species, a new form of molecular genetics and a revolution in the possibilities for studying the human genome were initiated.

The 1970s were filled with other important discoveries, including the technique developed by Joseph Gall and Mary Lou Pardue of in situ hybridization and the formulation of the two-hit hypothesis of Alfred Knudson to explain the inheritance of retinoblastoma and eventually other cancers. Chromosome banding, first described by Torbjorn Caspersson, expanded the use of somatic cell hybrids for gene mapping and allowed, for the first time, the precise identification of each human chromosome. Edward Southern described his blotting technique, which increased the efficiency of using recombinant DNA technology to isolate and clone DNA fragments. Methods were developed by Sanger and Allan Maxam and Walter Gilbert to sequence DNA molecules, allowing scientists to determine exactly the base sequence of any DNA molecule.

A major discovery that altered the understanding of the structure of genes occurred in 1977, with the isolation of the rabbit and mouse β-globin gene and chicken ovalbumin gene [10]. Using the newly developed Southern blotting and restriction endonuclease technology, scientists found a new structure for mammalian genes. The genes of a mammalian cell were not just simple base sequences in the DNA colinear with the protein coded for by the respective gene. Instead, the DNA of the genes had interruptions; that is, intervening sequences that interrupted the coding sequences. The terms *intron* to describe the intervening sequences that interrupted the coding sequences and *exons* to describe the DNA sequences that were eventually found in the mature mRNA were introduced.

This new understanding of the structure of mammalian genes led to the finding that the entire sequence of bases of a gene was transcribed into a RNA molecule that was then processed with the removal of the intron sequences and the production of a mature mRNA containing only exon sequences. The discovery of interrupted genes in mammalian cells was one of the first indications that the genetic concepts developed from the elegant studies of bacteria might not be the same in the more complicated world of the mammalian genome.

## Period of Human Genetics

From the late 1970s to the present, many additional important findings have altered the thinking and the ability of scientists to investigate and understand human genetics. All of the previous discoveries in genetics have led to the current knowledge and to a new and exciting time in the study of inherited human diseases. Nearly every day a new human gene is isolated and identified, opening up new approaches for the treatment of human disorders. Soon, the complete base sequence of the entire human genome will be known, which will eventually lead to the ability to identify and study every human gene and its function in cells.

New techniques are being developed for somatic cell gene therapy and for the alteration of fetuses in vivo to prevent or treat human diseases. The future of the treatment of human genetic diseases is just beginning, and every physician will need to deal with new medical molecular genetics as it begins to revolutionize the medical field.

# REVIEW QUESTIONS

**Directions:** The groups of questions below consist of lettered choices followed by several numbered items. For each numbered item, select the appropriate lettered option with which it is most closely associated. Each lettered option may be used once, more than once, or not at all.

### Questions 1–3

Match each of the following definitions to the term it correctly describes.

    **(A)** Principle of segregation
    **(B)** A centimorgan
    **(C)** One gene–one enzyme hypothesis
    **(D)** Transformation in bacteria
    **(E)** Conjugation in bacteria

1. Distance of map units on a chromosome

2. Type of gene transfer in bacteria in which a donor cell transfers DNA to a recipient cell by cell-to-cell contact

3. The concept that a gene has two alleles with each allele segregating separately in the gametes

### Questions 4 and 5

Match each of the following definitions to the term it correctly describes.

    **(A)** Regulator genes
    **(B)** Introns
    **(C)** Promoters
    **(D)** Structural genes
    **(E)** Operators

4. RNA polymerase binds to these DNA base sequences and initiates transcription

5. *Trans*-acting factors interact with these DNA base sequences, resulting in the repression of transcription of genes in bacterial operons

# ANSWERS AND EXPLANATIONS

**1–3. The answers are: 1-B, 2-E, 3-A.** Studies by Morgan showed that map distance on a chromosome is a reflection of the recombination frequency between two genes, that is, the closer together two genes are on a chromosome, the lower the recombination frequency is between them. The distance on a map can be expressed as centimorgans (cM), in which one cM is equivalent to 1% recombination frequency.

    Conjugation occurs between a male donor cell containing a fertility (F) factor that has the capacity to initiate gene transfer of the donor DNA and a recipient or female cell. This gene transfer requires cell-to-cell contact and involves the transfer of a single strand of donor DNA to the recipient cell. Genetic material from the donor DNA replaces homologous DNA sequences in the recipient cell and permanently alters the recipient's genotype.

Mendel's second principle—the principle of segregation—presented the concept of a gene with two alleles that segregated in the gametes. One copy of each allele is passed to the gametes at meiosis, and at fertilization, one allele from each parent is combined to form the diploid cell or F1 generation. If a mating occurs between two F1 parents, each of which is heterozygous for a particular allele, the F2 offspring show a 3:1 ratio of the dominant phenotype to the recessive phenotype.

**4 and 5. The answers are: 4-C, 5-E.** Promoter sequences, located 5′ to the structural genes, are the binding sites for RNA polymerase. RNA polymerase binds to specific base sequences within the promoter site and initiates transcription. In bacterial operons, a single polycistronic mRNA is produced, which contains information for the synthesis of all the enzymes coded for by a single operon.

Operator sites located 5′ to the structural genes in a bacterial operon contain base sequences that are recognized by specific repressor proteins. When a repressor protein is bound to the DNA at the operator site, RNA polymerase is prevented from initiating transcription. In this way, all of the structural genes within an operon can be turned off simultaneously. When the repressor is bound to small inducer molecules, the protein no longer can bind to the DNA at the operator site, and RNA polymerase initiates transcription.

# REFERENCES

1. Mendel G: Letter from Gregor Mendel to C. Nageli. *The Birth of Genetics.* Supplement to *Genetics* 35(5): Part 2, 1950.
2. Garrod A, Harris H (ed): *Garrod's Inborn Errors of Metabolism.* London, England: Oxford University Press, 1963, p 19.
3. Tatum EL: Nobel Prize lecture of Edward L. Tatum, 1958. The Nobel Foundation for Les Prix Nobel. *Science* 129:1715–1719, 1959.
4. Avery O, MacLeod C, McCarty M: Studies on the chemical nature of the substance inducing transformation of pneumococcal types. *J Exper Med* 79:137–158, 1944.
5. Watson JD, Crick FH: Genetic implications of the structure of deoxyribonucleic acid. *Nature* 171:964–967, 1953.
6. Biological Laboratory: Cellular regulatory mechanisms. *The Cold Spring Harbor Symposium on Quantitative Biology*, vol. XXVI. Cold Spring Harbor, NY: Biological Laboratory, 1961.
7. Jacob F, Monod J: On the regulation of gene activity. In *The Cold Spring Harbor Symposium on Quantitative Biology*, volume XXVI. Cold Spring Harbor, NY: Biological Laboratory, 1961, pp 193–211.
8. Englesberg E, Irr J, Power J, et al: Positive control of enzyme synthesis by gene C in the L-arabinose system. *J Bacteriol* 90:946–957, 1965.
9. Zubay G, Schwartz D, Beckwith J: Mechanism of activation of catabolite sensitive genes: a positive control system. *Proc Natl Acad Sci* 66:104–110, 1970.
10. Chambon P: Split genes. *Sci Am* 244:60–71, 1981.

# 2

# STRUCTURE OF NUCLEIC ACIDS AND PROCESS OF DNA REPLICATION

## CHAPTER OUTLINE

## INTRODUCTION OF CLINICAL CASE

Mary, a 23-year-old woman in her third month of pregnancy, and Joe, her husband, visited their gynecologist for a routine examination. The physician informed the couple that they were going to have twins, but he could not determine if the twins would be monozygotic (identical) or dizygotic (nonidentical). In addition, he explained that twins only occur at a frequency of 1/100 deliveries, with monozygotic twins making up about one-third of those delivered. Mary was very excited about having twins and laughed with her husband about dressing the twins alike and their possible role in a gum commercial. The physician, however, once again emphasized that there is only a 33% chance that the twins will look alike and a 67% chance that the twins will simply look like brothers or sisters. Mary and Joe wanted to know why. The physician handed the couple a small pamphlet explaining how monozygotic and dizygotic twins arise. After reading this chapter, the student should be able to explain monozygosity and dizygosity to Mary and Joe and also the molecular basis for these two alternatives.

## STRUCTURE AND PROPERTIES OF NUCLEIC ACIDS

### DNA: The Genetic Material of All Cells

Human cells, along with most organisms, store their genetic information in molecules of *deoxyribonucleic acid* or *DNA*. The information stored in a DNA molecule must be stable, must be able to resist change, and must be faithfully replicated and transmitted to progeny cells without modification. All of these properties are important to prevent the occurrence of mistakes or defects arising in the DNA and being expressed in the daughter cells. DNA is also the site of gene expression in cells, the site that determines whether a particular function is turned on or off during the growth and life of the cell. These features of the DNA molecule—stability, resistance to change, faithful replication, and the ability to transfer its information to the cell—are built into the structure of the molecule. This structure, which was determined by Watson and Crick in 1953, revolutionized the thinking in modern biology (Figure 2-1).

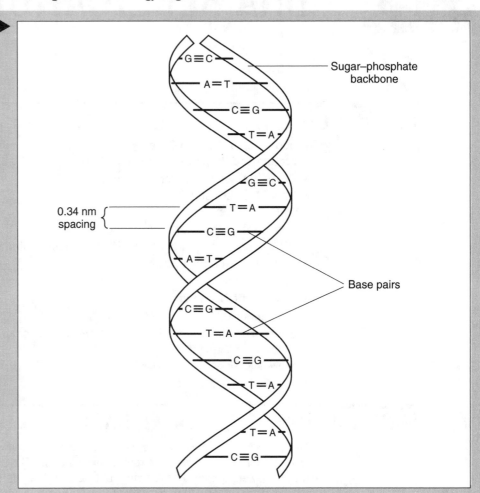

**FIGURE 2-1** ▶

**The Watson-Crick Model of the Structure of DNA.** *A double-stranded DNA helix has two polynucleotide strands of DNA coiled around a common axis and running in an antiparallel orientation. The two strands are held together by hydrogen bonding between adenine (A) and thymine (T) residues and guanine (G) and cytosine (C) residues. Each turn of the helix represents 10 bp.*

DNA is made up of defined bases: A, T, G, and C.

Each base is a strand of DNA, and RNA is joined to the next base by a phosphodiester bond.

All DNA molecules, whether they are found in prokaryotic or eukaryotic cells, are made up of four types of nucleotides linked together by phosphodiester bonds to form long chains or polynucleotides. Each nucleotide in DNA consists of a nitrogenous base, the sugar deoxyribose, and a phosphate residue. The nitrogenous bases found in DNA are of two types, purines (i.e., adenine [A] and guanine [G]) and pyrimidines (i.e., cytosine [C] and thymine [T]). These nucleotides are linked together by covalent phosphodiester bonds that join the 5' carbon of one deoxyribose to the 3' carbon of the next deoxyribose to form polynucleotide chains (Figure 2-2).

The DNA chain can exist as a single-stranded molecule but is found most often in cells as the double-stranded structure shown in Figure 2-1. This structure is helical in nature with two strands of DNA coiled around a common axis and running in antiparallel (opposite) orientation. This opposite orientation, or polarity of the molecule, is an important concept in the biochemistry of RNA and DNA. Figure 2-3 illustrates the polarity of two strands of DNA in the helix with the bases hydrogen-bonded in a complementary manner.

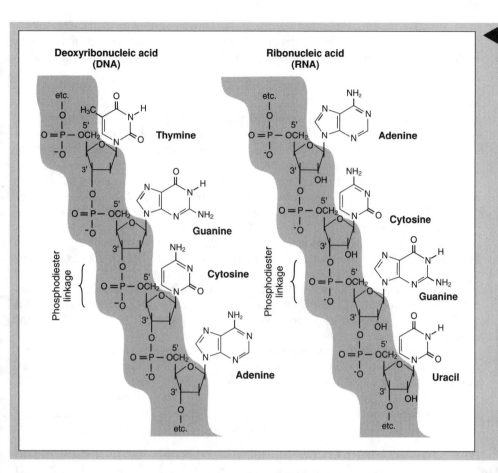

**FIGURE 2-2**
**Polynucleotide Chains Formed by 3'–5' Phosphodiester Bonds.** Both RNA and DNA are made up of four nucleotides linked together by 3'–5' phosphodiester bonds. DNA is made up of four bases (adenine [A], thymine [T], cytosine [C], and guanine [G]) and the sugar, deoxyribose. RNA contains the four bases (A, uracil [U], C, and G) and the sugar, ribose.

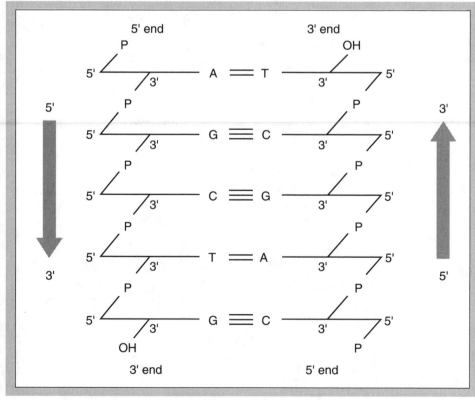

**FIGURE 2-3**
**Antiparallel Orientation and Base Pairing in DNA.** The two DNA strands in a double-stranded helix have opposite polarity with one strand running in a 5' to 3' direction and the complementary strand running in a 3' to 5' direction. The four bases (adenine [A], thymine [T], cytosine [C], and guanine [G]) are placed on the inside of the helix face-to-face to allow hydrogen bonding between purine and pyrimidine residues.

*The two strands of DNA in a DNA helix have opposite polarity, one strand going in a 5' to 3' direction, and the other strand going in a 3' to 5' direction.*

One DNA strand has a 5' to 3' orientation, while the complementary strand runs in a 3' to 5' direction. Conventionally, a double-stranded DNA molecule is written as follows:

$$5'\text{-----------------}3'$$
$$3'\text{-----------------}5'$$

The top strand of the molecule is written left to right, 5' to 3', whereas the bottom strand is in the opposite polarity and written left to right, 3' to 5'. This directionality is also seen in RNA molecules and is important to remember in order to understand many of the functions RNA and DNA have in the cell.

The most common form of DNA found in nature is the B-DNA structure (also known as the Watson-Crick double helix), where the deoxyribose-phosphate backbone is located on the outside of the molecule and the bases are stacked inside the molecule perpendicular to the axis of the helix [1]. The two strands of DNA, which wrap around each other in a right-handed turn, are held together by *hydrogen bonds* between the bases, which are face-to-face inside the molecule. A simple way to visualize the helix is to imagine a rope ladder with the rungs of the ladder formed by the base pairing A-T or G-C. If the ladder is given a right-handed twist, the result is a helix in the form of B-DNA. In B-DNA, each turn of the helix represents 10 base pairs (bp) and is equivalent to 3.4 nm in length. This number is fixed, enabling us to calculate the length of any B-DNA molecule from the number of base pairs it contains.

*The two strands of DNA in a DNA helix are attached to each other by the formation of hydrogen bonds between A-T and G-C.*

When purines and pyrimidines are placed in the double-stranded helix face-to-face, their pairing properties result in an A always pairing with a T and a G always pairing with a C (see Figures 2-1 and 2-3). This specific hydrogen bonding is diagrammed in Figure 2-4.

**FIGURE 2-4** ▶

***Hydrogen Bonding Between Adenine and Thymine and Between Guanine and Cytosine.*** *Adenine, a purine residue, forms two hydrogen bonds with thymine, a pyrimidine residue. Guanine, a purine residue, forms three hydrogen bonds with cytosine, a pyrimidine residue.*

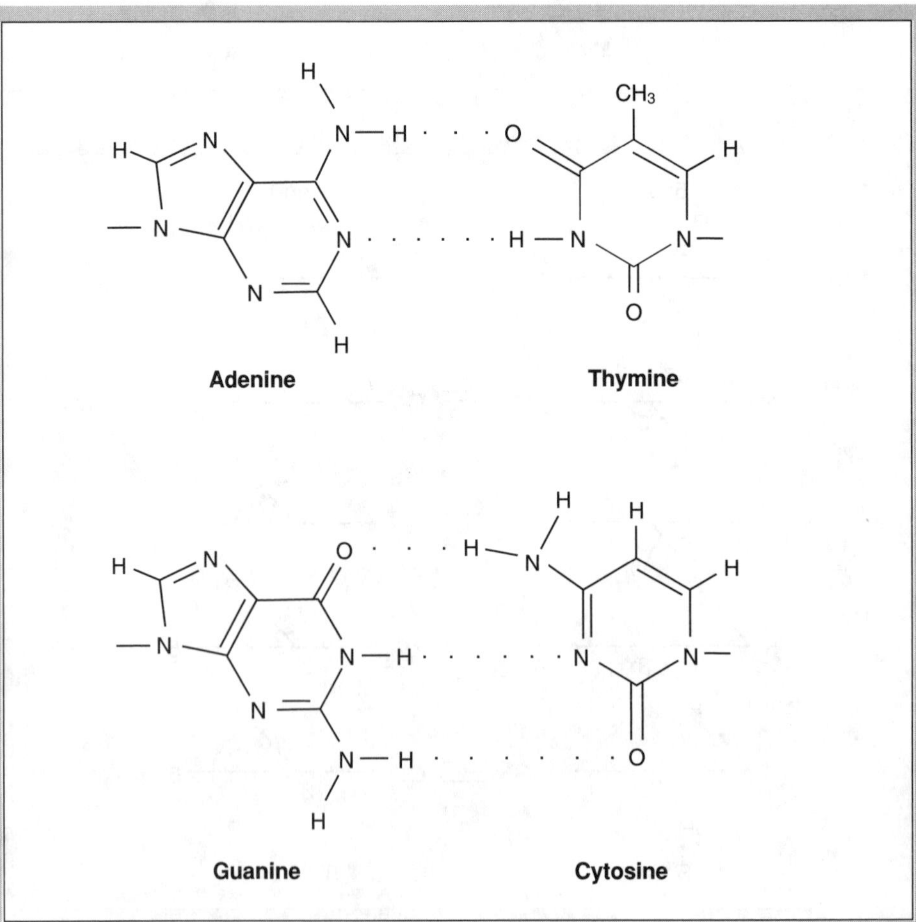

Notice that there are two hydrogen bonds formed in an A-T pair and three hydrogen bonds formed in a G-C pair.

The DNA helix is made up of successive A-T and G-C base pairs along the length of the polynucleotide chain. The two strands of the double helix are *complementary*; that is, A always hydrogen bonds to T, and G always hydrogen bonds to C. Therefore, given the sequence of bases on one strand of the DNA double helix, the sequence on the complementary strand is fixed and easily determined. For example, the sequence ATTAGGC-CATTG on one strand of DNA will have the complementary sequence TAATCCGGTAAC on the opposite strand of DNA in an antiparallel orientation. The variation of the sequence of nucleotides along the DNA strand determines the function of each section of the DNA molecule and its ability to transmit information to RNA and protein.

The total DNA content of a cell makes up the entire genetic information available to the organism and is referred to as its *genome*. DNA genomes come in various sizes, with the human genome containing approximately $3.3 \times 10^9$ bp per haploid nucleus, nearly 1000 times that found in a bacterial cell (Table 2-1).

◀ **TABLE 2-1**
*The DNA Content of Some Prokaryotic and Eukaryotic Genomes*

| Organism | Number of Chromosomes[a] | Size in Base Pairs (bp)[a] |
|---|---|---|
| Escherichia coli | 1 | $4 \times 10^6$ bp |
| Saccharomyces | 16 | $1.3 \times 10^7$ bp |
| Mouse | 20 | $3 \times 10^9$ bp |
| Human | 23 | $3.3 \times 10^9$ bp |

[a] Numbers represent haploid genomes

DNA molecules can exist in several different forms in cells. For example, bacterial cells and mammalian mitochondria have DNA molecules that are double-stranded and form covalently closed circular molecules. Some mammalian viruses, such as adenovirus, have linear double-stranded DNA molecules. Human cells, with the exception of the sperm and egg, contain in their nuclei 46 molecules of linear double-stranded DNA, each of which is combined with protein molecules to make up bodies called *chromosomes*. A normal human cell contains 23 pairs of chromosomes, 22 autosome pairs and 1 pair of sex chromosomes.

The 46 chromosomes of a human cell are found inside a small organelle called the nucleus. If we consider the number of base pairs of DNA in a human nucleus ($6.6 \times 10^9$) and the fact that these base pairs are arranged into 46 individual DNA molecules, it becomes obvious that the DNA would not be able to fit into the nucleus if it remained linear. Thus, we find that the DNA molecules inside the nucleus are compacted into much smaller forms.

One of the mechanisms by which this compacting can occur is by bending DNA molecules in regions where certain sequences are present. For example, when the sequence AAAAAXXXXXAAAAAAXXXXXAAAAA is found in a DNA molecule (where A represents adenine residues and X represents any other base), bending occurs in the molecule, giving an overall shorter structure.

A second mechanism that allows DNA to exist in a more compact structure is the process of supercoiling. In the supercoiled state, the double-stranded helix falls back on itself to give a less relaxed and more compact DNA molecule. If we have a DNA helix of 1000 bp, there will be 100 turns of the helix with 10 bp per turn. If we cut the DNA and remove 100 bp or 10 turns, the DNA backbone must compensate for the loss of turns and will form the supercoiled form. This type of superhelical DNA is referred to as a negative or right-handed superhelix, indicating that the loss of turns and supercoiling is to the right (Figure 2-5). The negative supercoiled form is the form in which DNA is almost always found in nature, allowing the DNA to fit into the small space of the nucleus. A further organization of the DNA occurs in the nucleus when the DNA forms a complex with basic proteins called *histones*. This complex, known as *chromatin*, is the natural

**DNA is compacted by bending, supercoiling, and histone binding.**

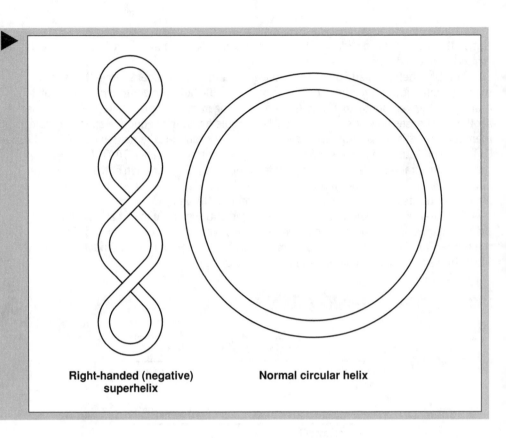

**FIGURE 2-5** ▶

**Supercoiling of DNA.** *The supercoiling of a DNA molecule allows it to become more compact and fit into the small space of the nucleus.*

**Right-handed (negative) superhelix**

**Normal circular helix**

form in which DNA occurs in a human nucleus and allows the DNA to become even more compacted and more ordered in structure.

## RNA Structure and Properties

Ribonucleic acid (RNA) molecules are also composed of nucleotides linked together by phosphodiester bonds, but they generally occur as single-stranded polynucleotides and are usually shorter than most DNA molecules (see Figure 2-2). As with DNA, RNA is made up of the bases A, G, and C but contains uracil (U) in place of T. The other major difference between RNA and DNA is the presence of the sugar ribose in RNA in place of the deoxyribose found in DNA. Since U has the ability to pair with A in much the same way as a T pairs with A, the four bases found in RNA—A, U, G, and C—can form complementary pairs with other bases in RNA as well as with the bases found in DNA. Although some viruses, such as retroviruses or poliovirus, contain RNA as their genetic material in place of DNA, the major function of RNA molecules in human cells is their involvement in the transfer of information from DNA to protein.

*RNA has U in place of T and ribose in place of deoxyribose.*

Because of the single-stranded nature of RNA and the pairing properties of the four bases, RNA often contains intramolecular hydrogen bonding and secondary structures (Figure 2-6). An important point to notice is the directionality of the RNA molecule, going in a 5′ to 3′ direction. If the molecule twists around itself, opposite polarity can occur in some regions of the molecule, resulting in the formation of base pairs between complementary purine and pyrimidine bases. This intrastrand base pairing allows a single-stranded molecule of RNA to form double-stranded portions. Such base pairing regions create structures known as stem-loop structures with the base pairing sections forming the stem and noncomplementary bases forming the loop as shown below.

*RNA can have intrastrand hydrogen bonding and form stem-loop structures.*

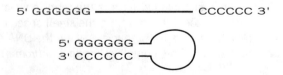

5′ GGGGGG ———————— CCCCCC 3′

5′ GGGGGG
3′ CCCCCC

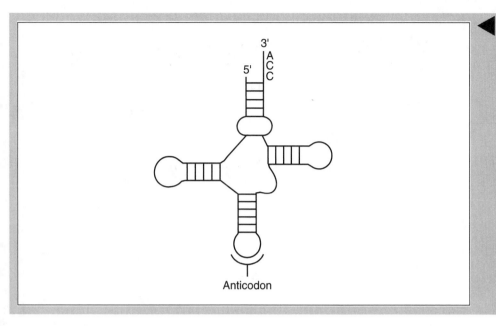

3'
|
A
C
C
5'

Anticodon

Intramolecular hydrogen bonding gives RNA a secondary structure, which makes the molecule very compact, more stable, and able to fit in a smaller space. In addition, the presence of a secondary structure allows the RNA molecule to carry out biologic functions more efficiently than a linear molecule, as we will see when we look at the process of protein synthesis in Chapter 3.

***Classes of RNA Molecule Found in Human Cells.*** Five main classes of RNA are found in human cells. These include *messenger RNA (mRNA)*, *transfer RNA (tRNA)*, *ribosomal RNA (rRNA)*, *heterogeneous nuclear RNA (hnRNA)*, and *small nuclear RNA (snRNA)*. mRNA makes up only a small percentage of the total RNA in the cell, is short-lived, and shows a large variation in base sequence from one mRNA molecule to another. They are the messengers that carry the information from the DNA to the protein-synthesizing machinery.

*mRNA molecules carry information from DNA to the protein-synthesizing machinery.*

tRNA molecules are small polynucleotides ranging from 74 to 95 nucleotides in length. Their major role is to carry specific amino acids to the ribosomes during protein synthesis. For each of the 20 amino acids found in proteins, there is a unique tRNA molecule that specifically recognizes that amino acid. In some cases, there is more than one tRNA species for a single amino acid.

*tRNA molecules carry specific amino acids to the ribosomes.*

rRNA is the most abundant RNA species in cells and is found associated with proteins in structures called ribosomes. rRNA varies in size with the specific rRNAs of eukaryotic cells designated by sedimentation coefficients (S values). Human ribosomes contain rRNA species of the following sizes: 28S, 18S, 5.8S, and 5S.

*rRNA molecules are found in the ribosomes.*

hnRNA and snRNA molecules are found in the nucleus of the human cell. hnRNA is the immediate product of transcription, is complementary to one strand of the DNA, and is the precursor to mRNA before it is processed by splicing (see Chapter 3). snRNA is found associated with specific proteins and is involved in the processing of the hnRNA to mRNA before the exit of mRNA from the nucleus to the cytoplasm.

The role of these RNA molecules in the process of *transcription* (the transfer of information from DNA to mRNA) and the process of *translation* (the transfer of information from mRNA to protein) [Figure 2-7] is discussed in more detail in Chapter 3.

# DNA REPLICATION

## Important Features of DNA Replication

For cells to pass on their genetic material to their progeny, they must replicate their entire genome before cell division occurs. Replicating an entire genome is no easy matter since this process must be carried out efficiently in a defined period and with extreme fidelity.

**FIGURE 2-7** ▶

*Information Transfer in Cells.* Information present in a DNA molecule is transcribed by RNA polymerase into a single-stranded RNA molecule with a base sequence complementary to one of the DNA strands. The information in the RNA molecule is translated into protein by the protein-synthesizing machinery.

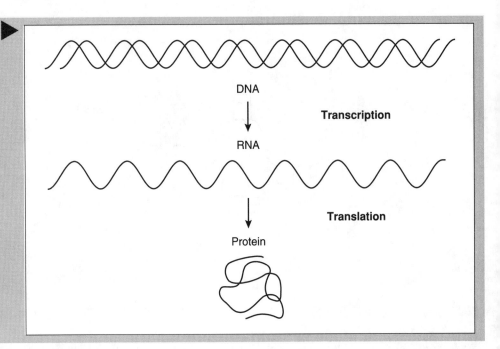

DNA

RNA

**Transcription**

**Translation**

Protein

Consider, for example, the human cell, which contains more than 6 billion bp of DNA. This DNA must be replicated once and only once at a precise time in the cell cycle. Furthermore, the DNA of a human cell is present as 46 individual chromosomes, each with multiple sites where replication begins. The period of DNA synthesis in a human cell with a generation time of 24 hours has a duration of approximately 7–8 hours. Imagine if you were given 46 long sheets of paper with millions of base pairs written in a particular sequence. How long would it take you to copy the sequence with complete accuracy? This is the job facing every human cell each time it replicates its DNA. Even in the bacterium *Escherichia coli*, replicating a single chromosome of 4 million bp is a complex and highly regulated process, which is just now being elucidated in detail. Knowledge about the process in human cells is understandably lagging behind but is being intensely investigated.

Human cells carry out the process of DNA replication at a particular time in their life cycle. All dividing human cells go through this *cell cycle*, which is made up of four stages, G1, S, G2, and mitosis as diagrammed in Figure 2-8. The stages of the cell cycle include *G1, gap phase 1*, which is characterized by active metabolic activity and protein synthesis in preparation for the *S phase*, or the period in which DNA synthesis (replication) occurs. S phase is followed by the *G2 phase*, during which chromosome condensation begins in preparation for the *M, or mitotic phase*. At M phase, the chromosomes are segregated, and cell division occurs, producing two daughter cells, each with a full content of DNA identical to that originating in the parent cell. The G1, G2, and S phases make up the time in the cell cycle between cell divisions; together, they are referred to as the *interphase* portion of the cell cycle. Cells not undergoing cell division, such as neuronal cells, are removed from the cell cycle and exist in a phase called G0. If cells in G0 are stimulated to grow, they move from G0 into the G1 phase.

As we will see in Chapter 5, entry and exit from one phase of the cell cycle to the other is a tightly controlled process. For this discussion, it is important to realize that if the DNA in the parent cell is not completely and accurately replicated, the cell is prevented from entering mitosis until the defect is remedied. Any mistakes made during DNA replication or any defects present in the DNA that are not corrected before mitosis are passed on to the daughter cells once the cell is allowed to proceed through M and undergo cell division. Thus, DNA is the only molecule found in human cells that is capable of repairing defects in its sequence. Any change not repaired has the potential of resulting in a mutation that will be expressed in all daughter cells derived from that parent.

The process of DNA replication must be very precise, and mistakes must be avoided at all costs. As we will see in Chapter 4, not only does the cell take precautions to make

DNA replication occurs in the S phase of the cell cycle.

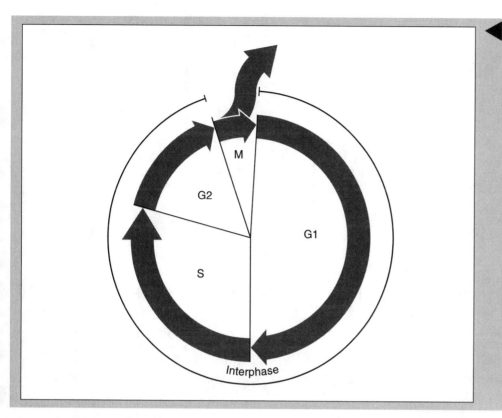

◀ FIGURE 2-8
**The Cell Cycle.** *The human cell cycle is represented as having a doubling time of 24 hours. The G1 phase lasts approximately 12 hours, the S phase about 7–8 hours, the G2 phase 3–4 hours, and mitosis about 1 hour. There are two major checkpoints in the cell cycle, one at the G1/S border and one at the G2/M border. The G1, S, and G2 phases make up the interphase portion of the cell cycle.*

sure the entire base sequence is correctly replicated during S phase, but the cell also has evolved numerous pathways to correct defects or mistakes in the DNA that escape the replication machinery or arise from damage to the DNA after replication. These repair processes are extremely important to maintain the proper DNA sequence during growth and reproduction of cells and to keep mistakes or mutations in the DNA to a very low frequency in all cells.

How does the cell manage to replicate 6 billion bp of DNA in a period of 7–8 hours without making mistakes, rereplicating, or missing any parts of the genome? Remember that the DNA molecule is a linear double-stranded structure with the two strands running in opposite polarity and held together by hydrogen bonding between A-T and G-C. During the process of DNA replication, the two strands separate, with each strand serving as a template for the synthesis of the two new daughter strands. This mode of replication is referred to as semiconservative replication and is diagrammed in Figure 2-9. After the first round of DNA synthesis and cell division, each of the two daughter cells contains one old strand of DNA and one newly synthesized strand of DNA. Because of the specificity of the hydrogen bonding, where A pairs with T and G with C, the sequence of bases in the template strand dictates the sequence of bases in the newly synthesized strands. In this way, the nucleotide sequence of the DNA molecule is maintained in both daughter cells and is identical to the original parental molecule.

## DNA Replication Starts at Specific Sites and Is Bidirectional

The chromosomes in human cells contain linear double-stranded DNA and are thought to have their ends fixed. The replication process on each chromosome starts at specific positions, referred to as the *origins of replication (ori)*. Ori, defined as the sites of initiation of DNA replication, must have certain properties that signal the replicating system to start replication. Each human chromosome has multiple ori placed every 150–200 kbp with 30,000 initiation sites localized over the entire human genome. The presence of many ori per chromosome results in multiple sections of the genome being replicated at the same time. Each small replicating unit, or *replicon*, has its own ori site where DNA replication begins. Once initiated, the replication complex proceeds bidirectionally on the chromosome until each replication bubble, or replicon, comes in contact

*The human genome has multiple sites for the initiation of DNA replication.*

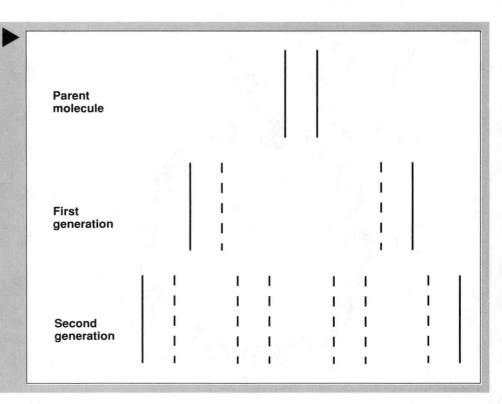

**FIGURE 2-9**

**DNA Replication Is Semiconservative.** The two strands of the double-stranded DNA helix are pulled apart with each parent strand serving as a template for the synthesis of a new daughter strand. After the first division, the daughter cells contain one new strand of DNA and one old strand of DNA. After the second division, two cells have one old and one new strand, while two cells contain only newly synthesized DNA.

Parent molecule

First generation

Second generation

*The leading DNA strand is synthesized continuously in a 5' to 3' direction. Lagging strand DNA synthesis is discontinuous with small fragments synthesized in a 5' to 3' direction.*

with the next one (Figure 2-10). In this way, the entire chromosome can be completely replicated during a single S phase.

## DNA Replication Is Discontinuous

In a short segment of DNA containing an ori site, as illustrated in Figure 2-11, there are two strands of the DNA, one strand running in a 5' to 3' direction and the opposite strand running in a 3' to 5' direction. As the two strands unwind and come apart, DNA synthesis begins at the ori and proceeds down the two strands. However, because of the properties of DNA polymerase, the enzyme involved in DNA replication, synthesis can only proceed from the ori in a 5' to 3' direction. One strand, therefore, is synthesized in a 5' to 3' direction, and the opposite strand is also synthesized in a 5' to 3' direction. Notice, however, that the two forks move away from each other. Since there is no known DNA polymerase that can synthesize DNA in a 3' to 5' direction, a DNA strand cannot be used as a template in the 5' to 3' direction. The cell resolves this problem by synthesizing short fragments of DNA, known as *Okazaki fragments*, using the 3' to 5' strand as a template. Each short segment, which is 200 nucleotides in length, is synthesized in a 5' to 3' direction. The resulting fragments are then joined together by an enzyme called DNA ligase to give one continuous strand of DNA. The strand of DNA that is synthesized continuously in a 5' to 3' direction is known as the *leading strand of DNA synthesis*, since it starts at a fixed point and leads DNA synthesis. The strand of DNA that is synthesized 5' to 3' in short pieces or discontinuously is called the *lagging strand of DNA synthesis*, since it was thought to lag behind the synthesis of the other strand. The current thinking is that the replication of both strands is coordinate.

Another important term used in describing the process of DNA replication is the *replication fork*. The replication fork refers to the part of the DNA that is being replicated at a given time, and it represents the region between the unreplicated portion of the DNA and the newly replicated portion of DNA. Since the DNA is synthesized bidirectionally, there are two replication forks in each replicon (see Figure 2-11).

What is the mechanism that signals DNA replication to begin at the site of initiation? Cells contain a protein, referred to as an *initiator protein*, which is coded by some other sequence on the DNA, not by the ori sequence. This initiator protein has the ability to recognize the origin sequence and signal the initiation of DNA replication. The basic mechanism by which initiator proteins recognize ori sequences in human cells remains to

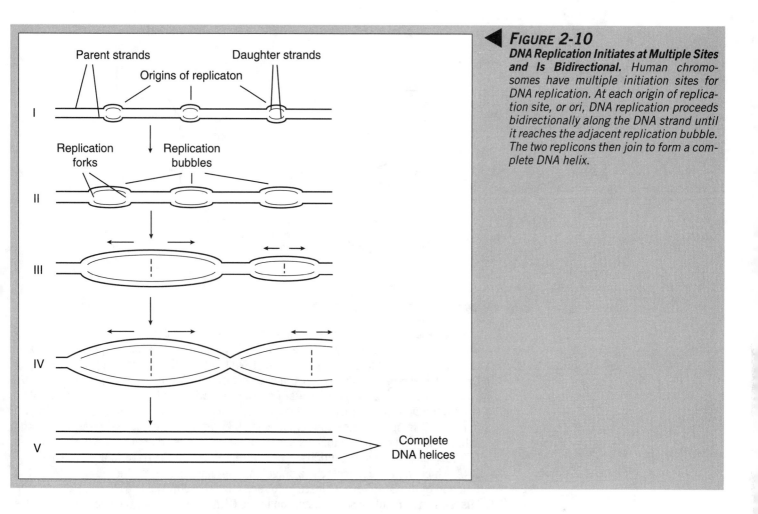

**FIGURE 2-10**
*DNA Replication Initiates at Multiple Sites and Is Bidirectional.* Human chromosomes have multiple initiation sites for DNA replication. At each origin of replication site, or ori, DNA replication proceeds bidirectionally along the DNA strand until it reaches the adjacent replication bubble. The two replicons then join to form a complete DNA helix.

be determined. In some other organisms, such as yeast, the details of this reaction are just being elucidated. A likely hypothesis is that this initiator protein binds the ori sequence and in some manner attracts the DNA replicating complex to the site on the DNA. This DNA replicating complex contains many proteins, some of which are listed in Table 2-2 and discussed further below.

## DNA Replication Requires a Primer

*DNA polymerase*, one of the key enzymes involved in DNA replication, is responsible for polymerizing deoxyribonucleotides to form the DNA strands. An important property of all known DNA polymerases is their inability to initiate DNA synthesis. All known DNA polymerases must have a *primer*—that is, a free 3' hydroxyl end of a polynucleotide—to which they can add another nucleotide. Without this primer, DNA polymerases will not synthesize DNA. The primer in DNA replication is not DNA. Instead, a small molecule of RNA, 5–10 nucleotides in length, which is synthesized by an enzyme called DNA primase, acts as the primer. DNA primase has the ability to initiate the synthesis of an RNA molecule at the origin of replication. Using this RNA primer, DNA polymerase adds deoxyribonucleotides to the free 3' hydroxyl group of the RNA and synthesizes a new strand of DNA complementary to the template strand. After the DNA is synthesized, the RNA molecule is removed from the DNA structure, and the resulting gap in the DNA is filled by DNA polymerase.

*DNA synthesis is initiated on both the lagging and leading strands by the synthesis of a short RNA fragment made by DNA primase.*

## Fidelity of DNA Replication

DNA polymerases, highly complex proteins made up of multiple subunits, serve several functions in the human cell. Some DNA polymerases, such as DNA polymerase-δ, in addition to catalyzing DNA synthesis, have a *3'–5' exonuclease* activity. An exonuclease is an enzyme that removes one base at a time from the end of a DNA molecule. A 3'–5'

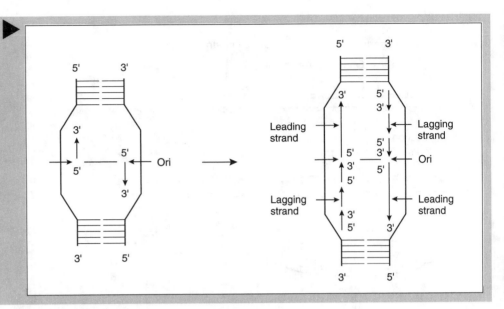

**FIGURE 2-11**

**DNA Synthesis Is Discontinuous.** *DNA polymerase can only synthesize DNA in a 5' to 3' direction. For both strands of the parental DNA molecule to serve as templates, one new strand, the leading strand, is synthesized continuously in a 5' to 3' direction, while the other strand, the lagging strand, is synthesized discontinuously in short pieces but still in a 5' to 3' direction. The short pieces, or Okazaki fragments, are then joined together to form the final DNA molecule.*

exonuclease removes the base at the 3' end of the DNA molecule. This nuclease activity is a very important property of all replicating DNA polymerases, serving as a safety valve to prevent mistakes in base pairing during DNA replication. With some low frequency, all DNA polymerases mistakenly put the wrong base into the new growing strand of DNA. When this occurs, a *mispair* results. For example, when an A appears on the template strand, the DNA polymerase occasionally places a C in the new DNA strand instead of a T. If this A-C mispair is not corrected, a mistake is made in the base sequence, and subsequent replications of this mistake result in a G-C base pair in the DNA instead of an A-T base pair. Such substitutions of one base pair by another lead to a *mutation* in the DNA, which might be reflected in a defect in a cellular function of the daughter cells. The function of the 3'−5' exonuclease is to recognize these mispairs as soon as they occur. Mispairs often result in some distortion in the DNA molecule that then acts as a signal to the exonuclease to remove the newly inserted but incorrect base. The DNA polymerase then has a second opportunity to insert the proper base into the growing chain. This exonuclease activity of DNA polymerase is referred to as the *proofreading function*, and it serves to reduce mutation rates in cells 100 to 1000 times. Even with this safety check, some mistakes are not corrected. To deal with these mistakes, the cell uses another process called DNA repair, which is discussed in Chapter 4.

## Proteins Involved in DNA Replication

Some of the proteins known to have an important function in the process of DNA replication are listed in Table 2-2. The known components acting at the replicating fork in a human cell include DNA polymerases, *DNA helicase, DNA primase, DNA ligase, RNase H1,* single-stranded DNA binding proteins, and other protein factors, which stimulate the action of DNA polymerases. In addition, the enzyme *DNA topoisomerase,* which acts downstream from the replicating fork, is essential for DNA replication.

***DNA Polymerases.*** DNA polymerases catalyze the formation of the phosphodiester bonds between adjacent deoxyribonucleotides in the DNA chain. DNA polymerase uses the template strand of DNA and the hydrogen-bonding properties of the purines (A, G) and the pyrimidines (C, T) to determine which deoxyribonucleotide is added to the 3' hydroxyl group of the growing chain. All DNA polymerases, as mentioned above, catalyze the synthesis of DNA only in the 5' to 3' direction. Human cells contain five known DNA polymerases, each of which serves a different function in the cell. The five known human polymerases are referred to as DNA polymerase-α, -β, -γ, -δ, and -ε. Some of the properties of these enzymes are given in Table 2-3.

DNA polymerase-δ is the major replicating protein in human cells and is involved in both leading strand and lagging strand replication. DNA polymerase-α is found in cells in a complex with another protein, the DNA primase. This complex is involved in the replication of the lagging strand. The role of DNA primase is to make the RNA primers

*Human cells contain five different DNA polymerases.*

◀ TABLE 2-2
Human Proteins Involved in DNA Replication

| Protein | Function |
|---|---|
| DNA helicase | Unwinds DNA and breaks hydrogen bonds |
| Single-stranded DNA–binding protein (RPA) | Binds single-stranded DNA to prevent hydrogen bonding |
| Proliferating cell nuclear activity (PCNA) | Stimulates DNA polymerase-$\delta$ activity |
| DNA polymerase-$\delta$ | Leading and lagging strand DNA replication and 3′–5′ exonuclease proofreading |
| DNA polymerase-$\alpha$/DNA primase complex | Synthesis of RNA primers and lagging strand synthesis |
| DNA ligase | Seals 3′ terminal hydroxyl and 5′ terminal phosphate groups of adjacent nucleotides in DNA |
| Ribonuclease H1 (RNase H1) | Removes RNA from a RNA–DNA hybrid |
| DNA topoisomerase | Relaxes DNA by breaking and resealing phosphodiester bonds |

Note. RPA = replication protein A.

discussed above with DNA polymerase-$\alpha$ then adding deoxyribonucleotides to the 3′ terminal of the primer for a short distance of 30 or so nucleotides. The DNA polymerase-$\alpha$/DNA primase complex then falls off the DNA and is replaced by DNA polymerase-$\delta$, which continues the synthesis of the growing chain. DNA primase also must synthesize the single RNA primer needed by the leading strands at each origin of replication.

DNA polymerase-$\beta$ and -$\epsilon$ function mainly in the process of DNA repair and do not appear to be directly involved in the process of replicating the entire genome. DNA polymerase-$\gamma$ is found exclusively in the mitochondria and is responsible for the replication of the circular double-stranded DNA found in this organelle.

***DNA Helicase and DNA Topoisomerase.*** During the process of DNA replication, the two strands of DNA must be pulled apart to form the single-stranded DNA required by the replicating proteins. The pulling apart of the two strands of DNA is catalyzed by an enzyme called DNA helicase. DNA helicase functions by breaking the hydrogen bonds holding the two DNA strands together, then unwinding the DNA and opening up the molecule at localized spots. This breaking of hydrogen bonds requires energy, which is obtained from the hydrolysis of adenosine triphosphate (ATP). It is important to realize that if the two strands are not kept separated by some means, they will simply come back together and form new hydrogen bonds. This function of keeping the stands separated is carried out in human cells by a protein called replication protein A (RPA). RPA is a single-stranded DNA binding protein that coats the two separated DNA strands, thus masking the base pairs and preventing base pairing from occurring.

◀ TABLE 2-3
Properties of Human DNA Polymerases

| DNA Polymerase | Size (catalytic subunit) [kilodaltons] | Location | Function in the Cell |
|---|---|---|---|
| $\alpha$ | 160–185 | Nucleus | Lagging strand replication |
| $\beta$ | 40 | Nucleus | DNA repair |
| $\gamma$ | 125 | Mitochondria | Replication of mitochondrial DNA |
| $\delta$ | 125 | Nucleus | Leading and lagging strand replication |
| $\epsilon$ | 210–230 | Nucleus | DNA repair (?) |

DNA helicase functions at the edge of the replication fork, unwinding and opening up the DNA as replication proceeds down the DNA molecule. However, as the DNA continues to be unwound at the replicating fork, a problem arises further down the DNA molecule. As the helicase unwinds the DNA at the fork, the helix downstream becomes more and more tightly wound. With this increase in supercoiling of the DNA, movement of the replicating fork becomes more and more difficult and finally stops. To release the tension on the DNA molecule requires an enzyme that can relax the DNA. This relaxing protein found in all cells is called DNA topoisomerase. DNA topoisomerase functions by breaking the phosphodiester bonds between adjacent base pairs and then, by holding onto one strand of the DNA, unwinds it to make the DNA less tight. After unwinding the DNA, the topoisomerase reseals the phosphodiester bond, resulting in a DNA molecule that is now more relaxed and no longer too tightly supercoiled. DNA topoisomerases and DNA helicases are crucial proteins in the cell, not only for the process of DNA replication but also for the process of transcription.

The major differences between DNA helicase and DNA topoisomerase activities are as follows. DNA helicases break hydrogen bonds and unwind the DNA, whereas DNA topoisomerases break phosphodiester bonds, unwind the DNA, and then reseal it by forming a new phosphodiester bond.

*Ribonuclease H1 (RNase H1), DNA Ligase, and PCNA.* The RNA primers used by DNA polymerases to synthesize DNA must be removed from the DNA molecule before replication is completed. Removal of the RNA primer is carried out in human cells by the enzyme *RNase H1*. RNase H1, H standing for hybrid, specifically degrades RNA present in a DNA–RNA hybrid, hence the name. After removal of the RNA primer from the RNA–DNA hybrid, lagging strand DNA synthesis is completed by filling in the gap function of DNA polymerase. Then the ligation of the 3′ hydroxyl terminus of the DNA of one Okazaki fragment with the 5′ terminal phosphate of the DNA of the adjacent fragment occurs to form a phosphodiester bond. The enzyme that carries out this final ligation step is called *DNA ligase*. DNA ligase activity is essential in the cell for many functions that involve DNA integrity. The enzyme is required for closing any nicks that occur in double-stranded DNA.

An additional protein involved in DNA replication in human cells is known as *proliferating cell nuclear antigen* (*PCNA*). This protein, which is part of the DNA polymerase-δ protein complex, acts by stimulating the activity of the DNA polymerase.

The functioning of these proteins in lagging strand DNA replication can be visualized in the model shown in Figure 2-12. In lagging strand DNA replication, the two DNA strands are pulled apart by DNA helicase, exposing single-stranded regions on the DNA. RPA, a single-stranded DNA-binding protein, binds the DNA to keep the strands apart. The DNA polymerase-α/DNA primase complex then binds to the DNA and begins the synthesis of the RNA primer. Using the RNA primer, DNA polymerase-α synthesizes a small stretch of DNA by the addition of deoxyribonucleotides to the 3′ terminus of the primer. The DNA polymerase-α complex is then dissociated from the DNA and replaced with DNA polymerase-δ, which continues replication of the lagging strand. After two Okazaki fragments have been replicated on the lagging strand, they each still contain a small RNA primer. Before DNA replication can be completed, the RNA must be removed from the DNA. This step is carried out by RNase H1. Once the RNA has been removed from the DNA, the gap that remains is filled by DNA polymerase-δ, using the opposite DNA strand as a template. Finally, the nick remaining between the two adjacent Okazaki fragments is closed by the action of DNA ligase to give a large, continuous, and newly synthesized DNA strand.

## Replicating the Ends of DNA: Telomeres and Telomerase

Located at the ends of each human linear chromosome is the randomly repeated base sequence, TTAGGG. These sequences, which are repeated 100 to 1000 times, form the ends, or *telomeres*, of all chromosomes and are important for maintaining the integrity of the chromosome. Replication of DNA at the ends of chromosomes proves to be a special problem in DNA replication. Remember that the lagging strand is copied in small pieces from the template strand. However, since the DNA polymerase-α/DNA primase complex

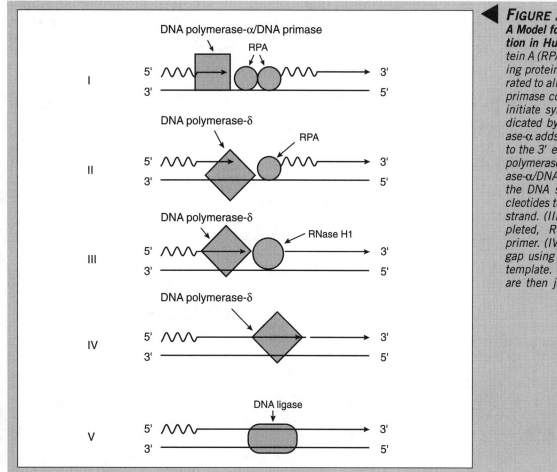

FIGURE 2-12
*A Model for Lagging Strand DNA Replication in Human Cells.* (I) Replication protein A (RPA), a single-stranded DNA-binding protein, keeps the DNA strands separated to allow the DNA polymerase-α/DNA primase complex to bind to the DNA and initiate synthesis of an RNA primer (indicated by the wavy line). DNA polymerase-α adds about 30 deoxyribonucleotides to the 3' end of the RNA primer. (II) DNA polymerase-δ displaces the DNA polymerase-α/DNA primase complex and extends the DNA strand by adding deoxyribonucleotides to the 3' end of the growing DNA strand. (III) When DNA synthesis is completed, RNase H1 removes the RNA primer. (IV) DNA polymerase-δ fills in the gap using the opposite DNA strand as a template. (V) The two Okazaki fragments are then joined together by DNA ligase.

cannot complete replication at the very end of a DNA molecule after removal of the RNA primer, the very ends of the chromosome remain unreplicated. If the ends of the chromosome were left unreplicated after each round of replication, the chromosomes would become shorter and shorter, and eventually the loss of important genes located near the ends would lead to a lethal event in the cell.

The cell overcomes this problem by having a special enzyme called *telomerase*. The role of telomerase in cells is to add bases to the 3' termini of the template DNA. This elongation of the template strand allows for continued synthesis of the DNA of the lagging strand by the normal activity of the DNA polymerase-α/DNA primase complex. The mechanism by which telomeres are replicated by telomerase is diagrammed in Figure 2-13.

Telomerase is an unusual enzyme, consisting of both a RNA component and a protein component. The RNA component of human telomerase is approximately 450 nucleotides in length and contains a template region of 11 nucleotides, which is complementary to the telomere sequence, TTAGGG. This complementary RNA sequence allows recognition and base pairing of telomerase with the 3' ends of telomeres. The protein component of telomerase, which functions like a polymerase, can then direct the synthesis of the telomeres by catalyzing the addition of deoxyribonucleotides to the 3' ends of the chromosome, using the RNA sequence as a template. Telomerase then moves along the DNA, repeating the process several times and extending the length of the telomeres on the template strand. The elongation of the template strand now allows for continued synthesis of the lagging strand by DNA polymerase-α/DNA primase.

Recent evidence suggests a crucial role of telomeres and telomerase in cellular aging and in the development of cancer. Studies with human cells have shown that telomerase is active in germline cells but is not detectable in many normal somatic cells. Consistent with this observation is the finding that somatic cell telomeres are much shorter than those found in sperm cells. As a normal somatic cell divides, it gradually loses telomeric sequences, suggesting that telomeric loss may play a role in determining

*The ends of human chromosomes contain randomly repeated sequences that make up structures known as telomeres. Telomeres are synthesized by a special enzyme called telomerase.*

the life span of a human cell. Other studies have shown that immortal or tumor cells have activated their telomerase activity, and the presence of this activity may be related to the uncontrolled growth of these malignant cells. (See Chapter 5 for more information on telomeres.)

***FIGURE 2-13*** ▶

**Replication of Telomeres by Telomerase.** *The enzyme telomerase contains a RNA component that is complementary to the human telomere sequence, TTAGGG. Telomerase binds to the telomeres on the ends of the chromosome and catalyzes the addition of deoxyribonucleotides to the 3' end of the template DNA strand. The enzyme then translocates in a 5' to 3' direction on the DNA and repeats the binding and addition of deoxyribonucleotides to the 3' end of the DNA. Once the template strand is elongated, the DNA polymerase-α/DNA primase complex can initiate complementary DNA synthesis on the lagging strand.*

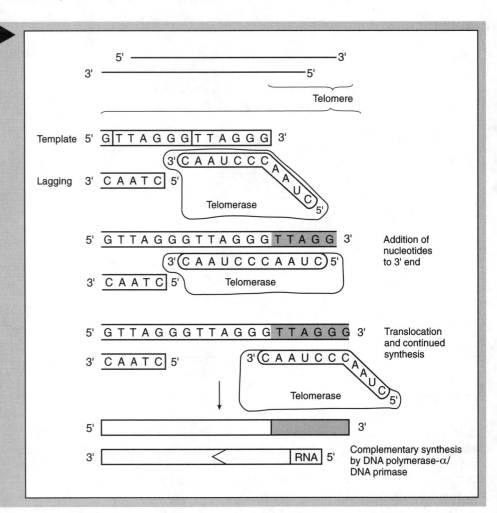

# RESOLUTION OF CLINICAL CASE

Mary got her wish. The twins were monozygotic, and the two girls were identical to one another. Monozygotic twins arise from a single zygote, which divides very early to form two identical embryos, each of which develops into a fetus. DNA replication is a faithful process, resulting in the production of two, new double-stranded DNA molecules that are identical in sequence to each other and to the parental molecule. Since both embryos developed from a single zygote, the two girls have identical DNA sequences, and their genotypes and phenotypes are the same (Figure 2-14).

In the case of dizygotic twins, two ova formed by a double ovulation are each fertilized by separate sperm cells. When this occurs, the resulting twins have on the average only one-half of their genes or DNA sequences in common. DNA replication is still faithful, but each fertilized egg begins development with a different complement of DNA. Thus, dizygotic twins look like siblings but are not identical.

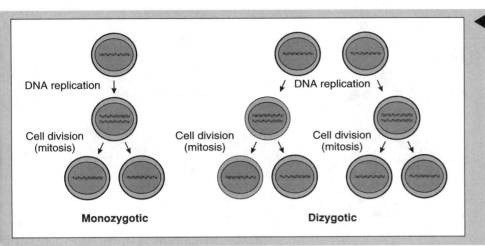

FIGURE 2-14
**Fidelity of DNA Replication Gives Two Identical Daughter Cells.** *Monozygotic twins arise from a single zygote, which undergoes DNA replication and then divides to form two identical embryos. Faithful replication of the genome assures that each fetus has an identical DNA complement and the resulting twins are identical. Dizygotic twins arise from two different zygotes with only half of their DNA content in common. Dizygotic twins are, therefore, not identical but look like siblings.*

# REVIEW QUESTIONS

**Directions:** For each of the following questions, choose the **one best** answer.

1. A sample of double-stranded DNA contains 20 mol percent adenine (A). What is its mol percent guanine (G)?

   **(A)** 20

   **(B)** 40

   **(C)** 60

   **(D)** 30

   **(E)** Insufficient information

2. Which of the following statements describes the replication of human chromosomes?

   **(A)** Replication of the lagging strand requires only a single RNA primer

   **(B)** Replication is unidirectional

   **(C)** Both strands are synthesized in a 3′ to 5′ direction

   **(D)** Replication requires a single type of DNA polymerase

   **(E)** Replication initiates at multiple origin sequences

3. Double-stranded DNA has which of the following characteristics?

   **(A)** It contains two strands that are joined to each other through phosphodiester bonds

   **(B)** It replicates in a semiconservative manner

   **(C)** It involves hydrogen bonds between purine–purine and pyrimidine–pyrimidine pairs

   **(D)** It always contains an equal number of adenine and guanine bases

   **(E)** It is exclusively found in the cell nucleus

4. Both DNA and RNA have which of the following characteristics in common?

   **(A)** They are exclusively found in the cell nucleus

   **(B)** They are bound to histone proteins in the nucleus of eukaryotic cells

   **(C)** They contain deoxyribose

   **(D)** They contain phosphodiester bonds

   **(E)** They contain adenine, thymine, guanine, and cytosine

**Directions:** The group of questions below consists of lettered choices followed by several numbered items. For each numbered item, select the appropriate lettered option with which it is most closely associated. Each lettered option may be used once, more than once, or not at all.

**Questions 5–7**

Match each site of action with the appropriate enzyme. The diagram pertains to the replication of double-stranded DNA. DNA shown with *arrows* indicates the newly synthesized strands and the direction of replication.

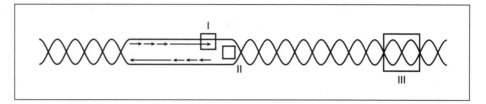

(A) Telomerase
(B) DNA topoisomerase
(C) DNA polymerase
(D) DNA helicase
(E) DNA ligase

5. This enzyme functions at the square marked I on the diagram.

6. This enzyme functions at the square marked II on the diagram.

7. This enzyme functions at the square marked III on the diagram.

# ANSWERS AND EXPLANATIONS

**1. The answer is D.** Double-stranded DNA has an equal number of A and T residues and an equal number of G and C residues. If A equals 20%, T must equal 20%, making A-T equal to 40%. The remainder is 60%, with 30% for C and 30% for G.

**2. The answer is E.** DNA replication, which is bidirectional, initiates at multiple sites on each chromosome to allow the total genome to be replicated during S phase. The lagging strand requires multiple RNA primers, one for each Okazaki fragment. Both the leading and lagging strands are synthesized in a 5′ to 3′ direction. Both DNA polymerase-$\alpha$ and DNA polymerase-$\delta$ are required.

**3. The answer is B.** The two strands of the double-stranded DNA molecule are pulled apart, each strand serving as a template for the synthesis of each daughter strand. In the daughter cells, the new DNA molecule is made up of one new strand and one old strand. The two strands are held together by hydrogen bonds; the hydrogen bonding occurs between purines and pyrimidines. The number of adenine bases equals the number of thymine bases. Double-stranded DNA is found in the mitochondria.

**4. The answer is D.** Both DNA and RNA contain phosphodiester bonds. In addition to being found in the nucleus, DNA is found in the mitochondria, and RNA is found in the cytoplasm and in the mitochondria. Only DNA is bound to histones. RNA contains ribose, not deoxyribose, and uracil, not thymine.

**5–7. The answers are: 5-C, 6-D, 7-B.** DNA polymerase acts at the replication fork and adds nucleotides to the 3′ end of the growing DNA strand. DNA helicase acts at the replication fork opening the double-stranded DNA to form single strands and allowing the DNA polymerase to move down the strand. DNA topoisomerase acts downstream of the replicating fork to release the tension on the supercoiled DNA.

# REFERENCES

1. Watson JD, Crick FH: Molecular structure of nucleic acid. A structure for deoxyribose nucleic acid. *Nature* 171:737–738, 1953.

# TRANSFER OF INFORMATION FROM DNA TO RNA AND PROTEIN

## INTRODUCTION OF CLINICAL CASE

Jessie and Susan Johnson brought their 6-month-old son, Charlie, to the physician for an examination to determine the basis for excess swelling of his feet and hands. They informed the physician that Charlie has had recurrent infections, often with associated pain. The physician, upon examination of the boy, found that Charlie's spleen was palpable 3.0 cm below his right costal margin. Charlie is of African-American descent, suggesting to the physician that with such symptoms the boy may have a blood disorder. She ordered a hemoglobin electrophoresis and a blood count and prescribed antibiotics for the boy.

Charlie's tests came back 2 days later with the following results:

|  | Patient | Normal |
|---|---|---|
| Hemoglobin (Hb) | 8.2 g/dL | 14–18 g/dL |
| Hematocrit | 24% | 40%–50% |
| Hb electrophoresis | HbS (70%) | < 1% |
| HbF | 24% | < 1% |
| HbA$_2$ | 5% | 1%–3% |
| HbA | < 1% | 98% |

The presence of HbS and the high percentage of HbF confirmed the physician's suspicion that Charlie has sickle cell anemia and will need constant evaluation and monitoring. As a final test, the physician ordered a DNA analysis on Charlie and his parents to determine if one or both parents are carriers of the disease.

# TRANSCRIPTION: RNA PRODUCTION FROM A DNA TEMPLATE

The biologic information stored in a DNA molecule must find a way to be expressed in a cell. This expression is accomplished by converting the information in DNA into protein molecules, using mRNA as an intermediate or messenger; hence the name, *mRNA*. The other forms of RNA, including tRNA, ribosomal RNA (rRNA), heterogeneous RNA (hnRNA), and small nuclear RNA (snRNA), also serve important functions in the final production of protein molecules and are involved in the process of protein synthesis or translation of mRNA into protein. All of these types of RNA molecules are synthesized directly from a DNA template by enzymes called RNA polymerases in a process known as *transcription*.

> **Transcription** is the process by which RNA polymerase synthesizes RNA directly from a DNA template.

## Important Aspects of a Typical Human Gene

A typical gene found in the nuclear DNA of a human cell has certain features that are common to all human genes. Figure 3-1 depicts such a gene and emphasizes these important features. Each gene is made up of a particular sequence of base pairs with the

**FIGURE 3-1**

***Diagram of a Typical Human Gene.*** *A typical human gene is made up of exon and intron sequences that are transcribed by RNA polymerase into a primary transcript. The primary transcript is then processed by the addition to the 5′ end of the Cap structure, the addition to the 3′ end of the poly A tail, and then the removal of the intron sequences and the splicing together of the exon sequences. The mature mRNA contains only exon sequences that have information for protein sequences as well as signals for the initiation and termination of protein synthesis. UTR = untranslated regions.*

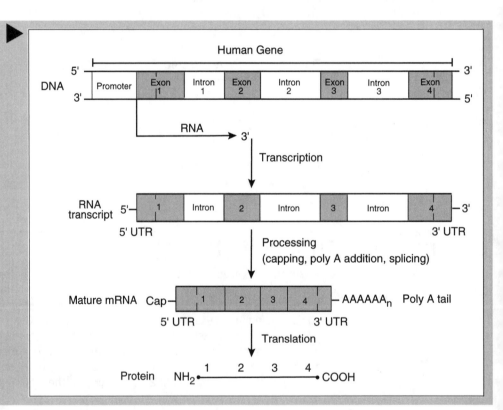

> **Exons** are sequences within a human gene that are conserved during the processing of the primary RNA transcript and are present in the mature mRNA.

> **Introns** are sequences within a human gene that are removed from the primary RNA transcript during processing and are not present in the mature mRNA.

size of different genes ranging from tens to millions of kilobase pairs (kbp) of DNA. The sequence of base pairs (bp) is specific for each gene and determines the specificity of the product coded for by that gene. However, not all of the bases present in a gene are expressed in the final product. Human genes are often split into: (1) sequences that remain in the final mature mRNA and code for information (*exons*) and (2) sequences that are removed from the primary RNA transcript during processing (*introns*). In addition to coding sequence information, exons contain other sequences necessary for the functioning of mRNA. Thus, an exon is defined as a sequence in a primary RNA transcript that is conserved during the processing of the transcript into a mature mRNA molecule.

Sequences that signal the initiation of transcription are present in each gene. One of these sequences is referred to as the *TATA sequence*; a second sequence is known as the initiator (Inr) sequence. These sequences, TATA and Inr, represent *promoter sequences* and determine the site at which the initiation of transcription begins on the DNA molecule. Transcription is initiated by the binding of transcription factors and the enzyme, RNA polymerase, to the promoter site. Once bound, RNA polymerase catalyzes the synthesis of an RNA molecule, starting at the promoter site and extending the length of the RNA, using the sequence of bases from one strand of the DNA as a template. Similar to the synthesis of DNA, RNA is synthesized starting at the 5' end of the DNA and moving in a 3' direction. The RNA molecule produced in transcription is a single-stranded molecule with a base sequence complementary to one strand, the template strand, of the DNA helix.

> **Promoter sequences** determine the site where the initiation of transcription begins on a DNA molecule.

The initial RNA transcript contains both intron and exon sequences and must be processed to the mature mRNA in a series of steps. These steps include the addition of a *Cap structure* at the 5' end of the mRNA, the addition of a *poly A tail* at the 3' end of the mRNA, and finally, the removal of the intron sequences and the *splicing* together of the exon sequences to produce the final mRNA product.

> RNA is synthesized by RNA polymerase in a 5' to 3' direction.

One additional feature of the mature mRNA molecule is the presence of untranslated regions (UTR) found both at the 3' and 5' ends of the molecule. These are referred to as 3' UTR or 5' UTR, and they represent sequences in the exons that remain in the mRNA but are not translated into protein. These regions contain signals that are necessary for both processing of the mRNA and its translation into protein.

> UTRs are sequences in exons found either at the 3' end or the 5' end of a mRNA that will not be translated into amino acids.

## Basic Steps in the Process of Transcription

### INITIATION OF TRANSCRIPTION

Every gene contains a base pair sequence in the DNA called a promoter. The function of the promoter is to set the location and the direction of transcription on a DNA template. RNA polymerase must be recruited to the promoter site, a task mediated by a large number of proteins called transcription factors. *Transcription factors* can bind to sequences in the promoter or can bind to one another in many different arrays to instruct RNA polymerase whether or not to transcribe a particular gene. Different cells in the body contain identical DNA sequences, but they may express different genes at different times and in response to different signals. Controlling the variation of expression of genes is the function of interaction of various transcription factors within different cells. The regulation of the transcription process by different transcription factors will be discussed in detail in Chapter 8. For this discussion, it is important to define some of the basic steps in the initiation of transcription of all genes. Foremost among these steps is the binding to the TATA sequence in a promoter of a gene by a protein known as the *TATA-binding protein (TBP)* [Figure 3-2].

> **Transcription factors** interact at sequences in the promoters of human genes and regulate transcription.

◄ **FIGURE 3-2**
**Initiation of Transcription at the Promoter Site.** *Promoter sites contain specific sequences such as the TATA sequence, which is recognized by the TATA-binding protein (TBP). The TBP attracts other transcription factors and RNA polymerase to the promoter. RNA polymerase then initiates transcription, using one of the DNA strands as a template.*

Once the TBP is bound to the TATA sequence in the promoter, additional transcription factors then bind either to the TBP or directly to the DNA, resulting in a protein-DNA complex that now recruits RNA polymerase to the promoter site. RNA polymerase binds to the DNA about 25 bp from the TATA box and initiates transcription in a specific direction (5' to 3'). Only the base sequence of one of the two DNA strands is copied; the directionality of the transcription process is determined by the presence of additional sequences in the promoter, such as the Inr.

> TBP binds to the TATA sequence in the promoter and, by interacting with additional transcription factors, recruits RNA polymerase to the promoter.

Figure 3-3 depicts the actions of RNA polymerase in transcription. Before transcription can begin, the two DNA strands must be pulled apart by the action of DNA helicases, an activity associated with some transcription factors. The separation of the two DNA strands to form single-stranded regions of DNA is necessary for the transcription process to begin and continue along a DNA molecule. Once the DNA strands have been pulled apart, only one strand, that running in a 3' to 5' direction, serves as the template for RNA synthesis. RNA polymerase catalyzes the formation of a phosphodiester bond by attach-

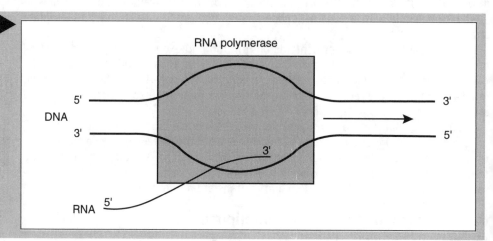

**FIGURE 3-3** ▶

*RNA Is Synthesized in a 5' to 3' Direction. The two strands of the double-stranded DNA helix are pulled apart by DNA helicase to give single-stranded regions. RNA polymerase begins synthesizing RNA in a 5' to 3' direction using the DNA strand, which is complementary and going in a 3' to 5' direction.*

ing the 5' phosphate of the incoming ribonucleotide to the 3' hydroxyl of the growing RNA chain. Multiple RNA molecules can be synthesized from a single DNA molecule with the binding of additional RNA polymerase molecules to the promoter sequence occupied by transcription factors.

In the nucleus of human cells, there are three different types of RNA polymerases referred to as RNA polymerase I, II, and III. *RNA polymerase II* is the polymerase discussed above whose function is to transcribe mRNA, starting from a TATA box within a promoter region. RNA polymerase II also synthesizes some of the snRNAs.

*Human cells contain in their nucleus three different RNA polymerases: I, II, and III.*

*RNA polymerase I* specifically transcribes the genes that code for three of the four ribosomal RNAs (28S, 18S, and 5.8S), while *RNA polymerase III* transcribes the genes that code for tRNAs and the 5S ribosomal RNAs. Each RNA polymerase is a complex molecule of high molecular weight and is made up of multiple subunits. The various RNA polymerases are attracted to their specific sites of RNA initiation by different transcription factors.

## RNA PROCESSING

The primary transcript of RNA made directly from the DNA template is not the final biologic form of the RNA. RNAs, including rRNA, tRNA, and mRNA, must be processed before they attain the form that functions in protein synthesis. rRNA is synthesized and processed in a small body in the nucleus known as the nucleolus. Processing of rRNA includes methylation of the RNA and removal of portions of the primary transcript by ribonuclease (RNase) [Figure 3-4].

A single RNA transcript produced by RNA polymerase I carries the information for

**FIGURE 3-4** ▶

*Processing of Ribosomal RNA. The primary RNA transcript synthesized by RNA polymerase I is processed in the nucleolus and is cleaved by ribonuclease (arrows) several times to produce the final RNA molecules of 18S, 5.8S, and 28S.*

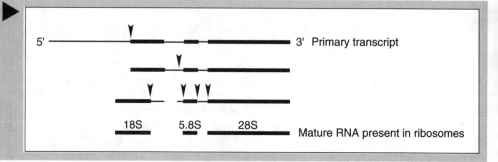

three of the mature rRNAs found in ribosomes, the 18S, the 28S, and the 5.8S rRNAs. Processing of the transcript occurs with cleavage of the primary transcript by RNase at the points marked by *arrows* in Figure 3-4. The initial cleavage yields an RNA product, which is then further cleaved to give the 18S, the 5.8S, and the 28S mature RNAs. Molecules of tRNA are processed in a similar way with the primary transcript being cleaved to smaller forms by RNase.

The major RNA processing in the cell occurs with mRNA. mRNA is synthesized by RNA polymerase II as a large primary transcript known as hnRNA, which contains both intron and exon sequences. Once the transcript is made by RNA polymerase, three independent processing events occur before the mRNA is competent to take part in protein synthesis. These events are diagrammed in Figure 3-5. The first step in the

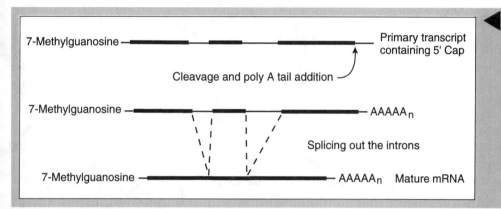

◀ **FIGURE 3-5**
**Processing of mRNA.** *mRNA is processed in the nucleus, where 7-methylguanosine is added to the 5′ end of the initial RNA transcript and poly A is added to the 3′ end of the initial RNA transcript. The introns are then removed, and the exons are spliced together.*

processing of hnRNA is the *capping*, or the attachment to the 5′ end of the RNA of a nucleotide called *7-methylguanosine* (*7-methyl G*). The 7-methyl G is attached through a 5′–5′ phosphodiester bond between the ribose on the 7-methyl G and the first ribose on the RNA. This bond is unique, since all the other phosphodiester bonds in RNA and DNA are 3′–5′. The addition of 7-methyl G or the *Cap structure* makes the resulting RNA molecule more resistant to exonucleases in the cell and increases the stability of the RNA.

*Capping is the attachment to the 5′ end of the primary RNA transcript of the nucleotide 7-methyl G.*

The second step in the processing of hnRNA occurs at the 3′ end of the molecule in an area referred to as the 3′ UTR. Within this 3′ UTR is the sequence, AAUAAA, known as the *poly A addition signal*. This sequence, located at the 3′ end of each primary RNA transcript, is recognized by a nuclease that cleaves the RNA 11–30 nucleotides past this site. Following the cleavage, an enzyme called poly A polymerase recognizes the poly A addition signal and adds residues of adenosine to the end of the mRNA to yield molecules of RNA that have 50–200 adenine residues (poly A tails) at their 3′ end.

*AAUAAA, a signal found at the 3′ end of the primary RNA transcript, is known as the poly A addition signal.*

The final step in the processing of hnRNA, which now carries a Cap at the 5′ end and a poly A tail at the 3′ end, is the removal of the intron sequences and the joining together of the exon sequences. This process, called *splicing*, occurs on small ribonuclear bodies known as *spliceosomes* (Figure 3-6). Spliceosomes function by recognizing the splice donor and splice acceptor sequences that occur at the beginning and end of each intron. The splice donor site, GU, begins each intron, and the splice acceptor site, AG, ends each intron. In addition, within each intron sequence is a site known as a branch point. Within the branch point is an adenosine residue which, in the presence of a spliceosome, combines with the G of the donor splice site. This reaction results in the release of the first exon and the formation of an intermediate structure known as the *lariat*. In the next step, exon 1 combines with the acceptor exon 2 to form a continuous mRNA that contains only the exon sequences, and the lariat is released.

*Splicing is the process whereby intron sequences are removed from the primary RNA transcript and the remaining exons are joined or spliced together.*

*The lariat is an unusual structure of RNA formed during the process of splicing.*

**FIGURE 3-6** ▶

**Splicing of the Precursor RNA to Form Mature mRNA.** *The removal of intron sequences from the primary transcript is carried out on spliceosomes, using the donor splice site, GU, and the acceptor splice site, AG, located at the exon–intron boundary. The A residue at the branch point combines with G at the donor exon–intron junction. This reaction results in the release of exon 1 and the formation of an intermediate lariat structure with exon 2. In step 2, exon 1 is joined or spliced to exon 2, and the lariat is released.*

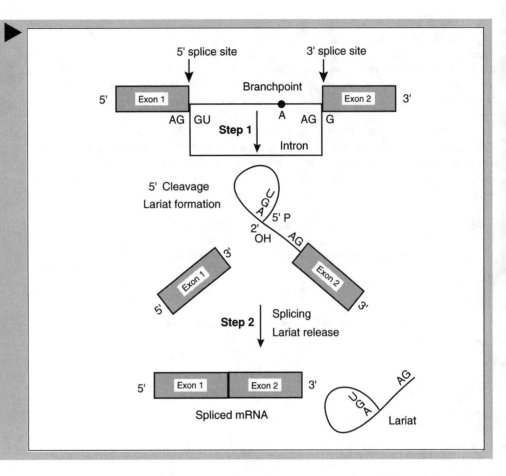

The RNA components of spliceosomes are the snRNAs discussed previously. The snRNA particle or spliceosome contains several snRNA molecules referred to as U-1 through U-6. (U refers to the high content of uridine found in the snRNA.)

The removal of introns and the splicing of exons must be an extremely precise process, considering that most genes are made up of multiple exons and introns, and the recognition sequences, GU and AG, can occur in many places in the RNA, roughly every 20 bp. Thus, there are additional sequences 5' and 3' to the GU and AG sequences, which play an important role in the splicing process. These sequences and their role in splicing are currently being elucidated. Different genes are made up of different numbers and sizes of both exons and introns, although there are some common features in the processing of mRNA. Notably, the order of exons in the DNA and in the primary transcript are preserved in the mature mRNA.

Even though the process of splicing is extremely accurate, mistakes do occur. Such mistakes can have disastrous effects on the protein produced, generally resulting in a mutation phenotype. Splicing mutations occur in the gene for dystrophin, the protein defective in Duchenne's muscular dystrophy. The dystrophin gene is made up of 79 exons and 78 introns spread over more than 2.5 million bp of DNA. The 79 exons are spliced together to give a mRNA of 14 kbp. If inaccurate splicing occurs, the resulting mRNA may be missing one or more exon sequences, resulting in a totally inactive protein and a child with Duchenne's muscular dystrophy.

On the other hand, splicing can give a cell flexibility in using the same DNA sequence to make different proteins. For example, if a gene is made up of 20 exons, the cell may determine which exon sequences are retained in the mature mRNA and will be coding sequences for the final protein product. The proteins will be similar but not identical, since they may share some exon sequences but differ in others. This process is referred to as differential or *alternative splicing*.

**Alternative splicing** *gives a cell flexibility to use the same DNA sequences to make different proteins.*

The final mature mRNA (see below) consists of a RNA molecule with: (1) a 7-methyl G Cap on the 5' end to give the mRNA stability; (2) a run of adenines or a poly A tail at the 3' end, which also contributes to stability as well as facilitates transport of mRNA from the nucleus to the cytoplasm; (3) exon sequences at the 5′ end and the 3′ end of the molecule, which are not translated into protein (UTR); and (4) exon sequences that are translated into protein.

```
           5' UTR                            3' UTR
5' Cap---------AUG-----------------UAA---------AAAAAAAAAA
```

At the beginning of the first translated exon, there is an AUG sequence, which is the signal for the initiation of protein synthesis. Toward the end of the last translated exon is a second signal (i.e., UAA, UAG, or UGA), one that signals protein synthesis to terminate.

## NUCLEAR TRANSPORT AND RNA STABILITY

The processes of DNA replication and transcription occur inside the nucleus. The site of protein synthesis, however, is located in the cytoplasm. Thus, the cell must transport to the cytoplasm, mRNA, and ribosomes, and transport back to the nucleus proteins such as DNA and RNA polymerases, transcription factors, and many others. The transport process, both in and out of the nucleus, is energy-requiring and driven by the hydrolysis of adenosine triphosphate (ATP). Proteins synthesized in the cytoplasm that are destined to the nucleus contain sequences known as nuclear localization signals, which enable them to be labeled for transport back to the nucleus. Likewise, the ribonuclear protein particle movement from the nucleus to the cytoplasm is mediated by a signal known as cytoplasmic localization signal.

Once the RNA molecule is made in the nucleus, it can have varying stabilities, unlike DNA, which is always very stable. (Stability refers to the degradation or loss of the nucleic acid inside the cell.) The binding of DNA to chromatin tends to protect the molecule from degradation by blocking the sites of action of deoxyribonucleases (DNases). RNA, on the other hand, is not protected in the same way and is much more subject to degradation by the existence of RNases in the cell. RNA is also single-stranded and, although it can form intrastrand bonding, the free ends are often exposed and degraded. RNA molecules differ from one another in their stability or so-called *half-life*. (Half-life is the time it takes to degrade half of the molecules present.) rRNA and tRNA form defined secondary structures that act to protect them from nucleases, making them more stable than most mRNA.

*The **half-life** of a molecule is the time it takes to degrade one half of the molecules present.*

Although generally less stable than tRNA and rRNA, mRNAs show major differences in stability from one to another. The significance of the stability of a mRNA molecule is reflected in its coding into protein. The amount of protein coded for by a mRNA is directly related to the stability of that mRNA. More stable mRNAs with longer half-lives make more protein simply because they are around longer and serve as messengers for longer periods of time. A mRNA with a shorter half-life will be degraded sooner and will code for fewer protein molecules. This variability of mRNA half-lives gives the cell another level of control over the amount of protein that is made from a particular mRNA molecule. If a cell requires large amounts of a particular protein, the mRNA coding for that protein tends to be more stable. If a particular protein is required only at certain times (e.g., proteins that function only in a particular phase of the cell cycle), it will generally be coded for by a less stable mRNA.

# TRANSLATION: mRNA-DIRECTED PROTEIN SYNTHESIS

## The Genetic Code

The mature mRNA molecules produced by the process of transcription and processing carry the information from the DNA sequence to the ribosomes, where the information is translated by the ribosomes to produce a protein of a determined amino acid sequence. The mRNA, which contains instructions for the synthesis of proteins in its structure, is composed of a single strand of ribonucleotides held together by phosphodiester bonds. At the 5′ end of mRNA is the Cap sequence, followed first by a sequence known as the 5′ UTR and then by a signal for translation initiation, the AUG codon. Toward the 3′ end of

*Translation is the process by which the ribosomes translate the information in mRNA into protein.*

the mRNA, there is a signal for translation termination (UAA, UAG, or UGA), followed by a 3' UTR. Finally, at the 3' end of the mRNA is the poly A tail. This mRNA is now ready for translation into protein.

```
            5' UTR                                      3' UTR
5' Cap----------AUG------------------UAA---------AAAAAAAAAA
```

The sequence of bases in the mRNA between AUG and UAA determines the exact sequence of amino acids in the protein. The mRNA at this point is colinear with the protein to be produced, meaning that there are no breaks or interruptions in the reading of the mRNA sequence. Protein synthesis starts at AUG initiator codon and proceeds in a 5' to 3' direction, each ensuing triplet coding for the next amino acid in the protein until a termination codon is reached, ending protein synthesis.

**Protein synthesis** *initiates at the first AUG codon (the initiator codon) present in the first exon.*

```
5'------ AUGUUUAAA -------------UAA---- 3'
         N-Met-Phe-Lys-----------stop
```

The information in the mRNA sequence is read by way of the *genetic code* shown in Figure 3-7. The genetic code is made up of 64 codons, each containing three bases: A, U, C, or G. All permutations of the four bases in groups of threes are represented in

**FIGURE 3-7** ▶

**The Genetic Code.** *The genetic code is made up of 64 codons, 61 of which code for amino acids and 3 of which code for stop codons. Each codon is made up of three nucleotides read in a 5' to 3' direction, starting with the first letter, then the second letter, and finally the third letter. For example, the codon AUG is the start or initiator codon for methionine. Phe = phenylalanine; Leu = leucine; Ser = serine; Tyr = tyrosine; Cys = cysteine; Trp = tryptophan; Pro = proline; His = histidine; Gln = glutamine; Arg = arginine; Ile = isoleucine; Met = methionine; Thr = threonine; Asn = asparagine; Lys = lysine; Val = valine; Ala = alanine; Asp = aspartic acid; Glu = glutamic acid; Gly = glycine.*

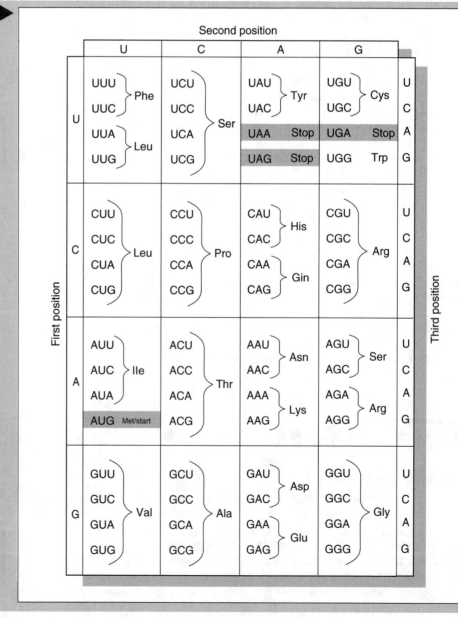

Figure 3-7. Each of 61 triplets codes for a particular amino acid. For example, when the codon UUU appears in a mRNA molecule, a phenylalanine (Phe) will be found at that position in the protein. Likewise, if the codon AAA appears in the mRNA, a lysine (Lys) will be found at the corresponding position in the protein.

Of the 64 possible codons, 3 of the codons, *UAA, UAG, and UGA*, do not code for an amino acid and are referred to as *nonsense* or *termination codons*. When one of these codons appears in the mRNA sequence, it is a signal to stop translation. The genetic code, with some minor exceptions, is universal, with the same codons always coding for the same amino acid. For example, UUU will always code for phenylalanine, and AAA will always code for lysine. An exception to this general codon usage occurs in the mitochondria, where UGA codes for tryptophan (Trp), AUA for methionine (Met), and CUA for threonine (Thr).

From the codons listed in Figure 3-7, it is clear that more than one codon can code for the same amino acid. A good example is found with the amino acid, valine (Val), which can be coded for by either GUU, GUC, GUA, or GUG. The ability of more than one codon to specify a single amino acid is known as the redundancy of the genetic code. The amino acids tryptophan and methionine, on the other hand, are specified by a single codon, UGG and AUG, respectively.

## tRNA: Translator of the Genetic Code

Figure 3-8 is a diagram of a tRNA molecule that was discussed in Chapter 2. tRNA is a critical component of protein synthesis. Like rRNA and mRNA, it is first made

> *The **genetic code** is made up of 64 codons, which code for 61 amino acids and 3 termination or stop signals.*

> *A nonsense or **termination codon** (i.e., UAA, UAG, UGA) does not code for an amino acid and signals the termination of protein synthesis.*

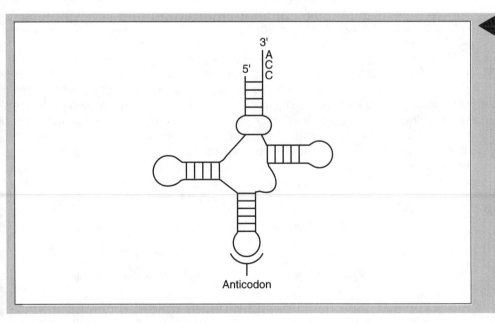

**FIGURE 3-8**
**The Structure of tRNA.** *Single-stranded RNA molecules can have intrastrand base pairing and form stem-loop structures. The ACC at the 3' end represents the site where amino acids attach to the tRNA. The anticodon loop recognizes complementary codons in the mRNA during protein synthesis.*

in the nucleus as a precursor RNA and then processed to form the final product. Note the 3' end of the molecule in the figure. With all tRNA molecules, the nucleotides CCA are added to the 3' end of the tRNAs after transcription has occurred. This CCA sequence is an important part of the tRNA molecule and represents the site at which specific amino acids attach. A second critical part of tRNA is indicated at the bottom loop of the structure. This site, the *anticodon site*, is required to recognize a specific codon found in the mRNA as indicated above.

tRNA has an important role in the process of taking the information in the mRNA and turning it into a protein molecule. Every amino acid is recognized by a specific tRNA molecule and becomes attached to the CCA end of the tRNA. The three base sequence on the anticodon is complementary to a specific codon found in the mRNA. The codon in the mRNA (e.g., GGC), is recognized by the anticodon of the tRNA, CCG, allowing the tRNA to attach to the mRNA by base complementarity (Figure 3-9). Since each tRNA carries

> *The **anticodon sequence** in a tRNA molecule recognizes specific codons in the mRNA.*

> *By way of its anticodon, a tRNA molecule binds complementary base pairs in the mRNA and adds the amino acid it carries to the growing polypeptide chain.*

FIGURE 3-9 ▶

**Interactions of Codons and Anticodons.**
*The anticodon on the tRNA recognizes a complementary codon on the mRNA and places the specific amino acid carried on the tRNA into the growing polypeptide chain. Met = methionine; Gly = glycine; Ser = serine; Ile = isoleucine; Ala = alanine.*

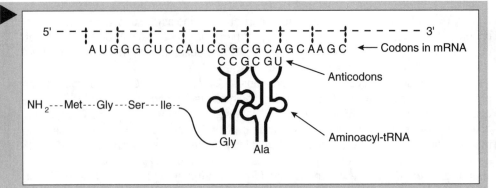

a specific amino acid, the binding of the tRNA to the mRNA by base pairing between the codon and the anticodon provides the specificity for placing a particular amino acid in the growing polypeptide chain. In the example shown, a glycine (Gly) is placed at the site of the GGC codon. UAA, UAG, and UGA, the three termination codons, do not have a tRNA with a complementary anticodon; therefore, they are not recognized by a tRNA, and no amino acid is added to the grown polypeptide chain. Their presence in a mRNA results in chain termination and signals the end of synthesis of a protein molecule.

One additional complication arises with the genetic code. There are 61 codons specifying amino acids, but there are less than 61 tRNA molecules found in cells. Therefore, some tRNA molecules must be able to recognize more than one codon, a phenomenon called "wobble." In some cases, a tRNA molecule, which has a fixed anticodon such as AAU, recognizes the sequence UUA in the mRNA. However, if the sequence in the mRNA is UUG, it may also be recognized by the same tRNA molecule; in both cases, leucine is added to the growing polypeptide chain. Wobble effects are found with the third base of the codon.

Once the reading of the message has begun at AUG, each triplet is read in order with no interruptions or gaps. If the sequence is changed even by a single base pair (e.g., UUU to UCU), the amino acid specified in the protein will be serine (Ser) instead of phenylalanine. Thus, just a single base change in the DNA sequence when transcribed into mRNA can result in a single amino acid change in a protein, producing a mutant protein. If that amino acid change occurs in a part of the protein necessary for activity, the activity will be lost, and the mutation will affect the phenotype of the cell. A single base change, or base pair substitution in the DNA resulting in the replacement of one amino acid in the protein by another, is referred to as a *missense mutation* (Figure 3-10).

**Missense mutations** *are due to a single base change in the DNA and result in the replacement of one amino acid in a protein by another.*

FIGURE 3-10 ▶

**Types of Mutations That Occur in DNA.**
*Mutations can arise in the DNA by single base pair substitutions or by the addition or deletion of one or two base pairs. Substitution of one base pair by another can lead either to a missense mutation, where one amino acid replaces another in the protein or to a nonsense mutation where a sense codon is changed to a nonsense codon. The presence of a nonsense codon in the middle of the coding sequence of an mRNA results in cessation of protein synthesis at the site of the nonsense mutation. Addition or deletion of one or two base pairs shift the reading frame of the mRNA, resulting in a protein with the wrong amino acid sequence from the site of the addition or deletion. Phe = phenylalanine; Ser = serine; Met = methionine; Lys = lysine; Val = valine; Glu = glutamic acid.*

### Base Pair Substitution

| Missense Mutation | | | Nonsense Mutation | | |
|---|---|---|---|---|---|
| | Normal | Mutant | | Normal | Mutant |
| DNA | AAA | AGA | DNA | AGC | ATC |
| RNA | UUU | UCU | RNA | UCG | UAG |
| Protein | Phe | **Ser** | Protein | Ser | **Stop** |

### Base Pair Addition or Deletion

#### Frameshift Mutation

| | Normal | | Addition |
|---|---|---|---|
| DNA | TAC AAA TTT CAA AGC | DNA | TAC AAA **C**TT TCA AAGC |
| RNA | AUG UUU AAA GUU UCG | RNA | AUG UUU **G**AA AGU UUCG |
| Protein | Met-Phe-Lys-Val-Ser | Protein | Met-Phe-**Glu**-**Ser**-**Phe** |

A single base pair substitution in the DNA also can lead to a second type of mutation. An example is when the codon UCG, which codes for serine, is changed to UAG, one of the termination codons. When a termination codon appears in the middle of a gene, protein translation stops at that point, and the resulting protein product is a truncated version of the normal product and is thus inactive. A single base pair substitution in the DNA that leads to the presence of a termination codon in the middle of the coding region of a mRNA is referred to as a *nonsense or termination mutation*.

A third type of mutation can result from an addition or deletion of one or two base pairs in the DNA. When this occurs in the coding region, the reading frame or sequence of triplets read on the mRNA is shifted, resulting in wrong amino acids being placed in the protein from the point of the addition or deletion. Such mutations, in which the mRNA is read out of frame, are known as *frameshift mutations*. Frameshift mutations usually have drastic effects on a protein, for it will now have multiple amino acids unlike those found in the normal protein.

*Nonsense mutations* are caused by a single base change in the DNA and result in the presence of a stop codon in the middle of the gene and premature termination of protein synthesis.

*Frameshift mutations* are due to additions or deletions of one or two base pairs in the DNA and result in shifting the reading frame of an mRNA.

## Steps in Protein Synthesis

The main components of the protein synthesis machinery include mRNA, tRNAs, amino acids, and ribosomes. mRNA provides the sequence of codons that are to be translated into protein. tRNAs bind specific amino acids and then transfer that amino acid into protein, using the specific codons in the mRNA. Finally, a particle consisting of RNA and protein, known as a ribosome, acts as the site of protein synthesis. The production of the final protein is a result of reactions that occur in a specific sequence as described below.

*Step 1.* Attachment of an amino acid to a tRNA molecule—a process called charging—is diagrammed in Figure 3-11. For each of the 20 amino acids found in a cell, there are 20 different enzymes called *aminoacyl-tRNA synthetases*. The function of these enzymes is to attach a specific amino acid to a specific tRNA molecule, the specificity of the reaction lying in the base sequence of the tRNA molecule. This reaction must be highly specific, since a mistake would result in the wrong amino acid being placed in the protein. In Figure 3-11, it is seen that a specific amino acid is initially attached to a specific aminoacyl-tRNA synthetase with the simultaneous hydrolysis of ATP to give an enzyme–amino acid adenosine monophosphate (AMP) intermediate. This intermediate complex now binds to a specific tRNA molecule with the corresponding transfer of the amino acid to the 3′ end of the tRNA and the simultaneous release of the enzyme and AMP. The aminoacyl-tRNA maintains the energy released in the hydrolysis of ATP in the ester bond formed between the amino acid and the tRNA molecule. This energy will be used later to drive the synthesis of a peptide bond.

*Charging of tRNA molecules is catalyzed by enzymes known as aminoacyl-tRNA synthetases.*

*Step 2.* The interaction of mRNA and tRNA with the ribosome—a process called the polyribosome cycle—is shown in Figure 3-12. *Ribosomes* are particles made up of both rRNA and specific proteins. They are found in the cytoplasm, where they serve as the site of protein synthesis. Each ribosome consists of two subunits, the 40S, or small subunit, and the 60S, or large subunit. When joined together, the entire particle is 80S. The 40S subunit is made up of approximately 30 different proteins and contains 18S rRNA. The 60S subunit has 50 different proteins along with the 5S, the 5.8S, and the 28S rRNA molecules. Each of the subunits carries out a particular function in the process of protein synthesis.

*Ribosomes* are made up of rRNA and protein and are the site of protein synthesis.

The 40S subunit, the 60S subunit, charged tRNA molecules, and a mRNA molecule come together to form a complex that, dictated by the codons in the message, directs the insertion of a specific amino acid into a growing polypeptide chain. The ribosome physically moves down the mRNA (5′ to 3′) with the sequential addition of amino acids from tRNAs to form the polypeptide. A mRNA molecule, which can be thousands of bases long, is translated by multiple ribosomes, each of which attaches at the 5′ end of the mRNA and moves down the message, producing a single protein molecule. In this way, multiple copies of a single protein can be produced from a single mRNA molecule. When the protein is completed, it is released along with the ribosome and tRNA molecules, which are free to begin the cycle again.

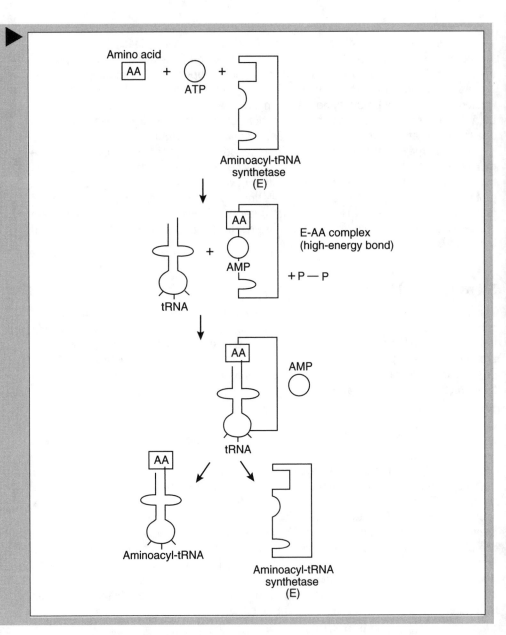

**FIGURE 3-11** ▶

*Charging tRNA Molecules. The enzyme aminoacyl-tRNA synthetase binds ATP and a specific amino acid to form an enzyme-AMP-amino acid complex containing a high-energy bond. A tRNA molecule then binds to the amino acid and is released from the complex with the amino acid attached to give a molecule of aminoacyl-tRNA. The enzyme is free and ready to start another cycle.*

## INITIATION OF PROTEIN SYNTHESIS

Protein synthesis involves three specific steps: *initiation, elongation, and termination.* Each of these steps involves a different set of proteins and the use of energy, either in the form of ATP or guanosine triphosphate (GTP). The process of initiation is considered first (Figure 3-13).

Specific initiation factors, IFs or proteins, are involved in the process of initiation. Even though the process of initiation involves more than ten proteins, for simplicity, only one of these, IF2, is considered in our discussion. IF2 binds to GTP to form a complex that binds a tRNA charged with methionine. (Remember that the methionine codon, AUG, is always the first codon, or initiator codon, on a mRNA molecule.) This complex, IF2–Met-tRNA–GTP, then interacts with the 40S subunit and the 5' end of a mRNA. After binding to the 5' end of the mRNA, the 40S subunit, along with the IF2–Met-tRNA–GTP complex, begins to move down or scan the mRNA. This scanning, which uses the 5' Cap for proper alignment and ATP as an energy source, continues until the complex reaches the first AUG, or the initiator codon, on the mRNA. Once the complex finds the first AUG codon, the 60S ribosome subunit binds to the complex to form the final ribosome structure. The attachment of the 60S subunit requires GTP as an energy source. With formation of this final structure, composed of a 40S subunit, a 60S subunit, Met-tRNA and a mRNA, the initiation step is completed. IF2 and guanosine diphosphate

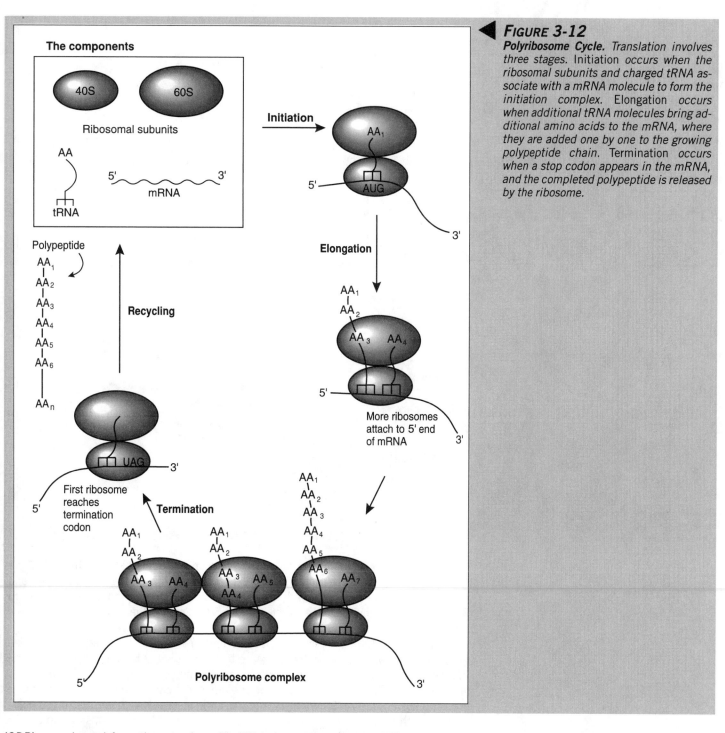

**FIGURE 3-12**

**Polyribosome Cycle.** *Translation involves three stages. Initiation occurs when the ribosomal subunits and charged tRNA associate with a mRNA molecule to form the initiation complex. Elongation occurs when additional tRNA molecules bring additional amino acids to the mRNA, where they are added one by one to the growing polypeptide chain. Termination occurs when a stop codon appears in the mRNA, and the completed polypeptide is released by the ribosome.*

(GDP) are released from the complex with IF2 being regenerated by GTP to reinitiate the cycle.

## PROCESS OF ELONGATION

Let us consider the mRNA molecule shown below to discuss the role of tRNA molecules in the process of elongation of protein synthesis.

5' Cap ------AUGUUUAAAGUUUCG------------UAG----AAAAAAA 3'
          Met Phe Lys Val Ser

Each codon in the mRNA is going to interact with a specific anticodon on the tRNA molecule. The first tRNA, Met-tRNA, with the anticodon UAC, has already interacted with the mRNA during the process of initiation. Thus, each protein begins with the amino

### FIGURE 3-13

***Initiation of Protein Synthesis.*** *The initiation factor 2 (IF2) in complex with GTP binds to a tRNA charged with methionine (Met). This complex interacts with the small 40S ribosomal subunit and the 5' end of a mRNA molecule. After binding to the 5' end of the mRNA, the 40S subunit scans the mRNA until it reaches the first AUG codon. At this point, the 60S subunit binds to the complex to form the final initiation complex.*

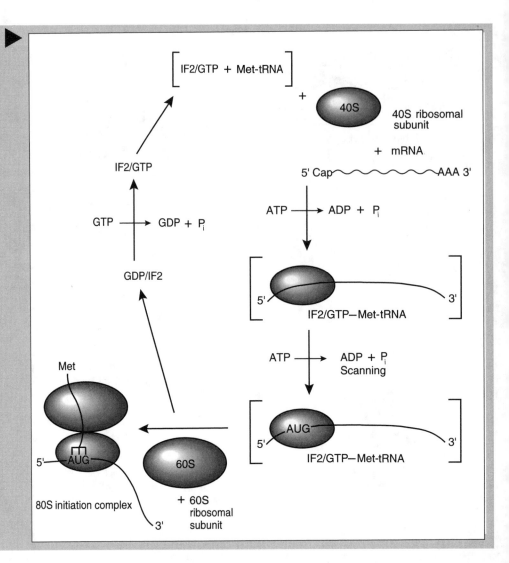

***Elongation of protein synthesis*** *involves the formation of a peptide bond between the carboxyl group of one amino acid and the amino group of an adjacent amino acid.*

acid methionine at its amino terminal end. The next two codons present in the mRNA are UUU and AAA. In the process of elongation, a second tRNA, with the anticodon AAA and with phenylalanine attached, binds to the mRNA. This binding is followed by a third tRNA with the anticodon UUU and the amino acid lysine attached and on and on until a termination codon is encountered.

In Figure 3-14, the process of peptide chain elongation is presented in a simplified sketch. Protein synthesis has been initiated by the binding of the Met-tRNA to the first AUG codon in a mRNA molecule bound to the ribosomes. Elongation begins when the next aminoacyl-tRNA enters the ribosome and binds to the second codon on the message directly following the AUG codon. Before the aminoacyl-tRNA enters the ribosome, it interacts with specific factors known as elongation factors (EFs), in this example, EF1 and the energy source, GTP. When the tRNA binds to its codon on the mRNA, GTP is hydrolyzed with the release of GDP and EF1. EF1 is regenerated in the presence of GTP and is able to initiate the cycle once more.

On the mRNA, Met-tRNA is now bound at AUG, and adjacent to it, the second aminoacyl-tRNA, Phe-tRNA, is bound at UUU. In the presence of the enzyme, *peptidyltransferase*, a peptide bond is formed between the carboxyl group of methionine and the amino group of phenylalanine. This reaction results in the attachment of methionine to phenylalanine. In the next step, the ribosome, in the presence of GTP, a second elongation factor, EF2, and an enzyme called *translocase*, moves (translocates) three nucleotides along the mRNA in a 5' to 3' direction. The uncharged tRNA is released, and a new codon on the mRNA is now exposed. This newly exposed codon on the mRNA is in position to accept the next aminoacyl-tRNA, and the cycle is repeated, each step resulting in a growing polypeptide chain with one additional amino acid.

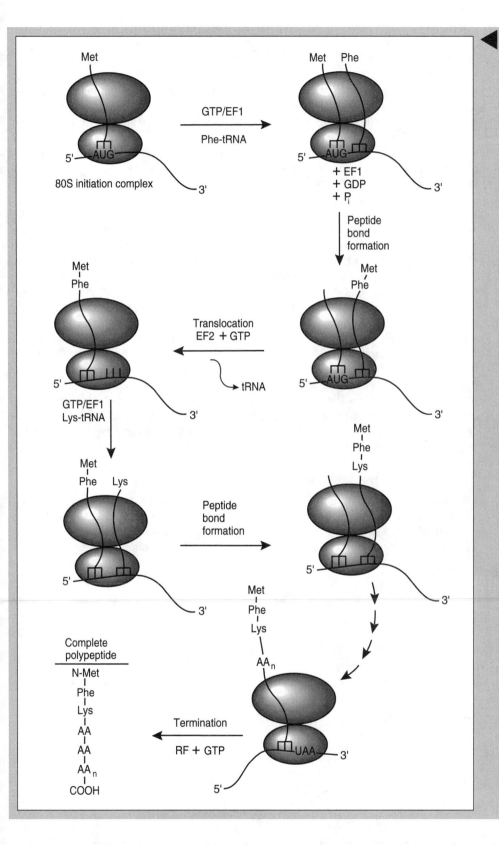

### FIGURE 3-14

**Elongation of Protein Synthesis.** *Protein synthesis has been initiated by the binding of methionine (Met)-tRNA to the AUG codon in a mRNA bound to the ribosome. A second aminoacyl-tRNA interacts with the elongation factor 1 (EF1) and GTP followed by the binding of this complex to the second codon on the mRNA. In the presence of peptidyltransferase, a peptide bond is formed between methionine and phenylalanine (Phe). The ribosome, in the presence of GTP, EF2, and an enzyme known as translocase, translocates one codon along the mRNA (5'–3') with release of the uncharged tRNA and the exposure of a new codon. The next aminoacyl-tRNA will now bind, and the cycle will repeat until a termination codon is reached. In the presence of release factor (RF), the completed peptide is released from the ribosome.*

When the moving ribosome comes to a termination codon (e.g., UAA, UGA, UAG), there is no tRNA that recognizes these codons. Instead, a factor known as release factor (RF), using GTP as an energy source, binds to the ribosome. In the presence of RF, the peptidyltransferase transfers the carboxyl terminal residue of the final amino acid to water, and the peptide is released from the ribosome. The ribosome then dissociates into its 40S and 60S subunits, and these subunits are now ready to begin a new cycle.

## Targeting Proteins for Their Proper Destination

Most of the proteins synthesized in a human cell are made on ribosomes that are located in the cytoplasm of the cell. Many of these proteins, however, must function in organelles such as the nucleus or the mitochondria or, in some cases, be secreted from the cell. In all cases, the protein must be labeled in such a way as to assure arrival at its proper destination. Thus, proteins are synthesized with certain sequences, known as signal sequences, which enable them to be transported into the proper organelle. For example, proteins bound for the nucleus have a specific sequence of amino acids at the N-terminal known as a nuclear localization sequence.

A different type of signal is found for proteins destined to be secreted from the cell. Inside the cell is the rough endoplasmic reticulum (RER), where proteins destined for secretion are synthesized. At the amino terminal end of secreted proteins is a sequence of 15–30 amino acids, 6–12 of which are very hydrophobic. This sequence is the signal marking the protein for the ER. Synthesis of these proteins begins on ribosomes in the cytoplasm, as we just discussed. However, as the signal sequence peptide is formed, it is recognized by a particle in the cytoplasm, the signal recognition particle (SRP). SRP binds the newly synthesized peptide still attached to the ribosome and transiently inhibits further translation of mRNA. The ribosome-SRP complex then moves to the ER membrane, where a second protein, the docking protein, binds to SRP and docks the complex to the membrane of the ER. Receptor proteins on the ER membrane now hold the ribosomes in place, and the SRP and docking protein are released. At this point, protein synthesis once again begins, and the signal sequence peptide with its hydrophobic properties enters the membrane. As the protein continues to be elongated, it is actually pushed through the membrane and into the ER lumen. Once the protein is completely synthesized, an enzyme known as a signal peptidase cleaves the signal sequence and releases the protein from the membrane.

## Synthesis of Mitochondrial Proteins

Mitochondria are very important organelles found in multiple copies in human cells. These organelles are the site of high energy production and are crucial to the well-being of a cell. Mitochondria have their own DNA genome, a circular DNA molecule of about 16,500 bp that contains 37 genes (Figure 3-15). The mitochondrial genes code for

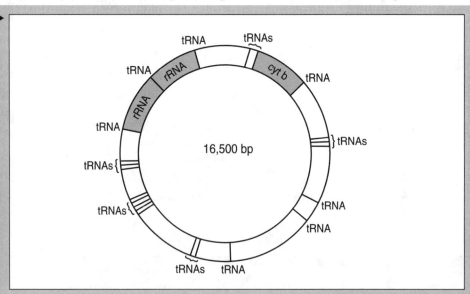

**FIGURE 3-15** ▶

**Human Mitochondrial Genome.** *Mitochondria contain a double-stranded closed circular molecule of DNA as their genome. The mitochondria genome is made up of 16,500 bp that code for tRNA molecules, rRNA molecules, and subunits of proteins required for the mitochondrial electron transport complexes such as cytochrome b (cyt b).*

*Mitochondria have their own DNA genome that codes for tRNAs, rRNAs, and polypeptides that make up the electron transport complexes.*

some of the tRNAs used in mitochondrial protein synthesis, rRNAs found in mitochondrial ribosomes, and polypeptides found in some of the proteins that make up the electron transport complexes. The RNA and protein made within the mitochondria only make up a small fraction of those required for the functioning of the mitochondria. All other RNA and proteins are made in the cytoplasm and transported into the organelle.

The processes of DNA replication, transcription, and translation all can occur within the mitochondria since mitochondria contain their own DNA polymerase, RNA polymerase, and ribosomes. These factors, however, resemble the RNA polymerase and ribosomes found in bacterial cells and differ significantly from the same factors found in the human cytoplasm and nucleus. The importance of mitochondria and their gene products are discussed later when we look at a number of human diseases that can result from mutations in mitochondrial genes (see Chapter 11).

# RESOLUTION OF CLINICAL CASE

Mutations within a gene, even a single base pair change, may cause a change in the way a particular fragment of DNA responds to enzymes that cut DNA at specific sites. These enzymes, known as restriction endonucleases, cleave DNA at specific sequences, resulting in the production of small fragments of DNA. These fragments can be separated by electrophoresis according to size and then examined for the presence of a particular gene by nucleic acid hybridization. (A discussion of this procedure is given in Chapter 6.) A point mutation or base substitution within a gene may cause the loss of two fragments and the appearance of a larger one if the cleavage site of a restriction enzyme is destroyed.

*Restriction endonucleases* are important tools in the study of DNA sequences.

This type of analysis was carried out by Chang and Kan [1] who showed that the β-globin gene in DNA from normal individuals can be distinguished from the same gene in individuals carrying the sickle cell anemia mutation. This mutation causes an amino acid change from glutamic acid to valine in the sixth position of the β-globin protein. Red cell sickling is related to the decreased solubility of deoxyhemoglobin carrying this mutation. A single base pair change of AT to TA in the β-globin gene causes the amino acid substitution by changing the Glu codon, GAG, to a Val codon, GUG, as shown below.

**Normal gene**: Protein   Leu-Thr-Pro-**Glu**-Glu-
          DNA     5'CTGACTCCT**GAG**GAG3'
                  3'GACTGAGGACTCCTC5'

**Mutant gene**: Protein   Leu-Thr-Pro-**Val**-Glu-
          DNA     5'CTGACTCCT**GTG**GAG3'
                  3'GACTGAGGACACCTC5'

Figure 3-16 demonstrates how this base pair change also destroys an *Mst II* restriction endonuclease cleavage site and alters the electrophoretic pattern of DNA fragments derived from the β-globin gene.

A single base pair change in the DNA in which an AT base pair is replaced by a TA base pair alters the cleavage site for the enzyme *Mst II*, which cleaves at CCTGAGG to give two β-globin gene fragments, one of 1.15 kb and one of 0.2 kb. The mutant gene loses the cleavage site and gives only one DNA fragment of 1.35 kb. As noted above, this single base pair change, resulting in a single amino acid change in the globin protein, has severe effects on the functioning of the protein and results in a life-threatening disease, sickle cell anemia. The analysis of Charlie's DNA confirms that he is homozygous for the sickle cell gene, since he has only the 1.35 kb band. The DNA pattern also shows that both parents are carriers of the disease with one normal gene, indicated by the 1.15 kb band and the 0.2 kb band, and one mutant gene, indicated by the 1.35 kb band. A normal noncarrier has only the 1.15 kb band and the 0.2 kb band. In this autosomal recessive disease, where both parents are carriers but do not have symptoms of the disease, there is a 25% chance that their next child will have the disease.

**FIGURE 3-16** ▶
**Use of the Restriction Endonuclease, Mst II, *in the Diagnosis of Sickle Cell Anemia.*** *(Source: Chang JC, Kan KW: A sensitive new prenatal test for sickle cell anemia.* N Engl J Med 307:31, 1982.*)*

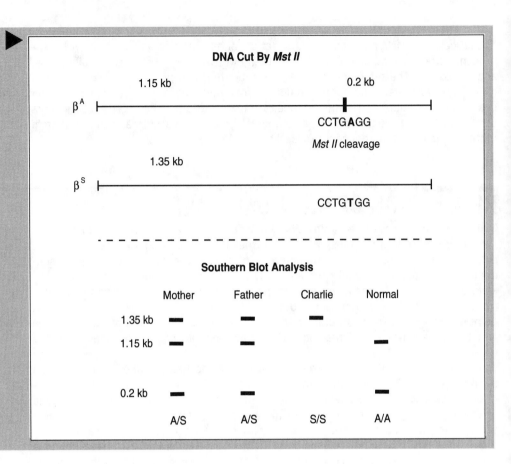

# REVIEW QUESTIONS

**Directions:** For each of the following questions, choose the **one best** answer.

1. The genetic code for amino acids is best described by which of the following statements?
    (A) It contains the signals that direct RNA polymerase to start and stop transcription
    (B) It is composed of code words (codons) containing three nucleotide letters
    (C) It is identical for protein synthesis occurring in the mitochondria and in the cytoplasm
    (D) It is translated into amino acid sequences by RNA polymerase
    (E) It contains the same number of code words as there are amino acids

2. Exons are best characterized by which of the following statements?
    (A) They are present in the same number in all protein-encoding genes
    (B) They contain sequences encoding the start and stop signals for translation
    (C) They code for sequences that are removed from RNA during processing reactions
    (D) They are required for the identification of a gene by DNA polymerase
    (E) They generally are much larger in size than introns

3. Messenger RNA production involves which of the following reactions?
    (A) The removal of introns from internal regions of the RNA transcript
    (B) Synthesis of the RNA in the 3' to 5' direction
    (C) The post-transcriptional addition of adenylate nucleotides to the 5' end of the RNA
    (D) The covalent attachment of an amino acid to the 5' end of the molecule
    (E) The binding of RNA polymerase III to the promoter sequence

4. Splicing introns out of a precursor mRNA involves which of the following actions?
    (A) It is directed by nucleotide sequences located in the 3' untranslated region
    (B) It has been rendered defective in some forms of sickle cell anemia
    (C) It requires the formation of a novel lariat structure
    (D) It occurs following translation of the mRNA
    (E) It is directed by RNA polymerase II

5. The sequence 5'-ATTGCCATGCTA-3' on a DNA template strand codes for an RNA molecule of which of the following sequences written 5' to 3'?
    (A) 5'-UAACGGUACGAU-3'
    (B) 5'-UAGCAUGGCAAU-3'
    (C) 5'-ATTGCCATGCTA-3'
    (D) 5'-TAACGGTACGAT-3'
    (E) 5'-TAGCATGGCAAT-3'

**6.** The coding sequence 5'-AGAUCAUGUUGCCUCGUUAUG-3' is located at the 5' end of a cytoplasmic mRNA molecule. Which of the following polypeptides is encoded by this portion of the mRNA molecule? (The genetic code is given in Figure 3-7.)

    **(A)** Arg-Ser-Cys-Cys-Leu-Val-Met

    **(B)** Asp-His-Val-Ala-Ser-Leu

    **(C)** Ile-Met-Leu-Pro-Arg-Tyr

    **(D)** Met-Leu-Pro-Arg-Tyr

    **(E)** His-Val-Ala-Ser-Leu

# ANSWERS AND EXPLANATIONS

**1. The answer is B.** The genetic code has 64 codons, each with three nucleotide letters. The signals that direct RNA polymerase are sequences in the DNA. The mitochondria has some codons that differ from those used by cytoplasmic mRNA. RNA polymerase transcribes RNA from DNA and is not involved in translation. There are 21 amino acids, but there are 64 code words, three of which do not code for an amino acid.

**2. The answer is B.** Exons have signals that are necessary for starting and stopping translation. Different genes contain different numbers of exons. The sequences removed during RNA processing are the intron sequences. DNA polymerase is involved in DNA replication and recognizes ori sites. Exons tend to be small, whereas introns can be very large in size.

**3. The answer is A.** Messenger RNA production requires processing that involves removal of the intron sequences from the initial RNA transcript. RNA is synthesized only in a 5' to 3' direction. The addition of a poly A tail is to the 3' end of the RNA. Processing involves the attachment of 7-methylguanosine to the 5' end of the RNA. RNA polymerase III specifically transcribes the genes for tRNA and 5S RNA.

**4. The answer is C.** Splicing involves the formation of a lariat structure as an intermediate. Splicing is directed by the donor splice site and the acceptor splice site at each intron–exon junction. Sickle cell anemia is due to a single base change in the hemoglobin (Hb) gene, which results in a single amino acid change in the protein and the production of HbS. Splicing occurs before the mRNA is translated by the ribosomes. RNA polymerase II carries out transcription and is not involved in RNA processing.

**5. The answer is B.** The RNA molecule is complementary to the DNA template strand, and is in the antiparallel direction. A is incorrect because it is not complementary to the DNA in the 3' to 5' direction. C is incorrect because the RNA will have U in place of T, and this sequence is not complementary to the DNA. D is incorrect because it contains T in place of U and is not complementary to the DNA in the 3' to 5' direction. E is incorrect because it contains T in place of U.

**6. The answer is D.** The mRNA is read starting at the first AUG codon, and then each codon is read in frame. The translation system scans the 5' end of the mRNA until it comes to the first AUG and then begins to translate. The other distractors are incorrect because AGA, GAU, AUC, and CAU are not initiator codons.

# REFERENCES

1. Chang JC, Kan KW: A sensitive new prenatal test for sickle cell anemia. *N Engl J Med* 307:30–35, 1982.

# ERRORS IN DNA: CAUSES AND REPAIR MECHANISMS

# INTRODUCTION OF CLINICAL CASE

An 8-year-old white girl named Caroline presented to a dermatologist. Caroline had extensive freckling, and her mother pointed out several small growths on the girl's forehead. The physician examined the girl thoroughly and found several more growths on areas of the skin that normally are exposed to sunlight. The mother stated that the family had just returned from spending the summer in Florida and that Caroline had played on the beach every day in the hot sun. The physician told the mother that the growths appeared to be small benign tumors, and they should be removed surgically. In addition, the physician warned Caroline that she should avoid the sun as much as possible and protect her skin from exposure to sunlight when outside.

A year after the surgery, Caroline and her mother returned to the dermatologist. At that time, Caroline had multiple skin tumors and, in addition, had developed neurologic abnormalities and problems with her vision and hearing. Upon further questioning, the physician learned that Caroline had not heeded his advice and had continued to play in the hot sun without protection. The physician explained to the mother that with Caroline's additional symptoms he suspected that she might be suffering from a rare genetic disease known as xeroderma pigmentosum. If she continued to ignore the physician and was exposed to sunlight, she would develop more malignant skin tumors, possible internal neoplasms, advanced neurologic problems, and be at risk of death at an early age. The physician then asked the mother if he could take a skin biopsy from her and from Caroline to verify his suspected diagnosis and to identify more accurately Caroline's problem. The mother agreed and assured the physician that she would make sure that Caroline avoided direct exposure to sunlight.

# MUTATION: STABLE CHANGES IN THE DNA BASE SEQUENCE

Mutations, defined as permanent changes in the base sequence of a DNA molecule, are the underlying cause of all human genetic disease. Many of the concepts concerning the types of mutations that occur in DNA and the potential mechanisms by which they occur were originally developed using bacterial cells as model systems. During the past 5 years, the molecular basis of mutations in many human genetic diseases has also been identified at the DNA level. This dramatic increase in knowledge about mutations and human disease is due mainly to the incredible advances in molecular techniques now available for studying DNA molecules and human chromosomes. Recent studies of diseased human cells have verified that all living cells using DNA as their genetic material have common mechanisms by which they undergo mutation and, more importantly, by which they can repair potential mutations [1, 2]. DNA is the only macromolecule found in cells that is capable of repairing itself. In this chapter, the ways in which DNA can be altered or damaged and the pathways human cells have evolved to counteract and correct this damage in their genome are examined.

> **Mutation** is a permanent change in the base sequence of a DNA molecule.

## Classes of Mutations Found in Human DNA

Table 4-1 lists the classes of mutation that occur in DNA molecules. It also lists how each mutation type can affect the structure of the product of a gene or the amount of the product that is coded for by a gene.

**TABLE 4-1** ▶
*Classes of Mutations Found in Human DNA*

| Class | Result |
|---|---|
| **Single base-pair substitutions (point mutations)** | |
| Altered structure of gene product | |
|     Missense mutation | Single amino acid replacement in the protein |
|     Nonsense mutation | A termination codon in the middle of the gene results in premature termination of protein synthesis |
|     RNA-splicing mutation | The protein may be missing part or all of an exon sequence |
| Altered quantity of gene product | |
|     Mutations in regulatory sequences | Transcription of the gene is altered, which can reduce or eliminate the gene product |
|     Mutations in RNA processing and translation | The stability of the messenger RNA is altered, which may reduce the amount of gene product |
| **Insertions or deletions** | |
| One or two base pairs (frameshift mutations) | The addition or deletion of one or two base pairs can affect the reading frame of the gene, resulting in a grossly altered or absent gene product |
| Large numbers of base pairs | Large pieces of the DNA may be lost or large segments of DNA may insert into the middle of a gene, resulting in loss of function |
| **Expansion of trinucleotide repeat sequences** | Unstable trinucleotide repeats can suddenly expand in number, resulting in the alteration of production or structure of a particular gene product |
| **Chromosomal alterations** | Inversions, translocations, duplications, or gene amplification may result |

## SINGLE BASE-PAIR SUBSTITUTIONS (POINT MUTATIONS)

Many of the mutations that occur in DNA are caused by single base-pair substitutions in which one base pair, such as an adenine–thymine (AT) pair, is replaced by a second base pair, such as a guanine–cytosine (GC) pair. The substitution of one base pair by a second base pair results in the change of a codon (see Chapter 3) and can lead either to a *missense mutation* in which one amino acid replaces another amino acid in a protein or to a *nonsense mutation* in which one of the termination codons appears in the middle of a gene. When a nonsense codon (UAA, UGA, or UAG) appears in the middle of the gene, there is no transfer RNA (tRNA) molecule to recognize these codons. Therefore, protein synthesis terminates at the site of the nonsense codon on the messenger RNA (mRNA), and a truncated polypeptide is produced. The effects of missense and nonsense mutations on a protein product are diagrammed in Figure 4-1.

***Missense or nonsense mutations** are single base-pair substitutions in the coding sequence of a gene.*

**Normal**

| | |
|---|---|
| DNA | ATGCAGGTGACCTCAGTG<br>TACGTCCACTGGAGTCAC |
| RNA | AUGCAGGUGACCUCAGUG |
| PROTEIN | Met-Gln-Val-Thr-Ser-Val |

| **Missense mutation** | | **Nonsense mutation** |
|---|---|---|
| DNA | ATGCAG**C**TGACCTCAGTG<br>TACGTC**G**ACTGGAGTCAC | ATGCAGGTGACCT**G**AGTG<br>TACGTCCACTGGA**C**TCAC |
| RNA | AUGCAG**CUG**ACCUCAGUG | AUGCAGGUGACC**UGA**GUG |
| PROTEIN | Met-Gln-**Leu**-Thr-Ser-Val | Met-Gln-Val-Thr-**stop** |

| **Frameshift mutation** | | **Insertion mutation** |
|---|---|---|
| DNA | ATGCAGGTG**A**ACCTCAGTG<br>TACGTCCAC**T**TGGAGTCAC | ATGCAGGTG-**LINE-3000bp**-ACCTCAGTG<br>TACGTCCAC-**LINE-3000bp**-TGGAGTCAC |
| RNA | AUGCAGGUG**AACCUCAGUG** | AUGCAGGUG-**LINE-3000bp**-ACCUCAGUG |
| PROTEIN | Met-Gln-Val-**Asn-Leu-Ser** | Met-Gln-Val ——— ? |

| **Deletion mutation** | | **Trinucleotide expansion** |
|---|---|---|
| DNA | ATGTCAGTG<br>TACAGTCAC | ATG-**(CAGCAGCAG)**$_{20}$CAGGTGACCTCAGTG<br>TAC-**(GTCGTCGTC)**$_{20}$GTCCACTGGAGTCAC |
| RNA | AUGUCAGUG | AUG-**(CAGCAGCAG)**$_{20}$CAGGUGACCUCAGUG |
| PROTEIN | Met-Ser-Val | Met-**(Gln-Gln-Gln)**$_{20}$-Gln-Val-Thr-Ser-Val |
| | | ↓ |
| | DNA | ATG-**(CAGCAGCAG)**$_{75}$CAGGTGACCTCAGTG<br>TAC-**(GTCGTCGTC)**$_{75}$GTCCACTGGAGTCAC |
| | RNA | AUG-**(CAGCAGCAG)**$_{75}$CAGGUGACCUCAGUG |
| | PROTEIN | Met-**(Gln-Gln-Gln)**$_{75}$-Gln-Val-Thr-Ser-Val |

*FIGURE 4-1*
***How Mutations in DNA Alter a Protein Product.** Mutations in DNA result from a single base-pair substitution (missense or nonsense mutations), one or two base-pair additions or deletions (frameshift mutations), insertion or deletion of a large number of base pairs (insertion or deletion mutations) and expansion of trinucleotide repeats (trinucleotide expansion mutations).*

Single base-pair substitutions can also occur in sequences other than those coding for amino acids. For example, a mutation that alters the splice acceptor or splice donor sequences can result in *aberrant splicing of the RNA.* The result of such a mutation is a mRNA that may be missing part or all of a particular exon and thus codes for a mutant protein. Other base-pair substitutions occur in regulatory sequences, which are necessary for binding of transcription factors or RNA polymerase. When mutations occur in regulatory sequences, the quantity of the product produced by the gene that is controlled by these sequences is altered. These mutations often lead to a complete absence of the gene product or to an increase in the amount of the gene product.

### INSERTIONS OR DELETIONS

***Frameshift Mutations.*** Frameshift mutations are caused by the addition or deletion of one or two base pairs within the coding sequence of a gene, which alters the reading frame of the mRNA. The sequence of codons in a mRNA is translated by the protein-synthesizing machinery, starting at the first AUG codon and then consecutively reading the mRNA in

*Frameshift mutation* is an addition or deletion of one or two base pairs within the coding sequence of a gene.

*Large deletions of 100–1000+ bp in the DNA are known to cause many human diseases. These deletions usually occur during meiosis.*

*The expansion of trinucleotide repeats (CAG, CTG, CGG, or GAA) gives rise to a number of expansion-disorder diseases.*

*Expansion of the trinucleotide CAG, which codes for glutamine, in the first exon at the 5′ end of the gene results in the production of a protein with an excess number of glutamines at the amino-terminal end.*

groups of three nucleotides without interruption (see Chapter 3). By adding or deleting one or two base pairs in the DNA, a mRNA is produced with an altered sequence and is translated out of frame from the site of the insertion or deletion of the base pair. The result of this frameshift mutation is the production of a protein that is altered in its amino acid sequence, starting from the point of the insertion or deletion of the base pair and continuing to the end of the protein, as shown in Figure 4-1. Often, the altered reading frame results in the production of a termination codon in the middle of the gene, which causes premature termination of protein synthesis.

Insertion and deletion of many base pairs also can occur within DNA molecules. *Deletion mutations* may occur in a chromosome with the loss of hundreds to thousands of base pairs from the DNA in which case the deleted genetic material is permanently lost from the cell. More than 100 human diseases result from such deletions, which occur during the process of recombination in meiosis. *Insertion mutations* (large insertions of DNA sequences) have been described, but they are rare occurrences in human cells. These insertions are caused by transposon-like elements, often repetitive DNA sequences such as the long, interspersed nuclear element (LINE) repeat. Transposon elements have the ability to move from one site on the genome to another, and if they insert into the DNA in the middle of a gene, the function of that particular gene is lost (see Figure 4-1).

## EXPANSION OF TRINUCLEOTIDE REPEAT SEQUENCES

Within the past 5 years, a new type of mutation that results in a number of human genetic diseases has been described [3, 4]. These mutations are the result of the expansion of trinucleotide repeats (CAG, CTG, CGG, or GAA), which are found throughout the human genome. Runs of these repeat triplets are found in exons at the 5′ end or the 3′ end of genes and, in some instances, even within an intron. The number of these repeats is unstable and can vary from one person to another. Normal, healthy individuals have a particular copy number range and, in most cases, have different numbers of repeats in their maternal and paternal alleles. Individuals affected with one of the expansion-disorder diseases have a sudden increase in the number of copies of the trinucleotide repeats. The expansion of the repeat sequences can alter either the expression or the structure of a particular protein, depending on where the repeats are located within a particular gene.

*Expansion-Disorder Diseases.* The trinucleotide CAG codes for the amino acid glutamine. In five human diseases—Huntington's disease, spinobulbar muscular atrophy, dentatorubral-pallidoluysian atrophy (DRPLA), and spinocerebellar ataxia (types 1 and 3)—the CAG repeat is located in the coding region of the first exon at the 5′ end of the gene. These repeats are translated and appear in the protein as a stretch of glutamines. A normal protein has a repeat of glutamines of fewer than 40 at its amino-terminal end, whereas the mutant proteins have a range of 40–100 glutamines at that site (Table 4-2; see also Figure 4-1).

All of the CAG repeat diseases involve late-onset neuronal loss and are inherited in an autosomal-dominant fashion. The various proteins involved in the five diseases above have no homology to one another and probably serve different functions in the cell, which suggests that the excess glutamine in the proteins gives them their mutant phenotype. The reason why the expanded glutamine region in these proteins results in a pathologic condition is still under investigation. Possible hypotheses include an alteration in transcriptional regulation (as suggested for the androgen-receptor protein) and an interaction among other proteins in the cell with the mutant proteins, which may be affected or inhibited by the expanded glutamine stretch. A recent novel idea suggests that the polyglutamine region can interact with the enzyme glyceraldehyde-3-phosphate dehydrogenase (G3PD). G3PD is an enzyme that functions in glycolysis and produces an important high-energy intermediate from the metabolism of glucose. Interference with this enzymatic activity could result in a cell with severe metabolic stress, particularly in neuronal cells, which obtain their energy from glucose. The complete connection between glutamine repeats and human disease is rapidly being elucidated and should provide important information, leading to a possible treatment of these severe diseases.

**TABLE 4-2**
*Human Diseases with Trinucleotide Expansions*

| Disease | Protein | Repeats | | Type |
|---------|---------|---------|---------|------|
| | | *Normal* | *Mutant* | |
| Huntington's disease[a] | Huntingtin | 11–34 | 36–120 | Glutamine (CAG) repeats |
| DRPLA[a] | Atrophin | 7–23 | 49–75 | |
| Spinobulbar muscular atrophy[a] | Androgen receptor | 11–33 | 40–62 | |
| Spinocerebellar ataxia (type I)[a] | Ataxin-1 | 6–44 | 40–82 | |
| Spinocerebellar ataxia (type 3)[a] | MJD-1 | 13–40 | 68–69 | |
| Fragile X syndrome[b] | FMR-1 | 6–50 | 60–1000 | CGG repeats |
| Myotonic dystrophy[c] | Protein kinase | 5–30 | 50–2000 | CTG repeats |
| Friedreich's ataxia[d] | Frataxin | 10–21 | 200–900 | GAA repeats |

[a] Glutamine repeats (CAG) in the coding region of the first exon of the gene.
[b] CGG repeats in the noncoding 5' end of the first exon of the gene.
[c] CTG repeats in the noncoding 3' end of the last exon of the gene.
[d] GAA repeats within the first intron of the gene.
DRPLA = dentatorubral-pallidoluysian atrophy; FMR-1 = originally for fragile X mental retardation; MJD = protein defective in spinocerebellar ataxia type 3, which is also known as Machado-Joseph disease.

Other trinucleotide expansions occur in noncoding regions; therefore, they are not translated into protein. The fragile X syndrome is associated with repeats of CGG in the 5' noncoding region of the gene. These repeats are hypermethylated on cytosine-phosphate-guanine (CpG), resulting in reduction of transcription activity and lowered mRNA levels. In myotonic dystrophy, the CTG repeats occur in the 3' UTR of the gene coding for a protein kinase. The RNA molecule produced, which has the expanded repeats at the 3' end, may not be polyadenylated properly, which prevents the movement of the RNA from the nucleus to the cytoplasm. Recent evidence suggests that the mutant RNA may affect other normal RNAs produced by the cell, leading to the pleiotropic and dominant effects seen with this mutation.

> **Fragile X Syndrome**
> *This syndrome is associated with repeats of CGG in the 5' noncoding region of the gene, leading to hypermethylation of cytosine and reduction of transcription of the gene.*

A recent addition to this growing list of trinucleotide expansion disorders is the disease Friedreich's ataxia [4]. This disease differs from those discussed above in that the trinucleotide repeat expansion GAA is located within the first intron of the gene and is not found in the mature mRNA. Cells from patients with Friedreich's ataxia contain little or no mRNA for the protein frataxin, which suggests that the excessive repeats of AG stretches within the intron may interfere with the normal splicing of RNA.

Increases in the number of repeat sequences probably occur by a slippage mechanism as described later for hereditary nonpolyposis colorectal cancer (HNPCC) or perhaps by unequal recombination of sister chromatids. However, there is no evidence to date to indicate that the expansion-disorder diseases are associated with a defect in DNA repair. Furthermore, there does not seem to be an increase in the incidence of cancers in the expansion-disorder diseases. The defect underlying the expansion of the trinucleotide repeats remains unknown.

## CHROMOSOMAL ALTERATIONS

A final class of mutations seen in human genetic diseases is the result of gross alterations of the chromosome structure. These mutations include inversions, translocations, duplications, and amplification of chromosome areas, which are discussed in more detail in Chapter 5.

## Effect of Mutation on Protein Structure and Function

The human globin genes provide many examples of the various classes of mutations found in DNA and of the effects on the normal functioning of a protein. Hundreds of point mutations have been identified in the globin genes, some of which are noted in Table 4-3.

*The human globin genes have many examples of missense, nonsense, frameshift, and regulatory mutations that lead to human disease.*

Missense mutations are quite common; the most well-known example is the transversion in codon six of an AT pair to a TA pair in the beta-globin gene. This simple base-pair substitution leads to the replacement of a glutamic acid with a valine residue in hemoglobin A, resulting in the production of hemoglobin S and sickle-cell anemia in the homozygous state (see Chapter 3).

**β-Thalassemia.** Many examples of β⁰-thalassemia, in which there is no production of the beta-globin protein, have been documented as nonsense mutations that occurred in the middle of the beta-globin gene. Other thalassemias that present with decreased amounts of the beta-globin protein result from RNA splicing mutations that reduce the amount of properly spliced mRNA and, therefore, the amount of the protein in the cell. RNA translation mutations (e.g., a mutation that alters the initiation codon AUG) may also result in thalassemia in which decreased amounts of alpha globin are produced because of reduced translation of the mRNA. With the increase in knowledge about the molecular structure of other genes involved in human disease, additional examples of mutation types and their effects on proteins are described frequently.

***TABLE 4-3*** ▶

*Examples of Human Diseases Resulting from Different Types of Mutations*

| Mutation | Disease |
|---|---|
| **Single base-pair substitutions**[a] | |
| Missense mutation | Sickle-cell anemia |
| | Glutamic acid (Glu) is changed to valine (Val) in hemoglobin A to produce hemoglobin S |
| Nonsense mutation | β⁰-Thalassemia |
| | Loss of beta-globin production |
| RNA splicing mutation | β⁺-Thalassemia |
| | Reduced amount of beta globin |
| Regulatory mutation | Hemophilia B |
| | Reduced transcription of factor IX |
| RNA translation | α-Thalassemia |
| | Reduced amounts of alpha globin |
| **Frameshift mutation**[b] | Duchenne's muscular dystrophy |
| | Altered reading frame from improper splicing of exons gives mutant dystrophin |
| | β⁰-Thalassemia |
| | Loss of beta-globin production |
| | Tay-Sachs disease |
| | Altered reading frame from a 4-bp insertion in exon 11 of the gene coding for hexosaminidase A leads to a deficiency of hexosaminidase A[a] |
| **Small deletion** | Cystic fibrosis |
| | Deletion of 3 bp at codon 508 in the cystic fibrosis gene (ΔF508) leads to a deletion of phenylalanine in the cystic fibrosis transmembrane conduction regulator (CFTR) protein[b] |
| **Large deletion** | Duchenne's muscular dystrophy |
| | No dystrophin protein |
| **Large insertion** | Hemophilia A |
| | Inactivation of the gene for factor VIII |

[a] This mutation is found in 73% of Tay-Sachs disease patients of Ashkenazi Jewish descent.
[b] This mutation, ΔF508, is found in 70% of cystic fibrosis patients of Northern European descent.

**Duchenne's Muscular Dystrophy (DMD).** Frameshift mutations in the globin genes leading to β⁰-thalassemia have been known for some time. More recently, frameshift mutations have been found in the gene coding for the protein dystrophin, which is defective in patients with DMD. The gene that codes for dystrophin has 79 exons, all of which must be accurately spliced to produce a normal protein. With so many exons involved, it is not surprising that defective splicing occurs, resulting in the joining of exons out of frame and a frameshift mutation. Deletions are also common in the dystrophin gene. The size of the dystrophin gene, which is more than 2.5 million base pairs (bp), results in a very high spontaneous mutation rate with many of these mutations being caused by large deletions within the gene.

**Hemophilia.** An interesting example of an insertion mutation has been described for the factor VIII gene, which is mutated in patients with hemophilia A. In one case, the LINE element of 3800 bp has inserted into the middle of the coding region, resulting in total inactivation of the gene and a loss of the protein in the cell. The gene coding for factor IX, a protein that is defective in patients with hemophilia B, has a mutation in the site for regulation of transcription, so its expression is decreased.

The discovery of different types of mutations and the demonstration that they can lead to human disease indicate the importance of keeping the rate at which mutations occur in DNA very low. However, in the normal growth and division of a cell, mutations occur in the DNA with regular frequency. If these mutations remain in the DNA, they become permanent changes and ultimately are passed on to daughter cells. It is important, therefore, to understand how mutations arise in the DNA and what pathways are available in the cell to correct them before they become inherited.

*The dystrophin gene, which is defective in people with Duchenne's muscular dystrophy, is 2.5 million bp in size and is subject to a high spontaneous mutation rate.*

# POTENTIAL MECHANISMS FOR THE PRODUCTION OF MUTATIONS

## Errors during DNA Replication

Replication of DNA (see Chapter 2) occurs semiconservatively with each strand of the parental DNA serving as a template for the synthesis of a new daughter strand. This process, which is carried out by DNA polymerases, involves recognizing a base in the template strand and adding a properly base-paired incoming nucleotide to the 3′ end of the growing strand. DNA polymerases catalyze the proper pairing of adenine to thymine and guanine to cytosine with very high accuracy, rarely making a mistake. However, at a frequency of approximately $10^{-5}$, the nucleotides are capable of forming an alternative structure, called a tautomer, in which a hydrogen atom is shifted within the molecule. This enables a base to form a mispair (e.g., an AC pair instead of an AT pair). If such a mispair remains in the DNA, the initial AT pair, which has become an AC pair, now gives rise to a GC pair during the next replication cycle (Figure 4-2). Such mispairs in the DNA occur at a frequency of 1 in 1 million bp replicated, which is a rather high frequency considering that the human genome has 6 billion bp. Yet, the mutation rate in a cell is not 1 in 1 million; it is closer to 1 in 1000–10,000 million bp. To keep the mutation rate this low, the cell must have a way of correcting these mispairs before they become a permanent part of the DNA.

*Frequency of mispairs in the DNA is 1 in 1 million bp replicated.*

## Damage and Modification of Bases in DNA after Replication

There are a number of ways in which bases that are present in already replicated DNA molecules undergo damage or modification (see Figure 4-2). One frequent modification occurs with the purine bases, adenine and guanine. At a rate of more than 10,000 times a day per cell, purine residues are lost from the DNA by a process called *depurination* in which the glycosidic bond between the sugar, deoxyribose, and the base is hydrolyzed. This hydrolysis leaves behind a free apurinic site or gap in one of the DNA strands. Such damage must be corrected before the DNA is replicated, or it may lead to a mutation.

Another spontaneous chemical event that occurs in the DNA at a rate of 100 times a day is the *deamination* of the bases cytosine, adenine, or guanine. In the process of deamination, the base loses an amino group, and its structure is changed. For example, when deaminated, cytosine becomes uracil. Thus, spontaneous deamination of cytosine residues in the DNA leads to the presence of uracil in the DNA instead of cytosine. Uracil, which resembles thymine in its base-pairing properties, now pairs with an adenine

## FIGURE 4-2 ▶

**Mistakes and Damage Leading to Mutation.** *DNA molecules are constantly subjected to the occurrence of mistakes in base sequence or to damage and alteration of bases. During DNA replication, mispairing can occur, giving rise to single base-pair substitutions. Bases in nonreplicating DNA molecules undergo alteration by depurination, deamination, or chemical damage. Ultraviolet irradiation (e.g., sunlight) causes the formation of pyrimidine–pyrimidine dimers in the DNA.*

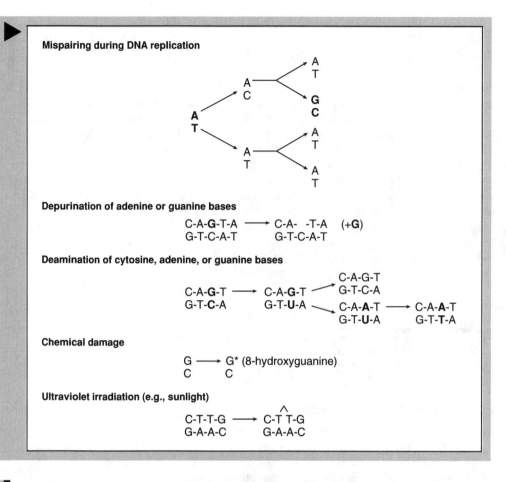

*Spontaneous deamination of cytosine residues in the DNA occurs 100 times a day per cell and gives rise to uracil residues.*

residue during the next replication cycle. If this occurs, the original GC pair (which after deamination is a GU pair) becomes an AT pair or a mutation.

Normal metabolism of many compounds in the cell often can lead to damage by producing oxidizing intermediates that have the ability to oxidize bases in DNA. In particular, guanine residues are commonly damaged by oxidation to give the altered base 8-hydroxyguanine. The presence of such oxidized bases can interfere with normal replication or transcription if not corrected.

In addition to spontaneous damage and modification of bases in DNA, damaging reagents present in the environment can have deleterious effects on DNA. Ultraviolet (UV) rays from sunlight are a common mutagenic agent that causes bonding between adjacent pyrimidines on the same DNA strand. The most frequent type of pyrimidine–pyrimidine dimer produced is the thymine–thymine (TT) dimer (see Figure 4-2). When a TT dimer is present in a DNA molecule, it blocks DNA replication and leads to the death of the cell if it is not corrected or removed. Other types of environmental mutagens include x-rays and many chemotherapeutic agents, which damage bases in DNA and interfere with normal DNA replication.

# DNA REPAIR MECHANISMS

## Correction of Mispairs during DNA Replication: Editing by DNA Polymerase

Considering that DNA polymerases must replicate $6.6 \times 10^9$ bp once every cell cycle, usually within an 8-hour period, the occurrence of errors is not unexpected. Yet, mutation rates are much lower than would be predicted from the rate at which mispairs are formed during replication (i.e., 1 per $10^6$ bp). This ability of a cell to keep the mutation rate low is attributed to the presence in all cells of specific mechanisms to deal with all known kinds of mistakes or damage to DNA. Cells have two lines of defense to deal with mispairs that occur when DNA polymerase adds the wrong nucleotide to the growing 3′ end of the new DNA strand.

A 3′–5′ exonuclease activity is associated with DNA polymerases that are involved in DNA replication. Human DNA polymerase-δ and -ε both have this activity. The 3′–5′ exonuclease activity is responsible for cleaving a mispaired nucleotide from the 3′ end of the newly replicated strand of DNA. When an incoming nucleotide in the altered tautomeric form mispairs with a base in the template strand, that mispair may result in a distortion in the DNA. The 3′–5′ exonuclease perceives this as a mistake and cleaves the nucleotide from the 3′ terminus. The DNA polymerase then has another chance to put the proper nucleotide in the new DNA strand, as shown in Figure 4-3.

*Proofreading function prevents mispairs from remaining in DNA during replication as a result of 3′–5′ exonuclease activity associated with DNA polymerases.*

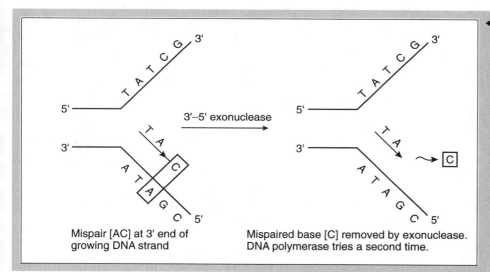

Mispair [AC] at 3′ end of growing DNA strand

Mispaired base [C] removed by exonuclease. DNA polymerase tries a second time.

◀ **FIGURE 4-3**
***Proofreading by the 3′–5′ Exonuclease Activity Associated with DNA Polymerases.*** *DNA polymerases responsible for DNA replication have an associated 3′–5′ exonuclease activity that can remove mispaired bases from the 3′-hydroxyl end of a growing DNA strand. When the polymerase places the wrong base in the newly synthesized DNA strand, the exonuclease recognizes the mistake and removes the mispaired base, allowing the polymerase a second chance to add the right base.*

The importance of the editing process associated with DNA polymerases is illustrated by what happens in a cell when this activity is lost. Strains of bacterial cells that have a dramatic increase in their mutation rate (e.g., an increase of 100–1000-fold) have been described and are known as mutator strains. These bacterial strains have increased mutation rates for all of the genes on the chromosome and have been shown, in some cases, to be deficient in the proofreading activity of DNA polymerase. Studies of these mutant bacteria and mutant bacteriophages have shown that the presence of the 3′–5′ exonuclease activity associated with the replicating DNA polymerase reduces the mutation rate approximately 1000-fold. Thus, with the accuracy of the DNA polymerase activity and the associated 3′–5′ exonuclease activity, the fidelity of DNA replication reduces the mutation rate during DNA replication to 1 in $10^9$ bp replicated.

## Correction of Mispairs after Replication: Methyl-Directed Mismatch Repair System

Even with the above accuracy, mispairs sometimes are not recognized in time, and the replication fork passes them by and leaves them in the newly replicated DNA. This occurs mechanistically when the altered tautomeric form of the nucleotide does not revert to the normal form until after the DNA polymerase complex has bypassed the mispair. Now we have a new double-stranded DNA molecule with a permanent mispair. If the mispair remains in the DNA, it leads to a mutation at the next DNA replication cycle. However, cells have evolved a mechanism to deal specifically with mispairs still present in the DNA immediately after replication.

The methyl-directed mismatch repair system was initially discovered during the study of *Escherichia coli* cells that exhibited a mutator (*mut*) phenotype, which means that a cell has an increased spontaneous mutation rate at all loci on its genome. Recently, a similar repair system has been described for human cells and probably is essential for reducing the spontaneous mutation rate in all known cells from bacteria to mammals [1,

*Mutator (mut) strains are characterized by an excessively high mutation rate.*

5]. The methyl-directed mismatch repair system scans the DNA and looks for mispairs as well as small insertions and deletions that are not properly paired to the template DNA strand. When a mismatch is detected, the repair system removes the defect from the DNA, and the molecule has its proper base sequence restored (Figure 4-4). How does the mismatch repair system know which strand needs to be repaired? For example, if the normal base pair is AT, and the mispair is AC, correction and removal of the adenine from the mispair leads to a GC pair or a mutation. On the other hand, removal of the cytosine from the mispair restores the original AT base pair. What is the mechanism of recognition?

DNA molecules in cells are methylated at specific sites, either on adenine or cytosine residues. In bacteria, a specific base sequence (5'-GATC-3'), when present in a DNA molecule, is recognized by a methylase, which adds a methyl group to the adenine residue. In human cells, cytosine residues located in CpG islands are methylated. It is important, however, to remember that methylation is a postreplication event, and newly synthesized DNA is not immediately methylated. Therefore, during the initial period of DNA replication, one strand (i.e., the template strand) is methylated, but the newly synthesized DNA strand is not methylated. When a mispair is detected, correction specifically takes place on the nonmethylated, or newly synthesized, DNA strand. This enables the repair system to correct the strand that has a normal base in the wrong place and prevents the mispair from leading to a mutation. The cells that are deficient in this repair system have a high mutation rate, because the mispairs are not corrected and thus lead to an increased number of mutations.

*The methyl-directed mismatch repair system specifically repairs the nonmethylated or newly synthesized strand of DNA.*

The proteins known to be involved specifically in methyl-directed mismatch repair are referred to as mutator (Mut) proteins. In *E. coli* these proteins are Mut S, Mut L, and Mut H. Human homologues are known for Mut S (hMSH2 and GTBP) and Mut L (hMLHI and hPMS2), but at this time, none are known for Mut H. The pathway for methyl-directed mismatch repair appears to be similar in bacteria and humans, although the human pathway seems to be more complex, as one might predict.

**FIGURE 4-4** ▶

**The Methyl-Directed Mismatch Repair System.** *Mispairs that remain in the DNA after the replicating complex has moved on are removed from the DNA by the methyl-directed mismatch repair system. The newly synthesized DNA strand remains unmethylated for a short time, enabling the repair system to identify the errant base. The mutator (Mut) S protein recognizes and binds to the mispair, followed by the recruitment of the Mut L and Mut H proteins. The Mut H protein, an endonuclease, cleaves the unmethylated DNA strand. An exonuclease, along with a DNA helicase, moves down the DNA strand, removing base pairs one at a time until the mispaired base and an additional 100 bases past the mispair are removed. The gap left in the DNA is repaired by the actions of DNA polymerase and DNA ligase to give a fully repaired DNA helix.*

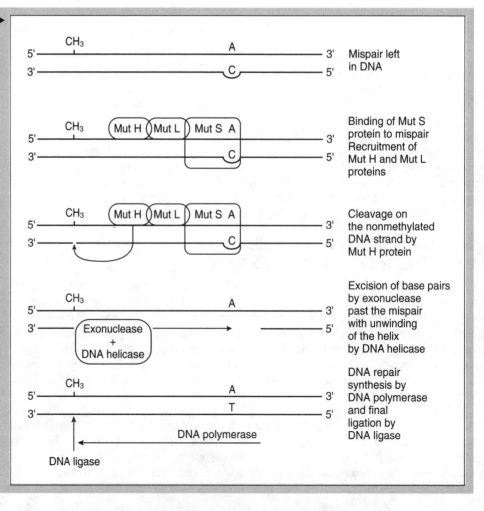

Figure 4-4 indicates the template DNA strand as having a methylated adenosine at the sequence dGATC. The newly replicated DNA strand is unmethylated but contains a mispair noted by the bulge in the DNA. This bulge is recognized and bound by the Mut S protein. The presence of the Mut S protein bound to the mismatch recruits the Mut L and Mut H homologues to the site. The next step is a cleavage of the nonmethylated DNA strand opposite the 5′-GATC-3′ site by Mut H, followed by the stepwise removal of nucleotides by an exonuclease to approximately 100 bp past the mismatch. This step is facilitated by a DNA helicase, which unwinds the DNA duplex. The resulting gap in the DNA is then repaired by DNA polymerase, using the base sequence in the template strand. The final step is the sealing of the final phosphodiester bond by DNA ligase.

In human cells, a mismatch is recognized by the protein hMSH2 or by a dimer composed of the proteins hMSH2 and GTBP. In addition, nucleotides that are not base paired and that loop out from the DNA are repaired in human cells by this methyl-directed DNA repair system. This loop repair is important in the occurrence of human diseases that result from a defect in methyl-directed mismatch repair.

*HNPCC: A Defect in Methyl-Directed Mismatch Repair.* Perhaps one of the most interesting and certainly one of the most important findings relating defective DNA repair to cancer has occurred just in the past few years with the discovery that one of the most common hereditary cancers, HNPCC, is caused by mutations in the methyl-directed mismatch repair system. HNPCC affects 1 in 200 people in the United States and accounts for 15% of all colon cancers. Evidence accumulated during the past 2 years indicates there are at least five genetic loci involved in the human mismatch repair process. These include the *hMSH2* gene, the *hMLH1* gene, the *hPMS1* and *hPMS2* genes, and the *GTBP* gene. The products of these genes remove either mismatches (see Figure 4-4) or loops of 2–4 bp that are displaced from the DNA helix.

> HNPCC can be caused by a defect in the methyl-directed mismatch repair system.

The hallmark of cells with HNPCC is microsatellite instability, which is presumably caused by the inability of these cells to remove displaced loops from the DNA. Microsatellites are repetitive nucleotide sequences (di-, tri-, or tetranucleotides) located throughout the human genome. The presence of these repeats in the DNA presents a unique problem to DNA polymerase during DNA replication. When the polymerase is confronted with a long sequence of repetitive DNA, it often undergoes slippage and produces a strand of DNA with extra bases, which are not base paired and which loop out from the helix. The mismatch repair system recognizes these loops as defective and removes them. However, if the repair system is defective, the loops remain, and the number of repeat units varies from the normal amount.

This microsatellite instability indicates that the cell has developed a *mut* phenotype and now has an increased rate of overall mutation. Such cells develop mutations in crucial genes (e.g., the p53 gene or other tumor-suppressor genes) at a much higher rate than normal. Therefore, they have a greater potential for becoming cancerous. Mutations in the *hMSH2* gene have been mapped to human chromosome 2. In families with HNPCC, a single defective gene is inherited. When a somatic-cell mutation destroys the function of the second allele, the cell is unable to carry out mismatch repair, and mutations occur at an increased rate. Other tumors also have been shown to have microsatellite instability and may be caused by defects in DNA repair.

## Repair of Altered or Damaged Bases in DNA

*Base-Excision Repair System.* In addition to errors occurring during DNA replication, another source of mutation in DNA is the alteration or damage of bases in a DNA molecule that is not undergoing DNA replication. Cells have evolved two major repair systems to deal with this type of DNA damage or alteration. The first system, the base-excision repair system, was investigated by asking a very simple question: Why is uracil not found in DNA? Uracil is known to be a normal component of RNA and to have normal base-pairing properties with adenine, similar to thymine. However, DNA found in most cells from bacteria to humans contains thymine, not uracil. This specificity became even more puzzling when the precursor for putting uracil into DNA (i.e., deoxyuridine triphosphate [dUTP]) was found to be a very good substrate for DNA polymerases and that, at least in vitro, it could be used to synthesize DNA that contained uracil. The key to this puzzle came initially in 1977, when a new enzyme called uracil-DNA glycosylase was

> The base-excision repair system removes altered or damaged bases from DNA, such as the uracil residues that arise from the spontaneous deamination of cytosine.

described in bacteria. This enzyme removes uracil residues from DNA by cleaving the glycosidic bond between the base and the deoxyribose residue. Uracil-DNA glycosylase was found to be present in human cells as well and is responsible for keeping uracil out of DNA molecules. Other enzymes have since been described that have a similar glycosylase activity; that is, they act in cells to remove specific altered or damaged bases from the DNA by the pathway diagrammed in Figure 4-5.

When a uracil residue occurs in a DNA molecule, it is recognized by uracil-DNA glycosylase and is removed from the DNA, leaving behind a gap. The lack of a base in the

**FIGURE 4-5** ▶

***Base-Excision Repair by Uracil-DNA Glycosylase.*** *Deamination of cytosine residues in the DNA results in the presence of uracil. If uracil remains in the DNA, a mispair will occur at the next replication cycle, leading to a base-pair substitution. To avoid high mutation rates resulting from deamination of cytosines, a specific enzyme, uracil-DNA glycosylase, specifically removes the uracil from the DNA. The apyrimidinic site is recognized by an endonuclease, which cleaves the DNA backbone at that site to allow for repair synthesis by DNA polymerase and DNA ligase. Other types of DNA glycosylases are also present in cells to remove other types of altered or damaged bases.*

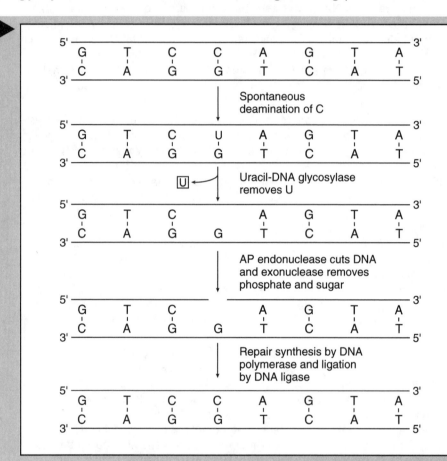

The uracil-DNA glycosylase specifically removes uracil from DNA.

DNA, which is called an apurinic or apyrimidinic site (AP site), is recognized by specific endonucleases known as AP endonucleases, which cleave the DNA strand at the site of the missing base. The resulting gap is then repaired by DNA polymerase, using the base present in the complementary strand as a template, followed by ligation by DNA ligase to restore a corrected DNA strand.

Why is it necessary for a cell to remove uracil from DNA? Cytosine residues in DNA are deaminated at a high spontaneous rate, giving rise to uracil in the DNA. If that uracil is not removed, it eventually results in a GU mispair, and the original GC pair becomes an AT pair or a mutation. Thus, cells have evolved a specific way to keep uracil out of DNA and to avoid possible mutation. An additional problem arises when cytosine bases in the DNA are methylated, as they often are. Deamination of methylated cytosine results in a thymine base in the DNA. The presence of thymine, although it will be mispaired to G at the next replication cycle, is not recognized as a wrong base. (Remember, this is happening in DNA that is already replicated and methylated, so the methyl-directed mismatch repair system will not correct this GT mispair.) As a result, sites in the DNA where cytosine is methylated are prone to high mutation rates and are referred to as "hot spots" for mutation. The presence of hot spots in DNA underscores the importance of the uracil-DNA glycosylase enzyme in reducing the number of mutations that would result from the spontaneous deamination of cytosine.

***Nucleotide-Excision Repair System.*** A more general repair system that is especially important for repairing many kinds of damage that occur to DNA molecules is known as the nucleotide-excision repair system. This repair system is necessary for the repair of bulky distortions in the DNA molecule. The nucleotide-excision repair system is the most complex system for DNA repair, and it involves a large number of proteins. Some of these proteins and their genes have been identified in human cells and are listed in Table 4-4. These genes have been classified mostly by the identification of patients with different mutations, all of which lead to a form of xeroderma pigmentosum (XP) or a related disease known as Cockayne syndrome (CS). Patients with XP are very susceptible to skin cancers upon exposure to sunlight because they are missing one or more proteins needed to carry out the process of nucleotide-excision repair (Figure 4-6). The association of a particular gene with its protein product now can be determined by in vitro assays, which can indicate the particular step that is defective in the repair pathway.

*The nucleotide-excision repair system removes bulky distortions from the DNA.*

UV light damages DNA by causing dimers between adjacent pyrimidines on the same DNA strand to occur (e.g., TT dimer). When a TT dimer is present in the DNA, distortion of the helix causes replication and transcription to stop at that point until the dimer is removed. The nucleotide-excision repair system removes these TT dimers, using a large number of proteins. The initial step is recognition of the damage by the protein XPA, which binds to the lesion along with the XPF-ERCC1 protein complex and the single-stranded DNA-binding protein (i.e., RPA). A complex of TFIIH, which contains the DNA helicases XPD and XPB as well as other polypeptides, is recruited to the damage by the presence of XPA. The helicase activity unwinds the helix and stimulates the excision activity of two endonucleases, XPF and XPG. The two nucleases make a dual excision, with XPG cutting the DNA 3 nucleotides 3′ to the damage, and XPF cutting the DNA 24 nucleotides 5′ to the damage. The result of dual excision is the release of an oligomer of 29 nucleotides containing the TT dimer, leaving behind a DNA molecule with a large gap and a free 3′-hydroxyl end. The 3′-hydroxyl is recognized by DNA polymerase-δ or -ε, which then carries out repair synthesis using the undamaged DNA strand as a template. The final nick is then sealed by DNA ligase, and the DNA is completely repaired (see Figure 4-6).

*The excision process in nucleotide-excision repair results in the release from the DNA of a 29-bp oligomer that contains the damaged site.*

The overall general scheme for nucleotide-excision repair resembles those seen for base-excision repair and methyl-directed mismatch repair. All systems have a specific protein (or proteins) that recognizes the damaged area of DNA as well as specific proteins that are involved in removal of the damage from the DNA. Following removal of the damage, the gap remaining in the DNA is filled by repair synthesis, which is catalyzed by DNA polymerases, and the final phosphodiester bond is sealed by DNA ligase. Other common proteins involved in DNA repair processes include DNA helicases and single-stranded DNA-binding proteins, which alter the structure of the DNA and facilitate the repair process.

**TABLE 4-4**
*Proteins and Genes Involved in Nucleotide-Excision Repair in Human Cells*

| Function of Gene Product | Gene |
|---|---|
| **Xeroderma pigmentosum (XP)-related functions and genes** | |
| Binds to damage on DNA | *XPA* |
| DNA helicase activity | *XPB* and *XPD* |
| 5′ single-stranded endonuclease activity | *XPF* |
| 3′ single-stranded endonuclease activity | *XPG* |
| Possible ATPase activity or DNA binding | *XPC* |
| **Other proteins and genes** | |
| Single-stranded DNA-binding protein | *RPA* |
| DNA polymerase-δ and -ε | *DNA pol* |
| DNA ligase | *DNA lig* |

*Note. A, B, C, D, F, G = complementation groups for XP. For example, XPA indicates that a defect in gene A results in the disease XP.*

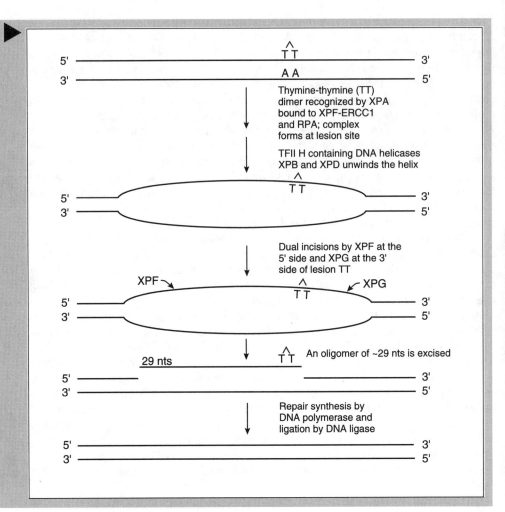

**FIGURE 4-6**

***The Nucleotide-Excision Repair Pathway.*** *The nucleotide-excision repair pathway is responsible for removing bulky distortions that can occur in the DNA, such as thymine–thymine dimers produced by ultraviolet irradiation. This repair pathway is quite complex and requires a large number of proteins, some of which have been identified in human cells. The xeroderma pigmentosum A (XPA) protein is a damage-recognition protein that binds to the lesion along with the XPF-ERCC1 protein complex and the single-stranded DNA-binding protein (i.e., RPA). This complex is then joined by the TFIIH complex, which contains two DNA helicase proteins, XPB and XPD. Two nucleases, XPG and XPF, are then recruited to the damage site to cut the DNA 3' and 5' to the lesion, releasing an oligomer of 29 nucleotides (nts). The remaining gap is filled by DNA polymerase and DNA ligase.*

# DEFECTIVE DNA-REPAIR SYSTEMS AND HUMAN GENETIC DISEASES

## Defects in Nucleotide-Excision Repair

*Xeroderma Pigmentosum (XP).* XP is a rare autosomal recessive disorder that has an incidence in the United States of 1 in 1 million. Patients with XP are especially sensitive to sunlight, and they develop severe sunburns with minimal exposure. At approximately 8 years, neoplasms on the skin develop with basal-cell carcinomas, squamous-cell carcinomas, and melanomas occurring at 2000 times the normal population. Other manifestations of XP can include eye problems and neurologic symptoms. Internal neoplasms develop in XP patients at 20 times the rate of the general population. Confirmed diagnosis of the disease can be made from biochemical and molecular analyses of cells cultured from the patients.

In the late 1960s, skin cells from XP patients were shown to be unable to repair DNA damage caused by exposure to UV light. Seven separate genes associated with XP were characterized and placed into the complementation groups *A, B, C, D, F,* and *G.* Since then, the genes of the seven complementation groups have been mapped to specific sites on human chromosomes, and the functions of most of the gene products as well as their role in the nucleotide-excision repair process have been identified (see Table 4-4).

Treatment of patients with XP involves early detection, education of the family and the patient, and constant monitoring of the skin for lesions. The disease is rare, but the risk of subsequent offspring having the disease is 25%. Most importantly, the patients

*The sensitivity to sunlight of patients with XP is caused by lack of the nucleotide-excision repair system.*

must take extreme precautions against sun exposure. In addition, these patients are more susceptible to chemical carcinogens and should avoid smoking or exposure to tobacco smoke.

*Cockayne Syndrome (CS).* In addition to XP, two other human diseases are associated with defects in the nucleotide-excision repair system. Patients with CS have growth and mental retardation and are sensitive to sunlight, but they do not develop skin cancer at an increased rate. These patients are of short stature and have progressive neurologic degeneration. Although cultured cells from CS patients are capable of carrying out nucleotide-excision repair, they have lost the repair coupling to transcription. This coupling of repair to transcription allows preferential repair of actively transcribed genes.

*Trichothiodystrophy (TTD).* TTD is an autosomal recessive disorder. Patients display brittle hair, short stature, mental retardation, and some neurologic problems. As with CS patients, they do not develop skin cancer at an increased rate. Mutations in the *XPB, XPD,* and *XPG* genes lead to TTD and may indicate a defect in coupling repair to transcription.

## Regulation of Cell Growth in Response to DNA Damage: Role of p53

The extensive systems of DNA repair emphasize the importance of protecting the integrity of the cell's genome. Otherwise, damage in the DNA would accumulate, leading to mutations, possibly to cancer, and sometimes to death of the cell. In addition to the many repair pathways, cells have other safeguards to ensure the repair of DNA damage before the DNA is replicated and passed on to daughter cells. In the cell cycle (see Chapter 2, Figure 2-8 and Chapter 5), two important checkpoints exist to signal the cell to stop progressing through the cycle if the DNA is not intact or if it is damaged. One of these checkpoints is at the G1/S border, and the second is at the G2/M border. The G1/S checkpoint is crucial for DNA repair and utilizes a signal protein known as p53, which is a tumor-suppressor protein (i.e., it acts in cells to prevent the development of tumors).

Normally, p53 functions in cells as a transcription factor and can regulate the transcription of a number of genes, either by stimulating or decreasing transcription of particular genes. When a cell is exposed to DNA-damaging agents, such as ionizing irradiation, levels of p53 protein are transiently increased, DNA replication is decreased, and the cell is arrested in the G1 phase of the cell cycle. With the cells temporarily halted in G1, the DNA repair systems have time to function, repairing the DNA damage before the damage can be replicated during the S phase. Cells that are mutant for p53 cannot stop damaged cells in G1, and they have a high incidence of developing into cancer cells. More than 50% of human cancers can be directly related to mutations in the p53 gene and to the production of mutant p53 proteins (see Chapter 14).

Recently, information about a human autosomal recessive disorder known as ataxia-telangiectasia (AT) further illustrated the role of the G1 checkpoint in preventing disease and facilitating DNA repair [6]. AT has an incidence ranging from 1 in 40,000 to 1 in 100,000 live births. Homozygous individuals have a 100-fold increase in the development of cancer when compared to the general population. Even heterozygous carriers, which represent approximately 1% of the population, have a three- to fourfold increase in cancer development. Phenotypic characteristics associated with AT include immune deficiencies, cerebellar degeneration, increased incidence of cancer, and sensitivity to ionizing radiation. Cells from AT patients have been shown to have a defect in the G1 checkpoint; they appear to be unable to use the p53 checkpoint pathway needed for proper cell-cycle control and DNA repair.

Cells from bacteria to mammals have evolved elaborate mechanims to protect their DNA. If something goes wrong with these mechanisms, a human cell is destined either to die or to lose control of growth and become a cancer cell.

*The p53 tumor-suppressor protein blocks cells at the G1/S border of the cell cycle in response to DNA damage.*

# RESOLUTION OF CLINICAL CASE

The fibroblasts derived from the skin biopsies of Caroline and her mother were grown in cell culture and tested for sensitivity to ultraviolet (UV) light and for the ability to carry out nucleotide-excision repair. Caroline's fibroblasts were 10 times more sensitive to killing by UV light when compared to her mother's cells or to control cells. The increased sensitivity to UV light and the knowledge that Caroline was showing neurologic complications indicated that her defect was probably in the gene coding for the *XPA* gene product. The *XPA* gene has been cloned and shown to code for a DNA-binding protein of 273 amino acids. The gene is comprised of six exons, which cover a span of 25 kb of DNA located on the long arm of chromosome 9.

Further testing involved measuring the ability of the cultured cells to remove TT dimers from the DNA of Caroline's cells, her mother's cells, and normal cells. After exposing the cells to UV light for increasing periods of time, samples of cellular DNA were analyzed for the number of remaining TT dimers. Both the normal cells and the mother's cells showed a decrease in the presence of dimers in the DNA. Cells taken from Caroline, however, were unable to remove any of the dimers, even 32 hours after irradiation. As a final test, the doctor performed complementation tests with cells that contained defects in the various *XP* genes. These studies involve fusing two types of cells and determining if the resulting hybrid is able to perform normal DNA repair. If two cells that form the hybrid are from the same complementation group, the resulting hybrid is defective. If the two cells that form the hybrid are from different complementation groups, the hybrid will be able to perform normal DNA repair (i.e., the two mutations complement one another). The results of the hybrid studies confirmed that the cells taken from Caroline had a defect in the *XPA* gene. The normal product of the *XPA* gene is responsible for the initial binding of the photoproduct formed after exposure to UV light. If the XPA protein is defective and unable to bind to the damage, the rest of the pathway for nucleotide-excision repair is not activated, and the dimers remain in the DNA.

As with most heterozygotes who are carriers of XP, the cells from Caroline's mother did not show any consistent abnormalities. However, the physician had to inform the mother that the risk of subsequent children having the disease was 25%. The physician arranged for the family to see a genetic counselor and recommended that Caroline not only be constantly protected from sunlight but also be examined on a regular basis so the development of any malignancies could be dealt with as early as possible. At the present time, there is no cure for XP.

# REVIEW QUESTIONS

**Directions:** For each of the following questions, choose the **one best** answer.

1. The defect in xeroderma pigmentosum (XP) is best described by which one of the following statements?
   - (A) XP results from a defect in methyl-directed mismatch repair
   - (B) XP results from a defect in nucleotide-excision repair
   - (C) XP results from a defect in cellular DNA polymerases
   - (D) XP results from a defect in DNA ligase
   - (E) XP results from a defect in topoisomerase I

2. The methyl-directed mismatch repair system is best described by which of the following activities?
   - (A) It corrects mistakes in the methylated DNA strand and not in the unmethylated strand
   - (B) It can remove lesions in the DNA caused by ultraviolet light and various chemicals
   - (C) It is impaired in hereditary nonpolyposis colorectal cancer (HNPCC)
   - (D) It involves mismatch correction by the proofreading activity of DNA polymerase-δ
   - (E) It is found only in bacterial cells

**Directions:** The group of questions below consists of lettered choices followed by several numbered items. For each numbered item, select the appropriate lettered option with which it is most closely associated. Each lettered option may be used once, more than once, or not at all.

### Questions 3–6

For each of the mutants given below, select the most appropriate corresponding sequence of amino acids.
   - (A) Normal      $H_2N-$Lys-Gly-Leu-Cys-Arg-Met-Thr− −COOH
   - (B) Mutant I     $H_2N-$Lys-Gly-Leu-Phe-Glu− −COOH
   - (C) Mutant II    $H_2N-$Lys-Gly-Leu-Ser-Arg-Met-Thr− −COOH
   - (D) Mutant III   $H_2N-$Lys-Gly-Leu-Cys-Pro-Asn-Asp− −COOH
   - (E) Mutant IV    $H_2N-$Lys-Gly-Leu-Cys− −COOH

3. This mutant has a single base-pair substitution that leads to a missense mutation

4. This mutant has a single base-pair substitution that leads to a stop codon

5. This mutant has a single base-pair deletion that shifts the reading frame, resulting in the formation of a stop codon

6. This mutant has a single base-pair addition that shifts the reading frame with each codon, following the addition being changed

# ANSWERS AND EXPLANATIONS

**1. The answer is B.** Xeroderma pigmentosum (XP) is a disease that is associated with the loss of the nucleotide-excision repair system. Patients with XP are extremely sensitive to sunlight because the thymine–thymine dimers formed by ultraviolet light cannot be repaired or removed in these patients. The methyl-directed mismatch repair system removes only normal bases from DNA. A defect in DNA polymerase would lead to a pleiotropic phenotype and probably be lethal. A defect in the enzymes DNA ligase or topoisomerase I would affect many pathways.

**2. The answer is C.** The methyl-directed mismatch repair system specifically removes the mistakes from the unmethylated or newly synthesized DNA strand. The template, which is methylated, is not repaired. In HNPCC, this repair system is defective, leading to a high spontaneous rate of mutation and the development of cancer. The methyl-directed mismatch repair system removes only normal bases from DNA. The proofreading activity of DNA polymerase is involved in correction of mismatches during DNA replication. Human and yeast cells, as well as bacterial cells, have this repair system.

**3–6. The answers are: 3-C, 4-E, 5-B, 6-D.** Mutant II is a missense mutant in which a single base-pair substitution leads to the codon for *Ser* replacing the codon for *Cys*. The normal code reads AAA-GGX-CUX-UGU-CGA-AUG-ACX. Mutant II reads AAA-GGX-CUX-AGU-CCG-AUG-ACX.

Mutant IV is a nonsense mutant in which a single base-pair substitution leads to the stop codon UGA. It reads AAA-GGX-CUX-UGU-UGA.

Mutant I is a frameshift mutation in which a single base deletion of a guanine shifts the reading frame, resulting in the formation of the stop codon of UGA. The code for mutant I reads AAA-GGX-CUX-UUC-GAA-UGA.

Mutant III is a frameshift mutation in which a single base addition of a cytosine in the codon CGA shifts the reading frame with each codon following the addition being changed.

# REFERENCES

1. Modrich P: Mismatch repair, genetic stability and cancer. *Science* 266:1959–1960, 1994.
2. Sancar A: Excision repair in mammalian cells. *J Biol Chem* 270:15915–15918, 1995.
3. Loeb LA: Microsatellite instability: marker of a mutator phenotype in cancer. *Cancer Res* 54:5059–5063, 1994.
4. Warren S: The expanding world of trinucleotide repeats. *Science* 271:1374–1375, 1996.
5. Leach FS, Nicolaides NC, Papadopoulos N, et al: Mutations of a Mut S homolog in hereditary nonpolyposis colorectal cancer. *Cell* 75:1215–1225, 1993.
6. Kastan MB, Zhan Q, El-Deiry WS, et al: A mammalian cell cycle checkpoint pathway utilizing p53 and GADD45 is defective in ataxia-telangiectasia. *Cell* 71:587–597, 1992.

# CHROMOSOME STRUCTURE AND THE MAMMALIAN CELL CYCLE

# INTRODUCTION OF CLINICAL CASE

Mrs. Clark and her husband brought their 7-year-old son, William, to the pediatrician. The mother claimed that the child is very argumentative and constantly throws tantrums when he does not get his own way. In discussing the child further with the physician, the mother said that when William was an infant, he was very difficult to feed, and he never wanted to eat. As a last resort, the parents fed William through a tube. William was born after a 43-week gestation; labor was induced because of decreased fetal movement. Prenatal ultrasound performed early in the pregnancy detected no apparent fetal abnormalities. William weighed 6 pounds, 14 ounces at birth and was very hypotonic, which most likely contributed to his feeding difficulties. His developmental milestones were somewhat delayed, sitting at 12 months, walking at 18 months, and talking by 24 months. William attends first grade in a regular school but has difficulty completing assignments and often demonstrates disruptive behavior in class.

The physician noticed that William is very fair with blond hair and clear blue eyes, although both parents have dark complexions, brown eyes, and brown hair. The physician asked the couple if any of their other children have similar problems. The parents said their two older sons who are 10 and 12 years of age are both quite well behaved and fairly tolerant of their younger brother's tantrums. The pediatrician asked to see pictures of the two older boys. As she expected, they both have dark complexions with dark eyes and hair. The pediatrician thought that perhaps William was adopted, but the parents assured her that he was not. The pediatrician gave William a physical examination after which she instructed the family to be patient and to work with William to help him control his unruly behavior. The pediatrician encouraged the Clarks to return in 6 months, unless something unusual occurred.

Four months later, the pediatrician received a frantic call from Mrs. Clark who stated that William had developed an unusual appetite. The mother explained that the child ate constantly, gaining 20 pounds in 3 months. The mother told the physician that William was in the kitchen day and night, madly stuffing his mouth with everything he could find. The parents finally resorted to locking the refrigerator to keep William from eating constantly. The mother said that she received several phone calls from William's teacher, who reported that he had been caught stealing food from the other children. Last night, the mother was very frightened when a neighbor knocked on the door at 2 A.M. with William in tow. The neighbor found the child digging in the trash cans and eating pieces of garbage. The pediatrician told the mother to bring William in immediately. Meanwhile, the pediatrician called the genetics department at the local hospital and spoke to a clinical geneticist about William's case.

# STRUCTURE OF HUMAN CHROMOSOMES

The human genome in its diploid state is composed of more than 6 billion base pairs (bp) of DNA arranged in linear structures known as chromosomes (*chromo* [colored] *soma* [bodies]). Chromosomes are unique in the nucleus of a human cell because they are the only objects that stain bright purple using the Feulgen-Rossenbach procedure. In a human diploid cell, there are a total of 46 chromosomes arranged as 22 pairs of autosomes and one pair of sex chromosomes. Human males have 22 pairs of autosomes, one X chromosome, and one Y chromosome, whereas human females have 22 pairs of autosomes and two copies of the X chromosome. The exception is the gametes—the egg and the sperm—both of which are haploid and contain only one copy of each pair of the 22 autosomes and one sex chromosome.

> **Normal Human Diploid Cell.**
> 46 chromosomes: 22 pairs of autosomes and 1 pair of sex chromosomes

## Chromatin

As mentioned in Chapter 2, large amounts of DNA can fit into the small space of the cell nucleus because the DNA strands are compacted into organized structures. The DNA molecules in the nucleus are not found as extended linear structures but are present in supercoiled structures, in which the double-stranded helix falls back upon itself to give a more compact form. Further organization of DNA molecules in the nucleus occurs when the DNA complexes with various nuclear proteins. DNA in the nucleus is always associated with proteins and is referred to as *chromatin*. At the various stages of the cell cycle, chromatin is compacted to different degrees. During cell division in the mitotic stage, chromatin is most tightly packaged. At mitosis, discrete chromosomes are visible under the light microscope after staining and can be accurately identified.

> In the nucleus, DNA is associated with proteins and is referred to as **chromatin**.

## Histones, Nucleosomes, and Packaging of DNA

At all stages of the cell cycle, DNA is associated with specific basic proteins, called *histones*, and forms structures known as *nucleosomes*. Five major classes of histones are associated with DNA in the nucleus: H1, H2A, H2B, H3, and H4. Histones contain high concentrations of lysine and arginine, which give them a positive charge and promote the formation of strong ionic bonds with negatively charged DNA. Under an electron microscope, chromatin is seen as repeating units of nucleosomes arranged with regular spacing to form long fibers with a bead-like structure [1].

Each nucleosome is made up of a core of eight histone proteins (two copies each of H2A, H2B, H3, and H4) with 146 bp of DNA wrapped around them to form a DNA-histone complex. This well-organized structure further reduces the length of the DNA. Adjacent nucleosomes are separated from each other by a stretch of 50–60 nucleotides of DNA referred to as spacer DNA or *linker DNA*. The linker DNA is generally considered to be part of the nucleosome, giving a total of approximately 200 bp of DNA per nucleosome (Figure 5-1).

> **Nucleosome**
> A structure made up of 200 bp DNA and 8 histone proteins.

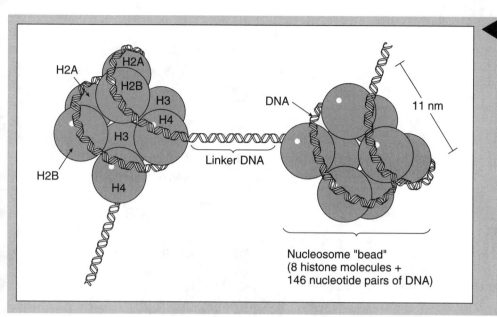

**FIGURE 5-1**
**Structure of a Nucleosome.** *The nucleosome is composed of eight histones (two each of H2A, H2B, H3, and H4) associated with 146 base pairs (bp) of DNA and 50 bp of linker DNA. (Source: Reprinted with permission from Becker W, Reece JB, Poenie M:* The World of the Cell, *3rd ed. Menlo Park, CA: Benjamin-Cummings, 1996, p 434.)*

Histone H1, which is composed of 220 amino acids, is the largest of the histones. Like the other histones, histone H1 is rich in the basic amino acids, especially lysine. Histone H1 is not part of the nucleosome structure, but it is important in assembling multiple nucleosomes into a complex fiber structure. This 30-nm chromatin fiber, sometimes called the *solenoid*, contains six nucleosomes per turn, as shown in Figure 5-2.

Additional packaging of the DNA occurs with the formation of domains or loops of chromatin held together by nonhistone proteins. These loops, containing 50,000–100,000 bp, are attached to a scaffold structure and represent regions of less compacted DNA, known as *euchromatin*, which are usually transcriptionally active. Topoisomerase II, an enzyme necessary for the uncoiling of DNA (see Chapter 2), is part of this loop structure and probably plays a role in the dynamic movement of the chromatin. Other areas of the DNA-protein structures are more highly compacted and make up *heterochromatin*, which are areas of DNA sequence believed to be transcriptionally inactive.

*Less compacted DNA (**euchromatin**) contains transcriptionally active DNA sequences. More highly compacted DNA (**heterochromatin**) contains areas of DNA sequence believed to be transcriptionally inactive.*

With further compaction of the loop-scaffold, the DNA-protein structure produces the chromatids that make up the chromosome structures seen only in the mitotic phase of the cell cycle, during which the chromosomes become highly condensed (see Figure 5-2). The final size of the compacted chromosomes is approximately 1.5–5 μm in length, although they contain nearly 75 mm DNA. This extensive compaction allows all of the chromosomes to fit into the nucleus, which is approximately 10 μm in diameter.

## DNA Replication Origins, Centromeres, and Telomeres

***DNA Replication Origins.*** The chromosomes, as we see them in mitosis, have three well-defined sequences or structures that are essential for their replication and distribution to daughter cells. These elements are DNA replication origins, centromeres, and telomeres (Figure 5-3).

*Important Elements of a Chromosome Replication origins, a centromere, and telomeres.*

As discussed in some detail in Chapter 2, the chromosomes in a human cell replicate only during the S phase of the cell cycle. The process of DNA replication is very precise and highly regulated. Each chromosome is replicated by simultaneous initiation of DNA replication from multiple origins within the chromosome. These multiple replicating blocks, or *replicons*, allow the total genome to be replicated during the S phase. The replication process is controlled to allow each chromosome to replicate completely but only once in a single S phase. In the event of improper or incomplete replication of a chromosome, the cell recognizes the mistake and may either block the cell from continuing in the cell cycle or direct that cell to a programmed death. Furthermore, as discussed in Chapter 4, the cell has evolved multiple repair systems to ensure the correctness of the DNA base sequences that are passed on to the daughter cells.

**FIGURE 5-2** ▶

**Packaging of DNA in the Nucleus.** *A model for the progressive stages of DNA coiling and folding in the nucleus. (A) Configuration of the nucleosomes. (B) The 30-nm chromatin fiber with H1 histone located in the interior of the fiber. (C) The looped domain of the 30-nm fiber. (D) The formation of heterochromatin made up of highly folded chromatin. (E) A duplicated metaphase chromosome with highly compacted DNA. (Source: Reprinted with permission from Becker W, Reece JB, Poenie M: The World of the Cell, 3rd ed. Menlo Park, CA: Benjamin-Cummings, 1996, p 435.)*

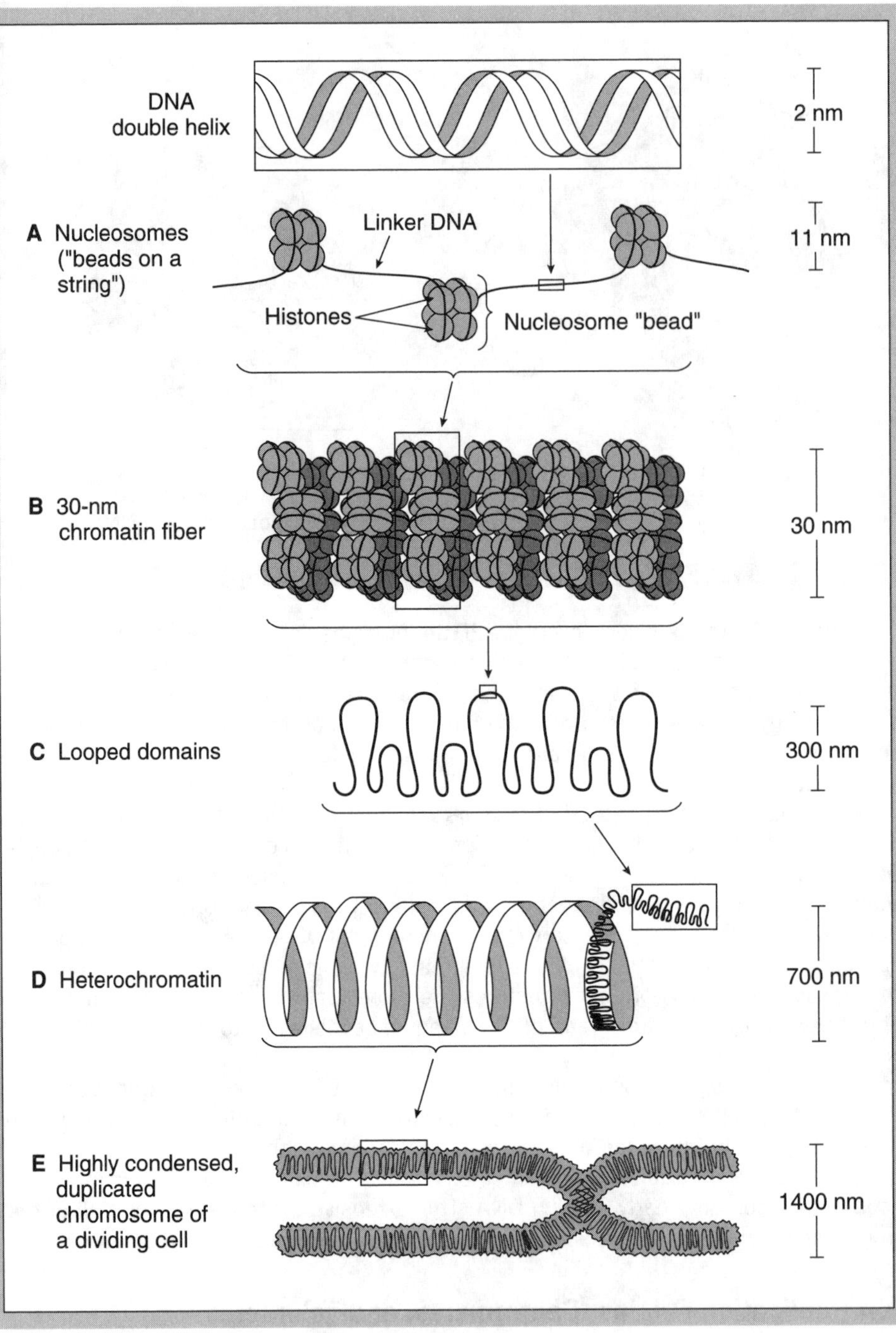

The **centromere** consists of highly repetitive DNA sequences.

**Centromeres.** A second important sequence found in all human chromosomes is the centromere. The centromere, which consists of highly repetitive DNA sequences, is the site where the two sister chromatids are attached. The constriction produced is used to define the arms of the chromosome as the short (p, or petite) arm and the long (q) arm. Centromeres are positioned at different sites on different chromosomes and are used to differentiate the chromosome structures seen in mitosis as metacentric, submetacentric, and acrocentric (Figures 5-3 and 5-4).

The function of the centromere is to ensure equal distribution of each chromosome to the daughter cells at cell division. During the process of mitosis, in prophase, proteins assemble at the centromere to form a DNA-protein complex known as a *kinetochore*. Each chromatid contains one kinetochore. Microtubules attach to each of the kinetochores and link the centromere of each chromosome to the spindle poles. The chromosomes move to the center of the cell until all of the chromosomes become aligned at the

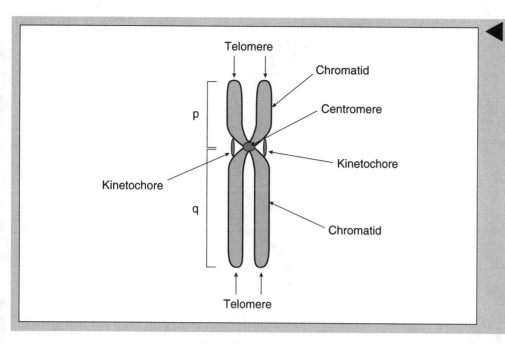

**FIGURE 5-3**
*Chromosome at Metaphase.* A duplicated chromosome is made up of two sister chromatids held together at the centromere. Each chromatid contains a kinetochore and telomeres at each end.

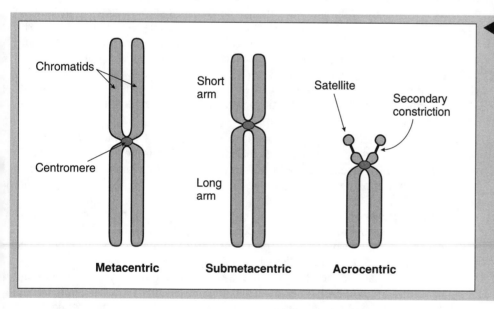

**FIGURE 5-4**
*Metacentric, Submetacentric, and Acrocentric Chromosomes.* The position of the centromere is used to describe a chromosome.

metaphase plate. When the appropriate signal is given, anaphase begins, and the sister chromatids are pulled apart. One chromatid from each chromosome moves to opposite poles of the cell, ensuring that one chromatid from each replicated chromosome is passed to each daughter cell (Figure 5-5).

A critical control point occurring in mitosis between metaphase and anaphase has been identified [2]. A protein, called an inhibitor of sister chromatid separation (ISS), appears to hold the two sister chromatids together until all of the chromosomes are properly aligned on the metaphase plate. At a crucial point, the protein ubiquitin directs the degradation of the ISS protein, and anaphase begins with the sister chromatids moving to each pole (see Figure 5-5).

In the absence of a centromere, control over the distribution of the chromosomes to the daughter cells is lost. One daughter cell may receive two copies of the acentric chromosome (a chromosome without a centromere), whereas the other daughter cell receives none. A well-known aberrant chromosome structure that does not contain a centromere is found in many tumor cells. These structures, known as *double-minute chromosomes*, are found in multiple copies as small circular DNA structures that often occur in pairs (Figure 5-6).

**ISS Protein**
*This protein holds the two sister chromatids together on the metaphase plate. When the ISS protein is degraded, anaphase begins.*

**FIGURE 5-5**

**Transition from Metaphase to Anaphase.** *The inhibitor of sister chromatid separation (ISS) protein holds the two sister chromatids together at the metaphase plate. When ISS is degraded by the ubiquitin system, anaphase begins with separation of the chromatids.*

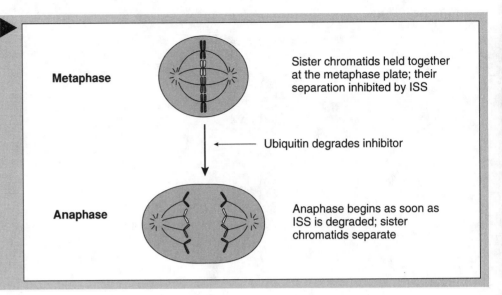

**Metaphase** — Sister chromatids held together at the metaphase plate; their separation inhibited by ISS

Ubiquitin degrades inhibitor

**Anaphase** — Anaphase begins as soon as ISS is degraded; sister chromatids separate

---

**Double-minute chromosomes** *do not contain a centromere and are distributed randomly to daughter cells.*

**Telomeres** *consist of multiple tandem repeats of the sequence TTAGGG and are located on both ends of each chromatid.*

**Functions of Telomeres**
*Maintain the stability of chromosomes, prevent chromosome degradation, ensure the proper replication of the DNA ends, and serve as signals that a chromosome is not broken or fragmented.*

In the absence of a centromere, these double minutes are randomly distributed to daughter cells. If these DNA elements contain genes that give a cell a selective growth advantage, cells containing multiple copies of the double minutes overgrow cells that have not received the element. Therefore, a single cell can accumulate hundreds of copies of the double minutes. When the double minutes carry a gene coding for a protein, a cell containing many copies of the double minutes overproduces that protein. These proteins, when present in high concentrations, often alter the growth properties of a cell and result in the cell escaping normal growth control and developing into a tumor cell. Thus, the presence of a centromere on a chromosome is an essential sequence that ensures normal distribution of all genetic material to each daughter cell and the maintenance of the normal state.

**Telomeres.** An important structure associated with the ends of all human chromosomes is the telomere. Telomeric DNA consists of multiple tandem repeats of a specific sequence—TTAGGG—located on both ends of each chromatid. Telomeres are known to serve several very important roles in human cells.

- Telomeres are essential for maintaining the stability of the chromosome and for preventing end-to-end fusions from occurring. In the absence of telomeric sequences, the chromosome ends appear very sticky and tend to fuse together. Such fusions can result in the formation of ring chromosomes, if the two ends of a single chromosome fuse, or they can result in a dicentric chromosome, if two different chromosomes fuse to each other. Fusion of chromosome ends interferes with normal distribution of the chromosomes at mitosis and often results in daughter cells with chromosome abnormalities as discussed later.

- The presence of telomeric sequences protects the ends of the chromosomes from being degraded by nucleases. Proteins that are just now being identified appear to interact with the telomeric DNA sequences to give a nucleoprotein protective cap on the ends of each chromatid, which makes the DNA much more resistant to enzymatic degradation.

- Telomeres are essential for ensuring the proper replication of the ends of the chromosomes. As discussed in Chapter 2, the ends of the DNA are not fully replicated by the normal DNA replication machinery and require the presence of a specific enzyme (i.e., telomerase) to add nucleotides to the very ends of the DNA. The presence of noncoding telomeric sequences at the ends of each chromosome protects the coding sequences of the DNA that might be located near the terminal ends of a chromosome from being lost during each replication cycle. Because the telomeric sequences are repeated hundreds of times and apparently do not code for any protein, loss of some of these sequences from the chromosomes can be tolerated at each cell division.

- Recently, a new role of telomeres in normal cell growth has been identified [3]. The presence of a telomere on the end of a chromosome can serve as a signal that a

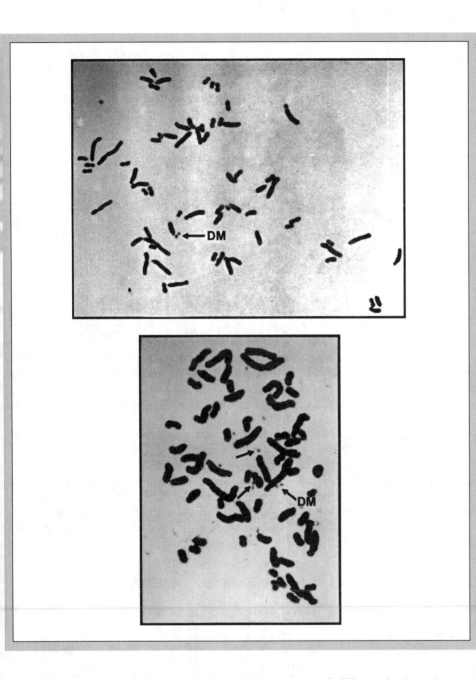

**FIGURE 5-6**
**Double-Minute Chromosomes.** These acentric circular DNA molecules, which usually occur in pairs, are randomly distributed to daughter cells and can accumulate to high copy number. DM = double minute.

chromosome is normal (i.e., not broken or missing an end). When a broken chromosome or a fragment of a chromosome arises in a cell, this abnormal structure is missing a telomeric sequence on at least one end. DNA damage controls that are present in all normal human cells recognize the abnormal chromosomal end and signal the cell cycle to stop temporarily. This interruption of the cell cycle allows the DNA repair systems described in Chapter 4 to repair the damage. In the event the damage cannot be repaired, the cell may be directed to a programmed death.

Each time the DNA undergoes replication in a normal human somatic cell, 50–100 bp of the terminal telomeric sequences of a chromosome are lost. This shortening of the chromosome ends can be prevented by the enzyme telomerase, which is normally present in embryonic stem cells and germ cells. Normal somatic cells, however, do not appear to have active telomerase. Thus, as a somatic cell continues to grow and divide, there is a constant shortening of the ends of the chromosomes. The continual shortening of chromosomes may function as a mitotic clock, signaling the occurrence of programmed death in aging cells before a chromosomal abnormality arises and is passed to daughter cells. If cell death does not occur, a chromosome that has completely lost its telomeric sequences becomes subject to end-to-end fusion, which produces an aberrant chromo-

Normal somatic cells lose 50–100 bp of their telomeric sequences at each replication.

some. Thus, the absence or shortening of telomeres can serve as a signal to damage control systems that a cell is no longer healthy and should be eliminated before it develops a mutation.

It has been observed that some cancer cells can escape the above control system and become immortalized [3]. The telomeric sequences on the chromosomes of these cancer cells are short but remain stable in length, resulting in uncontrolled growth of a cell that is now no longer subject to damage control. The stability of the telomeres in cancer cells is apparently caused by the reappearance or reactivation of telomerase, which is not active in normal somatic cells. What triggers this activation of telomerase in cancer cells is unknown, but the presence of telomerase activity in a somatic cell has become a potential marker of cancer development and is being considered as a possible diagnostic test for some cancers. Future research is aimed at developing pharmacologic agents that inhibit telomerase activity in cancer cells and promote their death.

*Active telomerase in a normal somatic cell may indicate a cancerous condition.*

# CYTOGENETIC METHODOLOGY AND CHROMOSOME NOMENCLATURE

Although the 46 chromosomes found in the human nucleus are unique, some of them resemble each other when stained by certain conventional methods. Prior to 1971, human chromosomes could be classified only by size. In a standardized karyotype, the 22 autosome pairs were ordered by placing them in seven groups (A through G) in descending order of size, as shown in Figure 5-7.

Within each group, the chromosomes are arranged again by size. Group A contains chromosomes 1, 2, and 3, the largest metacentric chromosomes. Group B contains chromosomes 4 and 5, which are large submetacentric chromosomes. Group C contains chromosomes 6, 7, 8, 9, 10, 11, and 12, which are the medium-sized submetacentric chromosomes. This group also includes the X chromosome, which is usually separated on the karyotype and presented at the end. Group D contains chromosomes 13, 14, and 15. These are medium-sized acrocentric chromosomes with satel-

*Chromosomes are placed in groups A–G based on their size and the location of the centromere.*

**FIGURE 5-7** ▶
*Standardized Human Karyotype. A designation of G-banding patterns, according to the 1971 Paris Conference on Standardization in Human Cytogenetics sponsored by The March of Dimes.*

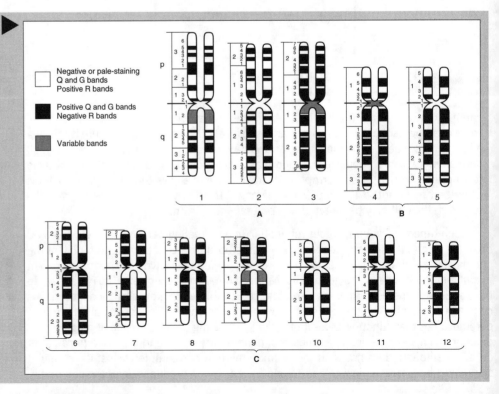

lites, which are regions of repetitive DNA that make up constitutive heterochromatin. Group E contains the relatively short chromosomes 16, 17, and 18, which have metacentric (16) or submetacentric centromeres (17 and 18). Group F contains the short metacentric chromosomes 19 and 20. Finally, Group G contains the very short acrocentric chromosomes 21 and 22, both of which contain satellite DNA. The Y chromosome is placed in group G, although the length of the Y chromosome can vary from male to male.

## Chromosome Banding

In 1971, a conference was held in Paris to establish a standardized system of nomenclature that would identify each chromosome as well as the regions of each chromosome. This system was made possible by Caspersson's discovery that when certain fluorescent dyes (e.g., quinacrine) are applied to a chromosome preparation, alternating light and dark bands result along each human chromosome. Each chromosome pair gives a specific banding pattern, which makes it possible to identify accurately every chromosome in a karyotype by this process known as *Q banding*. It was then possible to identify abnormal chromosome content and structure and determine their inheritance pattern and their relationship to human disease. Formal standardized nomenclature for human chromosomes was established and is maintained by a standing committee that publishes an International System of Human Chromosome Nomenclature (ISCN) to provide investigators with the accepted rules for noting karyotype designations. Some of these symbols are listed in Table 5-1.

> *Chromosome-banding techniques allow the identification of each chromosome in a karyotype.*

Following the discovery of Q banding, another procedure, Giemsa or *G banding*, was developed. This procedure combines the use of Giemsa stain with trypsin treatment and produces banding on human chromosomes that is identical to the bands produced by quinacrine. Regions of the chromosomes that give fluorescence with Q banding produce dark bands, or G bands, when treated with trypsin and Giemsa. G banding quickly became a standard procedure for identifying metaphase chromosomes in a human karyotype. An idiogram of the standardized G bands found for the 22 autosomes and the sex chromosomes is shown in Figure 5-7. A typical G-banded human karyotype is shown in Figure 5-8.

> *G banding, which combines Giemsa staining and trypsin treatment, is a standard procedure for identifying metaphase chromosomes in a human karyotype.*

Other banding procedures have been developed that can be used to reveal specific features of human chromosomes. Regions of chromosomes that contain constitutive het-

◀ **FIGURE 5-7**

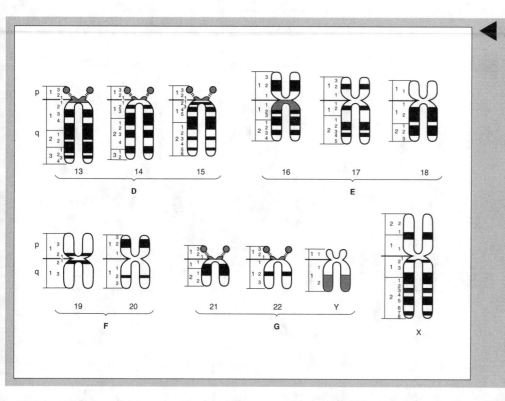

erochromatin are detected by *C banding*, which stains centromeric DNA as well as the satellite regions on chromosomes. Reverse banding, or *R banding*, is a technique that produces a banding pattern that is the reverse of G banding or Q banding. In this technique, the chromosomes are denatured at high temperature in the presence of salt and then stained either with Giemsa or acridine orange. R banding is particularly useful for studying the ends of chromosomes, which stain dark with G banding and light with R banding.

**TABLE 5-1** ▶

*Symbols and Abbreviations used in Cytogenetic Nomenclature*

| Symbol | Description |
|--------|-------------|
| p | Short arm of chromosome |
| q | Long arm of chromosome |
| ace | Acentric |
| cen | Centromere |
| del | Deletion |
| der | Derivative chromosome |
| dic | Dicentric chromosome |
| dmin | Double-minute chromosome |
| dup | Duplication |
| fra | Fragile site |
| hsr | Homogeneously staining region |
| i | Isochromosome |
| ins | Insertion |
| inv | Inversion |
| (−) | Loss of genetic material |
| (+) | Gain of genetic material |
| r | Ring chromosome |
| rcp | Reciprocal |
| rob | Robertsonian translocation |
| (;) | Semicolon; separates chromosome and chromosome regions when structural rearrangements involve more than one chromosome |
| t | Translocation |
| ter | Terminal |
| pter | End of the short arm |
| qter | End of the long arm |

**FIGURE 5-8** ▶

**G-banded Human Chromosomes.** *A G-banded karyotype of a normal human male. (Courtesy of Dr. Urvashi Surti, University of Pittsburgh School of Medicine and Magee Women's Hospital, Pittsburgh, Pennsylvania)*

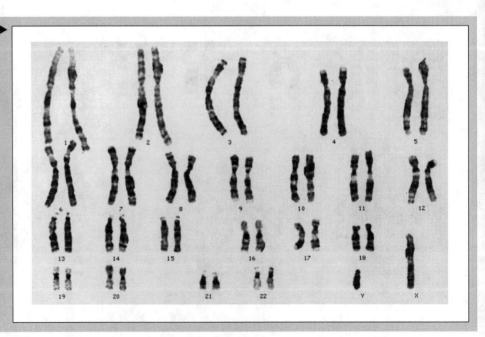

The dark bands, or G bands, produced with Giemsa and trypsin represent late-replicating heterochromatin bands, which are rich in adenine–thymine (AT) pairs. Approximately 20% of the human genes that have been mapped are located within the G bands (these are also the regions that give Q bands). The light bands, or R bands, represent areas of the chromosome that replicate early in S phase and seem to contain many of the sequences coding for known human proteins. Only recently has a satisfactory model been provided to explain the molecular basis for the various banding patterns seen on human chromosomes. Saitoh and Laemmli have attributed the difference in banding patterns to the chromatin loops within a metaphase chromosome [4]. Regions of DNA known as scaffold-associated regions (SARs) appear to define the base of the chromatin loops (see Figure 5-2). These regions are AT rich and are located along the length of the chromatids. Tighter coiling of the SARs occurs in the region of G/Q bands than occurs in the R-banded regions of the chromosome, giving the light and dark patterns seen along the chromosome.

The availability of banding procedures gave cytogeneticists, for the first time, the ability to identify individual regions within each human chromosome. Using the designation of p for the short arm and q for the long arm, each chromosome was further divided by regions as indicated in Figure 5-9. The bands and regions on each arm of a chromosome are numbered from the centromere to the telomere. A region is any area lying between two specified landmarks. The region adjacent to the centromere is numbered 1 in each arm followed by 2, 3, and so on going toward the telomere. Each band within a region is also numbered, once again starting proximal to the centromere and going toward the telomere. In a standard chromosome preparation, there are approximately 400 total bands seen per haploid chromosome number. Higher resolution banding is also now available to give 1000 or more bands. In this case, the bands are further subdivided.

In Figure 5-9, a normal chromosome 2 is represented. On the p arm there are two regions (1 and 2), and within each region are the bands (e.g., in region 1, bands are noted as 1, 2, 3, 4, 5, and 6). The *arrow* indicates a position on region 1 and band 6 and is noted as 2p16, meaning the location is on chromosome 2, the p arm, region 1, band 6. With the standardized nomenclature, a karyotype that deviates from normal can be clearly described and understood anywhere in the world.

Karyotype analysis is performed on growing cells that come from blood samples, skin samples, fetal amniocytes, chorionic villi, and tumor cells. When using peripheral blood samples, the cells are collected and stimulated to grow in cell culture for 72 hours, at which time a mitotic inhibitor, such as colchicine, is added. When a significant number

*Each band is named according to the chromosome, the region on the chromosome, and the position of the band within the region. For example, 2p16 indicates a position on chromosome 2, the p arm in region 1, band 6.*

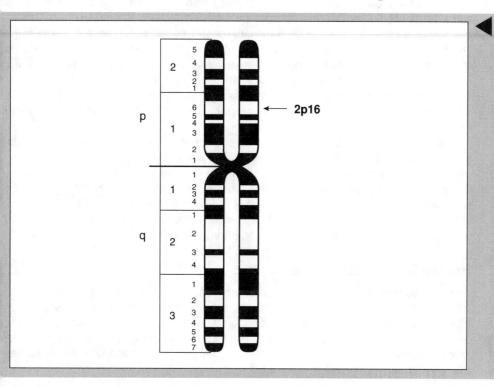

◀ **FIGURE 5-9**
**Nomenclature for Banded Human Chromosomes.** *A position of a gene sequence on the chromosome is named according to the chromosome (2), the arm of the chromosome (p), the region within the arm (1), and the band number (6), giving 2p16. The figure shows an R-banded pattern for chromosome 2.*

of cells are blocked in mitosis, the cells are collected and treated with hypotonic solution. After fixing, the cells are spread on a slide and stained with the appropriate stain. Mitotic cells are observed under a light microscope to identify metaphase spreads that have well-separated and well-defined chromosomes. Previously, the metaphase spreads were photographed and prints developed. The chromosomes from the print were painstakingly cut out and lined up by hand to create the standard karyotype shown in Figure 5-8.

Fortunately, most clinical laboratories are now equipped with an automatic karyotype machine designed to analyze the metaphase spreads. However, 1 to 2 weeks are still required to complete the analysis of a karyotype from the time the sample is taken to the time the results are obtained.

## In Situ Hybridization

A second and very important technique for evaluating karyotypes is a procedure known as in situ hybridization. With the increasing availability of labeled DNA probes, it is now possible to localize specific DNA sequences directly on a chromosome spread. Initial studies on in situ hybridization used radioactively labeled nucleic acid probes. These methods were adequate to detect multiple copies of genes on chromosomes such as the ribosomal RNA (rRNA) genes or genes that were amplified in tumor cells; however, single-copy genes were very difficult to detect simply because of the insensitivity of the method. Difficulty incorporating sufficient radioactive label into the probes and the length of time (months) needed to be able to see the radioactivity on the chromosome preparation limited the usefulness of the technique.

With the development of new labeling procedures, the technique is now much more efficient and is readily able to detect single-copy genes in a short time. Nonradioactive-labeling procedures, using a biotin-labeled probe that can be detected in combination with a fluorescent dye, is the current method of choice. This technique, known as fluorescent in situ hybridization (FISH), can be used to localize accurately a specific gene sequence within a chromosome. In the FISH procedure, metaphase chromosomes are denatured on a slide and then hybridized to a DNA probe labeled with biotin. The slide is then treated with the protein avidin, which binds very tightly to the biotin. With the addition to the slide of a fluorescent-labeled antibody to biotin, the site on the chromosome containing the sequence of interest fluoresces when placed under a fluorescent microscope, localizing its position.

> **Uses of FISH**
> *Accurately localizing a specific gene sequence within a metaphase chromosome and detecting numerical chromosome abnormalities in an interphase cell.*

Recent improvements in the FISH technique involve the use of different dyes on the same chromosome preparation to order probes with respect to one another. Current methodology allows two probes 2–3 megabases (Mb) apart to be ordered on the same chromosome. Additionally, the development of chromosome painting, in which mixtures of chromosome-specific DNA sequences are used, allows the identification of a specific chromosome within a metaphase spread. In this technique, DNA probes that are complementary to the entire length of a chromosome are used. These probes light up the specific chromosome and allow the detection of rearrangements and translocations with more precision than simple banding procedures.

Although initially used only with metaphase chromosome spreads, FISH is now being adapted for use with interphase cells. The advantage in being able to use interphase cells is the speed with which the technique can provide answers to the clinician. Because the procedure does not involve growing the cells in culture, results can be obtained in 24–48 hours. Interphase FISH is useful in the prenatal diagnosis of trisomies or sex-chromosome abnormalities in which the extra chromosome presents as an extra fluorescent spot in the nucleus of the interphase cell.

## Human Chromosome Abnormalities

With the availability of chromosome-banding techniques and FISH, it is now possible to recognize changes that occur in the structure or the number of normal chromosomes found in a metaphase cell. Chromosome abnormalities are common in humans, affecting as many as 7.5% of all conceptions. Most of these abnormal conceptions result in spontaneous abortions, reducing the live-birth frequency of chromosome abnormalities to approximately 0.5% or 1 in 200 births.

> *The frequency of human chromosome abnormalities is 1 in 200 live births.*

Chromosome abnormalities are classified as either numerical or structural abnormalities. Numerical abnormalities refer to a cell containing an abnormal number of normal chromosomes. Structural abnormalities involve a cell with one or more abnormal chromosomes that contain a structural rearrangement. Even cells that appear to have a normal karyotype by chromosome-banding analysis may in fact be abnormal. For example, sometimes all of the chromosomes in a diploid cell are derived from one parent (uniparental diploidy), or sometimes both copies of a single chromosome pair are derived from one parent (uniparental disomy). In these cases, a dramatic effect on the phenotype of the offspring can occur although the karyotype appears normal (see below).

## NUMERICAL CHROMOSOMAL ABNORMALITIES

Numerical chromosomal abnormalities occur when the normal human chromosome number of 46 is altered, resulting in euploidy (polyploidy), aneuploidy, or mosaicism. When a cell has balanced extra chromosome sets in multiples of n of 23 chromosomes, it is described as being polyploid or euploid. Cells that have a triploid (3n) number of chromosomes are identified as 69,XYY; 69,XXX; or 69,XXY (Table 5-2). *Triploidy* (3n) can result from fertilization of a single egg by two sperm (dispermy) or from a defective meiosis when a diploid gamete is produced. *Tetraploidy* (92 chromosomes or 4n) also occurs and is usually the result of a defect during the first zygotic division. Both triploidy and tetraploidy are lethal and generally result in spontaneous abortion of the fetus.

*Aneuploidy* is the term used to describe a cell in which an individual chromosome of one pair is found missing (monosomy) or a chromosome is present as an extra copy (trisomy). Monosomies, unless they involve a sex chromosome, are almost always lethal, as are most trisomies. However, three trisomies occurring in humans are compatible with life. These include trisomy 21 (Down's syndrome), trisomy 13 (Patau's syndrome), and trisomy 18 (Edwards' syndrome). Trisomy 21, the first trisomy associated with human disease, was described by Lejeune in 1959 when he identified three copies of chromosome 21 in patients with Down's syndrome.

Monosomy and trisomy can arise either during mitosis or meiosis by the process of nondisjunction, in which the two sister chromatids or two homologous chromosomes fail to separate during anaphase. Both meiosis I and meiosis II can give rise to nondisjunction as diagrammed in Figure 5-10. If nondisjunction occurs during meiosis I or II, one of

**Polyploidy or Euploidy**
*A cell that has a balanced extra chromosome set in multiples of n of 23 chromosomes.*

**Aneuploidy**
*A cell in which an individual chromosome of one pair is missing (monosomy) or a chromosome is present as an extra copy (trisomy).*

*Three trisomies are compatible with life: trisomy 13, trisomy 18, and trisomy 21.*

◀ **TABLE 5-2**
Notations for Chromosome Complement

| Condition | Notation |
|---|---|
| **Normal** | |
| Normal female | 46,XX |
| Normal male | 46,XY |
| **Numerical abnormalities** | |
| Triploidy | 69,XXX; 69,XXY; 69,XYY |
| Tetraploidy | 92,XXXX |
| Monosomy for 13 | 45,XX,−13 |
| Trisomy for 21 (Down's syndrome) | 47,XX,+21 |
| Mosaic for 21 | 47,XX,+21/46,XX |
| **Structural abnormalities**[a] | |
| Reciprocal translocation | 46,XY,t(1;12)(p35;p12) |
| Robertsonian translocation | 45,XX,der(14;21)(q10;q10) |
| Terminal deletion | 46,XY,del(5)(p15) |
| Interstitial deletion | 46,XX,del(2)(q31q33) |
| Ring chromosome | 46,XX,r(7)(p21q34) |
| Pericentric inversion | 46,XX,inv(6)(p12q23) |
| Paracentric inversion | 46,XY,inv(8)(q22q24) |
| Isochromosome | 46,X,i(X)(q10) |
| Dicentric | 46,X,dic(Y)(q13) |

[a] See Figure 5-12.

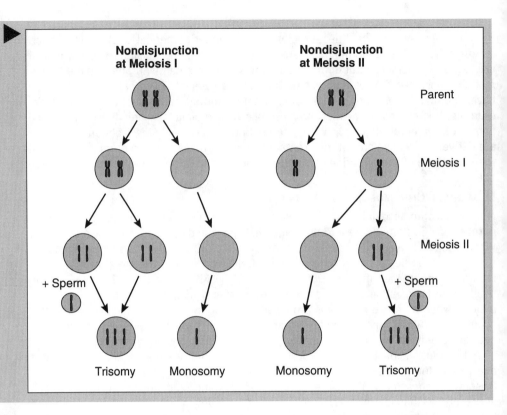

**FIGURE 5-10**

*Nondisjunction during Meiosis.* Meiotic nondisjunction results in two chromosomes migrating to one daughter cell with the second daughter cell receiving none. After fertilization by a normal sperm, a monosomic or trisomic cell results.

**Nondisjunction at Meiosis I**

**Nondisjunction at Meiosis II**

Parent

Meiosis I

Meiosis II

+ Sperm

+ Sperm

Trisomy    Monosomy    Monosomy    Trisomy

---

*Trisomy 21 (Down's syndrome) was the first trisomy associated with human disease.*

*Monosomies and trisomies arise during mitosis, meiosis I, and meiosis II by the process of nondisjunction.*

*Uniparental disomy occurs when one pair of chromosomes is derived totally from either the mother or the father.*

*Prader-Willi syndrome is caused either by a deletion of 15q11–13 on the paternal chromosome or by the absence of the paternal chromosome. Angelman's syndrome is caused either by a deletion of 15q11–13 on the maternal chromosome or by the absence of the maternal chromosome.*

the gametes produced will be missing one chromosome, whereas the other will have an additional copy of the chromosome. Fertilization by a normal sperm then gives rise to either trisomy or monosomy in the offspring. The frequency of meiotic nondisjunction increases with increasing maternal age, which is reflected in the increased risk in older women of having a child with Down's syndrome (trisomy 21).

When nondisjunction occurs during mitosis, the offspring produced is usually a mosaic and will have two or more cell lines with a different chromosome complement. Usually, a normal diploid cell line and a cell line with a chromosome abnormality are present. If mosaicism occurs early in development, multiple tissues may be affected. Mosaicism is discussed further in Chapter 9.

Monosomy and trisomy also can result from anaphase lag or the delayed movement of a single chromosome toward the poles during anaphase. Both nondisjunction and anaphase lag have the same consequence, producing one cell that is missing a chromosome and one cell with an additional copy of a chromosome. Human cells have evolved the mitotic control point described above to try to reduce the incidence of anaphase lag by delaying the initiation of chromatid separation until all of the chromosomes are properly aligned on the metaphase plate.

In some instances, the zygote receives a normal chromosome number, but the chromosomes all come from the father (uniparental diploidy), resulting in a complete hydatidiform mole (mass). The embryo is formed but dies early in gestation. A complete mole, with no fetus or normal placenta present, is the result of the loss of the female pronucleus in the fertilized egg followed by the duplication of the male pronucleus.

Ovarian teratomas arise when a zygote has only maternal chromosomes. This can occur when a female germ cell that is not fertilized is activated. These zygotes represent parthenogenomes because they appear to undergo partial embryogenesis without a paternal contribution. A fetus is never present; however, tissues derived from all three cell layers can be present and may give rise to some differentiated structures such as teeth, hair, and bone.

The importance of genomic imprinting in humans has become increasingly clear with the understanding of the molecular basis for two human diseases, Prader-Willi syndrome and Angelman's syndrome (Table 5-3). Both of these syndromes involve a critical region on the long arm of chromosome 15 (15q11–13). When the paternal chromosome 15 is missing or has a small deletion in the critical region, the child develops Prader-Willi syndrome. If the maternal chromosome 15 is missing or has a deletion in the critical area, the child develops Angelman's syndrome. Recent molecular studies have identified the candidate gene for Prader-Willi syndrome as a gene coding for a small nuclear ribonucleoprotein (snRNP)-associated peptide mainly expressed in the brain. This peptide is involved in the functioning of spliceosomes and is apparently expressed only from the paternal chromosome 15.

◀ **TABLE 5-3**
*Characteristics of Prader-Willi and Angelman's Syndromes*

|  | **Prader-Willi Syndrome** | **Angelman's Syndrome** |
|---|---|---|
| **Chromosome abnormality** | 70% have a deletion at 15q11–13<br>Chromosomes 15 are both of maternal origin (25%)<br>Imprinting error (~5%) | 70% have a deletion at 15q11–13<br>Chromosomes 15 are both of paternal origin (2%)<br>Imprinting error (3%) |
| **Suggested gene product(s) affected** | Small nuclear ribonucleoprotein-associated peptide N functions in spliceosomes<br>Other unidentified genes | Ubiquitin-protein ligase |
| **Clinical features** | Neonatal hypotonia<br>Hypogonadism<br>Short stature<br>Obesity with uncontrolled appetite<br>Mild mental retardation<br>Hypopigmentation (fair with blue eyes)<br>Behavioral problems | Severe mental retardation<br>Seizures<br>Inappropriate laughter and ataxic gait (happy puppet syndrome) |

The gene involved in Angelman's syndrome maps in proximity to the snRNP gene but is a separate gene. The gene involved in Angelman's syndrome and expressed only from the maternal chromosome 15 codes for an enzyme involved in ubiquitin-mediated protein degradation.

In summary, deviation from the normal chromosome number of 22 pairs of autosomes and 2 sex chromosomes generally results in spontaneous abortion or an abnormal fetus. Although some of these abnormalities are compatible with life, most are lethal. In addition, normal growth and development depends not only on the proper number of chromosomes but also on the presence of one homologue of each chromosome from the mother and one from the father.

## STRUCTURAL CHROMOSOMAL ABNORMALITIES

Structural chromosomal abnormalities occur at very high frequency (1 in 500 births) in the human population. All structural rearrangements of chromosomal material come from the breakage of chromosomes and the rejoining of the fragmented ends. Normal chromosomes contain telomere sequences at the ends of each chromatid to prevent fusion of one chromosome with another or with itself. When a chromosome breaks, a free end or free ends are produced. If these free ends are not repaired by rejoining to another free end, a signal to the damage control systems may eventually lead to the death of the cell. The rejoining of the ends of fragmented chromosomes is not always accurate, in which case a structural rearrangement of a chromosome results. However, if the rearrangement does not alter the content or the expression of the genetic material, no disease phenotype is observed. Nearly all possible structural rearrangements have been seen in human karyotypes and fall into one of the following classes.

*Translocations* occur when two nonhomologous chromosomes exchange genetic material. Translocations require that at least one break occurs on each of the two chromosomes. Reciprocal translocations are balanced exchanges between two chromosomes in which no genetic material is lost (Figure 5-11). The chromosomes carrying the translocations are referred to as derivative (der) chromosomes. Individuals with a balanced trans-

*A **reciprocal** or **balanced translocation** occurs when no genetic material is lost during an exchange of DNA between two nonhomologous chromosomes.*

**FIGURE 5-11**
*Structural Chromosomal Abnormalities.*
*Breakage of a chromosome and fusion of the broken ends leads to multiple structural chromosomal abnormalities.*

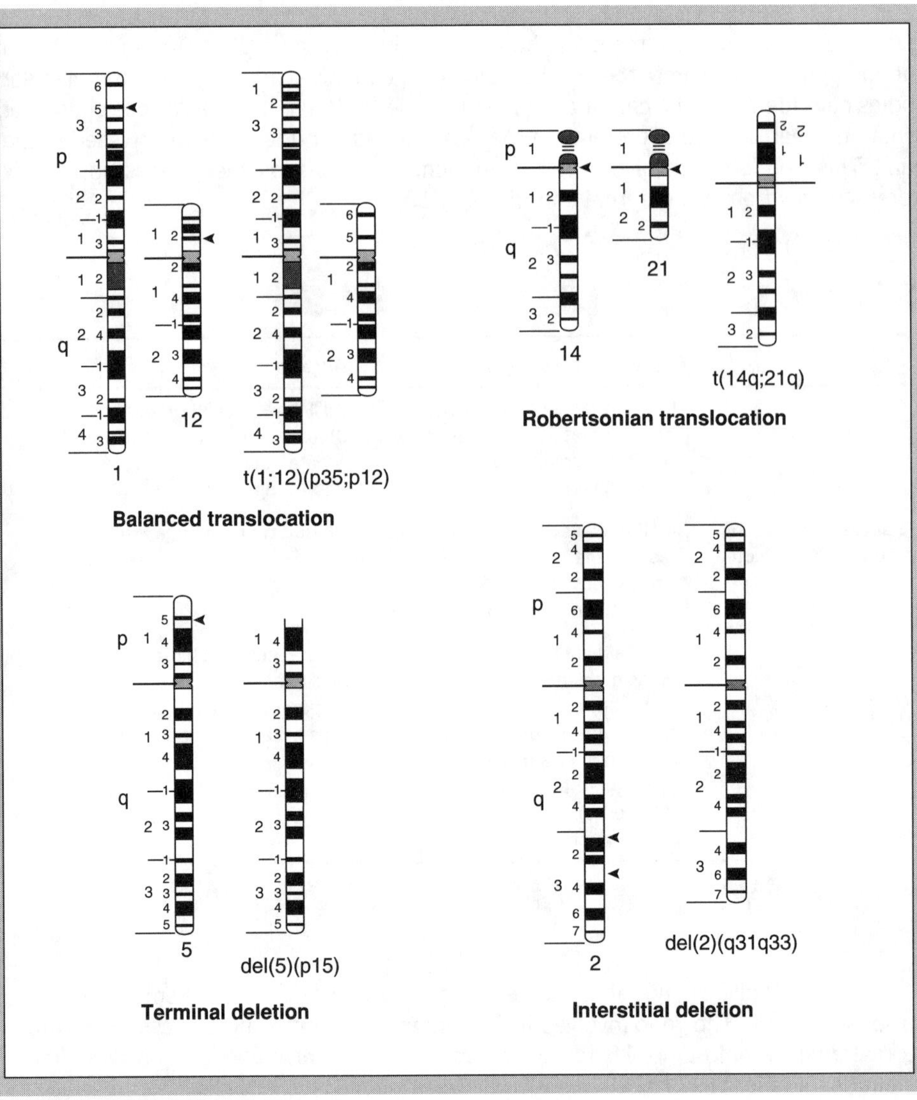

location are phenotypically normal; however, their offspring may express a genetic defect if they do not receive a normal complement of genetic material.

A Robertsonian translocation is the result of the fusion of two acrocentric chromosomes (i.e., chromosomes 13, 14, 15, 21, 22) at or near the centromeres. When this fusion occurs, the short (p) arms are lost from both chromosomes. This loss of the short-arm material from the acrocentric chromosomes is usually not lethal because this part of the chromosome contains either repetitive sequences or rRNA genes that are present in many copies. The problem with Robertsonian translocations arises with the offspring of the parents carrying such a translocation. A common example is the fusion of chromosome 14 to chromosome 21 to give t(14q21q). The mother carrying such a translocation will be normal having two copies of chromosome 14 and two of chromosome 21. However, if the child receives the translocation t(14q21q), he or she will have a normal chromosome 21 that comes from the mother, a normal chromosome 21 that comes from the father, and a third copy of chromosome 21 present on the translocation. This child is then trisomic for chromosome 21 and will have Down's syndrome. Approximately 5% of Down's syndrome patients result from this type of Robertsonian translocation.

*Deletions.* Deletions result when a break occurs within a chromosome and genetic material is lost. Large deletions (greater than 4000 kilobases [kb]) can be detected by G banding as a shortened chromosome with missing bands. *Terminal deletions* result in the loss of genetic material at either end of a chromosome and are indicated as pter or qter. The extent of a deletion and the genes involved determine the viability and the phenotype of an individual carrying the chromosome with the deletion. An example of a terminal deletion that is not lethal occurs in patients affected with cri du chat syndrome in whom the tip of the short arm of chromosome 5 is deleted (5p−).

A **Robertsonian translocation** *results from a fusion of two acrocentric chromosomes at or near the centromere.*

**Deletions** *result when a break occurs within a chromosome and genetic material is lost. Deletions can occur on the ends of chromosomes (*terminal deletions*) or within a chromosome (*interstitial deletions*).*

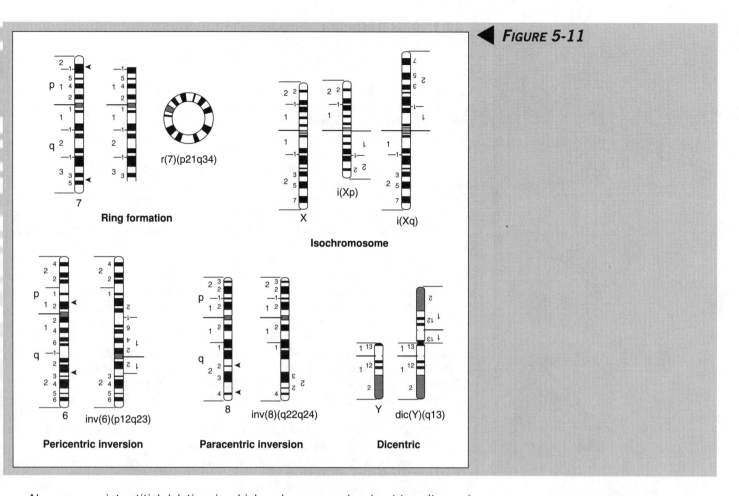

r(7)(p21q34)

**Ring formation**

i(Xp)

X    i(Xq)

**Isochromosome**

6    inv(6)(p12q23)

**Pericentric inversion**

8    inv(8)(q22q24)

**Paracentric inversion**

Y    dic(Y)(q13)

**Dicentric**

Also seen are *interstitial deletions* in which a chromosome breaks at two sites and the material between the breaks is deleted. Small deletions are not always visible by banding techniques and can only be detected by DNA probes. Small interstitial deletions (microdeletions) have been described for the long arm of chromosome 15. As discussed above, a crucial segment of chromosome 15 is related to imprinting. Approximately 70% of patients with Prader-Willi syndrome or Angelman's syndrome have been shown by DNA technology to have a common small deletion at 15q11–13, which localizes the genes involved in these two syndromes to a specific site on chromosome 15. Genes other than those discussed above may also be missing or defective in these patients and may account for the large number of clinical features observed.

Breakage of a single chromosome at the ends of the short and the long arms with loss of the telomere sequences can result in the formation of a ring chromosome in which the two ends of the chromosome fuse together. The presence of a centromere allows the ring chromosome to proceed through cell division. However, cell division sometimes proceeds with difficulty, in which case a cell without the ring (monosomy) is produced.

***Inversions.*** Inversions occur when a chromosome breaks at two positions and can involve the centromere (pericentric inversions) or not involve the centromere (paracentric inversions). Both breaks can occur on the same arm of the chromosome (paracentric inversion), or one break can occur on each arm of the chromosome (pericentric inversions). When the broken chromosome is rejoined, the fragment is attached in an inverted order (see Figure 5-11). If genetic material is not lost, gained, or altered in expression by the inversion, there are no obvious effects on the phenotype of the individual. However, because the DNA sequence has been inverted on one of the homologues, interference with chromosome pairing at prophase I in meiosis I may be affected. Inefficient or defective pairing can result in deletions or duplications of the genetic material, which is expressed later in the offspring.

***Isochromosomes.*** When division occurs along the wrong axis of a chromosome, the resulting chromosome has two copies of one arm and is missing a copy of the other arm. In most cases an isochromosome is lethal with the exception of an isochromosome

*Inversions can involve the centromere (**pericentric inversion**) or not involve the centromere (**paracentric inversion**).*

involving the X chromosome. Individuals with an isochromosome of the X chromosome resemble patients with Turner's syndrome; they are monosomic for one arm of X and trisomic for the other arm.

*Other Structural Abnormalities.* Other types of structural abnormalities seen in human karyotypes include chromosomes with duplications, chromosomes with amplifications of large segments, and chromosomes with insertions of genetic material. In addition, chromosomes with two centromeres or dicentric chromosomes are observed.

Two structural chromosomal abnormalities seen in many tumor cells result from the amplification of large pieces of DNA. One chromosomal abnormality of this type is the double-minute chromosome in which there are extrachromosomal elements that do not contain a centromere. Double-minute chromosomes usually have five to six copies of a particular gene, and a single cell may have hundreds of copies of the double minutes. The product of the gene carried on the double-minute chromosome may therefore be expressed at 500 times the normal level.

The other chromosomal abnormality associated with the amplification of genetic material involves the presence within chromosomes of 100–1000 copies of a particular gene sequence. These amplified sequences can be detected in a light microscope after G banding and appear as large areas within a chromosome or at the ends of chromosomes that stain very lightly. This nonbanded area is referred to as a homogeneously staining region (HSR). Typical HSR-containing chromosomes are shown in Figure 5-12. As with double minutes, a cell carrying a chromosome with an HSR may have 1000 times the number of copies of a normal gene, resulting in extremely high levels of that gene product being expressed in the cell. If that gene product alters the normal growth properties of a cell, the cell may lose growth control and develop into a cancerous cell. As opposed to genes on double-minute chromosomes, genes present on HSRs are not lost during division (and thus are stably inherited) because they are part of a chromosome carrying a centromere and are transmitted to both daughter cells.

The clinical phenotypes of many of the structural chromosomal abnormalities are further discussed in Chapter 9.

**FIGURE 5-12** ▶

*Chromosomes Containing a Homogeneously Staining Region (HSR). An HSR on a chromosome represents large sequences of DNA that have been amplified 100–1000 times. These sequences, found within a chromosome or at the ends, are equally distributed to each daughter cell and often are associated with cancer cells. A–E show different ways in which HSRs appear within chromosomes.*

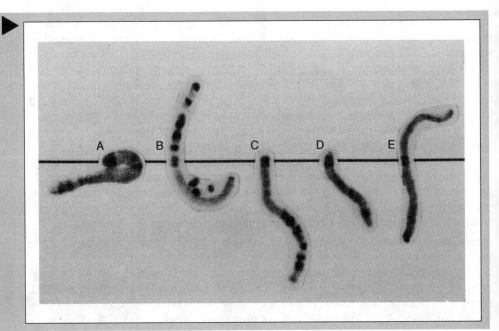

# CELL CYCLE, MITOSIS, AND MEIOSIS

The processes by which a cell replicates its genetic material, equally divides it, and transfers it to two new daughter cells are the keys to normal growth and development of any organism. In human cells, these processes take place during the well-regulated and ordered progression of the mammalian cell cycle. Control of the cell cycle ultimately determines if a cell will continue to cycle, grow, and divide; will undergo differentiation; or will become quiescent and stop cycling. Loss of control of the cell cycle leads to loss of control of normal growth and often results in abnormal cell growth (tumor cells or developmental defects) or programmed death of a cell (apoptosis).

As discussed in Chapter 2, the mammalian cell cycle consists of four distinct phases: G1, S, G2, and M. During the S phase, the total genetic complement of a cell is replicated, and during the M or mitotic phase, the duplicated DNA present in condensed chromosomes is equally distributed to two daughter cells, each of which then begins the cycle again. The G1 and G2 phases are used by a cell to prepare either for the S phase or the M phase, respectively. Cell-cycle progression is driven by biochemical reactions with multiple cyclin-dependent kinases (Cdk) being sequentially activated by the binding of proteins known as cyclins. Once activated, each Cdk-cyclin protein complex phosphorylates specific key proteins, which are required for the reactions that take place in each phase of the cell cycle.

*Cell cycle progression* is driven by the activities of Cdks. Multiple Cdk-cyclin complexes function in the human cell cycle.

Superimposed on this complex biochemical masterpiece are critical points, known as cell-cycle checkpoints, which monitor cell-cycle progression and maintain order. If there is a defect or any deviation from normal, the cell does not bypass the checkpoint, and the cell cycle is arrested until the defect is corrected. Thus, an organism can monitor the normal growth of a cell and eliminate any defective cells before the defect is passed to their progeny.

## Regulation of the Mammalian Cell Cycle

Orderly progression through the cell cycle depends on positive factors, which drive the cycle forward, and negative factors, which stop or arrest the cycle at a particular stage. The main positive factors are specific Cdks and specific cyclins, which function at each stage of the cycle. Negative factors act by blocking the activity of the specific Cdks and are known as cyclin-dependent kinase inhibitors (CKI). Table 5-4 lists the Cdks and cyclins that act at each stage of the human cell cycle as well as some of the negative-acting CKIs.

*Cdks and cyclins* act together to move the cell cycle forward, whereas *negative factors* (mainly CKIs) stop or arrest the cell cycle at various phases.

◀ **TABLE 5-4**
**Cyclin-dependent Kinases (Cdks), Cyclins, and Cyclin-dependent Kinase Inhibitors (CKIs) at Different Stages of the Cell Cycle**

| Cell-Cycle Phase | Cdk | Cyclin | CKI | |
| --- | --- | --- | --- | --- |
| | | | KIP[a] | INK[b] |
| G1 | Cdk4 | Cyclin D | p21, p27 | p15, P16 |
| G1/S | Cdk2 | Cyclin E | p21, p27 | |
| S | Cdk2 | Cyclin A | p21 | |
| G2/M | Cdc2 | Cyclin B | p21 | |
| M | Cdc2 | Cyclin B, cyclin A | | |

[a] KIP proteins (p21, p27) bind multiple cyclin-Cdk complexes and prevent activation or block kinase activity.
[b] INK proteins (p15 and p16) are specific for Cdk4/6 and cyclin D. They bind Cdk and prevent the binding of cyclin D.

Cdks, although present in a cell, are not enzymatically active until they form a complex with a particular cyclin. Cyclins are a class of proteins that are found to vary dramatically during the cell cycle. For example, the level of cyclin B increases continuously during interphase, then declines precipitously during M phase. These changes in the level of cyclin B are correlated with the activity of a specific Cdk called Cdc2, named for cell division cycle mutant 2. Cdc2 is active when cyclin B levels peak and becomes inactive as cyclin B levels suddenly decline. Thus, the phosphorylating activity of Cdc2 is modulated during the cell cycle by the availability of cyclin B. In addition, activation of Cdc2 also depends on phosphorylation of a specific threonine residue, which creates a second level of control over the kinase activity.

An active Cdk-cyclin complex can be inactivated by several mechanisms:

*Active Cdk-cyclin complexes can be inactivated by cyclin degradation, phosphate addition or removal, and the binding of CKIs.*

**CKI Classes: INK and KIP**
*Proteins of the INK class are specific for Cdk4/6–cyclin D complexes and act by binding the kinase and preventing its interaction with cyclin D. Proteins of the KIP class affect multiple Cdk-cyclin complexes and can prevent activation of the kinase or can inhibit the kinase activity directly.*

- The cyclin moiety can be degraded. This degradation of the cyclin protein involves an additional positive-acting factor necessary to move the cycle forward. A protein-degrading factor, known as the ubiquitin protein-degrading system, acts by specific proteolysis of certain marked proteins. The destruction of these proteins at key points in the cell cycle is necessary for the cycle to proceed.

- The critical phosphate needed to activate the kinase activity can be removed from the protein by a specific phosphatase. Also, the addition of more phosphate groups to the Cdk protein at thr-14 and tyr-15 blocks the kinase activity.

- CKI molecules can interact with Cdks or the Cdk-cyclin complexes and inhibit the kinase activity. Recent work has identified two classes of CKIs, the INK (inhibitor of Cdk) class and the KIP (kinase inhibitory protein) class. CKIs that fall into the INK class are the proteins p15, p16, p18, and p19, which are specific for the Cdk4/6–cyclin D complex. These inhibitors act by binding the kinase protein and preventing its interaction with cyclin D, thereby inhibiting the activation of the kinase activity. CKIs that fall into the KIP class are the proteins p21, p27, and p57, which can affect multiple Cdk-cyclin complexes and act by binding the complex and preventing activation of the kinase or inhibiting the kinase activity directly.

The interplay between the activation and deactivation of the Cdk activities at various steps of the cell cycle is key to the normal progression and regulation of the cell cycle, with each phase of the cycle offering a site or sites of specific control (Figure 5-13).

***G1 Phase of the Cell Cycle.*** The G1 phase of the cell cycle is often considered to be the start of the cycle. When human cells that are not dividing are stimulated to grow and divide, they enter the cycle at the G1 phase. New findings about the human cell cycle indicate that the functions of G1 phase proteins and their regulators are related to many biochemical processes that control cell growth. These processes include factors involved in differentiation, signal-transduction pathways, aging, cancer, and programmed death. The cells in the G1 phase react to environmental stimuli (e.g., growth factors, inducers of differentiation, negative factors) until the restriction point in G1 is reached. Once the cell passes this point, it is committed to the S phase and to dividing.

As indicated in Table 5-4 and Figure 5-13, the key positive regulators in G1 are Cdk4 and cyclins of the D family, which form a complex capable of phosphorylating various proteins required to function in G1. One key protein subject to phosphorylation by Cdk4–cyclin D in G1 is the retinoblastoma (pRb) protein. This protein, coded for by a gene that maps to chromosome 13 at 13q14, is mutated in some human cancers (e.g., those of children with retinoblastoma) and represents the first mammalian tumor suppressor identified at the molecular level.

**FIGURE 5-13** ▶
***Cyclin-dependent Kinases (Cdks), Cyclins, and Cdk Inhibitors (CKIs) Interact during the Cell Cycle.*** *Progression around the cell cycle is regulated by the interaction of positive and negative factors. Positive progression is directed by multiple cyclin–cyclin-dependent kinase complexes, which act by phosphorylating various proteins at different stages of the cycle. Negative factors include CKIs such as p16, p21, and p27, which block the phosphorylation of proteins by the kinase and arrest the cell cycle.*

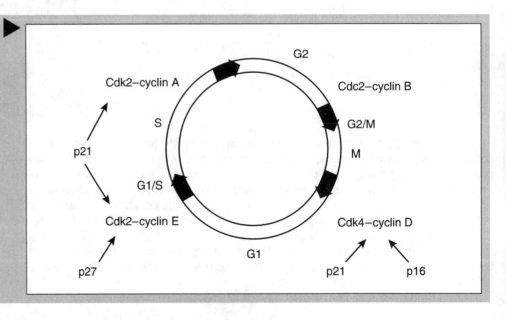

The pRb occurs in a nonphosphorylated form during the first two-thirds of the G1 phase before the restriction point is reached. Once past the restriction point in G1 phase and just prior to the transition from G1 to S phase, pRb becomes phosphorylated. Additional phosphorylation of pRb occurs in S phase and G2/M. Then, just before the M/G1 transition, pRb once again becomes dephosphorylated. When pRb is in the nonphosphorylated state, it functions by restricting cell growth. As pRb becomes phosphorylated, it loses its growth-inhibitory function, thus allowing the cell to proceed through the cell cycle (see Chapter 8) [5].

As long as pRb remains nonphosphorylated, it restricts further cell growth by blocking the progression of the cell and preventing it from moving past the restriction point in G1 phase and into S phase. If, however, environmental conditions are such that the cell is stimulated to grow and divide, pRb becomes phosphorylated, and the cell continues past the restriction point. When pRb function is lost by mutation, the mutated cell loses growth control and may become a cancerous cell. Other environmental stimuli (e.g., DNA damage) can interfere with pRb phosphorylation by way of the CKI protein p21-p53 pathway (see below). This results in the cell being arrested in G1 phase before the DNA damage can be replicated in S phase. If the damage cannot be repaired, the cell may be committed to apoptosis.

The key role of pRb is to serve as a regulator that can either repress or activate specific promoters, thereby linking the cell cycle to the transcription of a subset of genes. The pRb itself does not bind DNA. Instead, pRb interacts with and modifies the activities of other transcription factors, which bind DNA and regulate the expression of cell-cycle genes (see Chapter 8). The phosphorylation of pRb by the Cdk4–cyclin D complex allows previously repressed genes to be transcribed and allows the cell to progress from G1 to S phase.

A second important control that regulates the progression of a cell from G1 to S phase uses the Cdk inhibitor p27. This 27,000–molecular weight protein binds to the Cdk2–cyclin E complex and inactivates it. Quiescent human fibroblasts arrested in G1 phase have high levels of p27. Presumably, with p27 bound to the Cdk2-cyclin complex, the cells are unable to proceed into the S phase, and they remain arrested in G1. Growth-stimulation factors can result in the degradation of p27, activation of the Cdk2–cyclin E complex, and transition to S phase. Recent evidence has suggested that the same proteins that degrade cyclin B (i.e., the ubiquitin protein-degrading system) are responsible for the degradation of p27 [2].

Results obtained from studying transgenic mice with a disruption of the gene coding for p27 support the role of p27 in regulating cell growth. These mice, which are homozygously defective for p27, are larger-than-normal mice, have a 10-fold increase in Cdk2 activity, have an overall increase in cell proliferation, and have an increased incidence of developing pituitary tumors. This phenotype would be predicted for cells that have lost a protein that functions by blocking cells in the cell cycle and preventing uncontrolled cell growth.

**S Phase.** During the S phase of the cell cycle, the total complement of DNA must be replicated faithfully but only once. The process of DNA replication was discussed in detail in Chapter 2, and the reader is advised to review those concepts. The regulation of DNA replication in human cells is still not well defined, but two principles are clear. First, entry into the S phase is determined by a trans-acting cytoplasmic signal, most likely an active Cdk-cyclin complex. Secondly, there is a negative-controlling factor that prevents more than one S phase from occurring per cycle; that is, a second S phase cannot occur until the cell has passed through an M phase. This control, sometimes referred to as the S/M control, is not yet defined.

The entrance into S phase from G1 and progression through S phase to G2 depends on the functioning of specific Cdk-cyclin complexes. Cdk2 initially binds cyclin E as the cells proceed into S phase. Just before measurable DNA synthesis, cyclin A activates Cdk2, and phosphorylation of proteins essential for DNA replication occurs. Studies on simian vacuolating virus 40 (SV40) DNA replication demonstrate a role for the Cdk-cyclin complex at DNA-replication origins, where they may function to phosphorylate proteins that are involved in unwinding the DNA. Additionally, replication protein A

*pRb is a regulatory protein that can either repress or activate specific promoters by its interaction with various transcription factors.*

*Entrance into S phase from G1 and progression through S phase to G2 depends on the functioning of specific Cdk-cyclin complexes that phosphorylate crucial proteins needed for DNA replication.*

(RPA), which is a single-stranded DNA-binding protein, may be phosphorylated by the Cdk-cyclin complex, which regulates its function at DNA-replication origins.

The molecular controls that prevent a single replication origin from replicating twice in a single S phase are still unknown. Additionally, the connection between mitosis and the S phase remains to be elucidated. Several theories currently are being tested. These include the blocking of the RPA-DNA replication origin complexes by the mitotic Cdc2–cyclin B complex, thereby preventing S phase from occurring until the Cdc2–cyclin B complex is inactivated. Additionally, the breakdown of the nuclear membrane that occurs during M phase may be required for an S-phase signal to gain access to the chromatin. Such a control would prevent S phase from being initiated until after mitosis has occurred and the nuclear membrane barrier has been removed.

*G2/M Phase.* Once the DNA has been completely replicated, the cells proceed to G2 during which the chromatin begins to condense and form well-defined structures. A crucial point in the cell cycle occurs at the G2/M checkpoint, which is where the cells decide whether to enter mitosis. Critical proteins involved in the G2/M checkpoint include Cdc2 and cyclin B. The complex of Cdc2–cyclin B (previously known as M-phase promoting factor [MPF]), is essential for the entrance into and the exit from M phase. The activation and deactivation of the Cdc2–cyclin B complex involves a series of well-controlled phosphorylation and dephosphorylation steps coupled with the proteolytic destruction of the cyclin B protein. These reactions are catalyzed by a variety of protein kinases, protein phosphatases, and the ubiquitin protein-degrading system.

During G2 and prior to the entrance into M phase, Cdc2 binds to cyclin B but remains in an inactive state. Cdc2 is phosphorylated by a second kinase on three residues: thr-161, tyr-15, and thr-14. In this hyperphosphorylated state, the Cdc2–cyclin B complex remains inactive until dephosphorylation by a protein phosphatase occurs, removing the phosphate groups from tyr-15 and thr-14. At this point, the Cdc2–cyclin B complex is activated, and mitosis can begin.

*M Phase (Mitosis).* Three important transition stages are now recognized in the mitotic phase of the cell cycle. The first is the sudden activation of the Cdc2–cyclin B complex by dephosphorylation, which occurs at the G2/M border and results in the phosphorylation of many proteins required for mitosis. Mitosis requires a number of significant cellular changes, which include the condensation of the chromosomes, dissolution of the nuclear membrane, mitotic spindle formation, and changes in cell surface proteins. Proteins involved in these reactions, such as lamins, histones, actins, and cytoskeleton proteins, just to name a few, would be subject to phosphorylation by the activated Cdc2–cyclin B complex, preparing them for their role in mitosis.

Mitosis occurs in five distinct phases, each of which is defined by the action of the chromosomes. These five phases, diagrammed in Figure 5-14, are prophase, prometaphase, metaphase, anaphase, and telophase.

Prophase represents the stage at which the chromosomes have condensed to the point where they are visible in the light microscope as thin threads. Because the cells have just passed through an S phase, each chromosome has been duplicated once and consists of two sister chromatids connected by a centromere with each chromatid containing a kinetochore. Cytoplasmic microtubules are broken down during prophase and reorganized into the mitotic spindle. Prometaphase starts with the fragmentation of the nuclear envelope followed by the movement of the mitotic spindle into the area and attachment of some of the microtubules to the kinetochores on each chromosome. At this point, the chromosomes begin to move to the center of the cell where they will become aligned at the metaphase plate.

At metaphase, all the chromosomes are aligned at the metaphase plate, then the process is arrested for 20–60 minutes. As mentioned above, recent evidence has implicated a role for a protein (an inhibitor of sister chromatid separation [ISS]) that binds to the chromatids and keeps the sister chromatids from separating and entering anaphase until the right moment [2]. In this way, the cell can ensure that all of the chromosomes are perfectly aligned before the separation into daughter cells begins.

Anaphase starts suddenly with the separation of the sister chromatids and their movement to opposite poles of the cell. All of the pairs appear to separate simul-

---

*Entrance from G2 to M phase depends on the proper phosphorylation of Cdc2 and its binding to cyclin B.*

---

***Five Distinct Phases of Mitosis***
*These stages are based on the action of the chromosomes: prophase, prometaphase, metaphase, anaphase, and telophase.*

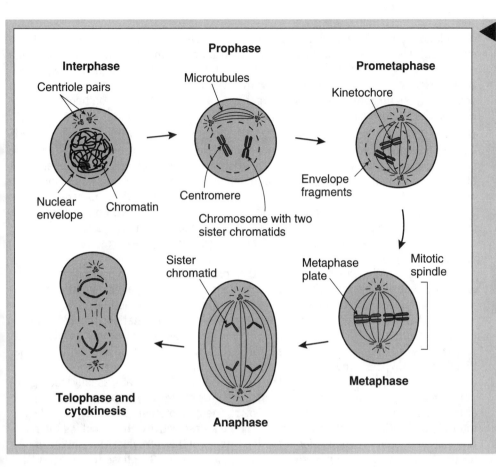

**Interphase**
Centriole pairs
Nuclear envelope
Chromatin

**Prophase**
Microtubules
Centromere
Chromosome with two sister chromatids

**Prometaphase**
Kinetochore
Envelope fragments

Sister chromatid

Metaphase plate
Mitotic spindle

**Metaphase**

**Telophase and cytokinesis**

**Anaphase**

**FIGURE 5-14**
**Stages of Mitosis.** In mitosis, one chromatid of each duplicated chromosome is transferred to each of two daughter cells to give two new diploid cells, which have a chromosome complement identical to the parental cell. (Source: Reprinted with permission from Becker W, Reece JB, Poenie M: The World of the Cell, 3rd ed. Menlo Park, CA: Benjamin-Cummings, 1996, pp 474–475.)

taneously, a process now believed to result from the sudden proteolytic destruction of the ISS protein, which releases the sister chromatids from one another. Additionally, at the onset of anaphase, cyclin B is rapidly degraded, which inactivates the Cdc2–cyclin B complex. This inactivation is required for the cell to exit mitosis.

The final phase, telophase, finds the chromatids present at the two poles. At this point, the processes that occurred at the beginning of mitosis are reversed with the beginning of chromosome decondensation, disassembly of the spindle apparatus, and reformation of the nuclear envelopes around the daughter chromosomes. Cytokinesis then occurs, resulting in two new daughter cells that are prepared to begin the cycle again.

In the above discussion three checkpoints are noted that are key to a properly ordered entrance into and exit from mitosis with each daughter cell receiving an exact copy of the parental genome. These three checkpoints are: (1) the transition from G2 to M concurrent with the activation of the Cdc2–cyclin B complex; (2) the M-phase checkpoint that occurs in metaphase, which is the point that regulates the timing of the separation of the chromatids and the initiation of anaphase; and (3) the sudden proteolytic destruction of cyclin B at the onset of anaphase with the simultaneous inactivation of Cdc2, which allows the cell to prepare for exit from M phase into a new G1 phase.

The latter two checkpoints entail the sudden degradation of two specific proteins (i.e., the ISS protein, cyclin B) and are regulated by a common pathway that involves the action of a proteolytic complex known as the ubiquitin-proteasome pathway. Recent evidence has implicated the ubiquitin-proteasome pathway not only in these two check-points in the cell cycle but also as a factor necessary for the transition step from G1 to S phase. With the proteolysis of p27 (a Cdk inhibitor) activation of the Cdk2–cyclin E complex occurs, which allows the G1/S transition to occur. The three checkpoints known to be regulated by the ubiquitin pathway are indicated in Figure 5-15.

The ubiquitin pathway, a complex multistep pathway involving three enzymes (E1, E2, E3), results in the covalent attachment of polyubiquitin chains to proteins subject to

**Three Checkpoints Controlling a Proper M Phase**
Activation of the Cdc2–cyclin B complex; separation of the chromatids and the initiation of anaphase; and proteolytic destruction of cyclin B to allow exit from M phase.

Ubiquitin directs the proteolysis of three critical cell-cycle proteins: ISS, cyclin B, and p27.

degradation. The polyubiquitinated protein is then degraded by a proteasome complex to small peptides. The advantages of using a destruction method to control the levels of a particular protein are the suddenness of the reaction, resulting in instant control, as well as the irreversibility of the action. Only with new protein synthesis can a cell once again accumulate significant levels of the target protein, giving the cell additional levels of control over the expression of certain proteins.

***Damage Control by p53-p21 Regulation.*** Crucial to normal growth and development is not only an ordered progression through the cell cycle but also the ability of a cell to sense some deviation from the normal. Problems arising in a cell, such as DNA damage, a lack of oxygen, a lack of sufficient energy or metabolites, infection by viruses, or any physiologic disturbance that can affect the normal growth and cycling of the human cell, must be dealt with, or an abnormal cell can result. To counteract such problems, cells have evolved negative regulatory mechanisms that can sense these abnormal conditions and either arrest a cell at a particular stage of the cell cycle or, in some cases, direct the cell to apoptosis.

> *The p53-p21 pathway of damage control is crucial for a cell to sense physiologic disturbances.*

The p53 protein, which is a DNA-binding protein that acts as a transcription factor, is the key controlling element that sits at the center of these negative regulatory mechanisms. Figure 5-16 illustrates the damage control system controlled by p53.

The p53 protein is a tumor-suppressor protein, and any mutations that result in the loss or alteration of its normal activity can lead to a high rate of cancer development. Nearly 50% of all human cancer cells have been found to contain an aberrant *p53* gene. When abnormal p53 levels exist, cells lose the ability to stop the progression of the cell cycle when adverse conditions arise. The cell continues to proliferate, usually leading to a defective cell type. The p53 protein exerts one of its controls by activating transcription of the gene coding for the Cdk inhibitor, p21. The p21 protein functions by binding to multiple cyclin-Cdk complexes and blocking the kinase activity, thus blocking the phosphorylation of proteins needed for the various stages of the cell cycle. The binding of p21 to the G1 cyclin-Cdk complexes is central to the arrest of G1 phase that follows DNA damage by ionizing radiation. An arrest of the cell in G1 allows time for the DNA repair

**FIGURE 5-15** ▶

***Cell Cycle Checkpoints Controlled by Ubiquitin.*** *Ubiquitin (ubi) directs the destruction of critical proteins in the cell cycle. (A) Cyclin B is degraded, resulting in the inactivation of the Cdc2 activity and exit from mitosis. (B) The inhibitor of sister chromatin separation (ISS) protein is degraded, resulting in the initiation of anaphase. (C) The p27 protein, which inhibits the Cdk2–cyclin E complex, is degraded to allow progression from G1 to S phase. (Source: Barinaga M: A new twist to the cell cycle.* Science *269:631, 1995.)*

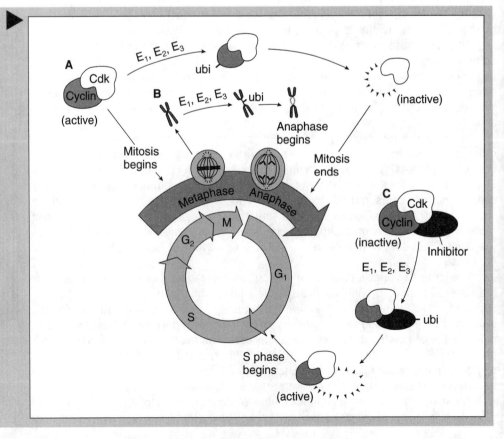

systems to correct the damage. In some cases, the damage can be repaired by systems described in Chapter 4; in other cases, the damage is too extensive, and the cell will be directed to apoptosis. Another function of p21 is to bind proliferating cell nuclear antigen (PCNA), which is a factor required for the activity of DNA polymerase δ. With p21 bound to PCNA, DNA replication is inhibited. New evidence suggests an additional regulatory role for p53 and p21 at the G2/M checkpoint in which a cell is prevented from entering M phase if there is chromosomal damage or a defect in the mitotic apparatus.

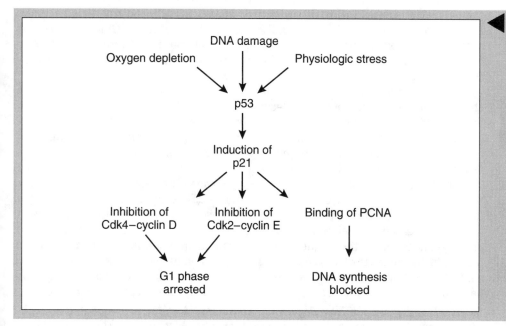

**FIGURE 5-16**
**Damage Control by p53 and p21.** *Damage or stress to the cell results in increased p53 activity. p53 acts as a transcription factor and induces the transcription of p21, a cyclin-dependent kinase inhibitor (CKI). p21 interacts with multiple cyclin-dependent kinase (Cdk)–cyclin complexes, inhibits the kinase activity, and arrests the cells in G1 phase. p21 also binds proliferating cell nuclear antigen (PCNA) and blocks DNA synthesis.*

The loss of p53 activity or the inability to express normal levels of p53 often leads to cancer or to other human genetic diseases (e.g., ataxia telangiectasia). Patients with ataxia telangiectasia are sensitive to ionizing radiation, and have a 100-fold increased risk of developing cancer, chromosomal and cell-cycle abnormalities, and other multiple dysmorphologies (see Chapter 8). The ataxia telangiectasia gene maps to 11q22 and appears to code for a phosphoinositol kinase (PIK) involved in signal transduction. In addition, cells from patients with ataxia telangiectasia express a suboptimal level of p53 and are less able to respond to DNA damage. In the absence of this damage control, DNA defects are not repaired but are passed on to daughter cells, accounting for the presence of many chromosomal abnormalities seen associated with this disease.

In summary, the human mitotic cell cycle is tightly regulated by a combination of positive and negative regulatory proteins, which can either allow the cell to move from one phase of the cycle to another on schedule or arrest the cell at a particular phase, if necessary. These regulators are coupled to the activity of Cdks, which act by phosphorylating other proteins necessary to function at each stage of the cycle. Multiple cyclins and Cdks, each of which has a specific role, have been identified in human cells. Further work is necessary to elucidate completely the details at the molecular level of the complex interactions of these positive and negative regulators and their ability to ensure that cells maintain their normal state.

## Meiosis

Preservation of the diploid chromosome number in human cells is accomplished during the process of meiosis, where gametes are produced with a haploid number of chromosomes. If meiosis did not occur, the process of fertilization would result in a doubling of the chromosome number in a cell at each generation. Meiosis, which occurs only in the germ cells, maintains the diploid chromosome number by having a single genome duplication, which is then followed by two successive division cycles, meiosis I and meiosis II.

Meiosis plays a major role in creating the genetic diversity seen among children born

**Meiosis**
*The process by which the diploid chromosome number in human cells is maintained after fertilization.*

of the same two parents. This diversity is accomplished by two means. First, various combinations of chromosomes (i.e., some of maternal origin, some of paternal origin) are randomly distributed to the gametes and passed on to the children. More than 1 million possible combinations of alleles can be produced by a single individual. Because the gametes receive only one of each pair of homologues, numerous combinations of the maternal and paternal chromosomes are possible. For example, one gamete may contain chromosomes 1, 3, 5, 7, 9, 11, 13, 15, 17, 19, 21, and X of maternal origin and 2, 4, 6, 8, 10, 12, 14, 16, 18, 20, 22, and Y of paternal origin, whereas a second gamete may contain chromosomes 2, 4, 6, 8, 10, 12, 14, 16, 18, 20, 22, and X of maternal origin and 1, 3, 5, 7, 9, 11, 13, 15, 17, 19, 21, and Y of paternal origin. Other gametes may contain any combination of the above. Because the paternal and maternal chromosomes have some genetic differences, the final chromosome complement of gametes demonstrates great genetic diversity.

Secondly, during meiosis I, recombination between synapsed (paired) homologous chromosomes occurs in prophase I. Recombination results from a physical breakage on the chromatids of two homologous chromosomes (one maternal chromosome and one paternal chromosome) with a reciprocal exchange of genetic material. The resulting chromosomes now contain genetic material from both parents and are referred to as recombinant chromosomes (Figure 5-17).

The frequency of recombination occurring between two genes on a single chromosome depends on the distance between the two genes. Thus, two genes that are very close together show a low frequency of recombination, whereas two genes that are far apart have an increased frequency of recombination. The ability to identify recombinant progeny and calculate the frequency of recombination allows the creation of genetic maps, which indicate the location of genes along a chromosome and their relationship to each other in terms of the distance between them. The process of linkage analysis is based on the determination of recombination frequency (see Chapters 7 and 10).

***Meiosis I: Segregation of Homologous Chromosomes.*** Meiosis I is unique in that duplicated chromosomes become linked to each other in homologous pairs (i.e., the maternal and paternal chromosomes pair). This pairing, termed synapsis, occurs during prophase I

> *Recombination between synapsed or paired homologous chromosomes occurs in prophase I of meiosis I.*

**FIGURE 5-17** ▶
***Recombination during Meiosis I.*** *During prophase I of meiosis I, two homologous chromosomes pair. Nonsister chromatids break and exchange genetic material, resulting in two recombinant chromatids and two nonrecombinant or parental chromatids. $M_1$ = maternal 1; $M_2$ = maternal 2; $P_1$ = paternal 1; $P_2$ = paternal 2.*

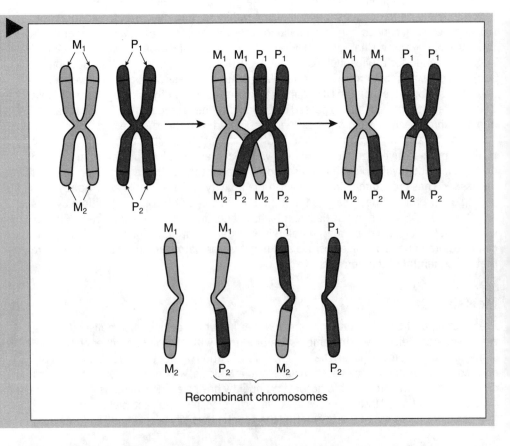

Recombinant chromosomes

of meiosis I and involves the alignment of two pairs of sister chromatids to produce a bivalent composed of four chromatids (Figure 5-18). The chromatids are now in position to allow crossing over between homologous chromosomes forming a chiasma (site of physical exchange of genetic material). The chiasma physically holds the two homologues together as the chromosome pair moves to the metaphase plate. At metaphase I, all of the chromosome pairs, or bivalents, become aligned at the metaphase plate with microtubules attached to their kinetochores. In meiosis I, each pair of chromosomes has a random orientation with respect to whether a maternal or paternal chromosome faces a particular pole.

Anaphase I begins when the two homologous chromosomes are pulled apart and begin to move to each pole. In meiosis I, it is the chromosomes (each of which is made up

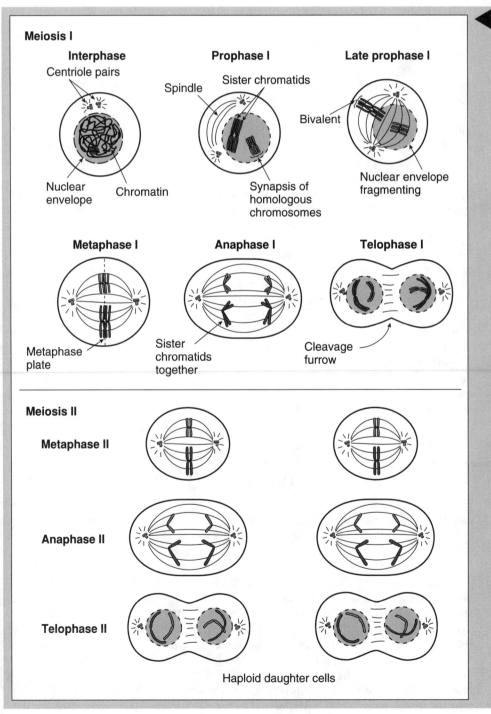

### FIGURE 5-18

**Stages of Meiosis.** *The process of meiosis reduces the number of chromosomes to one-half by going through two successive cell divisions following a single DNA replication. In meiosis I, homologous chromosomes align at the metaphase plate (metaphase I) and then separate at anaphase I to give two daughter cells, each of which has 23 chromosomes and 46 chromatids. During meiosis II, the sister chromatids of each chromosome separate and are transferred to the gametes to give four gametes, each of which contains 23 chromosomes and 23 chromatids. (Source: Reprinted with permission from Becker W, Reece JB, Poenie M:* The World of the Cell, *3rd ed. Menlo Park, CA: Benjamin-Cummings, 1996, pp 502–503.)*

of two sister chromatids) that separate and move to the poles, as opposed to meiosis II and mitosis, in which sister chromatids separate at anaphase and move to opposite poles. At telophase, one chromosome of each of the 23 pairs is found in the two daughter cells. The cells at this stage have 23 chromosomes and 46 chromatids. The combination of the 23 chromosomes is random with respect to their parental origin.

***Meiosis II: Formation of Haploid Gametes.*** Meiosis II begins shortly after the completion of meiosis I. The 23 chromosomes are present in the cells as duplicated DNA molecules, each of which contains two sister chromatids. Meiosis II resembles mitosis. During prophase II, the microtubules attach to the kinetochores and begin to move the chromosomes to the center of the cell. At metaphase II, the 23 chromosomes, some of which are recombinant chromosomes, are aligned on the metaphase plate. At the beginning of anaphase II, the centromeres divide, and sister chromatids move to opposite poles. The final stage of telophase II and cytokinesis produces four haploid cells with 23 chromosomes and 23 chromatids.

In human males, meiosis I begins at puberty and continues throughout adult life. Each diploid primary spermatocyte undergoes meiosis I to produce two haploid secondary spermatocytes, which then undergo meiosis II to form four spermatids. Each of these spermatids differentiate into mature sperm cells. Completion of the entire process, spermatogenesis, takes approximately 64 days.

In the human female, meiosis I of oogenesis begins prior to birth during the third month of prenatal development. At this time, the diploid primary oocytes begin to develop and enter meiosis I. These oocytes are arrested in prophase I of meiosis I and remain in this stage until puberty. At the time of birth, more than 2 million oocytes are present in the ovary, each of which is arrested in prophase I and can remain so for years. Approximately 400 of these eventually reach maturity. When ovulation begins, only one egg cell at a time completes meiosis I, giving one secondary oocyte and one polar body. Meiosis II begins immediately, is arrested at metaphase II, and is finally completed at the time of fertilization with the entrance of the sperm and the extrusion of the second polar body. In the overall process of meiosis in the female, only one of the four haploid cells produced during meiosis gives rise to a functional egg cell. The survival of the one egg cell is related to the amount of cytoplasm available because the egg must supply both cytoplasm and organelles to the fetus. The other cells—called polar bodies—degenerate and are lost. Once the sperm enters the egg and the second polar body is excluded, the male and female pronuclei fuse to form a diploid zygote. The zygote then proceeds to divide by mitosis to form diploid daughter somatic cells.

Many of the chromosomal abnormalities discussed earlier in this chapter occur during the process of meiosis. Nondisjunction can occur either during meiosis I or meiosis II. In addition, in meiosis I, mistakes in the recombination process can result in chromosomal deletions, inversions, translocations, and duplications.

# RESOLUTION OF CLINICAL CASE

Before the Clarks arrived with their son, William, the pediatrician met with the clinical geneticist, Dr. Parsons. Upon hearing the facts about William, Dr. Parsons suspected the child might be affected with a rare syndrome known as Prader-Willi syndrome. William had some of the characteristics associated with this syndrome, including his poor eating habits as an infant, his sudden uncontrolled craving for food, behavioral problems, and his hypopigmentation. Dr. Parsons explained to the pediatrician that Prader-Willi syndrome can result from three mechanisms: (1) a small deletion on the paternal chromosome 15 at 15q11–13; (2) the child receiving both copies of the maternal chromosome 15, a chromosome abnormality known as uniparental disomy; or (3) an imprinting defect that results from a mutation in the imprinting control region, which prevents erasure and re-establishment of the imprint in subsequent germ lines. Dr. Parsons stated that with current methodologies he could make a specific diagnosis if the pediatrician obtained a blood sample from William and his parents. Dr. Parsons explained that the blood samples are used for cytogenetic and FISH analysis to look for a deletion on chromosome 15. If no deletion is detected, then DNA analysis is performed, using chromosome 15 DNA-specific probes to look for the presence of uniparental disomy or an imprinting defect. Dr. Parsons also suggested that the pediatrician look for other characteristics of the disease, such as short stature, small hands and feet, and hypogonadism.

When William and his parents arrived at the office, the pediatrician was immediately struck by William's sudden weight gain. As she examined him, she found some of the characteristics of which Dr. Parsons spoke, such as small hands and feet. The pediatrician explained to the parents that William might have Prader-Willi syndrome and requested that they provide blood samples for the karyotype analysis. The results of the cytogenetic and FISH analysis would be available in 1 week, but additional DNA tests, if required, would take 2–3 weeks for the final diagnosis.

Ten days later, Dr. Parsons met with the pediatrician and explained the results of his tests. The cytogenetic analysis could not detect a deletion on chromosome 15 in William's cells; however, the FISH analysis revealed a microdeletion of chromosome 15q11–13. The deletion was detected using a fluorescently labeled DNA probe for the snRNP locus. Additional FISH analysis with several other DNA probes indicate that the deletion is quite large and probably includes a number of genes in that region of chromosome 15.

The region involved in Prader-Willi syndrome covers approximately 400 kb in the 15q11–13 area and includes a gene that codes for a snRNP. This gene is specifically expressed from the paternal chromosome 15. Because William has a deletion at 15q11–13 on the paternal chromosome 15, he does not express the snRNP protein, which functions predominantly in the neurons of the brain. It is possible that the lack of the snRNP protein, which is believed to play a role in RNA splicing, results in an interference with RNA processing or mRNA export from the nucleus and thus affects many genes. Additionally, hypothalamic development may be defective, accounting for the abnormal behavior phenotype, including the obsessive eating and behavioral problems.

The next day, the pediatrician met with William and his parents and explained the results of the tests confirming the diagnosis of Prader-Willi syndrome. The mother, who immediately blamed herself, wanted to know how her child could have this disease when she was perfectly normal. The pediatrician assured the mother it was not her fault and informed her that with every pregnancy there is a risk of a chromosomal abnormality occurring. There is nothing that could have been done to prevent the defect, and most likely it would not have been detected by normal karyotype analysis. In the case of uniparental disomy and microdeletions, the karyotype of the child would appear perfectly normal (i.e., 46,XY for William) and would not, at first examination, indicate any problems. Only after the child was born with phenotypic anomalies or later, when the child developed physical or behavior problems, would a genetic defect be suspected.

The pediatrician carefully counseled the parents that there is no cure for Prader-Willi syndrome but informed them of several clinics in the country that specifically deal with Prader-Willi patients. Patients are taught behavior modification with respect to diet control and exercise. With proper training and counseling, Prader-Willi patients can learn to control their eating habits and reduce their obesity. Unfortunately, even though they are only slightly mentally retarded, they usually cannot live outside a supervised group home because of their erratic behavioral problems and their obsession with food.

# REVIEW QUESTIONS

**Directions:** For each of the following questions, choose the **one best** answer.

1. Which one of the following individuals is expected to be phenotypically normal?
    - **(A)** A woman with 45 chromosomes, including a Robertsonian translocation between chromosomes 14 and 21
    - **(B)** A woman with 46 chromosomes, including a Robertsonian translocation between chromosomes 14 and 21
    - **(C)** A woman with the karyotype 47,XX,+18
    - **(D)** A man with deletion of a band on chromosome 4
    - **(E)** A man with deletion of a band on chromosome 5

2. Which one of the following cyclin-dependent kinase–cyclin complexes functions in the G2/M phase of the mammalian cell cycle?
    - **(A)** Cdk2–cyclin E
    - **(B)** Cdk2–cyclin B
    - **(C)** Cdc2–cyclin B
    - **(D)** Cdk2–cyclin A
    - **(E)** Cdk4–cyclin D

3. Which one of the following DNA elements ensures correct segregation of homologous chromosomes during meiosis and mitosis?
    - **(A)** Telomeres
    - **(B)** Centromeres
    - **(C)** Kinetochores
    - **(D)** Chromatids
    - **(E)** Satellites

4. Which one of the following cytogenetic notations indicates a man with trisomy 21?
    - **(A)** 47,XX,+21
    - **(B)** 45,X
    - **(C)** 47,XXX
    - **(D)** 47,XY,+21
    - **(E)** 45,XX,−21

5. Which one of the following cytogenetic notations indicates a woman with monosomy X?
    - **(A)** 47,XX,+21
    - **(B)** 45,X
    - **(C)** 47,XXX
    - **(D)** 47,XY,+21
    - **(E)** 45,XX,−21

6. Which one of the following cytogenetic notations indicates a woman with monosomy 21?
    - **(A)** 47,XX,+21
    - **(B)** 45,X
    - **(C)** 47,XXX
    - **(D)** 47,XY,+21
    - **(E)** 45,XX,−21

# ANSWERS AND EXPLANATIONS

**1. The answer is A.** The woman in option A will be normal; however, her offspring are at risk of developing Down's syndrome, or trisomy 21, if they receive the translocation. The woman in option B will have trisomy for chromosomes 14 or 21. The woman in option C will have trisomy for chromosome 18. The men in options D and E will both be deficient for the genes located on the deleted material.

**2. The answer is C.** Activation of the Cdc2–cyclin B complex at the G2/M border signals the cell to proceed into mitosis. Option A is incorrect because the Cdk2–cyclin E complex functions at G1/S. Option B is incorrect because Cdk2 does not complex with cyclin B. Option D is incorrect because Cdk2–cyclin A functions in S phase. Option E is incorrect because Cdk4–cyclin D functions in G1.

**3. The answer is B.** Centromeres are the constrictions on chromosomes that separate the long and short arms. They are essential for correct segregation of the homologous chromosomes in meiosis and mitosis. Option A is incorrect because telomeres, which are repetitive sequences located at the ends of chromosomes, function to maintain the integrity of the chromosome. Option C is incorrect because the kinetochore is a protein-DNA complex that binds microtubules. Option D is incorrect because a chromatid is a DNA strand. Option E is incorrect because satellites are tandem, repetitive sequences and are not involved in chromosome segregation.

**4. The answer is D.** 47,XY,+21 indicates a male (Y) with an additional copy of chromosome 21 (+21) with a total of 47 chromosomes.

**5. The answer is B.** 45,X indicates a female (X) who is missing a copy of the X chromosome to give only 45 chromosomes.

**6. The answer is E.** 45,XX,−21 indicates a female (XX) who is missing a copy of chromosome 21 (−21) [monosomy for 21] and has a total of only 45 chromosomes.

# REFERENCES

1. Becker WM, Reece JB, Poenie MF: *The World of the Cell*, 3rd ed. Menlo Park, CA: Benjamin-Cummings, 1996.
2. Barinaga, M: A new twist to the cell cycle. *Science* 269:631, 1995.
3. Holt SE, Shay JW, Wright WE: Refining the telomere-telomerase hypothesis of aging and cancer. *Nature Biotech* 14:836–839, 1996.
4. Saitoh Y, Laemmli U: Metaphase chromosome structure: bands arise from a differential folding path of highly AT-rich scaffold. *Cell* 76:609, 1994.
5. Sherr CJ, Roberts JM: Inhibitors of mammalian G1 cyclin-dependent kinases. *Genes and Development* 9:1149, 1995.

# MODERN METHODS FOR DETECTING DEFECTIVE GENES

## INTRODUCTION OF CLINICAL CASE

The Carson family, which has a family history of Duchenne's muscular dystrophy (DMD), was referred for prenatal diagnosis to a university molecular diagnostic service. After obtaining the family history, the family pedigree was established as shown in Figure 6-1.

DMD is a severe muscle-wasting disease that strikes 1 male in 3500 births. Symptoms are not noticeable in the first year or two of life, but between the ages of 3 and 6 years, the child begins to have difficulty rising from a sitting position on the floor and climbing stairs. By the age of 10–13 years, most patients are confined to a wheelchair. Death occurs usually by the age of 20 years, unless the patient is ventilated. The disease is X-linked, mapping to the Xp21 region of chromosome X, and shows expression mostly in males. The DMD gene has a very high spontaneous mutation rate (1 in 10,000 sperm or eggs), resulting in many cases being caused by new mutations. Because symptoms of the disease do not appear until 3–6 years, an unsuspecting family may have two or three boys, all of whom have inherited the disease, before the first child is diagnosed with DMD. Genetic counseling of families also is complicated by the high mutation rate within the disease gene. With the cloning of the gene, efficient tools for patient diagnosis, linkage and analysis, and mutation detection are now available for the disease.

An examination of the DNA available from the single affected male (III-1) showed no

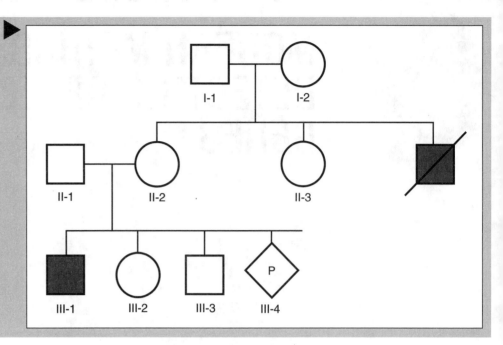

**FIGURE 6-1**
**Family Pedigree for the Carson Family.** *The pedigree shows the inheritance pattern of an X-linked recessive trait, as one would expect for Duchenne's muscular dystrophy. The solid symbols indicate an affected individual.*

deletion mutations in the dystrophin gene. An analysis was then performed using fluorescent cytosine–adenine (CA) repeat polymorphisms to determine the haplotypes of the members of the pedigree. There are four (CA)$^n$-repeat loci in the dystrophin gene with the following order:

$$5' - 5'Dys\text{II} - STR45 - STR49 - 3'CA - 3'$$

Polymerase chain reaction (PCR) analysis was carried out on DNA samples from the members of the pedigree (II-2, III-1, III-2, III-3, and III-4), using a fluorescent-labeled primer. The PCR products were analyzed on an automatic sequencer and gave the following traces shown in Figure 6-2. The alleles for each CA repeat were given as letters *a*, *b*, or *c* on the traces. A female has two letters for each allele because she has two X chromosomes. If she is homozygous at a particular allele, there is only one peak on the tracing, and her alleles at this locus are written twice (e.g., *a,a*). Males have only one X chromosome; therefore, they have a single allele at each site.

Analysis of these data gave the haplotypes for each individual tested and provided answers to the following questions asked by the family:

- The expected child (III-4) has been determined by ultrasound to be a male. What is the prenatal diagnosis for this fetus with respect to DMD?
- What are the carrier risks for the females in the pedigree (i.e., I-2, II-2, II-3, III-2)?

# MOLECULAR CLONING OF GENES AND PROBES

## Constructing Chimeric DNA Molecules Using Restriction Endonucleases and DNA Ligase

During the last 5 years, a major revolution has occurred in the field of human genetics, giving scientists an unprecedented ability to isolate and manipulate human genes. However, many of the keys to this success rely on findings made in the late 1960s and early 1970s, when bacterial studies led to the discovery of plasmids and enzymes that recognized and cleaved short, specific base sequences in DNA. These enzymes, known as restriction endonucleases, have been essential to the development of DNA cloning and have provided scientists with the ability to isolate and study genes from any species, including humans. A restriction endonuclease cleaves any double-stranded DNA mole-

*Restriction Endonucleases*
*These enzymes recognize and cleave short, specific base sequences in any DNA molecule.*

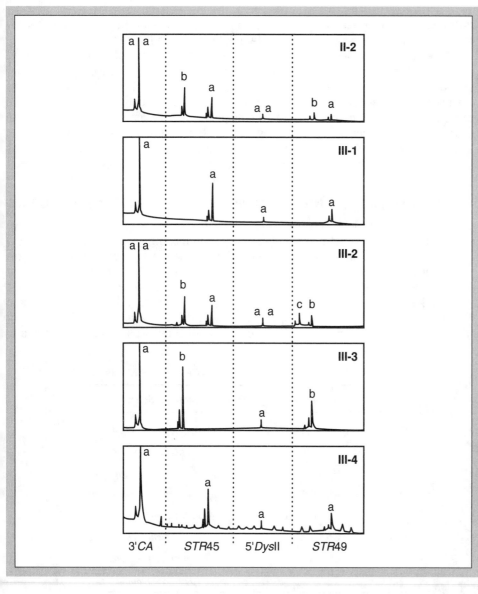

**FIGURE 6-2**
*Fluorescent Multiplex Linkage Analysis.*
*This type of linkage analysis is used for the prenatal diagnosis of Duchenne's muscular dystrophy. The electrophoretogram shows four polymorphic cytosine–adenine loci within the dystrophin gene. Males have a single marker at each locus, whereas females with two X chromosomes have two alleles at each locus. A single peak for a female indicates she is either homozygous for the marker or has a deletion.* (Source: Reprinted with permission from Schwartz LS, et al: Fluorescent multiplex linkage analysis and carrier detection for Duchenne/Becker muscular dystrophy. Am J Hum Genet *51:727, 1992.*)

cule that contains the proper base sequence, whether the DNA comes from a bacterial cell, a yeast cell, or a human cell. This unique property allows scientists to join together DNA fragments from any species to make a single chimeric molecule.

The various restriction endonucleases, of which there are now more than 100 known, are isolated from different bacterial species and are named accordingly. For example, *Eco*RI is from *Escherichia coli*, *Hae*III is from *Haemophilus aegyptius*, and *Pst*I is from *Providencia stuartii*. These enzymes, normally found in all bacteria, have evolved to protect the bacterial cells from infection by viruses. Viral DNA, when it enters a bacterial cell, is subject to degradation by the various restriction enzymes, which prevent the viral DNA from replicating and killing the host cell. The bacterial DNA, although it also contains the sites for cleavage, is protected from being degraded by its own enzymes by the presence of methyl groups at the site of cleavage.

Each restriction enzyme can be characterized by the 4- to 8-bp sequence recognized to be cut and the type of cleavage that occurs. The sequences recognized are generally palindromes, in which the sequence of bases on either strand of the DNA is the same when read in a 5′ to 3′ direction. In some cases, the DNA is cleaved to give blunt ends. However, with most restriction enzymes, cleavage of the two DNA strands is offset by several bases, giving either a 5′ overhang or a 3′ overhang. An overhang is a short single-stranded end on the DNA molecule, sometimes referred to as a sticky end. These sticky ends have the ability to hybridize to other DNA molecules that have a complementary sticky end, and they are crucial in the cloning of DNA fragments. The specificities of some of the known

**Sticky Ends**
*Most restriction endonucleases cleave DNA to give short, single-stranded ends sometimes referred to as sticky ends.*

restriction endonuclease are given in Table 6-1. The types of fragments produced by restriction endonuclease cleavage of DNA are illustrated in Figure 6-3.

When two DNA molecules (e.g., one from a plasmid DNA and one from human DNA) are cut by the same restriction endonuclease, fragments are produced with single-stranded overhangs. These single-stranded overhangs on the two cut DNA molecules have complementary base sequences and can hybridize, not only to themselves, but to the other type of DNA fragment as well. In the presence of the enzyme DNA ligase, these base-paired sequences are covalently ligated together to give a chimeric molecule; that is, a single DNA molecule composed of fragments of DNA from two different sources as illustrated in Figure 6-4.

> Single-stranded DNA overhangs with complementary base sequences can hybridize to each other to form a chimeric molecule composed of fragments of DNA from two different species.

## Vectors: Replicating DNA Molecules Designed to Carry Foreign Genes

DNA molecules to which a target DNA is linked are referred to as vectors. Vectors have several key properties that make them important in cloning genes. These include: (1) the presence of an origin of replication, (2) sites where restriction endonucleases cleave, and (3) the presence of a gene that allows the selection of specific cells that contain and replicate the vector DNA. The use of vectors allows the selection of a specific DNA fragment from a large number of DNA fragments. The particular fragment can then be amplified to a high copy number so it can be isolated in sufficient quantities for analysis. Commonly used vectors include bacterial plasmids and the bacte-

> **Vectors**
> Specific DNA elements to which a target DNA is linked are called vectors; they are used to amplify the number of copies of the target DNA.

**TABLE 6-1** ▶
*Examples of Restriction Endonucleases*

| Enzyme | Recognition Sequence | Bacterial Source |
|--------|---------------------|------------------|
| BamHI | G↓GATCC | Bacillus amyloliquefaciens |
| EcoRI | G↓AATTC | Escherichia coli RY 13 |
| HaeIII | GG↓CC | Haemophilus aegyptius |
| HindIII | A↓AGCTT | Haemophilus influenzae Rd |
| MstII | CC↓TN$^a$AGG | Microcoleus species |
| PstI | CTGCA↓G | Providencia stuartii |
| SmaI | CCC↓GGG | Serratia marcescens |
| TaqI | T↓CGA | Thermus aquaticus |

$^a$ N can be any nucleotide: adenine (A), thymine (T), cytosine (C), or guanine (G).

**FIGURE 6-3** ▶
*Restriction Endonucleases Can Generate DNA Fragments with a 3' or 5' Overhanging Single-Stranded End as well as Fragments with Blunt Ends. Each restriction endonuclease recognizes a specific base sequence that is cleaved to give single-stranded or blunt ends on the DNA.*

| | | |
|---|---|---|
| PstI (3' overhangs) | 5' CTGCA↓G 3' | — CTGCA 3' |
| | 3' G↑ACGTC 5' | 3' ACGTC — |
| EcoRI (5' overhangs) | 5' G↓AATTC 3' | 5' AATTC — |
| | 3' CTTAA↑G 5' | — CTTAA 5' |
| SmaI (blunt ends) | 5' CCC↓GGG 3' | — CCC 3'  5' GGG — |
| | 3' GGG↑CCC 5' | — GGG 5'  3' CCC — |

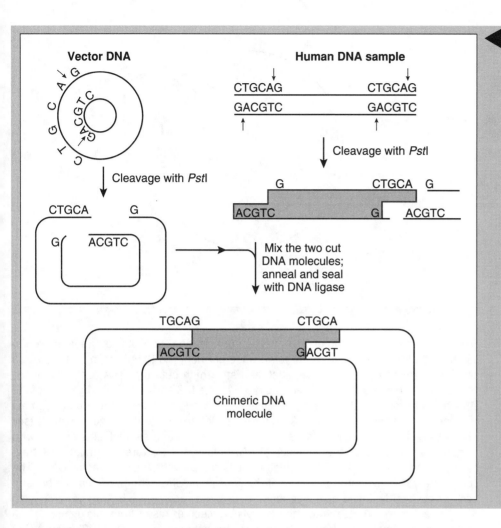

◀ **FIGURE 6-4**
*The Formation of Chimeric DNA Molecules Using Restriction Endonucleases. Chimeric DNA molecules are formed by cutting each of two DNA molecules with the same restriction endonuclease (e.g., PstI) and ligating them together in the presence of DNA ligase. The single-stranded complementary ends of the two DNAs can hybridize to produce a vector carrying a human DNA fragment.*

rial virus lambda. The important properties of a typical plasmid cloning vector are shown in Figure 6-5.

Plasmids are small, double-stranded, closed circular DNA molecules that exist extra-chromosomally in bacterial cells. These elements can replicate independently of the host chromosome, producing, in some cases, 50–100 copies of themselves in a single cell. Plasmids are also distributed to each daughter cell during cell division, and in the presence of selective pressure, a population of 1 billion bacterial cells, each of which carries 50 copies of the plasmid, can be produced. Plasmids are normal components of many bacterial cells, and often these DNA elements can confer on the host cell additional growth properties. One of the most important types of bacterial plasmids known is the resistance (R) factor. R factors are plasmids that carry genes for antibiotic resistance. When R factors are present in a bacterial cell, they make that cell resistant to antibiotics that would normally kill cells that do not have plasmids. The increasing spread of R factors in bacteria has become a major medical problem because they exist in many potentially pathogenic bacteria.

Molecular biologists have been able to use the antibiotic-resistant property of plasmids in DNA cloning by including on vectors genes that confer antibiotic resistance. The presence of these genes on the vector provides a method to select bacterial cells that carry the plasmid of interest (i.e., the vector carrying the target DNA). By attaching a foreign fragment of DNA to such a plasmid or vector, it is possible to amplify (greatly increase) the amount of the DNA fragment and clone it. The term "clone" refers to the production of a large population of DNA molecules, all of which are derived from a single DNA molecule and contain the same base-pair sequence. The basic method for using a vector to clone a DNA fragment is illustrated in Figure 6-6.

***Laboratory Example.*** Suppose a molecular biologist would like to clone a specific 4000-bp fragment of DNA from a cell. The first step in the process is purification of the DNA.

*Common vectors are derived from extra-chromosomal elements known as plasmids that exist and replicate autonomously in bacteria.*

*Plasmids often carry genes for antibiotic resistance, which can be used as a method to select bacterial cells that contain the plasmid.*

*The term "clone" can refer to the production of a large population of DNA molecules, all derived from a single DNA molecule and all containing the same base sequence.*

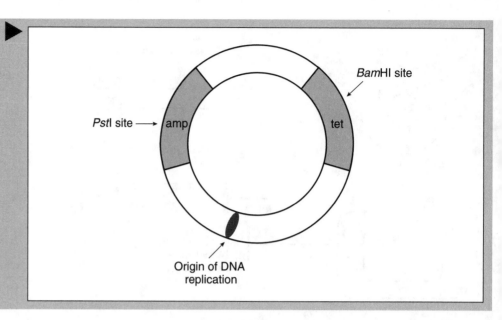

**FIGURE 6-5** ▶

*Typical Plasmid-Cloning Vector. A typical plasmid-cloning vector, such as pBR322, has an origin of replication, genes that confer antibiotic resistance, and sites for restriction endonuclease cleavage. These double-stranded, closed circular DNA molecules can replicate to high copy numbers in bacterial cells, allowing the in vivo amplification of any integrated DNA fragments.* amp = *ampicillin-resistant gene;* tet = *tetracycline-resistant gene.*

From each cell, 6 billion bp of DNA will be purified, and the 4000-bp fragment represents only a minute fraction of the total. Once purified, the DNA is then treated with a specific restriction endonuclease, such as *Pst*I, which cuts the DNA at the sequence 5'CTGCAG3' to give a 3'TGCA overhang (see Figures 6-3 and 6-4). This digestion of the DNA by *Pst*I produces thousands of fragments, only one of which is the *target fragment* that the biologist wants to clone. A vector is needed to isolate the target fragment (see Figure 6-5). The vector DNA also is cut with *Pst*I to give a linear molecule of DNA with single-stranded ends. In the presence of DNA ligase, the mixture of human DNA fragments containing the 4000-bp target DNA is mixed with the cut vector DNA. The complementary single-stranded ends of the DNA molecules hybridize to each other and become covalently linked by the action of DNA ligase (see Figure 6-4).

Three different types of recombinant DNA molecules result from this ligation: (1) a vector that has ligated back to itself and does not contain foreign DNA, (2) vectors that contain the various fragments of the human DNA produced by *Pst*I digestion, and (3) a few vectors that contain the 4000-bp fragment intended for cloning. With the use of selective antibiotic-resistant markers, it is possible to eliminate any vectors that have not integrated a fragment of human DNA. This is accomplished by transforming an *E. coli* strain with the entire population of vector DNA and selecting for bacterial cells that have developed antibiotic resistance.

It is important to note that the vector contains two genes for antibiotic resistance: one for ampicillin resistance and one for tetracycline resistance (see Figure 6-5). The *Pst*I site is in the middle of the gene for ampicillin resistance. If a vector has integrated a piece of foreign DNA at the *Pst*I site, the gene for ampicillin resistance is inactivated. Therefore, transformed bacterial cells that carry a vector with a human piece of DNA inserted into the *Pst*I site are sensitive to ampicillin but will remain resistant to tetracycline. By testing the transformed bacterial cells, one finds the following types of cells: (1) cells that are resistant to both antibiotics and contain a vector without human DNA, (2) cells that are sensitive to ampicillin but resistant to tetracycline and carry a vector that has integrated a piece of human DNA at the *Pst*I site, and (3) cells that are sensitive to both antibiotics and do not carry any vector DNA. Each of the ampicillin-sensitive, tetracycline-resistant cells must then be isolated and tested to determine which cell has a vector carrying the 4000-bp target fragment wanted for cloning.

The total collection of the cells carrying a vector with human DNA inserted into the *Pst*I site is referred to as a *DNA library* or recombinant DNA library. Two major types of DNA libraries are constructed. When the DNA library is made starting with complementary DNA (cDNA), which is complementary to messenger RNA (mRNA), it contains DNA sequences that represent only the coding regions of the genome. This is referred to as a *cDNA library*. When the DNA library is made starting with the total DNA isolated from cells, it contains DNA sequences that represent both coding and noncoding regions of

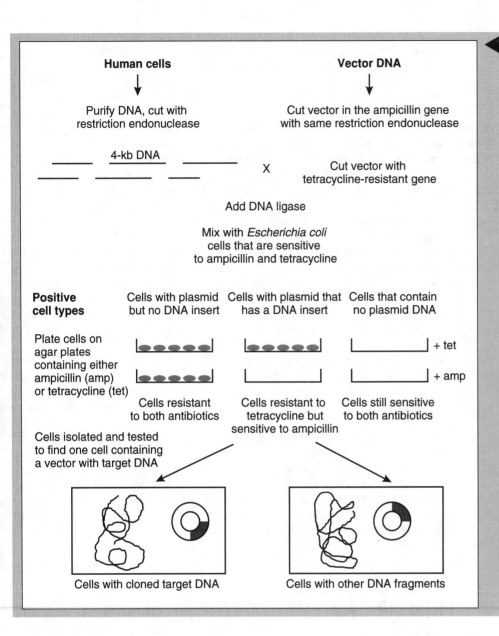

**FIGURE 6-6**
*Cloning of a Human DNA Fragment Using a Plasmid Vector. Insertion of a human DNA fragment into a restriction enzyme cleavage site within an antibiotic-resistant gene (PstI site in ampicillin gene) inactivates the gene. When vectors carrying DNA inserts are transformed into* Escherichia coli, *the cells that become tetracycline resistant and ampicillin sensitive are isolated by plating the bacteria on agar plates containing tetracycline and then testing the resulting colonies on agar plates with ampicillin. These cells are then isolated and further tested to determine if they contain the DNA fragment of interest.*

the genome and is referred to as a *genomic library*. Each cell in a library made from human DNA should theoretically contain a unique fragment of human DNA; in a genomic library, the entire library should contain the complete sequence of human DNA genome.

The normal human diploid cell contains 6 billion bp of DNA. The average size of a DNA fragment produced from a *Pst*I digestion of human DNA is estimated to be about 7000 bp. Therefore, for a library to represent all of the *Pst*I fragments, a minimum of approximately 1 million recombinant clones are necessary. However, only one of these clones will carry the 4000-bp fragment being sought. How can that one clone in a million be found? (The term "clone" here refers to a population of genetically identical bacterial cells, which have been derived from a single bacterial cell. In this discussion, the clone is considered a population of bacterial cells, all of which contain a vector with the same DNA insert.)

The two major approaches to identify clones with specific DNA fragments involve looking either at the DNA sequence or at the protein or RNA that might be produced from that DNA sequence. To utilize the sequence approach, it is necessary to have a *probe* available, which is a tagged RNA or DNA molecule that can be used to identify a clone carrying a particular DNA sequence. Probes must contain a base sequence that is complementary to the target DNA; then, utilizing nucleic-acid hybridization techniques, they are used to select the clone of interest.

**DNA Library**
*The total collection of bacterial cells containing a vector with a human DNA insert is referred to as a DNA library.*

*The term "clone" can refer to a population of genetically identical bacterial cells, all of which have been derived from a single bacterial cell.*

*A tagged RNA or DNA molecule used to detect another nucleic-acid molecule that has a complementary base sequence is called a **probe**.*

To identify the clone of interest, each cell in the DNA library is grown on agar plates into a bacterial colony; the clones are then transferred to a nitrocellulose filter. The filter is treated so that the DNA within the cells is denatured, and then the radioactive probe is added to the filter. If a cell contains a DNA fragment that is complementary to the probe, a radioactive spot is produced on the filter because of the hybridization of the radioactive probe to the DNA. These radioactive spots are detected by placing the filters on x-ray film. One then can go back to the original agar plate, select the clone that hybridized to the probe, grow the cells to high density, and isolate the vector DNA in large quantities.

If a nucleic-acid probe is unavailable but the protein product that is produced by the DNA sequence is known, techniques can be used that are aimed at finding a clone that makes the particular protein or RNA of interest. However, this type of approach requires that the vector contain regulatory sequences that allow the transcription and translation of the target DNA, as well as the availability of an antibody made to the specific protein.

Since the development of the first vectors, many new innovations have been developed to increase the usefulness of vectors in cloning genes. The inclusion of regulatory sequences on the vector DNA allows for proper transcription and translation of genes on vector DNA. This proper regulation of expression is particularly important when using vectors to analyze gene expression in mammals. In addition, it allows the production of large amounts of a protein to be produced from a mammalian gene and expressed in bacterial cells.

The addition of gene sequences to a vector DNA, which allows the vector to replicate not only in bacterial cells but also in mammalian cells, has also become very important for the study of gene expression in mammalian cells. Such a shuttle vector can be easily grown to high density in bacterial cells and then introduced into mammalian cells to study gene regulation.

With the current interest in sequencing the human genome, vectors have been developed that allow the insertion of very large (megabase) fragments of DNA into vectors. One of these vectors, the yeast artificial chromosomes (YACs), is composed of DNA sequences derived from centromeres, telomeres, and autonomous replicating sequences (ARS) of yeast. A fragment (up to 2 Mb) of foreign DNA can be ligated to the YACs; the vectors are then transformed into yeast cells where they can replicate as extrachromosomal elements (see Chapter 7 for more detail).

# DETECTION OF DNA FRAGMENTS BY NUCLEIC-ACID HYBRIDIZATION

*Under the proper environmental conditions, two single-stranded nucleic-acid molecules with complementary base sequences will hybridize to form a double-stranded molecule.*

Nucleic-acid hybridization is one of the most important tools available for studying and detecting gene sequences. The principles of hybridization form the basis for many of the current technologies used to isolate and study normal and defective human genes. Nucleic-acid hybridization occurs when two single strands of DNA or RNA (with complementary base sequences) are mixed together. Under the proper environmental conditions, the two strands will anneal to each other or *hybridize* to form a double-stranded structure.

For successful nucleic-acid hybridization, it is necessary that the nucleic-acid molecules be single stranded. In the case of a DNA molecule, which is generally present as a double-stranded structure, the first step is to treat or denature the DNA to separate the two strands. This treatment usually involves heating the solution of DNA or treating it with alkaline. As the temperature of the solution increases, a double-stranded DNA molecule denatures to form single-stranded structures. Salt concentration, such as the concentration of sodium ($Na^+$) ions, has the opposite effect and tends to stabilize the double-stranded helix. Thus, increasing the salt concentration favors the annealing process. Organic solvents, on the other hand, destabilize the helix and pull the strands apart. By controlling the temperature, the salt concentration, and the organic environment of a DNA solution, it is possible to favor either the denaturation of a double-stranded DNA molecule to single-stranded molecules or the hybridization of two complementary single-stranded molecules to form a double-stranded helix as shown in Figure 6-7.

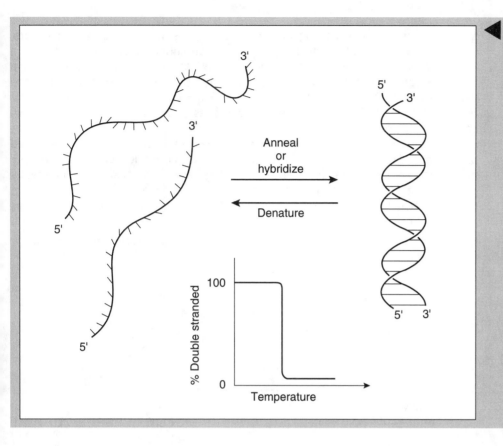

**FIGURE 6-7**
*Nucleic-Acid Hybridization.* At appropriate temperatures and salt concentrations, single-stranded nucleic-acid molecules that contain complementary base sequences will hydrogen bond to form a double-stranded helical structure. The double-stranded helix can be denatured to single strands by high temperature and low salt concentration.

When conditions are altered (e.g., lowering the $Na^+$ concentration, increasing the temperature), mismatched bases tend to denature. Conditions of *hybridization stringency* can be selected that will cause a duplex with one single mismatched base pair to be unstable. This concept of stringency is very important when trying to identify a single base change in a mutant DNA, such as needed for allele-specific oligonucleotide hybridization (ASOH) [see below].

The stringency of the hybridization reaction is controlled by the temperature, the salt concentration, and the organic environment of the solution.

The key to using nucleic-acid hybridization in the study of genes is labeling a particular molecule of DNA or RNA so it can be detected even if it is in the midst of 1 million unwanted molecules. This identification relies on the availability of *nucleic-acid probes*, which can be labeled and used to detect specific DNA fragments. Three major types of nucleic-acid probes used in such studies include the conventional DNA probes, RNA probes, and oligonucleotide probes. Conventional DNA probes are isolated by the cloning procedures described earlier in this chapter, using either plasmid vectors or virus-based vectors. RNA probes can be made if a DNA sequence is transcribed in the presence of regulatory sequences on the vector. More recently, oligonucleotide probes have become available. These probes are generally 30–50 nucleotides in length and are synthesized in vitro. To make an oligonucleotide probe, either some of the DNA or RNA sequence must be known, or the known amino acid sequence of a protein can be used with knowledge of the genetic code to synthesize the appropriate oligonucleotide. When using amino acid sequence information, it is necessary to use multiple oligonucleotides as the probe to compensate for the degeneracy of the genetic code.

In all cases, the probe must be tagged or labeled so that after hybridization the molecule can be detected. There are numerous methods available today to label nucleic-acid probes. These include incorporating radioisotopes such as phosphorus 32 ($^{32}P$) or hydrogen 3 ($^3H$) into the probe so that the probe and any nucleic-acid molecules to which it hybridizes can be detected by autoradiography. (Autoradiography is performed by placing a sample on x-ray film. The decay of the isotope causes a dark area on the film, which allows the detection of the labeled nucleic acid.)

Nonradioactive labeling systems are also available. In particular, labeling probes with fluorescent compounds coupled to biotin has become widely used. Nucleic-acid probes can be labeled using a biotinylated nucleotide such as deoxycytidine triphosphate (dCTP) then used in the hybridization reaction. A protein called avidin, which is derived

Biotin labeling of probes can be detected by forming a layered reaction:
1. *Target nucleic acid*
2. *Biotinylated probe*
3. *Avidin*
4. *Antiavidin antibody with fluorescent label*

from egg whites, has a very strong affinity for biotin and is added to the reaction mixture, where it binds to the biotin-labeled probe. Then, an antiavidin antibody containing a fluorescent label is added and binds the avidin. When the sample is placed under a fluorescent detector, it gives off a signal, which locates the position of the probe and detects a specific DNA fragment. This technique is used in fluorescent in situ hybridization (FISH) to locate DNA sequences at specific sites on a chromosome in a metaphase spread (see Chapter 5).

In Figure 6-8, the process of using a tagged probe is demonstrated. The tagged probe is mixed with a solution of unlabeled DNA molecules under conditions that favor hybridization of single strands with complementary base sequences. The probe specifically hybridizes to the DNA with the complementary sequence, and that DNA fragment becomes labeled, which allows it to be detected among the many millions of molecules that are not hybridized to the probe.

**FIGURE 6-8** ▶
**The Use of Nucleic-Acid Probes to Detect Specific DNA Sequences.** *Under conditions of high specificity, a tagged probe (\*) will hybridize only to a molecule that contains a complementary base sequence. Thus, a tagged probe can be used to detect one species of RNA or DNA in preparations of nucleic acid that contain thousands of molecules extracted from cells or tissues.*

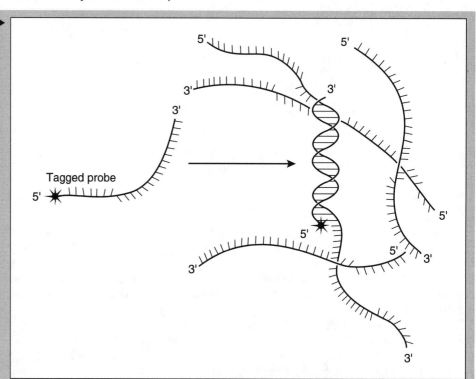

Tagged probe

## Southern and Northern Blot Hybridization

One of the first important applications of nucleic-acid hybridization in the study of molecular genetics was the development of the Southern blot, named after Edwin Southern, who first described the technique in 1975. Southern blotting has been used extensively during the past 20 years to detect target DNA fragments after size fractionation by gel electrophoresis. A typical Southern blot analysis is diagrammed in Figure 6-9.

The principle of the technique is quite simple, although the process often can be time-consuming. A purified DNA solution containing the particular gene of interest is treated with a restriction endonuclease to produce many DNA fragments of varying sizes. The solution of fragments is applied to an agarose gel, and an electrical current is added, which separates the gel fragments according to size. The smaller fragments move faster and farther down the gel. At this point, the Southern blot is performed by denaturing the DNA on the gel in an alkaline solution, then transferring the denatured fragments from the gel to a nitrocellulose membrane filter by blotting. The denatured DNA fragments, which are immobilized on the membrane filter, are then allowed to hybridize with a labeled nucleic-acid probe, which can detect the target DNA. The filter is placed on x-ray film to detect the site where the probe binds, which is indicated by a darkening of the film.

**Southern Blotting**
*This technique is used to detect DNA fragments after size fractionation by gel electrophoresis.*

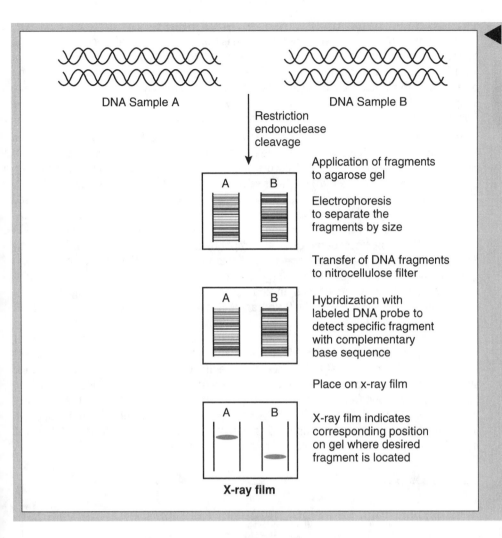

**FIGURE 6-9**
**Southern Blot Analysis of DNA Fragments.**
*After restriction endonuclease digestion of DNA and separation of the fragments by gel electrophoresis, the fragments are denatured on the gel and then transferred to a nitrocellulose membrane filter by blotting. The membrane is treated with a radioactive probe, which hybridizes to any DNA fragment with complementary base sequence. The localization of the fragment on the filter is detected by autoradiography.*

The position on the membrane filter where the probe binds indicates the presence of the target DNA. When this position is compared to the original gel and molecular weight standards, it can give an estimate of the size of the target DNA fragment. Results from a typical Southern blot are shown in Figure 6-10.

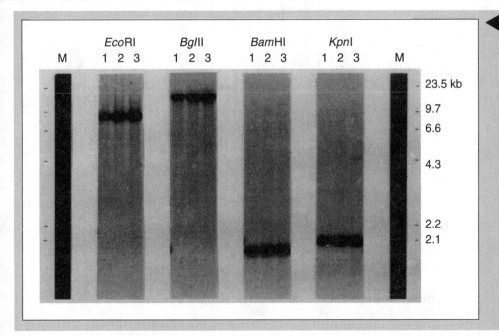

**FIGURE 6-10**
**Results from a Southern Blot.** *Five micrograms of DNA isolated from a mammalian cell were digested with the indicated restriction endonuclease. The resulting DNA fragments were separated by size using gel electrophoresis and then transferred by Southern blotting to a nitrocellulose filter. Filters were hybridized with a phosphorus 32-labeled probe, which detects the DNA sequence of the mammalian adenosine deaminase gene, and then placed on x-ray film. The figure is a picture of the x-ray film. The dark bands indicate hybridization of the probe to DNA fragments of specific sizes. Note that cleavage of the DNA by each restriction endonuclease gives a different size fragment hybridizing to the probe. Lanes 1, 2, and 3 indicate DNA samples isolated in three independent experiments. Lane M contains molecular-weight markers used to indicate the size in kilobases of the DNA fragments.*

A variation of the Southern blot technique is the Northern blot analysis, in which the target nucleic acid is a RNA molecule instead of a DNA molecule. A second variation of both Southern and Northern blotting analyses is a dot-blot (or slot-blot) assay. In this type of analysis, the nucleic-acid fragments are not separated by electrophoresis but are simply spotted either in a dot or in a slot directly on a nitrocellulose membrane. Then the fragments are treated with the probe and analyzed as above.

*Using Restriction Endonuclease Digestion and Southern Blotting to Obtain a Physical Map of DNA.* Figure 6-11 illustrates the use of restriction endonuclease digestion and Southern blotting to analyze the structure of a DNA fragment. The DNA fragment shown (approximately 15 kb in length) contains sequences for digestion by the enzymes EcoRI, BamHI and KpnI located throughout the fragment. A specific radioactive probe, located as shown in the figure, is available, which is complementary to only 1.5 kb of the DNA. The DNA is digested with each enzyme, alone or in combinations of two, to give DNA fragments that can be separated by electrophoresis on the basis of size. Testing by Southern blot analysis using the specific radioactive probe identifies the fragments and gives the approximate size of each fragment. For example, digestion with the enzyme KpnI gives a DNA fragment of 9 kb, indicating that the two sites for KpnI digestion lie 9 kb apart on the DNA fragment. The 9-kb fragment is located within the region of the DNA complementary to the probe. Digestion by EcoRI gives a fragment of 6 kb, and digestion by the BamHI gives two fragments of 7 kb and 3 kb. Digestion with a combination of KpnI and EcoRI produces a fragment of 2 kb. When the data are analyzed, a physical map of the DNA fragment of interest is obtained; the specific restriction endonuclease cleavage sites are located as shown in Figure 6-11. The fragments produced by digestion that do not overlap the probe (e.g., the 8-kb EcoRI fragment) are not detected on the Southern blot. This type of restriction analysis of DNA can be used to characterize the physical map of a DNA fragment. It also can be used to detect specific mutations in the DNA, if the mutation happens to alter or delete a restriction enzyme cleavage site within a gene.

**FIGURE 6-11**
*Obtaining a Physical Map of DNA. In determining the physical map of DNA, genomic DNA is treated with various restriction endonucleases, alone or in combinations of two. Southern blot hybridization is carried out to give the size of the resulting DNA fragments. The probe hybridizes only to a small portion of the total fragment and detects only those fragments that overlap the probe. Results can be used to create a restriction map of the DNA fragment.*

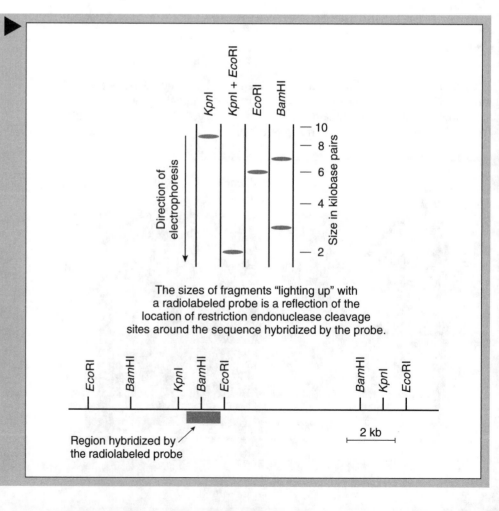

The sizes of fragments "lighting up" with a radiolabeled probe is a reflection of the location of restriction endonuclease cleavage sites around the sequence hybridized by the probe.

**Detection of Point Mutations in the β-Globin Gene by Southern Blot Hybridization.** In Chapter 3, the molecular basis for the human genetic disease sickle cell anemia was discussed. In sickle cell anemia, a single base-pair substitution in the β-globin gene results in the production of an altered β-globin protein. In this disease, a simple amino acid replacement of a glutamic acid residue by a valine residue produces a globin protein prone to sickling. This mutation in the β-globin gene can be detected by restriction endonuclease digestion and Southern blot analysis because the single nucleotide substitution of an A to T causes the loss of a *Mst*II restriction site in the β-globin gene. Thus, the DNA from a patient with sickle cell anemia when treated with *Mst*II and analyzed by Southern blotting gives a different pattern from that of a normal individual. A probe to the β-globin gene detects a single 1.35-kb *Mst*II fragment in the mutant DNA, whereas DNA from a normal individual shows two *Mst*II fragments, one 1.15 kb and one 0.2 kb (see Resolution of Clinical Case in Chapter 3).

> *Restriction endonuclease digestion and Southern blotting can be used to detect a mutation that alters or destroys a restriction enzyme cleavage site in the DNA.*

# POLYMERASE CHAIN REACTION

One of the most recent and innovative techniques developed for cloning and analyzing DNA is the polymerase chain reaction (PCR). With the use of PCR, it is now possible to make millions of copies of a specific DNA sequence in vitro in a few hours, and it can be done without first separating a specific DNA fragment from the rest of the genomic DNA. This extraordinary technique, developed by Kary Mullis in the mid-1980s, has revolutionized the way scientists analyze DNA in the detection and study of many human diseases. PCR is diagrammed in Figure 6-12.

> **PCR**
> *This technique is used to make millions of copies of specific DNA sequences in vitro in a few hours.*

To perform PCR, one begins with a double-stranded piece of DNA with the 5' and 3' ends as noted in Figure 6-12. Only a few molecules of starting DNA are necessary to begin, and they need not be purified away from other DNA molecules. In the first step of the reaction, the double-stranded DNA is denatured by heating to give two separate single strands. Added to the mixture of denatured DNA are two oligonucleotide primers (each 20–30 nucleotides long). One primer (the forward primer) is a 5'-to-3' oligonucleotide, which is complementary to one strand of the DNA; the second primer (the reverse primer) is a 5'-to-3' oligonucleotide, which is complementary to the other strand of the DNA. The two primers are added in excess so they will be present in sufficient concentration to complete 30 cycles of the reaction. Each primer has a 3' free end that serves as a substrate for DNA synthesis in the 5' to 3' direction, using each initial DNA strand as a template. After annealing the oligonucleotide primers to each template DNA strand, DNA polymerase is added along with the precursor deoxyribonucleotides. DNA synthesis proceeds down each DNA strand, making a new copy of the DNA and giving two double-stranded DNA molecules where there was initially only one. The 5' ends of the newly synthesized strands are fixed, but the 3' ends may vary in length. This ends the first cycle.

At this point, the two double-stranded DNA strands are again denatured by heating, which produces four single strands of DNA to serve as templates for new DNA synthesis. Once again, the primers bind to each single strand of DNA, and DNA polymerase begins synthesis at each primer, which results in eight single strands of DNA to serve as templates in the next cycle.

In the third cycle, the length of the newly synthesized DNA strands begins to become fixed; the 5' ends are defined by the primer, and the 3' ends are defined by the terminus of the opposite primer. The cycle is repeated again and again, resulting in the number of DNA molecules doubling at each cycle, with the short form that contains the target sequence beginning to predominate. After 30 cycles, nearly 1 billion copies of the DNA sequence located between the two primers are produced. This amount of DNA is sufficient to allow detection directly on an agarose gel, either by fluorescence, if a fluorescent-labeled primer is used, or simply by staining with ethidium bromide. A product produced from PCR and its detection by ethidium-bromide staining is shown in Figure 6-13.

Because the 30 cycles of the PCR reaction are carried out sequentially in a single reaction mixture, the DNA polymerase used in PCR, *Taq* polymerase, is isolated from a heat-resistant bacterium, *Thermus aquaticus*. The use of this heat-resistant polymerase allows the reaction mixture to be heated to 95°C for the denaturation step without

> *Taq polymerase, a heat-resistant DNA polymerase, is used in PCR.*

**FIGURE 6-12** ▶

**Polymerase Chain Reaction (PCR).** *The PCR is used to amplify DNA fragments in vitro without first separating out the target sequence from other DNA molecules. Two primers, which are complementary to the DNA sequence flanking the sequence to be amplified, are added to denatured DNA. The primers bind to each of the separated DNA strands, and, in the presence of DNA polymerase and deoxyribonucleotides, DNA synthesis proceeds to give two new DNA strands. In the second cycle, four single strands of DNA are available to act as templates to give eight strands of DNA containing the target sequence. In the third cycle, the short form with fixed ends begins to predominate. After 30 continuous cycles, nearly 1 million copies of the target sequence are produced. dNTP = deoxynucleoside triphosphates.*

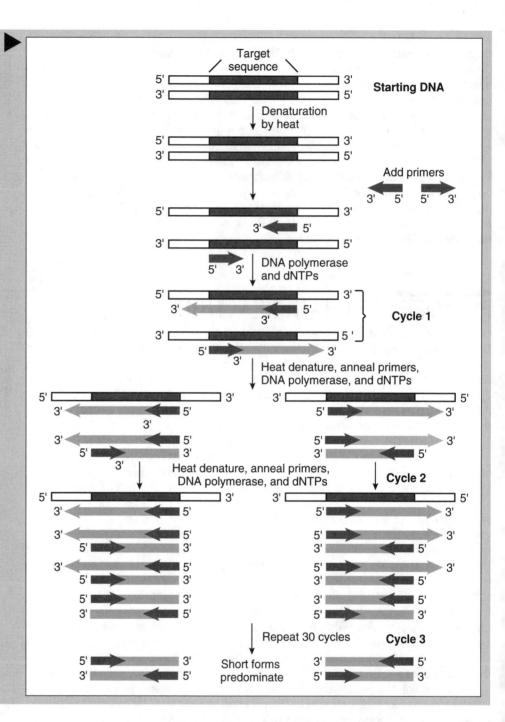

inactivating the polymerase activity. It is important to remember that one requirement of PCR is the prior knowledge of at least part of the base sequence of the DNA in order to make the necessary primers.

## Important Applications of PCR in Medical Genetics

**Detection of Mutations Using ASOH.** Single base change mutations can be detected in genes if the sequence around the potential mutation site is known, for example, when dealing with mutations in the β-globin gene (e.g., the mutation that results in sickle cell anemia). To detect these mutations, oligonucleotide primers that differ by a single nucleotide are synthesized in vitro and used with PCR-amplified DNA as illustrated in Figure 6-14.

Two allele-specific oligonucleotides (ASOs) are synthesized; one is complementary to the normal allele (ASO-N) base sequence, and the other is complementary to the

**ASOH**

*This technique is used to detect single base change mutations in a gene if the sequence change in the mutant phenotype is known.*

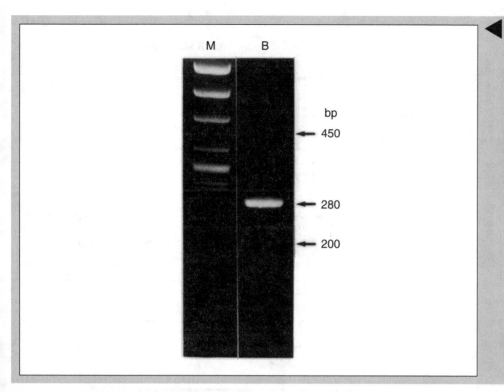

bp
← 450

← 280

← 200

**FIGURE 6-13**
**Product of a Polymerase Chain Reaction (PCR) Experiment.** *PCR amplification of a target sequence can produce $10^5$–$10^6$ copies of the desired DNA molecule. This amount of DNA is easily detected by applying the DNA solution to an agarose gel and separating the fragments by electrophoresis. Ethidium bromide staining is used to detect the fragment, which then can be isolated from the gel. This gel is stained with ethidium bromide to detect a 280-bp fragment produced in a PCR amplification.*

mutant allele (ASO-M) base sequence. The two oligonucleotides differ by a single base (adenine to thymine). The normal ASO hybridizes to the normal DNA allele but not to the mutant DNA allele because a mismatch occurs. However, the mutant ASO hybridizes to the mutant DNA but not to the normal DNA.

When using ASOH to detect mutations, DNA is amplified by PCR from a normal homozygote (NN), a heterozygote (NM), and a homozygous mutant (MM) and used as a control relative to unknown samples. The three control DNA samples along with the experimental samples are spotted on a nitrocellulose membrane to give a dot blot as described previously. Duplicate filters are individually treated with the radioactive ASOs. The ASO-N shows positive hybridization with the DNA obtained from the normal (NN) and the heterozygote (NM). The ASO-M, however, shows positive hybridization to the heterozygote (NM) and the mutant (MM) DNA samples. Therefore, if the mutant oligonucleotide hybridizes to a DNA sample, the patient from whom that sample was taken must be carrying the mutant DNA. Further analysis using the normal ASO distinguishes a heterozygote or carrier state from the homozygous mutant state.

PCR is useful in this assay because a very low amount of DNA (e.g., that present in a blood sample) can be tested by first using PCR primers to amplify the gene of interest and then testing the PCR-amplified DNA with the various ASO probes. In the example of a β-globin mutation, two PCR primers are created that are 20 bases long and complementary to the DNA sequence of the β-globin gene. These primers are used in PCR to amplify samples of DNA from various patients. The amplified DNA is then hybridized to specific ASO probes, which detect either a normal DNA sequence or a mutant sequence such as is found with sickle cell anemia.

When using ASOH to detect mutations in DNA, it is important to know the sequence change that has occurred to give the mutant phenotype, as well as the surrounding sequences, so that PCR primers can be synthesized.

**Detection of Short Tandem Repeat Polymorphisms by PCR.** Previously, the occurrence in the human genome of tandemly repeated di-, tri-, and tetranucleotides was discussed. Short tandem repeats (STRs) are stretches of DNA sequences composed almost entirely of repeating CA dinucleotides. These stretches of CA repeats are found throughout the human genome and are often polymorphic, meaning that the precise number of repeats at any particular CA repeat locus varies. The exact number of repeats in a person's two alleles can be measured at any particular CA repeat locus using PCR amplification. In utilizing STRs to type alleles, primers are synthesized from sequences that flank the

*CA repeats can be used to determine haplotypes in a family and can identify the allele pattern of a "disease gene."*

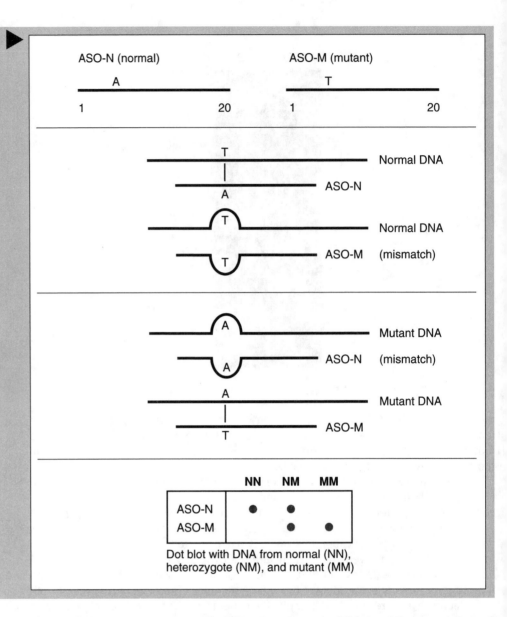

**FIGURE 6-14**

***Detecting Mutations Using Allele-Specific Oligonucleotides (ASOs).*** *Two short ASOs, one complementary to a normal gene (ASO-N) and one complementary to a mutant gene (ASO-M), are synthesized and tagged with phosphorus 32. DNA is amplified by the polymerase chain reaction from cells of normal (NN), heterozygous (NM), and mutant (MM) individuals and then dotted on a nitrocellulose filter. The two ASOs are then used in a hybridization reaction to detect complementary base sequence in the DNA. Stringent hybridization conditions are essential to ensure that a single mismatch will prevent hybridization of the ASO.*

Dot blot with DNA from normal (NN), heterozygote (NM), and mutant (MM)

STRs, allowing the amplification by PCR of alleles that differ by different numbers of repeat units as shown in Figure 6-15.

In the example shown, allele 1 has a CA repeat of 136 bp, and allele 2 has a CA repeat of 132 bp, each allele representing one of the two homologous chromosomes. During synthesis by PCR, the PCR products are labeled with a fluorescent marker (e.g., fluorescein), which allows their detection in an automated DNA sequencer. The sequencer is an electrophoresis system with a fixed fluorescence detector that measures fluorescent bands of DNA as they migrate past the detector. The type of data obtained in the analysis of CA repeats in the dystrophin gene using PCR is indicated in Figure 6-16. The automatic sequencer presents the data, where the x-axis represents the time of electrophoresis (an indication of the increasing size of the PCR DNA product), and the y-axis is the fluorescent intensity of the PCR DNA product.

The four repeat alleles found within the dystrophin gene differ in size by a few base pairs, with each allele detected as a major peak on the electrophoretogram. The different positions of the alleles are marked as 3'*CA*, *STR*45, 5'*Dys*II, and *STR*49. Two molecular-weight markers, 113 bp and 271 bp, are given to indicate the length of the fragments. The size of the allele to the left is the smallest, and the one to the right is the largest. For each position there can be more than one allele noted as *a*, *b*, or *c*. These alleles differ in size by a few base pairs. In the example shown, the female marked II-2 has a single peak at the 3'*CA* position, marked *a,a*, which represents her two dystrophin genes at this position and indicates that she is homozygous at this locus. A single peak at a locus in a female may also indicate a deletion at the site, in which case it is noted as *a*–.

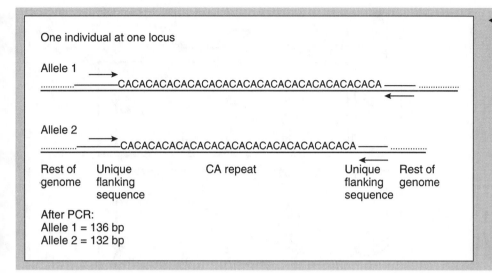

One individual at one locus

Allele 1

────────→
············——————CACACACACACACACACACACACACACACACACACACA ——————  ············——————
                                                                ←────────

Allele 2

────────→
············——————CACACACACACACACACACACACACACACACACACA ——— ············
                                                    ←────────

| Rest of genome | Unique flanking sequence | CA repeat | Unique flanking sequence | Rest of genome |

After PCR:
Allele 1 = 136 bp
Allele 2 = 132 bp

**FIGURE 6-15**
*Amplification of Cytosine–Adenine (CA) Repeats by the Polymerase Chain Reaction (PCR). The individual noted has two alleles, one located on each chromosome. The two alleles differ in size by a few base pairs, which represent polymorphisms in the CA repeats. Primers to the unique flanking sequence of the CA repeat unit are used in PCR to amplify the CA repeat unit. Alleles can be distinguished by the size of the CA repeat unit.*

At *STR*45, there are two peaks marked *a* and *b*, indicating the woman's alleles differ at this locus. The 5'*Dys*II and the *STR*49 locus each show two peaks *a* and *b*. Thus, this female has an allele array as follows:

| *a,b,* | *a,b* | *a,b* | *a,a* |
|--------|-------|-------|-------|
| 5'*Dys*II | *STR*45 | *STR*49 | 3'*CA* |

However, without information from other members of the pedigree, the linear array of the markers on each of the two chromosomes cannot be determined. For example, it is unknown whether the *a* allele of *STR*45 is on the same chromosome as the *a* allele of the 5'*Dys*II or if they are on opposite chromosomes. To determine this, the order of the markers on each chromosome and which alleles are linked together must be determined (setting phase). For example, is the order in female II-2 *b,b,a,a* on one chromosome and *a,a,b,a* on the other chromosome, or is some other combination present? To set phase with an X-linked disease, one starts with the males in the pedigree. Phase is already known in males in this pedigree because they have only one X chromosome and therefore only one dystrophin gene that is X-linked. A schematic of the two dystrophin genes of the mother (II-2) is shown in Figure 6-17 along with the haplotypes of both of her sons (III-1 and III-2).

Analysis of the mother's allele data indicates she has *a* and *b* alleles at positions *STR*45, 5'*Dys*II, and *STR*49. However, it cannot be determined if the *a* allele of *STR*45 is on the same dystrophin gene as the *a* allele of 5'*Dys*II or if they are on opposite dystrophin genes. The data from the sons are used to set phase. The affected son, III-1, has an *a* allele at each locus, giving a pattern of *a,a,a,a*. The unaffected son, III-2, has a pattern of *b,b,b,a*. Each son must have received one X chromosome from the mother. Because one son is affected, and one is unaffected, each son probably received different X chromosomes (thus different dystrophin genes) from the mother. Therefore, the mother is most likely *a,a,a,a* on one chromosome and *b,b,b,a* on the other. So, the disease gene must be the *a,a,a,a* pattern. Any female inheriting this gene is a carrier, and any male inheriting this gene is affected. A further example of setting phase using CA repeat data is given in the Clinical Case.

***Detection of Mutations Using Single-Strand Conformational Polymorphism (SSCP) Analysis.*** SSCP analysis of PCR-amplified DNA depends on the tendency of single-stranded DNA to form intramolecular hydrogen bonding. The electrophoretic mobility of a DNA molecule in a nondenaturing gel system is a function of its conformation as well as its length. If a point mutation affects the intramolecular hydrogen bonding and stem-loop structure of the molecule, the DNA has an altered electrophoretic mobility. Thus, two single-stranded DNA molecules that differ by even a single base can be differentiated by SSCP analysis (Figure 6-18).

A major advantage of SSCP analysis is that it can be performed without knowing what the mutation is in the gene to be analyzed (unlike ASOH, which requires prior knowledge of the base change that leads to the mutation). When a mutation is suspected in a particular region of a gene, the DNA for the suspect mutant gene and the DNA of the

**SSCP Analysis**
*This technique is used to detect a single base change in a DNA molecule if that base change affects the intramolecular hydrogen bonding of the DNA.*

**FIGURE 6-16** ▶

*Products of Polymerase Chain Reaction (PCR) Using Short Tandem Repeat (STR) Polymorphisms. In the dystrophin gene, four polymorphic cytosine–adenine (CA) loci are distributed throughout. The alleles at each locus differ by a few base pairs and can be detected by PCR amplification and electrophoresis. In the electrophoretogram shown, the positions are marked (3' CA; STR45; 5'DysII; STR49) along with the molecular weight markers, 113 bp and 271 bp. All males have a single allele at each locus. All females have two alleles at each position. (Source: Reprinted with permission from Schwartz LS, et al: Fluorescent multiplex linkage analysis and carrier detection for Duchenne/Becker muscular dystrophy. Am J Hum Genet 51:727, 1992.)*

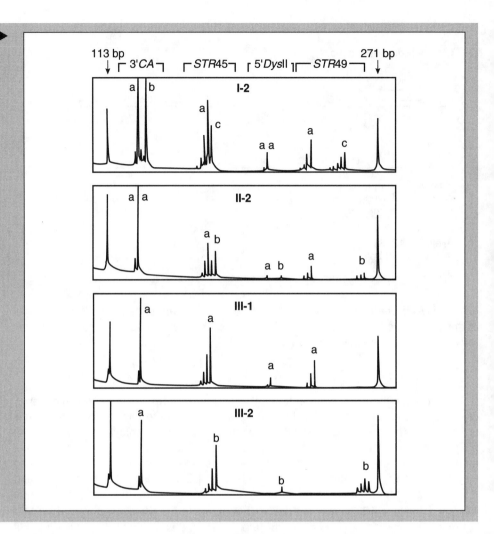

normal gene are amplified by PCR. The amplified DNAs are denatured and immediately applied to a nondenaturing gel to prevent the denatured strands from renaturing. On the gel, the single-stranded DNA molecules remain separated from each other and begin to migrate in the gel. The native environment allows the single-stranded DNA molecules to hybridize with themselves and form intramolecular hydrogen bonds. This intrastrand hydrogen bonding occurs because the double-stranded form of DNA is more stable than the linear single-stranded form. The folding that occurs depends on the base sequence, with the two complementary strands folding to make different forms with an altered mobility in the gel. A single base change in 500 bases on either strand can be detected as an altered conformer or a mutant form of DNA.

Further analysis of the altered DNA conformer can be carried out by isolating the DNA fragment from the gel and using automatic DNA sequence analysis to give the exact base change that has occurred. The base change identified in the DNA may or may not be related to the cause of the disease, because any polymorphism that has occurred will give an altered DNA conformer. To determine if the base change seen is the cause of the disease, several criteria must be fulfilled.

First, one would expect the base change to result in an altered amino acid in the protein or the appearance of a stop codon in the mRNA. This prediction can be tested by examining the genetic code and determining if the mutation seen leads to a change in a codon that would result in an altered amino acid or in a termination codon. Second, if the disease is recessively inherited, one would expect to find both alleles or both copies of the gene to be altered in a patient with a mutant phenotype. For a dominantly inherited disease, only one of the alleles would be expected to change. In the review questions at the end of this chapter, there is a problem given that demonstrates how SSCP is used in detecting a mutation and how those data are used to relate a mutation to a specific disease.

With the discovery of PCR, the analysis and detection of mutations in human genes has become a commonly used technique in the laboratory as well as in the clinic. The identification and cloning of additional human genes every day is making the detection of many human diseases increasingly possible using this technique. This increased ability to dissect a person's genetic makeup also is raising many important ethical questions, which will be addressed later in this text.

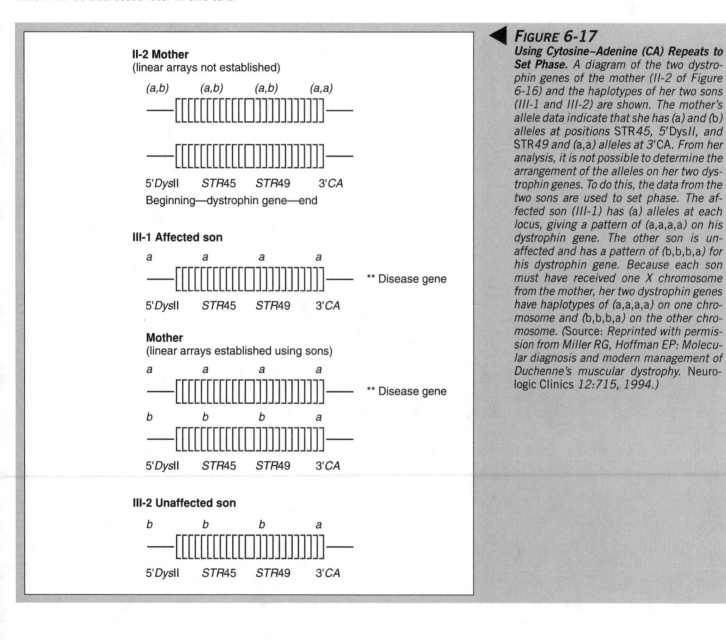

**FIGURE 6-17**

***Using Cytosine–Adenine (CA) Repeats to Set Phase.*** *A diagram of the two dystrophin genes of the mother (II-2 of Figure 6-16) and the haplotypes of her two sons (III-1 and III-2) are shown. The mother's allele data indicate that she has (a) and (b) alleles at positions STR45, 5'DysII, and STR49 and (a,a) alleles at 3'CA. From her analysis, it is not possible to determine the arrangement of the alleles on her two dystrophin genes. To do this, the data from the two sons are used to set phase. The affected son (III-1) has (a) alleles at each locus, giving a pattern of (a,a,a,a) on his dystrophin gene. The other son is unaffected and has a pattern of (b,b,b,a) for his dystrophin gene. Because each son must have received one X chromosome from the mother, her two dystrophin genes have haplotypes of (a,a,a,a) on one chromosome and (b,b,b,a) on the other chromosome. (Source: Reprinted with permission from Miller RG, Hoffman EP: Molecular diagnosis and modern management of Duchenne's muscular dystrophy. Neurologic Clinics 12:715, 1994.)*

**FIGURE 6-18** ▶
**Detecting Mutations Using Single-Strand Conformational Polymorphism Analysis.** *Single-stranded DNA molecules form sequence-specific intrastrand hydrogen bonding, which determines their migration in a gel under nondenaturing conditions. A single-base substitution can alter the mobility of the DNA molecule in a gel, allowing the detection of a single-base substitution in 500 bases. The two complementary DNA strands have different base sequences and give two different conformers on the gel.*

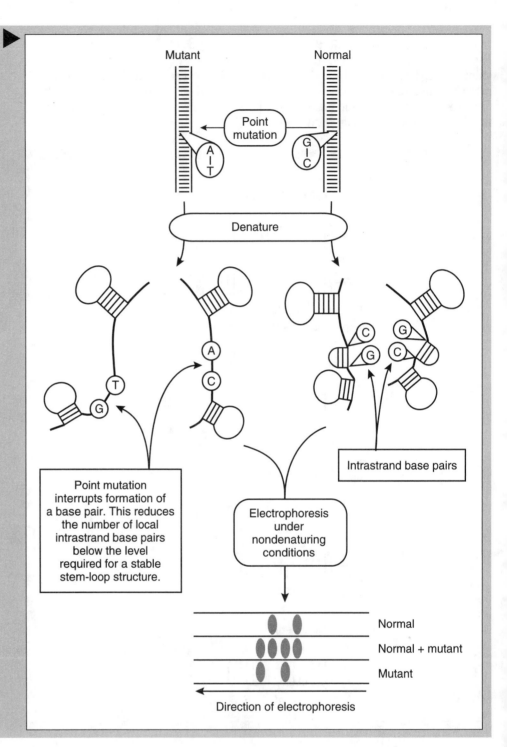

## RESOLUTION OF CLINICAL CASE

Additional members of the pedigree from the Carson family were tested for CA repeat polymorphisms (traces not shown), and the following results were obtained:

|  | I-1 | I-2 | II-3 |
|---|---|---|---|
| 5' *Dys*II | *a* | *a,b* | *a,a* |
| STR45 | *b* | *a,c* | *a,b* |
| STR49 | *b* | *a,c* | *a,b* |
| 3' CA | *a* | *a,b* | *a,a* |

To analyze the CA repeat data from the members of this pedigree for linkage, the first step is to determine the haplotype for each individual and note it on the pedigree using the order as above. This is done as described in the text using the haplotypes of the two sons, one affected with the disease (III-1) and one unaffected (III-3). As indicated from the traces, the affected son has *a* alleles at each locus, and his single dystrophin gene pattern is defined as *a,a,a,a*. The unaffected son (III-3), has a pattern of *a,b,b,a*. Each son must have received one X chromosome from the mother. Because one son is affected and one is unaffected, they probably received different X chromosomes and different dystrophin genes from the mother. We use the two sons to define the two patterns of the mother, assigning one haplotype as *a,a,a,a* and the other as *a,b,b,a* to represent her two dystrophin gene patterns. Once the mother's pattern is established, we can deduce the haplotypes of the rest of the members of the pedigree and note them as shown in Figure 6-19.

The "disease gene" is associated with the *a,a,a,a* haplotype. All males receiving this pattern are affected, and all females with this pattern are carriers.

- The expected male child received the "at-risk" haplotype (*a,a,a,a*) and has a 95% chance of being affected with DMD. Because the risk of recombination is approximately 5% for this region of the dystrophin gene, there is a small chance (approximately 5%), that the fetus might not receive the mutation because of a recombinational event. Therefore, the chance of being unaffected is 5%, and the chance of being affected is 95%.

- What are the carrier risks to the females (i.e., I-2, II-2, II-3, III-2) in the pedigree? Any female with the *a,a,a,a* haplotype is a carrier. Therefore, the risk for I-2, II-2, and II-3 is 100%. III-2 did not inherit the disease gene. Therefore, her risk of being a carrier is 5%, but only if a recombinational event occurs.

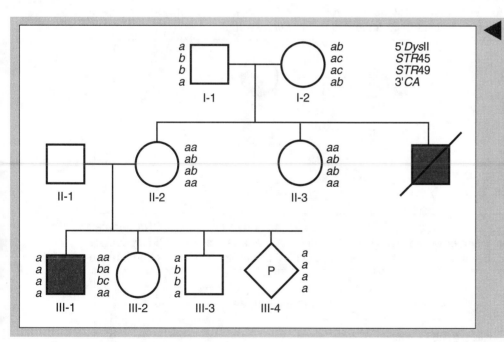

**◀ FIGURE 6-19**
**Notation of Haplotypes on the Carson Family Pedigree.** *The haplotypes of each individual of the pedigree are determined as described in the text and noted on the diagram. The unborn male (III-4) has inherited the (a,a,a,a) haplotype, which is the disease gene, and has a 95% risk of having Duchenne's muscular dystrophy.*

# REVIEW QUESTIONS

**Directions:** For each of the following questions, choose the **one best** answer.

### Questions 1 and 2

The questions that follow are designed to illustrate how modern techniques can be used to follow a mutation in a pedigree and to provide information about the change in DNA sequence caused by a mutation. The diagram below is a hypothetical gene containing three exons and two introns. The transcription unit is depicted by adjacent boxes symbolizing the exons (*darkened and numbered boxes*) and the introns (*open boxes*). This diagram contains information that is necessary to answer questions 3–6 as well.

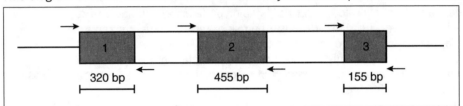

The polymerase chain reaction (PCR) primers, represented by *arrows*, are 20 nucleotides long and are derived from DNA sequences located directly outside the boundary of each exon. The size of each PCR product is 40 bp larger than the size of the exon (i.e., exon 1 PCR product = 360 bp).

The following pedigree is derived from a family at risk for an autosomal recessive disease with the proband (II-2) indicated by the *filled symbol*.

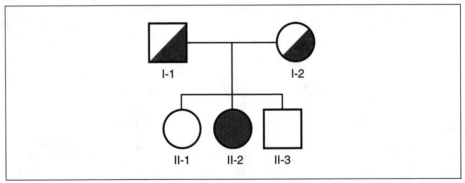

PCR is carried out using the three exon-specific primer pairs and a DNA template from each member of the pedigree. Electrophoretic analysis of the PCR products is illustrated below as a sketch of the polyacrylamide gel stained with ethidium bro-

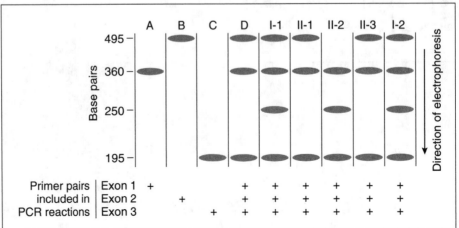

mide to detect the bands of DNA. Lanes A through D contain samples prepared from control DNA, using the indicated primer pairs. The size, in base pairs, of the DNA in each band is indicated to the left of the gel.

1. From the data given above, the mutation in the gene can be localized to which one of the following sequences?

    **(A)** Exon 1

    **(B)** Exon 2

    **(C)** Exon 3

    **(D)** Intron 1

    **(E)** Intron 2

2. The mutation will alter the protein encoded for by the gene in which one of the following ways?

    **(A)** The mutant protein will have a single amino-acid change from the normal

    **(B)** The mutant protein will be 20 amino acids larger than the normal protein

    **(C)** The mutant protein will be 85 amino acids smaller and translated in frame

    **(D)** The mutant protein will be the same size with multiple amino acid changes

    **(E)** The mutant protein will be 81 amino acids smaller and translated out of frame

**Questions 3 and 4**

The following is a three-generation pedigree of a family at risk for an autosomal dominant disease with moderate expressivity. The proband is III-2 (indicated by the *arrow*). The paternal grandmother (I-2) displays several of the less debilitating symptoms of the disease.

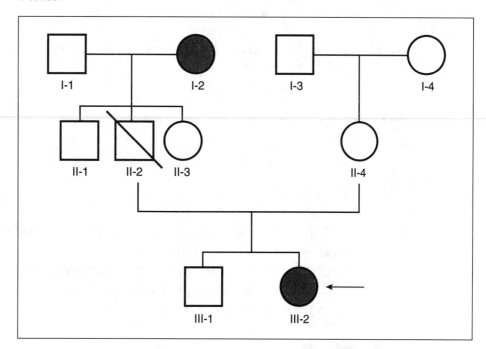

Electrophoretic analysis of the three exon-specific polymerase chain reaction (PCR) products is shown below.

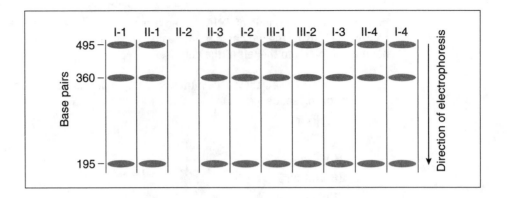

Single-strand conformational polymorphism analysis of the PCR products is shown below. Lanes A, B, and C contain DNA amplified from a normal individual using exon 1, exon 2, or exon 3 primer pairs, respectively.

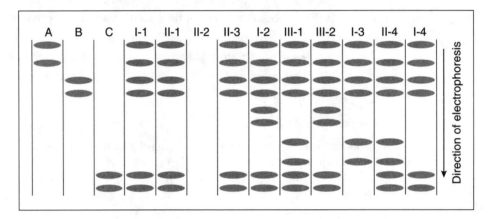

3. From the data given above, the mutation in the gene can be localized to which one of the following sequences?

   **(A)** Exon 1

   **(B)** Exon 2

   **(C)** Exon 3

   **(D)** Intron 1

   **(E)** The data are insufficient to determine the location

4. From the information given, the genotype of the deceased father (II-2) would most likely be which one of the following?

   **(A)** The father was heterozygous for the defective allele and affected

   **(B)** The father was homozygous for the defective allele

   **(C)** The father was homozygous for the defective allele but unaffected

   **(D)** The father was not carrying the defective allele

   **(E)** The data are insufficient to determine the genotype

**Questions 5 and 6**

The following is a pedigree of a family at risk for an X-linked recessive disease. The proband (III-2), indicated by the *filled symbol*, is the only member of the pedigree presenting with the disease. The maternal grandmother died from heart disease, and the maternal uncle died at 1 month of age from trauma caused by an automobile accident. Electrophoretic analysis of the polymerase chain reaction (PCR) products, using the three primer pairs for all three exons, is shown below the pedigree.

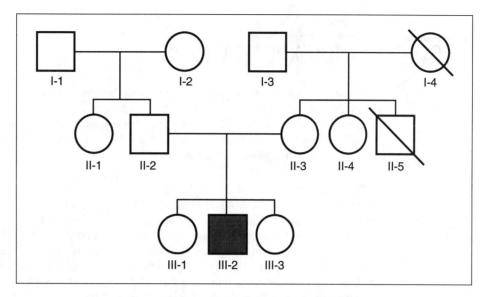

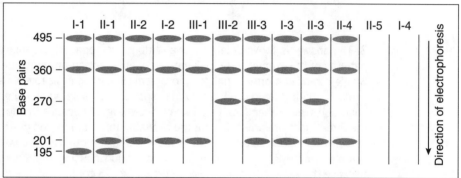

5. From the data given above, the mutation in the gene can be localized to which one of the following sequences?

  (A) Exon 1

  (B) Exon 2

  (C) Exon 3

  (D) Intron 1

  (E) The data are insufficient to determine the location

6. The change in DNA sequence in the proband could alter the expression of the gene in which one of the following ways?

  (A) The protein product will be missing a single amino acid and be inactive

  (B) The protein product will be missing 23 amino acids and be inactive

  (C) The protein product will not be detected because of a lack of an initiation codon in the messenger RNA (mRNA)

  (D) The protein product will not be detected because of a trinucleotide expansion in the 3' untranslated region of the mRNA

  (E) The protein product will not be detected because of a defect in the promoter sequence

# ANSWERS AND EXPLANATIONS

**1. The answer is B.** The data from the electrophoretic analysis indicate that exon 2 is 245 bp smaller in the DNA obtained from the proband as compared to the normal. When polymerase chain reaction (PCR) is carried out using the primers for exon 2, the proband has a PCR product of 250 bp instead of the normal 495 bp. Thus, the mutation lies within exon 2 of the gene.

**2. The answer is E.** The 245-bp deletion in exon 2 will remove 81 codons (243 bp) of amino acid sequence information plus two additional base pairs, causing a translational frameshift mutation. Option A is incorrect because a single amino acid change results from a single base-pair substitution, so the size of the polymerase chain reaction (PCR) product would not change. Option B is incorrect because the PCR product would be larger by 60 bp. Option C is incorrect because the PCR product would be smaller by 255 bp. Option D is incorrect because this type of mutant protein would result from a single base-pair addition or deletion, so the size of the PCR product would change by one base pair.

**3. The answer is E.** No deletions or additions are apparent because the electrophoretic patterns of the exon-specific polymerase chain reaction products are the expected size. Single-strand conformational polymorphism (SSCP) analysis reveals a unique conformer in the proband sample (III-2), which suggests a base-pair change is related to the condition. The proband appears to be a heterozygote because the normal conformers for the three exons are present. With the available information, the mutation in the proband cannot be mapped to one of the exons. Another unique SSCP conformer is present in samples from I-3, II-4, and III-1. The mutation causing this conformer is in exon 3 because the sample from I-3 has no normal exon-3 conformer. The mother is a carrier of this silent mutation. This result emphasizes the point that not all changes seen with SSCP analysis cause a disease.

**4. The answer is A.** The deceased father (II-2) should be heterozygous for the defective allele and have a banding pattern identical to the samples for the proband and I-2. He was most likely affected if the disease is an autosomal dominant disease. Option B is incorrect because both children would be affected if the disease is an autosomal dominant disease and the father had two defective alleles. Option C is incorrect because the disease gene seems to be passed from the father's mother to his son. Option D is incorrect because one would predict that even a single defective allele would give an affected phenotype. Option E is incorrect because the pedigree and the single-strand conformational polymorphism data give sufficient information.

**5. The answer is C.** The proband has a mutation in exon 3 for the following reasons. The inheritance pattern of the 201-bp and 195-bp bands on the paternal side demonstrates that the larger band is derived from a 161-bp exon 3 (201 − 40 = 161). The band derived from this exon is not present in the sample from the proband. Rather, exon 3 in the proband is 270 bp or 69 bp larger. Thus, in the proband, exon 3 has additional base pairs.

**6. The answer is D.** The last exon encodes the 3′ untranslated region of the messenger RNA (mRNA) and most likely codes for the C-terminal end of the protein as well. If a trinucleotide expansion within the 3′ untranslated region of the exon has occurred, the size of exon 3 would be larger (in this case 69 bp or 23 trinucleotide repeats). The presence of additional trinucleotide repeats at the 3′ end of the mRNA could destabilize the mRNA or block its processing, resulting in a dramatic decrease in the cytoplasmic levels of mRNA and a lack of detectable protein. Another possibility (but less likely) is that the additional 69 bp of DNA occur before the translational termination codon, and the protein product contains 23 additional amino acids, which could destroy its function. Option A is incorrect because a single amino acid change would not change the size of the PCR product. Option B is incorrect because the PCR product is larger not smaller. Option C is incorrect because the mutation is in exon 3 not in exon 1, where the initiation codon would be located. Option E is incorrect because a defect in the promoter sequence would not affect the size of exon 3.

# REFERENCES

1. Miller RG, Hoffman EP: Molecular diagnosis and modern management of Duchenne's muscular dystrophy. *Ped Neurogenet* 12:699–724, 1994.
2. Schwartz LS, Tarleton J, Popovich B, et al: Fluorescent multiplex linkage analysis and carrier detection for Duchenne/Becker muscular dystrophy. *Am J Hum Genet* 51:721–729, 1992.

# MAPPING AND IDENTIFICATION OF HUMAN DISEASE GENES

# INTRODUCTION OF CLINICAL CASE

A young Jewish couple brought Bruce, their 5-year-old son, to a physician for evaluation. After a recent vacation in Florida, the parents noticed that the boy's face had a butterfly rash that was exacerbated each time the boy was exposed to sunlight. Upon examining Bruce, the physician found additional rashes on other exposed areas of his skin. The physician also noted that Bruce was unusually small in stature for his age and had a narrow facies. Upon questioning, the parents stated that Bruce had been very small at birth, leading them to believe initially that he might have a growth deficiency. However, he had developed normally but remained small for his age. In addition, the mother told the physician that Bruce constantly had a cold or some other type of infection. Lately he seemed to be unusually tired and listless and was unable to play very long without having to stop and rest.

The physician had a blood sample drawn and suggested that the parents keep Bruce at home and quiet until he received the results of the blood studies. He sent the blood sample to the laboratory and ordered several tests, including a white blood cell (WBC) count, a determination of possible viral or bacterial infection, and a karyotype.

# PHYSICAL MAPPING

A *physical map* is the representation of the position of genes on each chromosome expressed in base pairs, with the genes placed in a known order and at a known distance from other genes. A physical map for the human genome is currently being constructed, using results obtained from both cytogenetic and molecular analyses. The final goal in creating this physical map is the knowledge of the entire linear sequence of nucleotides for each human chromosome.

## Cytogenetic Technology

### SOMATIC CELL HYBRIDS AND THEIR USE IN HUMAN GENE MAPPING

In the early 1970s, a method was developed for creating cells known as somatic cell hybrids, which have two entire mammalian genomes present in a single nucleus. In this technique, fusing agents (e.g., Sendai virus, polyethylene glycol) are added to cultured somatic cells to promote the formation of multinucleated cells known as homokaryons or heterokaryons. When cells of the same species are fused together, the resulting homokaryon is made up of an additive number of chromosomes from both parents (e.g., 92 human chromosomes in a human–human homokaryon). After nuclear fusion, the resulting hybrid is reasonably stable for many generations. When cells from two different species are fused together, the initial heterokaryon contains the full chromosome complement from both species. However, after nuclear fusion, the formed hybrid specifically loses chromosomes of one species and retains the chromosomes of the other species. Fusions of mouse or hamster cells with human cells generally result in the specific loss of human chromosomes and the retention of the rodent chromosomes in the hybrid nucleus. After 20–30 cell generations, a stable hybrid cell that contains the entire chromosome complement of the rodent but a reduced number of human chromosomes is formed. Because the loss of the human chromosomes occurs randomly, different hybrid cells in the population retain different human chromosomes.

This unexpected random loss of human chromosomes from human–mouse hybrids was exploited as a new and innovative method for mapping human genes to specific human chromosomes [1]. The assumption made in using somatic cell hybrids for mapping human genes is that a particular gene or gene sequence always is present in a hybrid when the chromosome on which it is located is present, and it is absent when the chromosome is absent from the hybrid. Therefore, a gene that is always associated with the presence of a particular chromosome is said to be syntenic with and localized to that chromosome. The methodology for somatic cell hybridization is diagrammed in Figure 7-1.

Two cell populations, one derived from a mouse with 40 chromosomes and one derived from a human with 46 chromosomes, are fused together in the presence of polyethylene glycol (PEG). After 24 hours, multinucleated cells arise in the population of cells at a frequency of approximately 30%. These multinucleated cells represent a large percentage of homokaryons (i.e., a fusion between two mouse cells or a fusion between two human cells) and approximately 1% heterokaryons (a fusion between a mouse cell and a human cell). To isolate pure populations of human–mouse heterokaryons, a selection procedure must be used that kills both of the parental cells and the homokaryons but allows the human–mouse hybrids to survive and grow. The most common procedure used for the selection of somatic cell hybrids is hypoxanthine-aminopterin-thymidine (HAT) medium selection or HAT-ouabain selection. In a medium that contains the drug aminopterin, the de novo purine and pyrimidine biosynthetic pathways of cells are blocked. However, if hypoxanthine and thymidine are present in the medium, the cells can overcome the block by synthesizing their purines and pyrimidines using salvage pathways (Figure 7-2).

For cells to grow in HAT medium, two key enzymes must be functional; that is, hypoxanthine-guanine phosphoribosyltransferase (HGPRT) and thymidine kinase (TK). In the somatic cell hybridization illustrated in Figure 7-1, the mouse cell is deficient in TK (TK−), and the human cell is deficient in HGPRT (HGPRT−). Neither mouse cells, human cells, or homokaryons formed from these cells can grow in HAT medium because

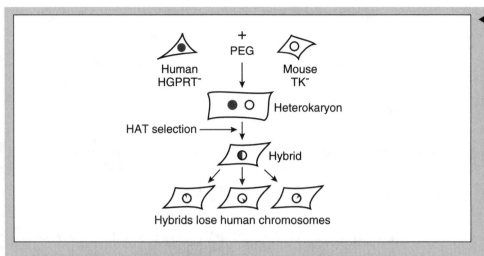

**FIGURE 7-1**
**Somatic Cell Hybridization.** Somatic cells derived from a mouse are fused in the presence of polyethylene glycol (PEG) with somatic cells derived from a human. The mouse cells are deficient in thymidine kinase (TK⁻), and the human cells are deficient in hypoxanthine-guanine phosphoribosyltransferase (HGPRT⁻). Neither can grow in hypoxanthine-aminopterin-thymidine (HAT) medium. Only heterokaryons, which contain the mouse HGPRT gene and the human TK gene, grow in HAT medium. After nuclear fusion, the resulting hybrid begins to segregate human chromosomes specifically. The result is a panel of hybrid cells, each of which contains a full complement of mouse chromosomes and only a few human chromosomes. Each hybrid contains different human chromosomes, allowing the assignment of human genes to specific human chromosomes.

they are missing one of the enzymes needed for nucleotide synthesis. Heterokaryons, on the other hand, have a normal TK gene that comes from the human genome and a normal HGPRT gene derived from the mouse genome. Therefore, in HAT medium, only human–mouse hybrids survive and grow.

Between 24 and 48 hours after fusion by PEG, the population of cells is plated in HAT medium. When the nonfused cells and homokaryons die, the surviving human–mouse hybrids are grown in nonselective medium to a pure population of cells that can be isolated and analyzed. As each isolated hybrid cell continues to grow and divide in nonselective medium, daughter cells arise that have randomly lost human chromosomes. Eventually, stable hybrid cells can be isolated, each of which has five to eleven human chromosomes remaining in the nucleus. However, each isolated stable hybrid has a different set of human chromosomes remaining. In this way, a hybrid cell panel is established in which each hybrid retains only a few but different human chromosomes.

Several methods can be used to determine which human chromosomes are present in each hybrid clone. Chromosome-banding procedures, such as G-banding, were once used exclusively to identify the human chromosomes present in the hybrid (Figure 7-3).

*A **hybrid panel** containing different human chromosomes is used for mapping human genes.*

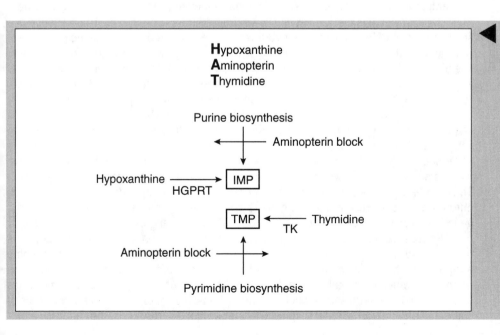

**FIGURE 7-2**
**HAT Selection.** A selection medium containing hypoxanthine, aminopterin, and thymidine (HAT) is used to select somatic cell hybrids. Aminopterin inhibits de novo purine and pyrimidine biosynthesis, forcing a cell to use exogenous metabolites to survive. A cell that has a functional thymidine kinase (TK) gene uses exogenous thymidine as a source of pyrimidines. A cell with a functional hypoxanthine-guanine phosphoribosyltransferase (HGPRT) gene uses exogenous hypoxanthine as a source of purines. Therefore, HAT medium selects for a cell with both an active TK gene and an active HGPRT gene. IMP = inosine monophosphate; TMP = thymidine monophosphate.

**FIGURE 7-3** ▶

*Analysis of Somatic Cell Hybrids. Somatic cell hybrids are analyzed for the presence of human chromosomes by G-banding. Cell-free extracts are made from each hybrid, and the proteins in the extract separated by gel electrophoresis. The location of a particular protein on the gel is determined by specific enzyme staining or by specific antibodies. If an assay or an antibody is not available to detect the protein, specific DNA probes can be used in Southern blotting to identify the presence of a specific DNA fragment in the hybrids that is complementary to the DNA probe. All the data are then used to correlate the presence or absence of a gene or gene product with the presence or absence of a specific chromosome. In the example given, the gene and the gene product are present when human chromosome 1 is present and absent when human chromosome 1 is absent. If additional hybrids indicated synteny between the gene and human chromosome 1, this gene would map to human chromosome 1.*

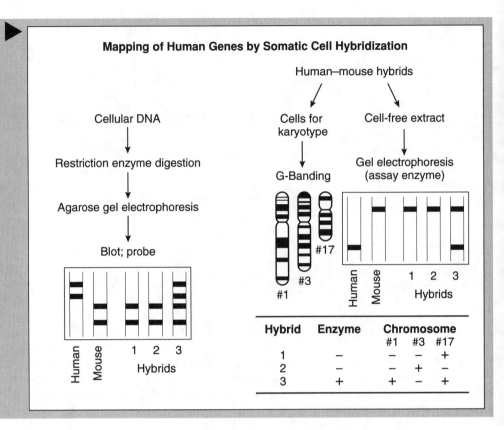

Mapping of Human Genes by Somatic Cell Hybridization

| Hybrid | Enzyme | Chromosome | | |
|--------|--------|-----|-----|------|
| | | #1 | #3 | #17 |
| 1 | − | − | − | + |
| 2 | − | − | + | − |
| 3 | + | + | − | + |

*The presence of a particular human gene in a hybrid is determined either by looking for the expression of the gene product or by looking for specific human gene sequences complementary to a DNA probe.*

More recently, sets of chromosome-specific primers have become available that allow polymerase chain reaction (PCR) screening to be used for the identification of human chromosomes that remain in the hybrids.

Once the specific human chromosomes present in each hybrid are identified, a number of approaches can be used to determine if a particular human gene or gene sequence is present or absent from the hybrid. If the function of the particular gene is known (e.g., an enzymatic activity) or if a specific antibody to the protein, which is coded for by that gene, is available, the hybrids can be analyzed for the expression of the human protein. This analysis uses electrophoretic separation of cell-free extracts isolated from the hybrids and then staining for enzyme activity or testing for a reaction to a specific antibody (see Figure 7-3). If the function of the gene is unknown, but a DNA probe is available, then the presence of a specific gene sequence can be determined by Southern blot analysis, which is carried out on DNA extracted from each hybrid. The presence of a specific human DNA fragment that hybridizes to the DNA probe indicates the presence of that gene in the hybrid, although it might not be expressed (see Figure 7-3). Primers specific for the gene sequences to be mapped now can be isolated, allowing the use of PCR analysis to detect the presence of human gene sequences in the hybrids (see Chapter 6).

Once each hybrid in the panel has been analyzed with respect to the presence or absence of a particular chromosome and the presence or absence of a particular gene sequence, the assignment of that sequence to a specific chromosome can be made. In Table 7-1, data from such an analysis for the test gene Z is presented. Five hybrids are identified in which chromosome 2 is present, and the gene is present (+/+), and five hybrids are identified in which chromosome 2 is absent, and the gene is absent (−/−). These (+/+ and −/−) hybrids are *concordant hybrids*. No hybrids are found in which chromosome 2 is present and the gene is absent (+/−) or vice versa (−/+); thus, the gene is syntenic with chromosome 2 and is assigned a map position on chromosome 2.

Because somatic cell hybrids retain more than one human chromosome, it is necessary to analyze hybrid panels as above to avoid ambiguity in the chromosome assignment of a gene. A modification of the above procedure is available that uses *microcells*, which are small human-derived cells that contain one to three chromosomes (rather than 46) to form somatic cell hybrids. When cells are exposed to very low doses of colcemid (an inhibitor of mitotic spindle formation) for 24–48 hours, they undergo a pro-

◀ **TABLE 7-1**
*Mapping of Gene X to Chromosome 2 Using Somatic Cell Hybrids*

| Hybrid Clone | Gene X DNA | 1 | 2 | 3 | 4 | 5 | 6 | 7 | 8 | 9 | 10 | 11 | 12 | 13 | 14 | 15 | 16 | 17 | 18 | 19 | 20 | 21 | 22 | 23 | X | Y |
|---|---|---|---|---|---|---|---|---|---|---|---|---|---|---|---|---|---|---|---|---|---|---|---|---|---|---|
| 1 | + | − | + | − | − | − | − | − | − | + | − | − | − | + | + | − | + | − | − | − | − | + | − | + | − | − |
| 2 | − | − | − | − | − | − | + | − | + | − | + | + | − | − | − | − | − | + | − | − | + | − | + | − | + | − |
| 3 | − | − | − | + | + | − | − | + | − | + | − | − | + | − | − | − | + | + | − | − | − | − | − | + | − | + |
| 4 | + | − | + | − | − | + | − | − | − | − | − | − | − | − | − | + | − | − | + | − | + | − | − | − | + | − |
| 5 | − | + | − | + | − | − | + | − | − | − | + | − | + | − | + | − | − | − | − | − | − | + | − | − | + | − |
| 6 | − | − | − | − | + | − | − | − | + | − | + | − | − | − | − | + | − | − | + | + | − | − | − | − | + | − |
| 7 | + | + | + | − | − | − | − | + | − | − | − | + | − | − | − | − | − | − | − | + | − | − | + | − | + | − |
| 8 | + | − | + | − | − | − | + | − | − | − | − | − | − | − | + | + | − | − | + | − | − | − | − | − | − | + |
| 9 | − | + | − | − | − | + | − | + | − | − | − | − | + | − | − | − | − | − | + | − | − | − | − | − | + | − |
| 10 | + | − | + | − | − | − | + | − | − | − | + | + | − | − | + | − | − | − | − | − | − | − | + | − | + | − |

longed mitotic arrest. This treatment gives rise to *micronuclei*, which have a few chromosomes enclosed within a nuclear membrane. Centrifugation of the micronuclei in the presence of cytochalasin B produces microcells in which the micronuclei with one to three chromosomes become enclosed within a plasma membrane. Just as somatic cells can be fused together in the presence of PEG, a microcell population can be fused to a somatic cell population, with the resulting hybrids containing only one to three human chromosomes.

The use of this technique was initially quite limited because the only microcell hybrids that could be selected involved those that retained a chromosome with a selectable marker (e.g., the X chromosome, which carries the HGPRT gene, or chromosome 17, which carries the TK gene). However, with the development of procedures for transferring exogenous DNA into mammalian cells (see Chapter 14), it is now possible to tag each human chromosome with an exogenously added selectable marker, the neo-resistant gene (*neo* for neomycin). The presence of a neo gene in a cell makes that cell resistant to the drug G418 and can be used to select hybrids that contain any neo-tagged human chromosome. This procedure provides a selection procedure for the isolation of microcell hybrids that retain a single copy of any human chromosome.

**Microcells** are small cells with one to three human chromosomes enclosed in a plasma membrane that can be fused to mouse somatic cells to form human-mouse hybrids with one to three chromosomes.

Human chromosomes tagged with a selectable marker, the **neo gene**, can be selected for in medium containing G418.

## SUBCHROMOSOMAL GENE MAPPING USING HYBRID CELLS

***Use of Translocations, Deletions, and Fragments.*** By using human cells that contain chromosomes with translocations or deletions to form somatic cell hybrids, a gene can be localized to a specific region within a chromosome. Balanced translocations occur at a high frequency in the human population and, in most instances, do not alter the phenotype of the individual carrying the translocation (see Chapters 5 and 9).

When using *translocations* for regional mapping, human cells that have a different portion of the X chromosome translocated to an autosome are used to form somatic cell hybrids (Figure 7-4). The translocated piece of the X chromosome carries the gene for HGPRT, and any hybrid with an autosome having this piece of the X chromosome can be selected for by plating in HAT-ouabain and keeping the hybrids under HAT selection as they segregate chromosomes. In this way, a hybrid that carries only the X–autosome translocated chromosome can be isolated. For these studies, the initial human cell is ouabain sensitive and HGPRT+, and the mouse cell is HGPRT− and ouabain resistant. The hybrids selected in HAT-ouabain retain ouabain resistance from the mouse and must retain the piece of the human X chromosome that carries the HGPRT gene or the translocation. As shown in Figure 7-4, hybrids can be selected that retain varying lengths of the X chromosome translocated to an autosome. These hybrids are then tested for other genes known to be located on the X chromosome so they can be mapped with respect to the HGPRT gene. Using this method, the relative positions on the long arm of the X chromosome of the genes for phosphoglycerol kinase (PGK) and glucose-6-phosphate dehydrogenase (G6PD) were determined.

Other types of balanced translocations, in which parts of autosomes are translocated

Translocations that carry the HGPRT gene are useful for the regional mapping of genes on a chromosome.

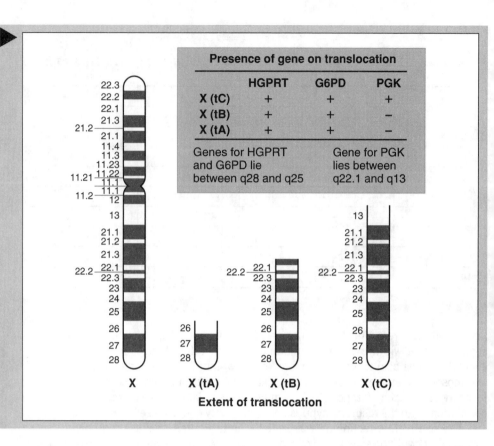

**FIGURE 7-4**

*Regional Mapping Using Balanced Translocations. Hybrid cell lines having human X–autosome translocations containing various lengths of the long arm of the X chromosome are analyzed for the presence or absence of a particular gene. Growth of the hybrids in hypoxanthine-aminopterin-thymidine (HAT) medium selects for retention of the translocation because the hypoxanthine-guanine-phosphoribosyltransferase (HGPRT) gene is located on the long arm of chromosome X. In the example given, the gene for phosphoglycerol kinase (PGK) must lie between q22.1 and q13, because the gene is present only when that portion of the X chromosome is translocated. The genes for HGPRT and for glucose-6-phosphate dehydrogenase (G6PD) must lie between q28 and q25, because they are present even on the smallest fragment of chromosome X translocated.*

**Presence of gene on translocation**

|        | HGPRT | G6PD | PGK |
|--------|-------|------|-----|
| X (tC) | +     | +    | +   |
| X (tB) | +     | +    | −   |
| X (tA) | +     | +    | −   |

Genes for HGPRT and G6PD lie between q28 and q25

Gene for PGK lies between q22.1 and q13

**Extent of translocation**

*Subchromosomal mapping can be carried out by using the chromosomal location of microdeletions or by analyzing chromosome fragments produced by radiation treatment of microcell hybrids.*

to the X chromosome, can also be used in regional mapping of a gene. Once again, growing the hybrids in HAT-ouabain selects for the presence of the chromosome that is made up of the long arm of chromosome X (which carries the HGPRT gene) and a translocated piece of an autosome. Hybrids are then tested for the presence of genes that might be located on the autosome piece. This type of translocation was important in the regional mapping to the long arm of chromosome 14 of the gene coding for purine nucleoside phosphorylase. Translocations serve another important role in gene mapping when using flow cytometry to isolate individual human chromosomes, as discussed later in this chapter.

*Deletions* that occur in human chromosomes and lead to a disease phenotype also have been used for regional mapping of genes. If a known phenotype is always associated with a small deletion detectable by banding techniques, the gene associated with that phenotype can be localized to that position on the chromosome, either through the use of somatic cell hybrids or by molecular analysis as discussed later in this chapter. Several examples of deletions that have been important for regional mapping include the genes responsible for Prader-Willi and Angelman syndromes, which have been localized to 15q11–13, and the gene or genes responsible for cri du chat syndrome, which was localized to the terminal end of 5p (see Chapters 5 and 9).

Subchromosomal mapping also can be carried out by *fragmenting chromosomes* into small pieces by the process of radiation-reduced hybrids. When a somatic cell hybrid with only a single human chromosome (obtained by microcell fusion) is treated with controlled doses of x-rays, fragmentation of all of the mouse chromosomes, as well as the single human chromosome, occurs. To rescue the human chromosome fragments, the irradiated cell is fused with a second viable rodent cell that integrates the human gene fragments into its chromosomes. Screening for integrated human chromosome fragments with specific DNA probes allows for the identification of genes or markers that are present on a single fragment. This information can produce a linear map order for the genes, assuming that the closer together two genes or markers are on the chromosome, the more likely it is that they will be found on the same fragment. As the distance between the two genes increases, the chances increase that they will be found on separate fragments.

**Fluorescence In Situ Hybridization (FISH).** The technique of FISH, in which nonradioactive-labeled DNA probes are used to localize a DNA sequence directly on a chromosome in a metaphase spread, was described in Chapter 6. The sensitivity of the FISH technique makes it possible to identify single-copy genes and map them to a specific region of a human chromosome. The resolution of this procedure is 2–3 Mb, but it can be increased by using more extended chromosome preparations. Therefore, if a DNA probe is available for a gene, the location and approximate position of that gene on a particular chromosome can be determined directly on a metaphase spread. Once this information is available, the application of more detailed molecular mapping can localize the gene more precisely.

*Isolation of Individual Chromosomes by Flow Cytometry.* In the early studies of somatic cell genetics, attempts were made to isolate pure preparations of human chromosomes and transfer them to recipient cells. Bulk preparations of total chromosomes could be isolated by various centrifugation methods, but it was never possible to transfer intact chromosomes into cells because the chromosomes were always fragmented into small pieces during the uptake procedure. Currently, individual human chromosomes are isolated in pure form by the process of flow cytometry and then used in molecular studies [2]. In this process, a population of human chromosomes is isolated intact from cells and then labeled with a fluorescent DNA-binding dye that can be detected by a laser beam. Each chromosome absorbs a different amount of the dye, depending on its size. Once labeled (Figure 7-5), the chromosomes are passed through a laser beam, and the intensity of the fluorescence is measured. The chromosomes are then automatically sorted according to the intensity of the fluorescence and thus isolated according to size. Because some human chromosomes are very similar in size, the separation is not always ideal, as illustrated in Figure 7-6. However, certain modifications of the procedure can be used to achieve a better separation of chromosomes of a similar size. Somatic cell hybrids that contain only a few human chromosomes can be used, simplifying the starting material and enabling a more precise separation of chromosomes of the same size. Particularly useful in this regard are human-Chinese hamster hybrids because Chinese hamster DNA has a different base content than human DNA. This difference results in different dye binding and better separation of the rodent and human chromosomes.

> **Uses for Human Chromosomes Purified by Flow Cytometry**
> Gene mapping
> Creating DNA libraries for each human chromosome

Additionally, chromosomes with translocations are valuable because they usually differ in size and intensity of fluorescence from the normal chromosome. This size difference results in an altered separation of the translocated chromosome in the flow cytometer and can be used to verify the position of a gene on a translocated chromosome.

Once the human chromosomes are isolated in pure form, they are used in Southern blot assays or dot blot assays for the direct localization of human genes or gene fragments to the specific chromosome (see Figure 7-6).

An additional advantage of having purified individual chromosomes available is the ability to construct specific DNA libraries for each human chromosome. No longer is it necessary to begin with the total DNA content of a human cell when preparing a DNA library. Instead, the libraries can be made starting with DNA from each isolated chromosome, which reduces the number of clones that must be assayed. This is particularly true if the gene of interest has already been mapped by somatic cell hybridization or FISH to a particular chromosome. DNA libraries for each human chromosome are now available and are being used in conjunction with yeast artificial chromosomes (YACs) and cosmids to determine the complete nucleotide sequence of the human genome.

## Molecular Technology

*DNA Libraries and YAC Clones.* Human DNA libraries (discussed in detail in Chapter 6) are made by cutting the complete human genome into pieces of defined size with restriction enzymes and then cloning the DNA fragments into bacterial plasmids or YACs, which are DNA elements that contain telomeres, a centromere, an autonomous replication sequence (Ars), and human DNA pieces ranging in size from 100,000 to 1 million bp (Figure 7-7).

The Ars enables the YAC to replicate in yeast cells, whereas the telomere sequences give stability to the YAC, and the centromere serves to promote distribution of the YAC to

> **YACs,** vectors being used for mapping the human genome, are DNA elements that contain telomeres, a centromere, an autonomous replication sequence, and pieces of human DNA ranging in size from 100,000–1 million bp.

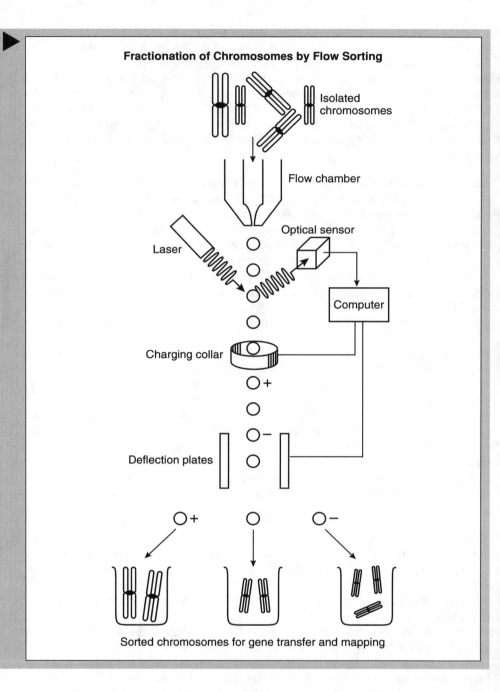

**FIGURE 7-5**

*Isolation of Human Chromosomes by Flow Cytometry.* *The process of flow cytometry enables mammalian genomes to be fractionated into single chromosomes or groups of chromosomes according to size. The isolated chromosomes are used for gene mapping and for the creation of specific DNA libraries containing DNA from each human chromosome. (Source: Reprinted with permission from Shows TB, Sakaguchi AY: Gene transfer and gene mapping in mammalian cells. In Vitro 16:65, 1980.)*

**Fractionation of Chromosomes by Flow Sorting**

Isolated chromosomes

Flow chamber

Laser

Optical sensor

Computer

Charging collar

Deflection plates

Sorted chromosomes for gene transfer and mapping

each daughter cell at cell division. The construction of human YAC DNA libraries has simplified the analysis of the human genome by reducing by approximately 10-fold the total number of clones required to contain the entire human genome sequence in the library. Many of the YAC libraries have been constructed by partial digestion of isolated human chromosomes with a restriction enzyme to produce DNA fragments that have overlapping sequences. The DNA fragments are separated using pulse gel electrophoresis to select for larger pieces of DNA and cloned into YAC vectors, which are propagated in yeast cells.

Each isolated YAC clone contains a specific segment of DNA of 100–1000 kb that must be analyzed and ordered with respect to all other DNA segments to form a composite map of each chromosome. The DNA fragments are analyzed for specific landmarks or sequences and then compared to each other to look for overlapping regions on the different fragments. Ordering of the overlapping fragments into a continuous piece of DNA (i.e., a contig) allows for the construction of the final physical map. One landmark used to construct the physical map is the location of restriction endonuclease cleavage sites that are used to make a restriction map (see Chapter 6).

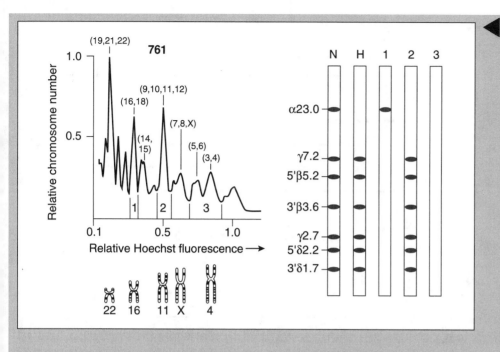

**FIGURE 7-6**

**Use of Sorted Human Chromosomes for Mapping Genes.** *A flow karyotype for human chromosomes at the top shows the separation of human chromosomes according to size. The sorted chromosomes are then used to map the location of specific genes using Southern blotting. Chromosomes sorted from regions 1, 2, and 3 are identified by chromosome banding. Region 1 contains chromosomes 16 and 18, region 2 contains chromosomes 9 through 12, and region 3 contains chromosomes 3 through 6. DNA from chromosomes in each region is treated with a restriction endonuclease, and the fragments are run out on a gel (bottom). The location of the various globin genes are detected by Southern blotting with a complementary DNA (cDNA) probe. Lane N is total DNA from a normal individual. Lane H is total DNA from an individual lacking the α-globin gene. Lanes 1, 2, and 3 represent DNA from the corresponding region of the sorted chromosomes. From these data, the α-globin gene appears to map to chromosome 16 or 18, and the β-globin genes map to chromosomes from region 2. More precise mapping of each gene can be done by sorting chromosomes containing translocations of the above chromosomes and looking for a new location of the genes in the flow karyotype. (Source: Reprinted with permission from Gray JW, Langlois RG: Chromosome classification and purification using flow cytometry and sorting. Ann Rev Biophys Biophys Chem 15:224, 1986.)*

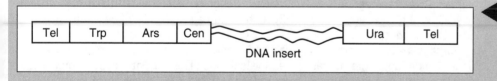

**FIGURE 7-7**

**Typical Yeast Artificial Chromosome (YAC) Cloning Vector.** *A YAC cloning vector contains yeast telomere sequences (Tel), a centromere sequence (Cen), an autonomous replication sequence (Ars), and selective markers for tryptophan (Trp) and uracil (Ura). The wavy line indicates a double-stranded DNA molecule of 100–1000 kb derived from a human cell.*

More recently, the use of sequence-tagged sites (STSs) has greatly enhanced the physical mapping of human DNA and has provided a large number of landmarks on each chromosome. STSs are short fragments of DNA, approximately 200–500 nucleotides in length, which have a unique known sequence and thus can be amplified by PCR using forward and reverse primers with a specific oligonucleotide sequence. Once the position and order of STSs on a chromosome are known, they can be used to order YAC clones. The presence or absence of various STSs on different YAC clones is determined by PCR, and these data are used to construct a content map of STSs. The assumption in putting together the map is that if the same STS is present on two YAC-cloned DNA fragments, these fragments overlap. Overlapping fragments are then arranged to give a final physical map as shown in Figure 7-8. STSs can be known sequences derived from isolated genes or gene fragments, or they can be derived from CA repeat sequences used in linkage analysis. By combining the use of restriction endonuclease digestion sites, STSs, and

**Sequence-Tagged Sites**
*STSs are used to order YAC clones and to form a content map.*

**FIGURE 7-8** ▶

**Sequence-Tagged Sites (STSs) Content Map.** *STSs are short DNA pieces (200–500 bp in size) of known sequence that can be amplified by polymerase chain reaction (PCR) and mapped to specific sites on a chromosome. The position of a STS can be used to order yeast artificial chromosome (YAC) clones by first determining whether or not a specific STS is present on the DNA fragments from different YAC clones. If the DNA fragment from two YAC clones (e.g., YAC clones 1 and 4) both have STS B, the DNA fragments are assumed to overlap. If the DNA fragments (e.g., from YAC 1 and 3) do not contain the same STS, they are assumed not to overlap. In this way, fragments of DNA cloned into YAC vectors can be ordered with respect to one another on a specific human chromosome.*

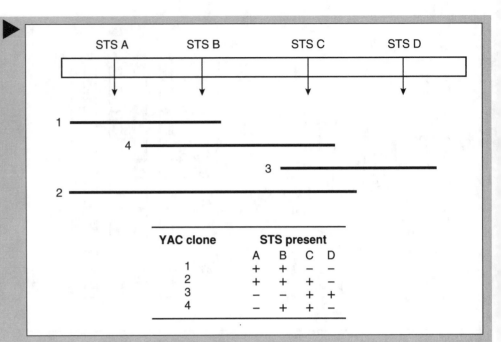

YAC DNA libraries, a low-resolution physical map has been constructed for each human chromosome [3].

A high-resolution physical map currently is being constructed by cutting each DNA fragment in each YAC clone with a restriction endonuclease and subcloning these smaller DNA fragments into cosmid vectors, which contain inserts of approximately 40 kb. The cosmid DNA library is analyzed as above for landmarks to produce an overlapping sequence map now made up of smaller DNA segments. DNA fragments isolated from individual cosmid clones are then cut with a restriction endonuclease and subcloned into vectors that carry smaller fragments of DNA (e.g., plasmids, M13 phage). These smaller DNA fragments can then be purified and sequenced to give the complete DNA base pair sequence for each fragment.

One continuing problem with YAC clones has been their instability, resulting in the isolation of clones with deletions and rearrangements, or the presence of more than one human DNA fragment in one YAC clone. Recently, bacterial artificial chromosomes (BACs), which can carry inserts of 350 kb of human DNA, have been introduced. The BAC DNA clones, which use the F factor (or fertility plasmid) of *Escherichia coli* as a vector, appear to be more stable than YAC clones and may become more useful in the final elucidation of the human gene physical map. The currently used approaches for constructing a physical map are summarized in Table 7-2.

**TABLE 7-2** ▶

*Constructing a Physical Map*

| Approach | Technique | Size Resolution |
|---|---|---|
| Chromosome | Flow cytometry, microcell hybrids | $2 \times 10^8$ kb |
| Cytogenetics | Banding, FISH | $2 \times 10^3$ kb |
| YACs | Fragments cloned, overlapping order | 100–1000 kb |
| BACs | Fragments cloned, overlapping order | 150–350 kb |
| Restriction sites | Order restriction sites | 200 kb |
| STSs | Order STSs on YAC clones | 100 kb |
| Cosmids | YAC clones cut and recloned | 40 kb |
| Plasmid/M13 | Cosmid clones cut and recloned | 1–5 kb |
| Sequencing | Each isolated fragment sequenced | 1 bp |

*Note. BACs = bacterial artificial chromosomes; FISH = fluorescent in situ hybridization; STSs = sequence-tagged sites; YACs = yeast artificial chromosomes.*

***Direct DNA Sequencing.*** The final goal of the Human Genome Project is to determine the complete nucleotide sequence of the human genome. This goal necessitates actual sequencing of many large DNA fragments, which is a technique that is rapidly being improved and further automated with fluorochrome-labeled bases. By the year 2005, or sooner, the entire DNA sequence of each human chromosome is expected to be known, raising many ethical and moral questions (see Chapter 14).

# GENETIC MAPPING

## Recombination Analysis

A genetic map is a representation of the relative position on a chromosome of a gene or gene loci based on the frequency of recombination between two segments of DNA during meiosis. This frequency is expressed in centimorgans (cM), where 1 cM equals a recombination frequency of 1%. As noted in Chapter 5, during meiosis I, two homologous chromosomes can pair and undergo recombination, which results in an equal exchange of genetic material between the two chromosomes and produces recombinant chromosomes in the gametes. The probability that two gene sequences on a single chromosome will undergo recombination depends on the distance between the two sequences; that is, the closer the two sequences are together, the less the probability that a recombinational event will occur between them and the less the probability that they will be separated from one another during meiosis I. As the distance between the two genes increases, the probability increases that a recombinational event will occur between them and that they will be separated from one another during meiosis I. Recombination frequency, therefore, is a relative measure of the distance between two loci on a chromosome as shown in Figure 7-9.

> ***Genetic Map***
> *A representation of the relative position of genes on a chromosome based on the frequency of recombination between two DNA segments at meiosis.*

> *The probability that two gene sequences on a single chromosome will undergo recombination is dependent on the distance between the two sequences.*

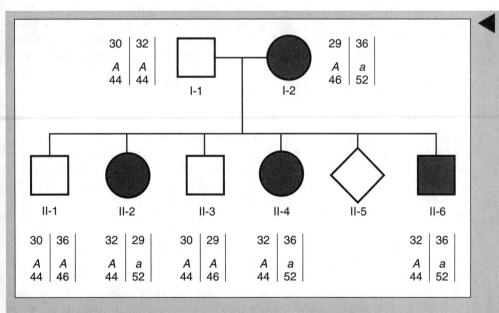

◀ ***FIGURE 7-9***
***Recombination Frequency Is a Reflection of the Distance Between Two Loci.*** *The mother (I-2) in this family is affected with a dominantly inherited disease, a, and is heterozygous at the disease locus (Aa). She is also heterozygous at two other loci, 29/36 and 46/52. The father (I-1) is unaffected and is homozygous normal (AA) at the disease gene locus and at one of the other loci (44/44). He is heterozygous at the second locus (32/30). Of the five children, three are affected with the disease, a, and have inherited the a allele and allele 52 of the marker locus from the mother. The two unaffected children received the normal allele A and the marker 46 from the mother. All five children are nonrecombinant for the disease gene locus and the marker locus, which indicates the close linkage of the two loci. However, two of the children, II-1 and II-2, are recombinant for the disease gene locus and the second genetic marker (29/36), which indicates that this genetic marker is probably not linked to the disease gene locus. Because the marker 52 is linked to the disease gene locus, a, if the unborn child (II-5) receives the 52 allele from the mother, there is a 95% chance it will also receive the disease allele, a, and be affected.*

In the figure, three positions are indicated on each of the two homologous chromosomes derived from the mother and from the father. The father is unaffected with the disease *a* and carries both normal alleles of gene *A*. The mother is affected with the disease and carries a normal allele, *A*, and a mutant allele, *a*. Two marker loci are indicated, one close to the disease gene, and one further away. The marker closest to the disease gene does not show any recombination in the mother because the children receive either *A*/46 or *a*/52. Because the mother is heterozygous, and thus informative at the marker locus, an evaluation of linkage of the marker to the disease gene can be made. The father is homozygous for the close marker and the disease gene and thus is noninformative. The marker located further away from the disease gene shows recombination in this small sample, indicating that crossing over has occurred between the marker and the disease gene. This high rate of recombination indicates that this marker is most likely not closely linked to the disease gene. Examination of the markers in the unborn child can be used to give a risk estimate of whether the child will be affected with the disease. If the unborn child receives the 52 marker from the mother, there is a 95% probability it will also receive the defective allele and have the disease.

The measurement of recombination frequency, known as genetic linkage analysis, is based on the use of genetic markers as shown in Figure 7-9. Genetic markers are sequences in the DNA that are polymorphic, that are heterozygous in a random population, and that allow the detection of recombinational events that occur between two chromosomes. In mapping disease genes by linkage analysis, the marker used is generally not within the disease gene itself, but it is a DNA sequence that is closely linked to the disease gene (e.g., disease gene, *a*, and marker 52 in Figure 7-9).

Several types of markers are used in genetic mapping. Originally, restriction fragment length polymorphisms (RFLPs) were used. These polymorphisms arise in the population when single base-pair mutations in the DNA alter the cleavage site for a restriction endonuclease. When the cleavage site is altered, cutting the DNA with a restriction endonuclease generates DNA fragments of different lengths (i.e., RFLPs). The use of RFLPs in linkage analysis has been replaced largely by the use of variable number tandem repeats (VNTRs) and short tandem repeats (STRs). Tandem repeats are stretches of base sequences composed entirely of repeating di-, tri-, or tetranucleotides, which tend to be quite polymorphic in human DNA, in which the precise number of repeats at any locus varies greatly. Currently, CA repeats are used in linkage studies in combination with PCR analysis, in which PCR primers that flank the tandem repeat sequences can be used to detect and analyze each repeat sequence (see Chapter 6).

## Linkage Analysis

Construction of genetic maps by linkage analysis began with the identification by Bell and Haldane in 1930 of the linkage of color blindness and hemophilia on the X chromosome. Linkage of genes on the X chromosome is often determined by linking a specific disease to X-linked inheritance and following the pattern of inheritance of the disease in a single family. Linkage of genes on autosomes, however, is much more difficult and depends on the availability of multigeneration families with large numbers of children, living parents, and grandparents. Unfortunately, for most inherited human disease genes, such families are not available. Therefore, multiple smaller families must be studied, and the data must be analyzed statistically.

These analyses are based on LOD score (*l*ogarithm of the *od*ds ratio for linkage) calculations. A determination of a LOD score is a statistical expression of confidence that a specific theta ($\theta$) value between two markers has been identified. Theta is defined as the distance between two regions of DNA expressed in units of recombination fraction. A 50% recombination between two regions is equivalent to a theta value of 0.5 and is equal to no linkage. A 1% recombination between two loci is a theta value of 0.01 and is equivalent to 1 cM or approximately 1 million bp present between the two loci. Therefore, markers separated by more than 50 million bp ($\theta = 0.5$) show no linkage.

An illustration of how LOD scores are determined is shown by examining 10 meiotic events in which no recombination between two markers has occurred, giving a theta value of 0.00 (0% recombination). Statistically, the confidence of this theta value is similar to the likelihood of flipping a coin 10 times and always getting tails; that is, $2^{10}$ (1:1024, or 1

---

**Genetic Linkage Analysis**
*This analysis is based on measuring the recombination frequency between two genetic markers.*

**Genetic Markers Used for Linkage Analysis:**
*RFLPs*
*VNTRs*
*STRs*

**LOD Score**
*A statistical method for analyzing linkage between two genetic markers.*

in $10^3$, or a log of odds of 3, LOD = 3). Thus, a LOD score of 3 or greater is considered significant that two genetic markers are closely linked.

If a marker is found that is linked to a particular inherited human disease gene, it can be analyzed and followed through many families. How is linkage between a marker and a disease gene determined? If 10% recombination is found between two loci, *A* and *B*, in the families studied, that means 90% of the time there is no recombination between *A* and *B*. The recombination frequency, θ, is 10%, and the genes are roughly 10 cM apart. Are these two loci linked? To answer this question, one must calculate statistical support for the 10 cM distance. This calculation comes from the available data, the LOD score. As in the example above of flipping a coin 10 times, if one head and nine tails are found in 10 flips (i.e., a recombination of 10%), the statistical support for tails and flipping having an association is weaker than if tails are flipped 10 out of 10 times; the LOD falls below 3. We can still make 10 cM statistically significant; however, we must flip the coin many more times and always observe the 9:1 ratio to obtain a LOD score of 3 or greater. In a study of markers *A* and *B*, a large number of individuals, perhaps from many families, must be analyzed. If a LOD score of 3 or greater is found, the two markers *A* and *B* are in fact linked (see Chapter 10 for details on LOD scores). The ideal situation is to find no recombination between the two markers under study, so there is greater confidence that the two loci are linked.

## Setting Phase

To assess the recombination frequency between two loci, it is necessary to know phase information (i.e., is allele *A* on the same chromosome as allele 2 [in phase or in coupling], or are the two alleles on different chromosomes [not in phase and in repulsion]?). An illustration of knowing phase is diagrammed in Figure 7-10.

In this pedigree, phase is known because the grandparents are available for analysis. The maternal grandmother has the disease gene, *a*, and must have passed it to her daughter along with the marker 1 allele. The mother's *A*2 combination must have come from the father, who is homozygous at both alleles. Phase in the mother is then *a*1 on one chromosome and *A*2 on the other chromosome. Of the four children, two are affected with the disease, one is unaffected, and one's genotype is unknown. Both children who received the *a*1 haplotype are affected, whereas the child receiving the chromosome with

> **Setting Phase**
> *A determination of whether two alleles are on the same chromosome or on different chromosomes.*

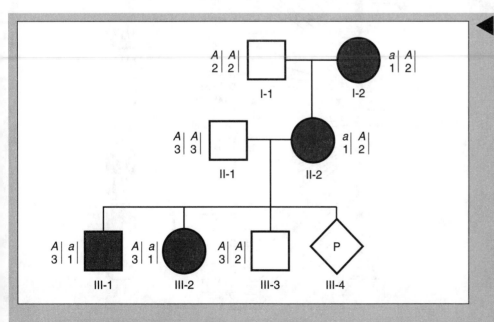

◀ **FIGURE 7-10**
**Knowing Phase in Linkage Analysis.** *An accurate determination of linkage in a pedigree depends on knowing the phase of two genetic markers; that is, whether the two markers are on the same or different chromosomes. In this pedigree, phase can be determined because the grandparents are available for analysis. The grandmother, I-2, is affected with a dominantly inherited disease, a, and is heterozygous at the disease gene loci (Aa) and at the marker locus (1/2). The grandmother passed the disease gene, a, and the marker locus to the daughter; thus, the daughter must be a1 on one chromosome and A2 on the chromosome received from the father. Therefore, we have set phase in the mother (II-2). We can now analyze the children, and see that they all are non-recombinants, having received either the chromosome containing a1 and being affected or receiving the chromosome containing A2 and being unaffected. Knowing phase also allows a determination of whether the unborn child will be affected or unaffected, depending on whether the child receives marker 1 or 2.*

*A2* is normal. No recombination events between the marker and the disease gene are seen in this small sample consistent with linkage of the two loci.

When phase is known, the probability that an unborn child will have the disease can be predicted. If the child receives marker 1, the chances are the child will also receive the disease gene and be affected. However, if the child receives marker 2, the chances are the child will receive the *A* allele and not be affected, unless a recombinational event occurs between the two chromosomes. If phase could not be set in this pedigree, which would be the case if neither grandparent were available for analysis, it would be impossible to determine if any of the children were recombinants. In that case, the linkage between the marker and the disease gene could not be determined. Thus, the unborn child could not be diagnosed.

Knowing phase in a pedigree, therefore, is important in determining linkage of two markers and underlies the importance of obtaining as much data about a family pedigree as possible. For X-linked human disease genes, phase can be set by the haplotype of an affected son. Because the single X chromosome in a son comes from the mother, phase can be automatically set in the mother by determining the phase in the affected son (see Chapter 6).

***Linkage Map of the Human Genome.*** A comprehensive human linkage map consisting of 5840 loci, 970 of which have been ordered, recently has been constructed [4]. This linkage map has markers every 0.7 cM and forms the basis of a high-density genetic map of the human genome. A typical linkage map for chromosome 14 is shown in Figure 7-11.

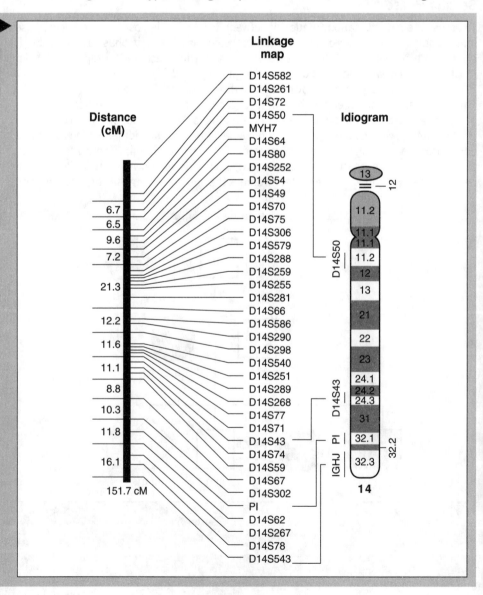

**FIGURE 7-11 ▶**

***Linkage Map of Human Chromosome 14.*** *The figure represents a linkage map of chromosome 14 as of July 1994. Genetic distances are indicated in centimorgans (cM). The linkage markers are short tandem repeat polymorphisms, and the idiogram shows the standard G-banding pattern. (Source: Reprinted with permission from Murray JC, Buetow KH, Weber JL, et al: Human genetic map—genome map V. Science 265:2055, 1994.)*

The map is created with the use of reference markers, STRs, and other polymorphisms. Markers (e.g., D14S74) are noted as D for DNA, 14 for chromosome 14, and S74 for site of marker 74. Recombination rates generally are higher in females as compared to males; therefore, the distances in cM between markers on a human chromosome are greater for females than males.

The exact correlation between the physical map and the genetic map for the human genome is not yet complete. However, as more information becomes available, and the amount of data on linkage and physical mapping increases, it becomes more certain that most inherited human disease genes eventually will be mapped, identified, and isolated for further study.

# MODEL SYSTEMS TO IDENTIFY HUMAN DISEASE GENES

One of the major goals of medical molecular genetics is the identification and cloning of all inherited human disease genes. The attainment of this goal began with the advent of recombinant DNA technology and the utilization of a strategy called *functional cloning*, which relies on knowing the biochemical structure of the protein involved and using the protein as a means of isolating the corresponding gene. Moreover, previous identification of the responsible protein meant that the underlying cause of the disease was already known, which generally is not the case.

Fortunately, as more genes were localized to specific human chromosomes by somatic cell hybridization, linkage analysis, and the visualization of chromosomal abnormalities in the light microscope, a technique known as *positional cloning* began to be used for identifying human disease genes. In positional cloning, a gene can be identified and cloned simply by knowing its map position within the genome. Neither the function nor properties of the protein product coded for by the gene need to be known. Within the past 10 years, the use of positional cloning approaches has enabled the identification and cloning of more than 50 inherited human disease genes with more genes being identified every day [5]. The combining of positional cloning with the identification of candidate genes—an approach called *positional candidate gene cloning*—is expected to result in an even greater increase in the number of human disease genes cloned during the next few years.

## Functional Cloning

In some inherited disorders, the basic protein defect in the disease was known at the biochemical level, and this subsequently enabled the identification of the corresponding gene. In these cases, the genes were isolated directly by recombinant DNA technology with no information about the map position of the gene. In functional cloning, the protein product of the gene must be known, purified, and usually sequenced. Knowledge about the amino acid sequence of the protein can be used to make in vitro synthetic oligonucleotides, which are used as probes to screen complementary DNA (cDNA) libraries (see Chapter 6). When a cDNA clone that is complementary to the oligonucleotide probe is isolated, it is used to screen genomic libraries in an attempt to isolate the complete gene sequence. Alternatively, antibodies directed against the protein can be used to look in an expression library for bacterial clones that produce the specific human protein. A good example of a gene isolated by functional cloning is the β-globin gene, which when mutated can give rise to sickle cell anemia or β-thalassemia (see Chapter 3).

*Functional Cloning*
*This cloning approach requires knowledge about the function and structure of the protein product of a gene but requires no prior information about the map position of a gene.*

## Positional Cloning

Positional cloning requires knowing only the chromosomal map position of a gene. No biochemical or functional information about the gene product is necessary. With positional cloning, the location of a gene on a chromosome is determined either by pedigree and linkage analysis, as described earlier in this chapter, or by the identification of chromosome abnormalities similar to those described in Chapter 5. A very important type of chromosome abnormality used in positional cloning is a *translocation*, which breaks

*Positional Cloning*
*This cloning approach requires knowing the map position of a gene, but no prior information about the function or structure of the gene product is necessary.*

the chromosome within a gene and disrupts the gene's function. If an inherited human disease is consistently associated with a common breakpoint in a chromosome, that point of breakage becomes an important clue to the position of the gene on the chromosome. Such a translocation was crucial in the identification and cloning of the gene for Duchenne's muscular dystrophy (DMD), as discussed later in this chapter.

*Microdeletions* that are found to be consistently associated with an inherited human disease are also important clues to the position of a gene. In the case of Prader-Willi syndrome (see Chapter 5), a microdeletion at 15q11–13 of the paternal chromosome is consistently associated with the disease. Using this information, the gene for the small nuclear RNA-associated polypeptide SmN was mapped to this position and shown to be associated with the disease.

If no chromosomal abnormalities are found associated with the disease of interest, linkage analysis data must be used to localize the gene to a specific map position. Linkage information can resolve the map position usually to only within 1–3 cM or 1–3 Mb of DNA. Thus, additional information is needed before the desired gene can be definitely identified and isolated. Experimental techniques used to identify the candidate gene in a large genomic fragment of 1–3 million bp are not simple and can be very time-consuming, as discussed later in describing the isolation of the DMD gene. Fortunately, more information is becoming available on the sequence and order of YAC clones containing DNA fragments from each human chromosome. As the sequence of these YAC clones becomes known, the ability to identify and isolate a candidate gene from a particular position within a chromosome will become much more feasible.

## Positional Candidate Gene Cloning

**Positional Candidate Gene Cloning**
*This cloning approach relies on determining the map position of a gene and then a molecular analysis of the region for a predicted candidate gene.*

Genes coding for many proteins have been localized within the genome, although many of these genes have no specific human disease associated with them. The localization of genes has increased rapidly over the past few years because of the advent of expressed sequence tags (ESTs). ESTs are small cDNA sequences generated from mRNA molecules that have been isolated from specific human tissues. Currently there are 450,000 human sequences in the available EST collection, and they may represent nearly 80% of the total human genes. With the use of ESTs, many or all genes can be mapped to a specific region on a human chromosome. When a family segregating a disease gene is studied by linkage analysis and a locus is identified, then it can be immediately determined if a previously mapped gene (i.e., EST) is located in the same region of the chromosome. In this way, the gene identified by linkage analysis can be mapped to the EST site. This technique has been successful in a number of cases and is predicted to become the major strategy in the future for the isolation of inherited human disease genes. The success of the positional candidate gene approach is directly related to the successful creation of a high-density EST map for the human genome.

Recently, several genes have been isolated by this approach, including the genes implicated in hereditary nonpolyposis colorectal cancer (HNPCC) and the gene implicated in Bloom's syndrome [6, 7]. Some examples of inherited disease genes isolated by positional cloning and the positional candidate gene approach are listed in Table 7-3.

## Genes Isolated by Positional Cloning

### THE GENE FOR DUCHENNE'S AND BECKER'S MUSCULAR DYSTROPHIES

One of the first examples of the use of positional cloning for identifying and cloning an inherited human disease gene was the isolation in 1987 of the DMD gene [8]. DMD is a common and devastating disease that before implementation of molecular diagnostics had an incidence of approximately 1 in 3500 live male births. The disease has an exceptionally high spontaneous mutation rate of $10^{-4}$ in both sperm and eggs. This high mutation rate, which is at least 10–100 times that found for most genetic diseases, implies that every male with a sperm production of $10^7$–$10^8$ per day will produce at least one sperm every day with a mutation in the DMD gene. Because the disease is inherited as an X-linked recessive trait, as a rule only males are affected with the disease, and females are carriers. There is no specificity to the disease; all populations as well as many animal species can be affected.

| Disease | Location | Aberrant Protein |
|---|---|---|
| Alzheimer 1 | 21q | β-Amyloid precursor protein |
| Alzheimer 2 | 14 | S182 membrane protein |
| Alzheimer 3 | 1q31–42 | STM2 membrane protein |
| Ataxia telangiectasia | 11q22–23 | Phosphatidylinositol 3-kinase |
| Bloom's syndrome | 15q26.1 | DNA helicase |
| Duchenne's muscular dystrophy | Xp21 | Dystrophin |
| Early onset breast cancer 1 | 17q21 | BRCA1 (nuclear protein) |
| Early onset breast cancer 2 | 13q12–13 | BRCA2 |
| Cystic fibrosis | 7q31.2 | CFTR |
| Familial melanoma | 9p21 | p16 |
| Fragile X syndrome | Xp27.3 | FMR1 |
| Hereditary nonpolyposis colorectal cancer (HNPCC) | 2p15–22 | hMSH2 (Mut S) |
|  | 3p21.3 | hMLH1 (Mut L) |
|  | 2q31–33 | hPMS1 (Mut L) |
|  | 7p22 | hPMS2 (Mut L) |
| Huntington's disease | 4p16.3 | Huntingtin |
| Li-Fraumeni syndrome | 17p13 | p53 |
| Marfan syndrome | 15q21.1 | Fibrillin |
| Myotonic dystrophy | 19q13 | Protein kinase |
| Neurofibromatosis type 1 | 17q11.2 | Neurofibromin |
| Neurofibromatosis type 2 | 22q12 | Schwannomin |
| Retinoblastoma | 13q14 | Rb protein |
| Werner's syndrome | 8p12 | DNA helicase |

*Note.* CFTR = cystic fibrosis transmembrane conductance regulator.

Boys who inherit the disease gene are asymptomatic at birth and appear to show few if any symptoms until the age of 4–5 years. By the age of 2 years, they may show some calf hypertrophy with noticeably large muscles, but this abnormality often goes undetected by the parents. Because of the high mutation rate, many cases of the disease are sporadic, often occurring in families with no history of the disease. By the time the child is in first grade, he begins to show muscle weakness. A blood test at this age shows that creatine kinase levels are extremely high (50–100 times the normal level), which indicates muscle destruction. By the age of 11 years, the boys are confined to wheelchairs, and by age 15 years, they have little muscle left. Patients generally die of respiratory failure by age 20 years if they are not ventilated.

Another form of muscular dystrophy, known as Becker's muscular dystrophy (BMD), has a similar phenotype but is much milder. Patients with BMD are still walking at 16 years and can live for many years. Prior to 1987, it was not possible to determine if the same gene or if different genes were involved in the two forms of muscular dystrophy. The answer to this question came only after the isolation and analysis of the DMD gene.

Before 1987, there was little or no information about the molecular basis for the muscle wasting in DMD. The muscle from DMD patients showed muscle fibers that appeared to die and then grow back, only to die again and finally be replaced by scar tissue. Nothing was known about the protein that was defective in DMD. This paucity of knowledge made the isolation of the gene by functional cloning highly improbable. The alternative approach was to use gene-mapping information and look directly for the gene without knowing anything about the properties of the gene or the gene product (i.e., positional cloning). If the gene could be isolated, then the function of its encoded protein could be determined. Some of the important steps used in the isolation and identification of a human disease gene (e.g., the DMD gene) by positional cloning, as well as steps used

**TABLE 7-4** ▶
*Identification of a Disease Gene by Positional Cloning*

| Goal | Steps |
|------|-------|
| Localize gene to a specific site | Linkage analysis using multiple families<br>Chromosomal abnormalities:<br>    Translocations<br>    Deletions<br>Fluorescence in situ hybridization (FISH) |
| Identify exon (expressing) sequences | Isolation of genomic DNA clones:<br>    Enlarge DNA size by chromosome walking<br>    Ensure that clone is correct (analyze patient<br>      DNA)<br>Analyze DNA clones for the presence of exons:<br>    Carry out a zoo blot<br>    Perform Northern blot analysis on mRNA<br>      samples<br>    Screen a cDNA library<br>    Find an open reading frame by sequencing the<br>      cDNA |
| Identify function of gene | Use the cDNA base pair sequence to determine an<br>    amino acid sequence of the protein<br>Explore data banks to find homology to a known<br>    protein<br>Use the cDNA in an expression vector to express<br>    the protein<br>Make antibodies to the expressed protein<br>Use the antibody to test normal and patient<br>    samples |
| Correlate the gene with the disease | Correlate mutational defects in the patient's genes<br>    with the disease phenotype and inheritance<br>    pattern<br>Create a "knockout" mouse and reproduce the<br>    disease phenotype<br>Correct the defect in vitro with a normal,<br>    functional gene |

*Isolation of the **DMD gene** by positional cloning was facilitated by the identification of a female who had a translocation at Xp21 that cut in the middle of one of her DMD genes.*

to show the association of that gene with a particular inherited human disease, are presented in Table 7-4 and described in more detail below.

***Localization of the DMD Gene to a Specific Site on the Genome.*** The isolation of the DMD gene was simplified by knowing that the disease was inherited as an X-linked recessive trait (Figure 7-12). Location of the DMD gene on the X chromosome immediately reduced the amount of DNA to be analyzed to 10% of the genome. However, this amount of DNA still represents approximately $3 \times 10^8$ bp. By comparison, a normal gene represents only $3 \times 10^4$ bp, or less than 0.01% of the total DNA of the X chromosome. Therefore, the next step was to localize the DMD gene within a smaller region of the X chromosome and reduce the amount of DNA that needed to be analyzed.

DMD is an X-linked recessive disorder; the mother is usually unaffected but often is a carrier of the disease with one defective gene and one normal gene. In sons, the only copy of the X chromosome must come from the mother, whereas daughters receive a copy of the father's only X chromosome and one of the mother's X chromosomes. Because the mother has two copies of the X chromosome, there is a 50% chance her son will receive the disease gene and express the disease and a 50% chance that her daughter will receive the disease gene and become a carrier. In the pedigree in Figure 7-12, the grandmother (I-2) must be a carrier because all three of her sons (II-1, II-3, II-5) have the disease. In this case, both daughters (II-2, II-4) are carriers who pass the defective gene on to 50% of their sons (III-1, III-4).

***Advantages of Having Chromosomal Abnormalities Associated with a Disease.*** A very important piece of information that led to the positional cloning of the DMD gene was the observation of a series of young girls who had a disease that looked like DMD clinically

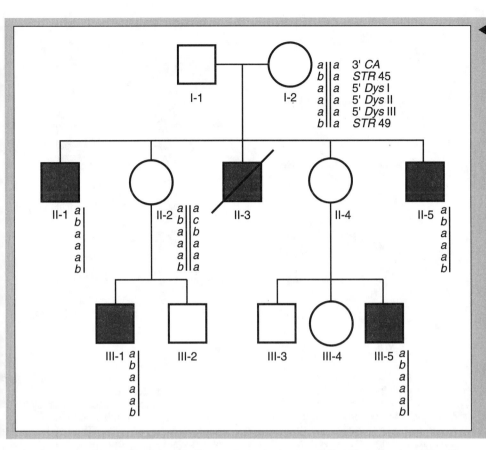

**FIGURE 7-12**
*Inheritance Pattern of X-linked Duchenne's Muscular Dystrophy (DMD).* DMD is inherited as an X-linked recessive disorder. Six genetic markers (CA repeats) are linked to the DMD gene and can be used to follow the disease in a family. All of the affected children are male and carry the (abaaab) haplotype, which came from the grandmother (I-2), who is a carrier. The daughter, II-2, also is a carrier with the abaaab haplotype, and she transmits the disease to 50% of her sons. The other daughter must also be a carrier because she transmits the disease to 50% of her sons. Note that every son who receives the abaaab haplotype is affected with DMD.

and who were shown by cytogenetic analysis to have an X–autosome translocation. Analysis of their karyotypes showed the presence of one normal X chromosome; however, part of their other X chromosome was translocated to an autosome (Figure 7-13). All of the girls showed the breakpoint in the X chromosome in the same place (i.e., Xp21); however, the autosome involved was different for each girl.

It was hypothesized that each of these girls with an X–autosome translocation had DMD, with the function of the gene responsible for DMD being destroyed by a break at Xp21 that cut within the gene. However, only one of the DMD genes should be affected by the translocation, leaving the other DMD gene intact. Thus, these girls should not show symptoms and should resemble nonsymptomatic carriers. Why then were these girls affected with the disease? The answer lies in the process known as X inactivation, which is discussed in Chapter 9. In a human female, only one of the X chromosomes is functionally active; the other X chromosome undergoes inactivation, and most of the genes on the inactive X chromosome are not transcribed. Normally, this process is random. In some cells, one X chromosome is inactivated, and in other cells, the other X chromosome is inactivated. A problem occurs when part of the X chromosome is translocated to an autosome. X inactivation shuts down one X chromosome as usual. However, if the X chromosome with the translocation is the one shut down, the autosome piece also shuts down, and the cell likely dies. A cell survives only if the normal X chromosome is preferentially inactivated. In that case, the cell has only one complete X chromosome active, both copies of the autosome are active, and the cell is viable. If the normal X chromosome is preferentially inactivated, the normal DMD gene located on the normal X chromosome also is inactivated. Therefore, if the breakpoint at Xp21 occurs in the middle of the DMD gene, the female will have both DMD genes inactivated (one by X inactivation and the other by a disruption occurring in the middle of the DMD gene), and she will be affected by the disease.

***Isolation of the Genomic DNA Clone Containing the DMD Gene Sequence.*** The eventual isolation of the DMD gene sequence was accomplished in the laboratory of Louis Kunkel with the use of a technique called *subtraction cloning* [8]. Dr. Kunkel used DNA from a young male patient with the initials BB, who was affected not only with DMD but with several other diseases, including retinitis pigmentosa and chronic granulomatous

## FIGURE 7-13

**Xp21/12 Translocation in a Female with Duchenne's Muscular Dystrophy (DMD).** *In this female who is affected with DMD, one DMD gene is inactivated by a break at Xp21 followed by a translocation of that part of the X chromosome to the autosome chromosome 12 to give the translocation 12/X. The second DMD gene, which is located on the normal X chromosome, is inactivated when the viable cells preferentially inactivate the normal X chromosome through X inactivation. Only cells with one copy of the genetic material from active chromosome X and both copies of the genetic material from active chromosome 12 survive.*

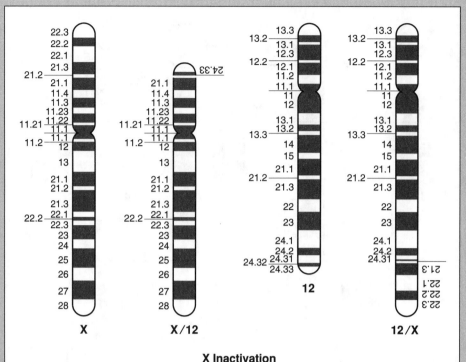

**X Inactivation**

**Inactive X/12**
The cell is *not viable*. The cell has too many X chromosomes. One complete X chromosome is active, and part of the other X chromosome, which is on 12/X, is also active. The cell does not have two copies of chromosome 12. Part of one chromosome 12 is on X/12 and is inactivated.

**Inactive normal X**
The cell is *viable*. The cell has only one active X chromosome: part on 12/X and part on X/12. The cell has two normal active chromosomes 12: one normal and the other active on X/12 and 12/X.

**Viable cells have an inactivated normal X chromosome.**

**Subtraction Cloning**
*This is a technique that looks for a piece of DNA that is present in the genome of a normal individual but is missing in a patient's deleted DNA.*

disease. Analysis of the boy's DNA by cytogenetic chromosome banding revealed a microdeletion of approximately 10 million bp localized to Xp21. These data suggest that the deletion at Xp21 resulted in the loss of a large piece of DNA that contained all of the missing genes leading to his disease phenotype, including the gene for DMD. Dr. Kunkel and his colleagues surmised that if the DNA missing from the patient's deleted X chromosome could be isolated from the DNA of a normal individual, one would have the 10 million bp that included the sequence for the DMD gene (subtraction cloning). Initially, fragmented DNA isolated from the patient, BB, was denatured and then reannealed in the presence of DNA fragments, which came from the restriction endonuclease digestion of DNA from an individual with a normal X chromosome. The patient's DNA is used in excess so that all of the normal DNA fragments preferentially hybridize to the patient's DNA. DNA fragments that do not find a hybridization partner (e.g., DNA present in the normal DNA and not in the patient DNA) are then cloned into vectors. In this way, a DNA library containing the DNA sequences on the X chromosome that correspond to the deleted area of X chromosome in the patient is constructed.

The DNA fragments in the library are then used as probes in Southern blotting with DNA isolated from normal individuals and the patient with DMD to look for those that hybridize to normal DNA but not to DNA that came from patient BB. Any such fragment that can hybridize to normal DNA but not to DNA from patient BB represents a DNA sequence that comes from the 10 million-bp–deleted region. Thus, the fragment represents a potential candidate gene sequence for the DMD gene. Of the clones isolated and tested, one clone of 180 bp was found to be missing, not only in patient BB but also in 10% of male patients affected with only DMD. The DNA fragment in this clone was then used in chromosome walking to isolate a larger DNA fragment, which resulted in the isolation of a genomic DNA fragment of 140,000 bp believed to contain sequences from the DMD gene.

DNA isolated from patients with different deletions covering the DMD gene were tested with the cloned DNA fragment to try to localize the position of the cloned fragment within the DMD gene. This analysis showed that the cloned fragment was located in the middle of a gene, which turned out to span 2.5 million bp of DNA and to represent one of the largest human genes known.

### Analyzing the Cloned DNA Sequences for the Presence of Exons.

Remaining to be demonstrated, however, was whether the cloned DNA fragment represented the coding sequence for the DMD gene. To use the cloned DNA fragment to investigate the function of the DMD gene product, the DNA fragment must contain exon sequences, which code for amino acids in the final gene product. Potential exon sequences can be identified in several ways.

One technique is to look for DNA sequences conserved through evolution. Exon sequences tend to be conserved from one species to another, whereas intron sequences generally are not. Conserved sequences are assayed by performing a *zoo blot*, in which DNA isolated from many species is tested by Southern blotting, with the isolated DNA fragment as a probe to look for complementarity in human and animal DNA samples. Finding hybridization with DNA from many species indicates that the isolated genomic DNA sequence has been conserved and probably contains exon sequences. The zoo blot was positive for the 140,000-bp DNA gene fragment isolated as described above.

*The presence of potential exon sequences in cloned genomic DNA fragments is determined by analyzing zoo blots, looking for complementarity to mRNA sequences, and isolating complementary sequences from a cDNA library.*

The second requirement of an exon sequence is that a complementary sequence should be found in messenger RNA (mRNA). Thus, a Northern blot was performed to look for a mRNA species present in muscle tissue that could hybridize to the cloned DNA fragment. In the case of the DMD clone, an RNA molecule of 14,000 bp was found to hybridize to the cloned DNA fragment.

If the isolated genomic fragment contains exon sequences, it can be used as a probe to detect and clone complementary DNA (cDNA) sequences (sequences complementary to mRNA) from a cDNA library constructed from human fetal muscle mRNA. The isolation of cDNA fragments that can hybridize to the genomic DNA indicates that the genomic DNA contains exon sequences. These experiments with the putative DMD gene sequence resulted in the isolation of fragments of cDNA complementary to the entire 14,000-bp RNA. When the cDNA was sequenced, the sequence was analyzed for open reading frames (ORF), which are large DNA base sequences without a stop codon that generally represent protein coding sequences. A single ORF of 11,000 bp, which would code for a protein of 3600 amino acids, was detected.

### Identification of the Function of the Gene.

This information allowed investigators to look at this potential gene product of 3600 amino acids and to determine its function in muscle tissue. With the amino acid sequence and nucleotide sequence already known for many proteins and genes (and stored in data banks), information about an unknown protein can be obtained by looking for sequence homology to other known proteins. When this was done with the putative DMD protein sequence, homology was found to two cytoskeletal proteins: α-actinin and spectrin.

*Information about the putative gene product of an isolated cDNA sequence can be obtained by looking for homology to known proteins and by expressing the protein in vitro and making antibodies that are used to compare normal and diseased tissue samples.*

A further analysis of the protein function was carried out by placing the isolated cDNA sequence in an expression vector, transforming the vector into bacteria and allowing the bacterial cells to synthesize the protein. The protein was isolated and used to produce antibodies in rabbits. Fluorescent-tagged antibodies specific for the putative DMD gene product were used to look for the presence of the protein in muscle tissue. These experiments revealed that the DMD protein, now known as dystrophin, was present in muscle biopsies of normal individuals but absent in patients with DMD.

### The Defective Gene in DMD and BMD.

Further studies revealed that the level of dystrophin also was reduced in patients with the milder form of muscular dystrophy, BMD. Thus, it was clear for the first time that both DMD and BMD were the result of a defect in the same gene, and differences in the severity of the disease were apparently related to the amount of dystrophin present in the muscles of the patient.

Additional experiments using muscle samples from many patients with either DMD or BMD verified that the 140,000-bp genomic DNA isolated represented sequences from the gene that coded for a protein (i.e., dystrophin) and that portions of the gene were deleted in approximately 60% of patients tested. The entire DMD gene is now known to have 79 exons that span more than 2.5 million bp of DNA and is nearly 100 times the size of most other human genes. The size of this gene explained the very high mutation rate associated with the disease, because the frequency of mutation is usually related to the size of the gene.

One puzzle remained concerning the differences in the severity of DMD versus BMD. Both types of muscular dystrophy are associated with deletions of the gene. Surprisingly, the mild BMD patients often have larger sections of the gene missing when compared to the more severely affected DMD patients. Analysis of muscle tissue using antibodies to dystrophin indicated that BMD patients have dystrophin present, but it is abnormal in molecular weight. DMD patients, on the other hand, have no dystrophin protein in their muscle.

The answer to the puzzle involves knowing the relationships between exons and introns, splicing of RNA molecules, and the translation of mRNA in the proper reading frame (see Chapter 3). The dystrophin gene is made up of 79 exons, all of which must be properly spliced to produce a normal mRNA and a normal protein. When a deletion occurs in the gene, the RNA molecule transcribed from the DNA is missing the base pairs corresponding to the deletion. During the process of splicing, introns are removed, and the remaining exons are spliced together. In the normal splicing process, the exons are joined together so that the reading frame of the mRNA is conserved. If an exon or exons are deleted, most deletions appear to be tolerated in BMD patients, providing that the remaining exons are spliced in frame. Correct splicing of the remaining exons occurs in BMD patients and allows them to produce a modified, partially functioning dystrophin molecule. In DMD patients, however, the deletion occurs in a manner that results in the remaining exons being spliced out of frame. This aberrant splicing produces a mRNA that is translated out of frame, resulting in the encounter of a termination codon in the middle of the mRNA and premature termination of protein synthesis. When this occurs, the result is a complete absence of the dystrophin protein. Thus, although both forms of the disease result from deletions occurring in the same gene, the type of deletion rather than the extent of the deletion determines the severity of the disease [9].

With the DMD gene now identified and cloned, many advances have been made in the prenatal diagnosis of the disease. Prenatal diagnosis is especially critical for DMD with its high mutation rate and its lack of expression until the affected boys are 4 or 5 years old. Short terminal repeat (CA) sequences are particularly useful in prenatal diagnosis of DMD (see the Clinical Case in Chapter 6). Although there are now serious attempts being made to develop gene therapy protocols for this disease, the unusually large size of the gene has resulted in significant problems in any attempts to find a suitable vector or a successful technique by which the gene can be delivered to muscle tissue.

## Genes Isolated by Positional Candidate Gene Cloning

### THE GENE FOR HEREDITARY NONPOLYPOSIS COLORECTAL CANCER

The positional candidate gene approach for identifying and isolating inherited human disease genes is illustrated by the isolation of the gene responsible for one of the most common cancer disposition syndromes, HNPCC. HNPCC affects 1 in 200 individuals in the Western world, with those affected developing tumors of several organs, including the colon, the ovary, and the endometrium, before they reach the age of 50 years.

*Localization of the HNPCC Gene to a Specific Site on the Genome.* HNPCC is inherited as an autosomal dominant disease and is diagnosed when three or more family members in at least two successive generations present with colorectal cancer, with at least one affected member being younger than the age of 50 years. In 1993, linkage analysis performed on two large families identified a microsatellite (short repetitive DNA sequence) DNA marker on chromosome 2 that was linked to HNPCC [10].

The autosomal dominant inheritance pattern of HNPCC initially suggested that the candidate gene causing the disease was probably a tumor suppressor gene. Other tumor suppressor genes, such as *p53* or *Rb*, show autosomal dominant inheritance and have been linked to some inherited forms of cancer. However, when the suspected region on chromosome 2 was analyzed at the molecular level, no apparent deletions that might be associated with tumor suppressor genes were found. Instead, an unusual DNA sequence

---

*DMD and BMD are both a result of deletions in the same gene, a gene that codes for the protein dystrophin. The type of deletion rather than the extent of the deletion determines the severity of the disease.*

*The gene responsible for HNPCC was isolated by the positional candidate gene approach. The gene product, hMSH2, was shown to be homologous to a yeast and bacterial gene, mut S, which is necessary for methyl-directed mismatch repair.*

abnormality was detected. The short repeat DNA sequences at the marker position on chromosome 2 varied in length from one tumor to another. This microsatellite instability was not unique to the marker on chromosome 2 but was found in the DNA of many tumor samples, which indicated that this instability was present throughout the genome and was perhaps a hallmark of HNPCC. These data led to the hypothesis that the gene located on chromosome 2 was probably not a tumor suppressor gene but was somehow involved in maintaining the accuracy of DNA replication. Loss of the gene function would result in a high rate of DNA replication errors and lead to an increase in mutation rate or to a mutator phenotype.

***Identifying Potential Candidate Genes.*** Years earlier, similar types of mutations had been described in bacteria and yeast and had been called *mutator strains* or *mut* strains. These *mut* strains were shown to be defective in a DNA repair system known as the methyl-directed mismatch repair system. The function of this repair system is to remove incorrectly paired bases from newly replicated DNA to prevent mutations from arising (see Chapter 4). Based on the linkage analysis data, the similarity of the phenotypes of *mut* strains in yeast and the microsatellite instability found associated with HNPCC, Dr. Bert Vogelstein and his colleagues began a search for the gene on chromosome 2 that was responsible for HNPCC [6].

***Isolation of the Genomic DNA Clone Containing the HNPCC Gene Sequence.*** In linkage studies described by Peltomaki and colleagues in 1993, the analysis of 345 microsatellite markers showed a single marker located at position 2p15–16 that was linked to HNPCC [10]. Vogelstein and his colleagues set out to isolate that fragment of chromosome 2. They began by forming somatic cell hybrids containing chromosome 2 fragments using either microcell-mediated cell fusions or radiation-reduced hybridization, as described earlier in this chapter. Hybrids that contained a piece of chromosome 2 with the previously mapped markers were then used to narrow the relevant portion of chromosome 2 that contained the gene they were seeking. Additional polymorphic markers were obtained and used to isolate genomic clones by PCR screening of YAC libraries. A further detailed study of HNPCC families finally led to the identification of a 0.8 Mb sequence from chromosome 2 that contained the HNPCC locus [6, 11].

Various candidate genes, including an SOS gene from yeast, a tumor suppressor gene, and the *mut S* gene from yeast, were then tested to see if they were present within the 0.8 Mb fragment. Neither the SOS gene nor the tumor suppressor gene mapped to the correct position. A human homologue of the *mut S* gene was isolated by PCR amplification of cDNA from colon cancer cell lines and named *hMSH2*. This human cDNA gave a positive Southern blot when tested with DNA isolated from somatic cell hybrids that contained the fragment of chromosome 2p believed to have the HNPCC locus. The cDNA from this positive clone was used as a probe to isolate additional cDNA fragments from cDNA libraries generated from human colon cancer cells or human fetal brain tissue. A final composite of the cDNA fragments revealed a single ORF of 2802 bp that could code for a protein of 934 amino acids. When this DNA sequence was further analyzed, it showed homology to the yeast *hMSH2* or *mut S* gene, indicating that HNPCC potentially resulted from a defect in mismatch repair.

***Identifying the Disease Gene for HNPCC.*** To prove that the *hMSH2* gene was responsible for the disease, additional studies were performed to show that mutations within this gene were associated with the disease and occurred in the germline of patients with HNPCC. Analysis of DNA from various families with HNPCC showed that there were alterations in highly conserved regions of the DNA that would be expected to alter significantly the protein product coded for by the gene. All of the data then strongly support the hypothesis that mutations in *hMSH2* result in HNPCC.

Defects in *hMSH2* may account for as many as 60% of all cases of HNPCC. At the molecular level, the defect in the human homologue of *mut S, hMSH2*, interferes with mismatch repair and results in a high mutation rate in cells. Knowledge about the other proteins involved in mismatch repair in yeast and bacteria strongly suggested that there may be additional human genes associated with HNPCC, which would account for the remaining 40% of cases. This assumption was correct, and further studies have identified at least three more genes present in humans that are required for normal mismatch repair (see Table 7-3).

In summary, the positional candidate gene approach involves the initial mapping of a disease gene to a chromosomal region using linkage analysis and then an analysis of that region for the presence of a potential candidate gene based on the phenotype or the molecular description of the genetic defect. As more markers are placed on the human gene map and more cDNA sequences for genes become available, the number of genes isolated by this approach will undoubtedly increase. Eventually, the physical map and the genetic map of the human genome will be correlated, and the complete map of all of the human chromosomes will be delineated. As we have seen in this chapter, numerous methods are available for carrying out physical and genetic mapping, many of which are often combined finally to isolate and identify an inherited human disease gene. In Table 7-5, a list of the important terms and methods used in both physical and genetic mapping is presented for review.

**TABLE 7-5** ▶

*Important Terms Used in Mapping Human Genes*

| Term | Meaning |
| --- | --- |
| BAC | Bacterial artificial chromosome |
| cM | centimorgan: 1% probability of recombination = 1 Mb |
| Contig | DNA fragments mapped in overlapping order |
| Cosmid | Lambda vector with 30–50 kb of insert DNA |
| EST | Expressed sequence tag (a sequence of a cDNA clone that can be assayed by polymerase chain reaction [PCR]) |
| FISH | Fluorescence in situ hybridization |
| HAT | Hypoxanthine-aminopterin-thymidine selection medium |
| LOD | Logarithm of the odds: measure of the probability of linkage between two loci at a specified rate of recombination |
| ORF | Open reading frame: codes for amino acid information |
| RFLP | Restriction fragment length polymorphism |
| STR | Short tandem repeats |
| STS | Sequence-tagged sites (any DNA with known sequence that can be assayed by PCR) |
| SSCP | Single stranded conformation polymorphism |
| Theta ($\theta$) | Distance between two loci expressed in recombination units; 50% recombination is equal to a theta value of 0.5 |
| VNTR | Variable number tandem repeats |
| YAC | Yeast artificial chromosome with 1 Mb of insert DNA |
| Zoo blot | Southern blot using DNA of different species |

# RESOLUTION OF CLINICAL CASE

The test results of the patient presented at the beginning of the chapter indicated that Bruce was in the first stages of acute leukemia. Additionally, his karyotype was abnormal, showing structurally rearranged chromosomes that contained multiple gaps and breaks in the DNA. The physician suspected that Bruce was suffering from a rare autosomal recessive disease and notified the parents immediately. The physician then called a clinical geneticist, Dr. Smith, for a consultation.

When the physician described the patient's symptoms and the abnormal karyotype, Dr. Smith suggested that the boy might be suffering from Bloom's syndrome, which is characterized by proportional dwarfing, erythremia, chronic infections, and craniofacial abnormalities. Furthermore, the presence of leukemia at such an early age was characteristic of the syndrome, which was associated with early onset of neoplasia. Dr. Smith asked if she could be present when the family arrived for their appointment.

Dr. Smith informed the parents of the potential diagnosis of Bloom's syndrome; she explained that the syndrome resulted in many breaks and gaps occurring in the patient's DNA, as well as an increase in recombination frequency, all of which lead to an increased mutation rate in the cells. With the very high mutation rate, the cell's tumor suppressor genes are inactivated at a higher-than-normal rate. When the function of these tumor suppressor genes is lost, the cells become malignant. Therefore, the presence of Bloom's syndrome would explain Bruce's development of leukemia at an early age. In addition, his risk for other cancers, such as lymphomas and carcinomas, is greatly increased. To complicate the matter further, chemotherapeutic agents that would normally be prescribed are often not tolerated by patients with Bloom's syndrome because of the damaging effect these agents can have on DNA. Dr. Smith asked the parents if they would permit her to take a small skin biopsy from Bruce so that she could do further analysis of his cells and verify the diagnosis.

The parents, of course, were very upset with the diagnosis, but they agreed to the biopsy if the skin sample would provide more information about Bruce's disease. Dr. Smith informed the parents that she would not have the results from the sample for at least 6–8 weeks.

Six weeks later, Dr. Smith called the physician and told him that she had cultured Bruce's cells, and they demonstrated the classic hallmark of Bloom's syndrome, a 10–15-fold increase in sister chromatid exchange. Furthermore, she had carried out a collaborative experiment with a colleague using a recently isolated cDNA probe to verify that Bruce had a mutation in the gene for Bloom's syndrome, which has recently been isolated and identified using positional cloning. Although there had been many candidate genes, none of the predicted ones turned out to be the gene for Bloom's syndrome. The gene now has been identified as coding for a DNA helicase, and it is believed that its loss interferes with DNA replication, DNA repair, and normal DNA recombination.

Dr. Smith asked the physician if the parents were of Ashkenazi Jewish ancestry. The physician said he knew they were Jewish but did not know if they were of Ashkenazi descent. Dr. Smith explained that in a recent study, a specific mutation in the gene for Bloom's syndrome had been associated with at least four Ashkenazi Jewish families. Bruce had the same mutation in his gene. The mutation is a result of a 6-bp deletion and a 7-bp insertion at position 2281 bp in the gene. The result of the mutation is the creation of a frameshift mutation and a premature stop codon in the middle of the gene. A person who originally had this mutation was probably a founder of the Ashkenazi Jewish population, and all Ashkenazi Jews who inherit Bloom's syndrome inherit this mutation from a common ancestor, a phenomenon known as the *founder effect*.

Dr. Smith and the physician met with the family and told them of the findings. The family was in fact of Ashkenazi Jewish descent, although they did not know of any other family members or friends with the disease. The physician explained that currently there is no cure for the disease. He would initially give Bruce low doses of chemotherapeutic agents to see if they would be tolerated, and he would follow Bruce very closely in the event he developed any other forms of cancer. Dr. Smith further counseled the parents that they were both likely carriers of the disease, and the risk of their future children inheriting the disease was 25%.

# REVIEW QUESTIONS

**Directions:** For each of the following questions, choose the **one best** answer.

1. Which one of the following strategies was used in the identification of the gene for Duchenne's muscular dystrophy (DMD)?
   - **(A)** Functional cloning
   - **(B)** Positional candidate gene cloning
   - **(C)** Identification of point mutations by restriction enzyme digestion
   - **(D)** Identification of a female DMD patient with a preferentially active normal X chromosome
   - **(E)** Identification of a male patient with multiple X-linked disorders

2. Which one of the following strategies was used in the identification of the gene for hereditary nonpolyposis colorectal cancer (HNPCC)?
   - **(A)** Functional cloning
   - **(B)** Positional candidate gene cloning
   - **(C)** Identification of a deletion mutation on chromosome 2 consistently associated with HNPCC
   - **(D)** Identification of a female HNPCC patient with a preferentially inactive normal X chromosome
   - **(E)** Identification of a male HNPCC patient with an X translocation

**Directions:** The groups of questions below consist of lettered choices followed by several numbered items. For each numbered item, select the appropriate lettered option with which it is most closely associated. Each lettered option may be used once, more than once, or not at all.

### Questions 3–5

Match each description with the appropriate term.
   - **(A)** Allele
   - **(B)** Locus
   - **(C)** Phase
   - **(D)** LOD
   - **(E)** Theta

3. Alignment of two genetic markers on the same chromosome

4. A measure of the probability of linkage between two loci

5. An expression of distance between two loci expressed in recombination units

### Questions 6–8

Match each description with the appropriate term.
   - **(A)** Sequence-tagged sites (STSs)
   - **(B)** Yeast artificial chromosome (YAC)
   - **(C)** Short tandem repeat (STR)
   - **(D)** Open reading frame (ORF)
   - **(E)** Hypoxanthine-aminopterin-thymidine (HAT)

6. A selection medium used to isolate human–mouse somatic cell hybrids

7. A DNA base sequence that codes for amino acid information

8. A cloning vector that can carry a human DNA insert up to 1 Mb in size

# ANSWERS AND EXPLANATIONS

**1. The answer is E.** DNA from patient BB, who had a deletion of the X chromosome that deleted several genes, including the DMD gene, was used in substraction cloning to isolate the DNA fragment carrying the DMD gene from a normal individual. Functional cloning cannot be used if there is no information about the function and properties of the gene product. In this case, there was no candidate gene for DMD. The mutations in the DMD gene are usually caused by large deletions, not point mutations. The female DMD patient has a normal X chromosome that is preferentially inactive.

**2. The answer is B.** The gene for hereditary nonpolyposis colorectal cancer (HNPCC) was identified first by mapping the gene to chromosome 2 by linkage analysis and then looking at the region for a gene involved in DNA repair. The function and properties of the HNPCC gene product were not known, so functional cloning could not be used. No known deletion mutations on chromosome 2 were consistently associated with HNPCC. Options D and E are incorrect because the HNPCC gene is not on chromosome X.

**3–5. The answers are: 3-C, 4-D, 5-E.** Setting phase is the determination of whether two genetic markers are on the same or different chromosomes. A LOD (log of the odds) score of 3 or greater is an indication that two genetic loci are closely linked. A theta ($\theta$) value of 0.5 is equivalent to 50% recombination between two loci. A theta value of 0.01, or 1% recombination, is approximately 1 million bp and equal to 1 centimorgan.

**6–8. The answers are: 6-E, 7-D, 8-B.** Hypoxanthine-aminopterin-thymidine (HAT) is a selection medium used to isolate human–mouse somatic cell hybrids. An open reading frame (ORF) is a long base sequence in DNA that codes for amino acids and has no termination codons. A yeast artificial chromosome (YAC) is a cloning vector that contains large fragments of DNA (100–1000 kb) linked to sequences necessary for the vector to replicate in yeast cells.

# REFERENCES

1. Shows TB, Sakaguchi AY: Gene transfer and gene mapping in mammalian cells in culture. *In Vitro* 16:55–76, 1980.
2. Gray JW, Langlois RG: Chromosome classification and purification using flow cytometry and sorting. *Ann Rev Biophys Biophys Chem* 15:195–235, 1986.
3. Hudson TJ, Stein LD, Gerety SS, et al: An STS-based map of the human genome. *Science* 270:1945–1954, 1995.
4. Murray JC, Buetow KH, Weber JL, et al: A comprehensive human linkage map with centiMorgan density. *Science* 265:2049–2055, 1994.
5. Collins F: Positional cloning moves from perditional to traditional. *Nat Genet* 9:347–350, 1995.
6. Leach FS, Nicolaides NC, Papadopoulos N, et al: Mutations of a *mut S* homologue in HNPCC. *Cell* 75:1215–1225, 1993.
7. Ellis NA, Groden J, Ye T-Z, et al: The Bloom's syndrome gene product is homologous to rec Q helicase. *Cell* 83:655–666, 1995.
8. Koenig M, Hoffman EP, Bertelson CJ, et al: Complete cloning of DMD cDNA and preliminary genomic organization of the DMD gene in normal and affected individuals. *Cell* 50:509–517, 1987.
9. Hoffman EP, Kunkel LM: Dystrophin abnormalities in Duchenne/Becker muscular dystrophy. *Neuron* 2:1019–1029, 1989.
10. Peltomaki P, Aaltonen LA, Sistonen P, et al: Genetic mapping of a locus predisposing to human colorectal cancer. *Science* 260:810–812, 1993.
11. Fishel R, Lescoe MK, Rao MRS, et al: The human mutator gene homologue, *MSH2* and its association with HNPCC. *Cell* 75:1027–1038, 1993.

# 8

# REGULATION OF GENE EXPRESSION

## CHAPTER OUTLINE

## INTRODUCTION OF CLINICAL CASE

Marie and George brought their son Jerry to the pediatrician for a well-care visit. Jerry, who is 14 months of age, has been a normal, active child. Marie informed the pediatrician that recently she noticed that Jerry's left eye had begun to cross and that he often rubbed the eye. The pediatrician examined Jerry's eyes and found a white, pupillary reflex in the left eye but not in the right. Other than his eye problem, Jerry was in good health. The pediatrician recommended that Marie and George take Jerry to an ophthalmologist for a thorough examination. Marie became concerned and asked for more information. The pediatrician explained that she was unable to see a red reflex in Jerry's left eye, indicating that something may be blocking his retina; therefore, it is very important that the problem be diagnosed as soon as possible. Marie became very concerned and asked the pediatrician if it was possible that Jerry has a tumor in his eye. The pediatrician, seeing Marie's distress, tried to calm her down and told her that it was possible, but it was too soon to know for sure. More tests would be necessary to diagnose the problem accurately. She then asked Marie and George if they or anyone in their families had ever had an eye tumor. They both said they were unaware of anyone, but they would check further.

The next day, Marie called the pediatrician to tell her that Jerry had an appointment with the ophthalmologist in 2 days. She also told the pediatrician that no one they contacted in their families had ever had an eye tumor. Then, somewhat hesitantly, she asked the pediatrician why that would be important. The pediatrician explained to Marie that eye tumors, which are known as retinoblastomas, can run in families and can be inherited. However, because there is no incidence of retinoblastoma in the parents' families, either Jerry does not have retinoblastoma or, if he does, it is caused by a new mutation. If the eye examination indicated that Jerry has retinoblastoma, they would be able to determine the genetic basis of the disease with some straightforward molecular tests. Marie then hesitantly told the pediatrician that there was an additional problem that she did not initially want to reveal. George is not Jerry's biologic father, and Marie is not sure where his real father is. She asked the pediatrician to please keep that information confidential unless it became a factor in treating Jerry's condition. The pediatrician assured Marie that the information would be held in confidence.

# OVERVIEW OF CONTROL POINTS IN THE REGULATION OF GENE EXPRESSION

Every cell in a human body contains the same DNA sequence, a sequence that carries information for 60,000–100,000 different RNA or protein molecules. Yet, not every cell in the body expresses the same genes. Even within a single cell, different genes are expressed at different times in response to external stimuli, developmental controls, or cell cycle signals. Crucial to normal growth and development of all cells, as well as to the tissue or organism as a whole, is the proper programmed expression of every gene in every cell. All cells continually express the proteins necessary for providing energy and for synthesizing the components that are required for growth. In contrast, some proteins are expressed only in particular cells. For example, hemoglobin synthesis is generally confined to erythroid cells, and immunoglobulin synthesis occurs predominantly in B lymphocytes.

Cells must have the capacity to regulate the turning on and turning off of certain genes at the right time. For example, some proteins are required only at a certain phase of the cell cycle; their expression at other times would be detrimental. Progression through the cell cycle and the expression of genes in the various phases of the cell cycle are tightly regulated. Any deviation from normal control leads either to uncontrolled cell proliferation and malignancy or to programmed cell death. Additionally, cells must be able to regulate the expression of specific genes in response to hormones or growth factors to carry out certain functions required in the life and development of an organism.

*Cells use elaborate and complex control mechanisms to control the expression of every gene in the human body.*

*The predominant step in all cells for regulating gene expression is at the level of transcription initiation.*

It is not surprising that cells have evolved elaborate and complex mechanisms by which they can specifically control the expression of each gene. Every step in the production of a protein product coded for by a specific DNA sequence is subject to control. The predominant step in all cells for the regulation of gene expression is at the level of *transcription initiation*, the process in which RNA polymerase binds to a specific site on the DNA and begins to transcribe an RNA molecule from a DNA template. Transcriptional control, which mainly occurs by controlling the access of RNA polymerase to the promoter site of a gene, is an amazingly complex process that involves many different proteins interacting with DNA and with each other in very specific ways.

Additional levels of control of gene expression include the following:

- Controlling the processing of messenger RNA (mRNA) by determining which exons present in the precursor mRNA are retained in the mature functional mRNA. This control mechanism can involve either alternative splicing of exons or differential polyadenylation of the precursor mRNA.
- Controlling the stability or the rate of degradation of the mature mRNA.
- Controlling the process of translation of the mRNA into protein.

# TRANSCRIPTIONAL CONTROL OF GENE EXPRESSION

This discussion of transcriptional control of gene expression concentrates on the regulation of RNA polymerase II, the enzyme responsible for transcribing protein-encoding genes and producing mRNA. The machinery necessary for the production of each mRNA molecule in a human cell involves more than 40 different proteins that interact in very complex ways at specific sites on the DNA. Transcriptional control involves the interaction of proteins (i.e., trans-acting factors) with specific sequences on the DNA (i.e., *cis*-acting elements). The interaction of these proteins with DNA and with each other serves to regulate transcription of the right genes at the right time to produce the proper amount of a protein.

## *Cis*-acting Elements

*Cis*-acting elements exist as short base-pair sequences that can be adjacent to or within a gene, or they can be distal sequences that occur thousands of base pairs away from the gene. These sequences are required for the recognition of a gene by RNA polymerase II, and they act as binding sites for the proteins that regulate the rate and the specificity of transcription. The major types of *cis*-acting elements of a gene are illustrated in Figure 8-1 and include the following:

> **Cis-*acting elements*** *are short base-pair sequences in DNA that act as binding sites for the proteins that regulate the rate and specificity of transcription initiation.*

- The *core promoter element* is located 5′ to the gene and comprises the site where the transcription complex containing RNA polymerase II assembles on the DNA. Two sequence elements with a more or less fixed location are present at the core promoter site: the *initiator element (Inr)*, which determines the transcription start site, and the *TATA element*, located 25–30 bp upstream of the Inr.
- The *promoter proximal elements* are composed of two types of *cis*-acting sequences that are found 50 to a few hundred base pairs upstream of the start site. The first type of promoter proximal element is a common class of base sequences that occurs in many genes. These sequences (e.g., CAAT or GC) act as binding sites for proteins, called *upstream transcription factors*. The second type of promoter proximal element is the *response element (RE)*. These sequences occur in promoters under common control (e.g., the genes that respond to glucocorticoid stimulation).
- The *promoter distal elements* are *cis*-acting sequences that can occur thousands of base pairs away from the transcription start site. These distal sites, known as enhancers or silencers, exert their action from large distances and may be localized either upstream or downstream of the gene they regulate.

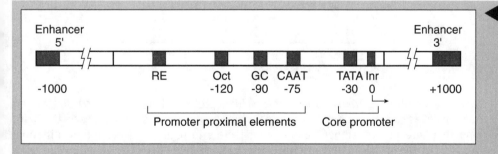

**FIGURE 8-1**
***Localization of* Cis-*acting Sequences in a Typical Human Gene.*** *The core promoter is where the TATA and initiator (Inr) sequences are located. The TATA sequence, located 30 bp upstream of the Inr sequence, is the binding site for the TATA-binding protein (TBP), and the Inr sequence is where RNA polymerase II binds and starts transcription. Promoter proximal elements are located 50 to several hundred base pairs upstream of the Inr site and include the common sequences CAAT, GC, and Oct. These sequences are the binding sites for upstream transcription factors. The second base sequences in the promoter proximal regions are the response elements (RE), which are the binding sites for inducible transcription factors. Located thousands of base pairs away, either 5′ or 3′ to the gene, are enhancer elements that bind activators.*

## Trans-acting Transcription Factors

Trans-acting transcription factors are proteins that bind to *cis*-acting elements on the DNA and interact with other transcription factors. These proteins function to control the initiation of transcription and can be classified into the following five groups:

> **Trans-*acting transcription factors** are proteins that bind to cis-acting sequences on the DNA and then interact with other transcription factors to regulate gene expression.*

- *General transcription factors (GTFs)* are a set of 30 or more polypeptides that assemble at the core promoter site and recruit RNA polymerase II to that site to form a preinitiation complex.
- *Common transcription factors*, sometimes called upstream transcription factors, are proteins that bind the common *cis*-acting sequences proximal to many promoters, such as the sequences CAAT and GC.
- *Inducible transcription factors* are proteins that are not always active in a cell. They respond to external stimuli that activate them and potentiate their binding to the RE sequences, resulting in increased transcription of genes containing the RE sequence. Only when the inducible transcription factor is activated in a cell is the transcription of a gene containing the particular RE activated.
- *Activator proteins* are transcription factors that bind enhancers and increase transcription initiation of a gene.
- *Repressor proteins* are transcription factors that bind silencers and inhibit transcription initiation of a gene.

Table 8-1 lists some of the *cis*-acting elements and the transcription factors that they bind in the control of transcription initiation by RNA polymerase II.

**TABLE 8-1** ▶
*Types of Cis-acting Sequences and the Transcription Factors They Bind*

| Cis-*acting Sequence* | *Transcription Factor* |
|---|---|
| **Core promoter elements** | |
| Inr | RNA polymerase II* |
| TATA | TATA-binding protein (TBP)* |
| **Promoter proximal elements** | |
| Common elements | |
| CAAT | CTF |
| GC | Sp1 |
| Octamer | OTF |
| Response elements (RE) | |
| GRE | Glucocorticoid receptor |
| TRE | AP1 |
| CRE | CREB |
| **Promoter distal elements** | |
| Enhancers | Activators |
| Silencers | Repressors |

* GTFs bind to RNA polymerase and TBP at the core promoter.

***Common Features of Transcription Factors.*** Most proteins that function as transcription factors have two major features in common: they bind to specific sequences on the DNA, and they interact with other transcription factors to regulate transcription initiation. Analysis of the structure of transcription factors indicates that they generally are composed of two functional domains as shown in Figure 8-2. One part of the protein, the *DNA-binding domain*, is the portion of the molecule that recognizes and binds specific DNA sequences. The ability of proteins to bind DNA is a reflection of particular amino acid sequences in the protein and the formation of specific motifs. Several common motifs involved in DNA binding are the zinc finger motif, the helix-turn-helix motif, the leucine zipper motif, and the helix-loop-helix motif.

> *Many transcription factors contain a **DNA-binding domain** and a **transactivation domain**.*

The zinc finger motif illustrated in Figure 8-3 is found in several transcription factors, including the common transcription factor (Sp1), the general transcription factor (TFIIA), and the glucocorticoid receptors. This motif is named for the characteristic

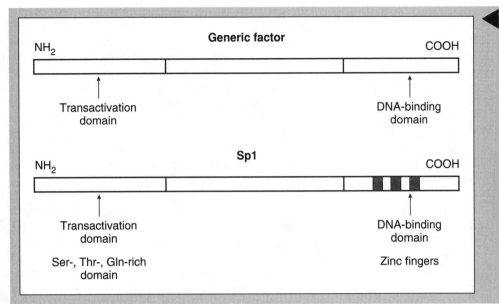

**FIGURE 8-2**
**Common Functional Domains in Transcription Factors.** *Many transcription factors contain two common functional domains. The transactivation domain is the amino acid sequence of the protein, which interacts with other protein factors and is responsible for activating the transcription of genes. The DNA-binding domain is made up of amino acid sequences, which are responsible for interacting with and binding to specific DNA sequences. The upstream transcription factor Sp1 binds to GC sequences by way of its DNA-binding domain, which includes three zinc finger motifs. The transactivation domain of Sp1 is rich in the amino acids serine (Ser), threonine (Thr), and glutamine (Gln) and interacts with the TAFIID 110 subunit of TFIID.*

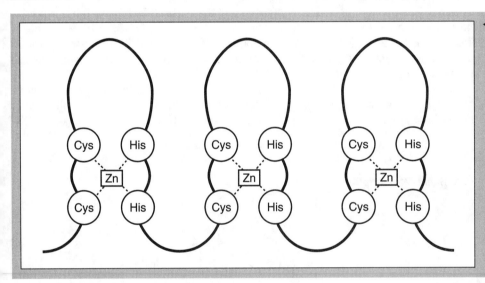

**FIGURE 8-3**
**Zinc Finger Motif for DNA Binding.** *The zinc finger motif is a common motif found in the DNA-binding domain of many transcription factors, such as Sp1, TFIIA, and the glucocorticoid receptor. The characteristic finger-like protrusions of the zinc finger motif are formed by the binding of zinc (Zn) to cysteine (Cys) and histidine (His) residues in the DNA-binding domain.*

structures that are formed when zinc binds to cysteine and histidine residues to form finger-like protrusions that can interact with DNA.

The second important functional domain of transcription factors, known as the *transactivation domain*, is the amino acid sequence in the protein that is necessary for the activation of transcription. The transactivation domain, which is a serine-, threonine-, and glutamine-rich amino acid sequence, is the part of the protein that interacts with other protein factors such as the TFIIs or RNA polymerase II, the proteins that make up the basal transcription complex. Transcription factors may contain similar DNA-binding domains but have different transactivating domains. In this case, the factors bind the same DNA sequence but activate transcription in a different way. Alternatively, transcription factors may have different DNA-binding domains but similar transactivating domains. This results in the binding of the factors to different DNA sequences, but the activation process is similar. Many transcription factors have been isolated and studied in vitro; however, the exact mechanisms by which these factors regulate transcription initiation in vivo is still unknown.

## Initiation of Transcription by RNA Polymerase II

***Formation of the Basal Transcription Complex.*** Figure 8-4A illustrates the interaction of the *cis*-acting core promoter sequences and the GTFs involved in the formation of the basal transcription complex during the process of initiation of transcription by RNA polymerase II. Many of the GTFs known to be involved in the initiation of transcription have been purified and characterized, and their putative function in transcription initiation has been studied using in vitro transcription systems [1]. In an in vitro transcription system, the GTFs assemble at the core promoter site in sequential order to form a complex that recruits RNA polymerase II to the start site. Table 8-2 lists the known human GTFs and some of their putative functions. These factors are named TF, for transcription factor; II, for RNA polymerase II control; followed by a letter (A, B, D, E, F, or H) to indicate the different factors.

TFIID is the first factor to bind to the core promoter sequence. TFIID is made up of at least 10 polypeptides, one of which is the TATA-binding protein (TBP). TBP, a protein of 38 kD, specifically recognizes the TATA sequence at all promoter sites. When TBP binds to the DNA, the protein forms a "saddle-like" shape around the DNA double helix. This distorts and bends the DNA and is believed to facilitate the binding of additional proteins. The remaining polypeptides in TFIID are referred to as TBP-associated factors (TAFs). The TAFs range in molecular weight from the largest of 250 kD (TAFIID 250) to the smallest of 10 kD (TAFIID 10) and interact with TBP to form a complex at the TATA-binding site. For simplicity, only some of the TAFs are illustrated in Figure 8-4. TAFs have multiple functions, one of which is their interaction with specific transcription factors that are bound to upstream DNA sequences.

Once TFIID is bound at the TATA sequence, the other GTFs bind to the complex in the following order to form the preinitiation complex: TFIIA, TFIIB, TFIIF/RNA polymerase II, TFIIE, and TFIIH. When this complex is stimulated by other transcription factors, mRNA synthesis begins with the movement of RNA polymerase II away from the promoter and the elongation of the mRNA transcript.

Functions attributed to the various TFII proteins (in addition to their ability to interact with transcription factors, activators, and repressors to control the initiation of transcription) include helicase activity, protein kinase activities, and cyclin-like activity.

> The **TBP** is a subunit of the general transcription factor TFIID and is the first protein to bind to the core promoter site at the TATA sequence.

## FIGURE 8-4 ▶

**Model for the Initiation of Transcription by RNA Polymerase II.** (A) Basal transcription complex. This figure illustrates the binding of the general transcription factors (GTFs) and RNA polymerase II (RNA pol II) to the core promoter. The TATA-binding protein (TBP), a subunit of TFIID, binds to the TATA sequence and facilitates the binding of the TBP-associated factors (TAFs). TBP and some of the TAFs are indicated as 250, 110, 150, and 60. Once TFIID is bound to the TATA sequence, the other GTFs (A, B, F, E, and H) and RNA polymerase II bind to the core promoter to form the basal transcription complex. Also indicated is the SWI/SNF multiprotein complex, associated with RNA polymerase II and believed to be necessary for disruption of the chromatin structure. (B) Activation of basal transcription complex. This figure illustrates the activation of transcription initiation by Sp1 bound to the GC sequence and interacting with TAFIID 110. Further activation is caused by the binding of an activator protein (ACT) to an enhancer sequence located 1000 bp away from the core promoter. The ACT is brought into proximity with the basal transcription complex by looping out of the DNA between the enhancer sequence and the core promoter to allow the activator to interact with TAFIID 250.

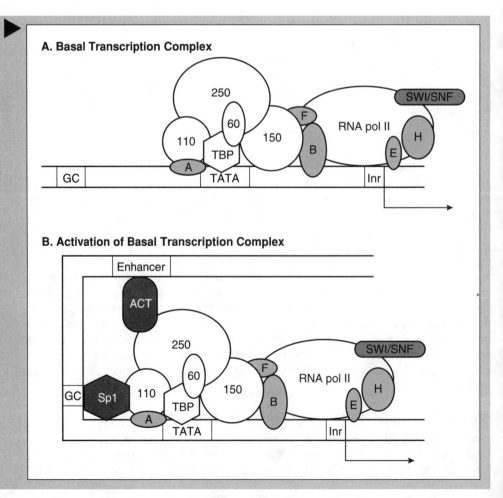

| General Transcription Factor | Number of Subunits * | Putative Function/Properties |
|---|---|---|
| TFIIA | 3 | Stabilizes TFIID-DNA complex; responds to activators |
| TFIIB | 1 | Promotes TFIIF-RNA polymerase II interaction; start site selection |
| TFIID | 10 | TBP binds TATA sequence; TAFs interact with activators and repressors; TAFIID 250 is a serine kinase |
| TFIIE | 2 | Stimulates TFIIH CTD kinase activity |
| TFIIF | 2 | Protein kinase; escorts RNA polymerase II to the preinitiation complex |
| TFIIH | 9 | Helicase activity; cyclin-dependent kinase activity; cyclin; CTD kinase activity |

*Note.* CTD = carboxyl-terminal domain.

◀ **TABLE 8-2**
*General Transcription Factors (GTFs) Associated with the Basal Transcription Initiation Complex for RNA Polymerase II*

TFIIH is of particular interest in this regard. This complex is made up of nine polypeptides, two of which have helicase activity and are involved in the process of nucleotide excision repair discussed in Chapter 4. Additionally, this complex contains one polypeptide that is identical to cyclin H and one that is a cyclin-dependent kinase involved in regulating the cell cycle. Thus, TFIIH couples transcriptional control both to DNA repair and to the regulation of cell cycle progression.

***Activation of the Basal Transcription Complex.*** The efficient and specific recognition of the core promoter site by the basal transcription complex depends on the short *cis*-acting sequences that are located upstream of the TATA sequence [2, 3]. These sequences include the common sequences found in RNA polymerase II promoters such as CAAT, octamer (Oct), and GC (see Figure 8-1 and Table 8-1). Specific upstream transcription factors recognize these sequences and bind to the DNA by specific interaction of amino acid sequences in the protein factor, the DNA-binding domain, and the base sequences in the DNA. As indicated in Figure 8-4B, the transcription factor Sp1, which contains a zinc finger DNA-binding domain, binds to GC sequences known as GC boxes. Once bound to the DNA, Sp1 can interact directly with the basal transcription complex through protein–protein interactions by way of its transactivating domain. The transactivating domain of Sp1 recognizes the TAFIID 110 subunit of TFIID bound at the TATA box and, by some still unknown mechanism, appears to enhance the assembly of the GTFs and RNA polymerase II at the promoter site. Other upstream-binding factors, such as AP1 and OTF (a protein that binds to Oct), presumably work in a similar manner; that is, once bound to the DNA, they can activate transcription, although each protein may interact with a different factor in the basal transcription complex.

***Upstream RE and the Inducible Transcription Factors They Bind.*** Some classes of genes are under common control in the sense that their transcription is activated in a similar manner under specific conditions. For example, the presence of glucocorticoids or phorbol esters in the external milieu of a cell results in a specific induction or an increased transcription of all the genes induced by these compounds. This inducible response is attributed to the RE sequences. These upstream *cis*-acting sequences occur only in special promoters and act as binding sites for specific inducible transcription factors. Some of these REs and the transcription factor they bind are indicated in Table 8-1.

One example of inducible control is the binding of the factor AP1 to the TRE sequence (TGACTCA) in genes that are turned on in the presence of phorbol esters, as well as growth factors, cytokines, and neurotransmitters. In the absence of phorbol esters, AP1 is phosphorylated and inactive in the sense that it cannot bind DNA. The activation of AP1 involves the protein kinase C cascade, which leads to the dephosphorylation of the AP1 molecule, altering it in such a way that it becomes capable of binding to promoters containing the TRE sequence. When AP1 is bound to the DNA, it interacts with the basal transcription complex and increases the rate of transcription initiation. A second example of inducible transcription is the response of cells to steroid hormones, a process that involves the binding of steroid receptors to specific REs, as discussed later in this chapter.

The **upstream transcription factor Sp1** binds to GC sequences, then interacts with TAFIID 110, a subunit of TFIID, to activate transcription.

***Binding of Transcription Factors to Enhancers and Silencers.*** The regulation of transcription initiation by enhancer and silencer sequences poses an interesting question of how sequences located thousands of base pairs away from a gene can control the transcription of that gene and do it in such a way that the sequence is independent of orientation (i.e., it can be upstream or downstream of the initiation start site). To regulate transcription initiation, the transcription factors bound at enhancer sequences must be brought into proximity with the basal transcription complex. The most likely hypothesis to explain the mechanism by which a distant sequence can interact at a promoter site is the process of DNA looping, in which the DNA located between the promoter site and the enhancer sequence loops out to bring the two sequences close together (see Figure 8-4B). In this way, the proteins bound at the enhancer sequences and the transcription factors bound at the promoter site can interact directly to activate transcription initiation. Silencers are thought to act by a similar mechanism, except the binding of a repressor protein to a silencer inhibits the transcription initiation process.

## Receptor-Mediated Control of Transcription

***Hormonal Control of Transcription.*** Proper growth, development, and homeostasis in humans is controlled in many instances by steroid and steroid-like hormones. Steroid hormones, such as glucocorticoid and mineralocorticoid, are important for regulating glycogen and mineral metabolism. The sex steroids (estrogen, progesterone, testosterone) are essential for normal sexual differentiation and reproduction. Other important biologic processes require steroid-like elements. For example, vitamin D is necessary for bone development, and retinoic acid functions as a regulator of differentiation.

Steroid and steroid-like hormones carry out their functions by binding to specific proteins—the steroid receptors—to form an activated complex that is capable of binding to RE sequences found in specific genes. The activated steroid-receptor proteins are essentially transcription factors that, when bound to RE sites in the DNA, activate transcription of a specific class of genes. Steroid-receptor proteins have several common features, which include a DNA-binding domain and a hormone-binding domain as indicated in Figure 8-5. As seen in many other transcription factors, the DNA-binding domain of steroid receptors contains zinc finger motifs.

**FIGURE 8-5** ▶
*Common Features of Steroid Receptor Proteins.* Steroid receptor proteins, such as the glucocorticoid receptor, contain a ligand-binding domain that binds glucocorticoid and results in activation of the receptor. Once the activated receptor enters the nucleus, the DNA-binding domain, which contains two zinc finger motifs, binds to response element sequences, such as the glucocorticoid response element, and activates transcription initiation of all of the genes regulated by glucocorticoids.

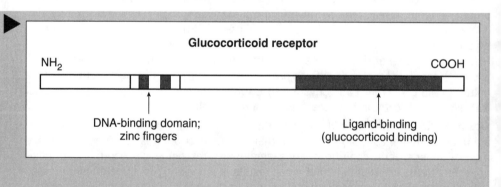

Steroids are synthesized in the endocrine cells and move to their target cells through the blood. Because of their lipid solubility, they can enter cells by diffusion across the plasma membrane. Once inside the cell, the steroid binds to a steroid receptor present either in the cytoplasm or in the nucleus. Some steroid receptors (e.g., the glucocorticoid receptor) are present in the cytoplasm, whereas others (e.g., the androgen receptor) are found in the nucleus. The general pathway of steroid hormone action is diagramed in Figure 8-6.

As the model indicates, a glucocorticoid enters a cell and is bound by the glucocorticoid receptor protein present inside the cytoplasm of the cell; then, the complex moves to the nucleus. Sex hormones, however, must enter the nucleus before they can bind to the androgen receptor. In both cases, the binding of the hormone activates the receptor

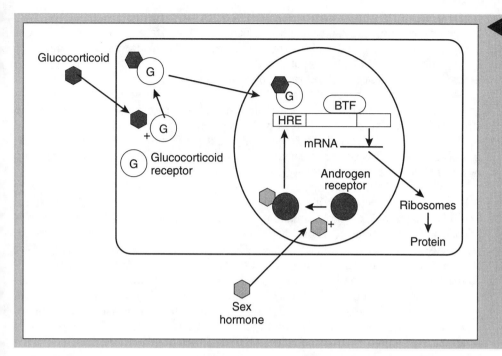

**FIGURE 8-6**
*Model for Steroid Hormone-Induced Transcription.* Glucocorticoids enter the cell by diffusion and are bound by the cytoplasmic glucocorticoid receptor (G). The activated complex moves to the nucleus and binds to hormone response elements (HRE), allowing the complex to activate the basal transcription factors (BTF) to stimulate transcription. Sex hormones must first enter the nucleus to activate the androgen receptor, which is located in the nucleus. Binding of the sex hormone to the ligand-binding site of the androgen receptor activates the receptor and stimulates its binding to the HRE. Once bound to an HRE sequence, the receptor interacts with the BTF and activates transcription initiation of all of the genes that have the binding site for this transcription factor.

protein to enable its binding to DNA. Once in the nucleus, the activated complex binds to the RE sequences on the DNA known as the hormone response elements (HREs). The HRE sequences may be located upstream of the promoter or may be in enhancer sequences. The binding of the activated complex to the HRE and the subsequent interaction of the receptor protein with factors in the basal transcription complex result in activation of the initiation of transcription by RNA polymerase II. All genes that have a common HRE sequence are simultaneously stimulated by a specific activated complex and show increased transcription. In this way, a cell can coordinate the inducible expression of multiple genes in response to specific hormone signals.

*Hormones regulate gene expression by binding to steroid receptors that act as transcription factors and by binding to HREs in promoters, resulting in the activation of transcription of a class of genes.*

*Signal Transduction and Transcriptional Regulation.* When a water-soluble ligand, such as a growth factor or a cytokine, is involved in activating transcription, the receptor protein for the ligand is localized within the plasma membrane. One important class of membrane protein receptors has tyrosine kinase activity and reacts to growth factors or cytokines to regulate cell growth. Important in this class are the signal transducers and activators of transcription (STATs). STATs are transcription factors present in an inactive form in the cytoplasm. When cytokines bind to receptors present in the plasma membrane, a tyrosine kinase associated with the receptor phosphorylates the receptor, providing a binding site for STAT proteins. The bound STAT proteins undergo phosphorylation on tyrosine residues and form dimers, which then migrate into the nucleus. Once inside the nucleus, they act as transcription factors by binding to specific sequences on the DNA located upstream of the TATA sequence.

*STATs are transcription factors located in the cytoplasm that are activated after tyrosine phosphorylation.*

Although each cytokine or growth factor has a different set of genes that it activates, they all act through a limited number of STATs. (There are currently six STAT family members known). How this differential expression of genes occurs is still unknown and will require a knowledge of how the different STATs interact with each other and with other transcription factors bound at specific *cis*-acting sequences.

## Role of Chromatin Structure

Current studies on transcription regulation are carried out in vitro using linear molecules of DNA and purified proteins. However, inside the cell, the DNA is not in a simple linear structure but is found wrapped around histones in the form of nucleosomes and ordered chromatin. For transcription initiation to occur within a natural chromatin structure, the protein factors necessary to carry out the process must gain access to the promoter sites on the DNA. The position of a DNA sequence in the chromatin can dramatically affect whether that sequence is transcribed or silenced. If the promoter sites are accessible to proteins, then transcription can occur, but if the promoter sites are inaccessible and blocked by histones, the genes remain in a nontranscribed state. Recently, the effect of chromatin structure on the regulation of transcription has been addressed. Two important findings have provided a glimpse of how the chromatin structure might be disrupted to expose promoter sites and allow transcription to occur [4].

When histones are acetylated on lysine residues at the amino-terminal end, an allosteric change in the nucleosome structure occurs, resulting in a destabilization of the structure and exposure of the DNA to proteins involved in transcription initiation. The TBP-associated protein TAFIID 250 (which is part of TFIID) contains histone acetyltransferase activity and may be one of the factors important in altering the chromatin structure by acetylation to facilitate transcription initiation.

In addition, RNA polymerase II contains a multiprotein complex known as the SWI/SNF complex (SWI = switch [a mating signal in yeast]; SNF = sucrose nonfermenting). The SWI/SNF complex, which is made up at least 10 proteins, includes a protein with adenosine triphosphatase (ATPase) activity that, in the presence of adenosine triphosphate (ATP), disrupts the nucleosome array. This disruption can facilitate transcription of DNA that was previously unavailable to the transcription complex. Although the SWI/SNF complex was initially found in yeast, similar complexes that disrupt the nucleosome structure are associated with RNA polymerase II in human cells. The human complexes appear to be more diverse and more complex than those found in yeast, and continuing studies are necessary to determine how these complexes function in transcription and what role the chromatin structure plays in the regulation of gene expression.

> *RNA polymerase II contains the multiprotein complex **SWI/SNF**, which helps disrupt the chromatin structure and facilitates access of transcription factors to promoter sites in the DNA.*

## Developmental Control of Globin Genes

The understanding of the regulation of hemoglobin synthesis has been fundamentally important to human biology and human genetics. Studies on hemoglobinopathies, the most common single-gene disorders found in humans, have contributed a wealth of knowledge to the molecular basis of human disease. In addition, the unique regulated pattern of synthesis of the globin genes during human development has features of special interest that are just now being determined [5].

Normal hemoglobin is a tetramer made up of two $\alpha$- and two $\beta$-subunits and functions to carry oxygen in red blood cells (RBCs). The genes that code for the $\alpha$- and $\beta$-subunits and of hemoglobin are found in two gene clusters that are located on separate human chromosomes, as illustrated in Figure 8-7. The genes coding for the $\beta$-like globin chains are found on the short arm of chromosome 11 (11p15.5) and span approximately 50,000 bp of DNA. The $\beta$-like globin genes code for the $\varepsilon$-, G$\gamma$-, A$\gamma$-, $\delta$-, and $\beta$-subunits of hemoglobin. The genes coding for the $\alpha$-like globin chains are located on the short arm of chromosome 16 (16p13.11-13.33) and cover approximately 30,000 bp of DNA. The $\alpha$-like globin genes code for the zeta and alpha chains. The two alpha genes, $\alpha_1$ and $\alpha_2$, are expressed equally and code for identical polypeptides. In addition, this cluster includes three pseudogenes (genes that do not code for a protein) and a theta gene with an unknown function.

**FIGURE 8-7** ▶

*α-Globin and β-Globin Gene Clusters Under Control of a Locus Control Region (LCR). The genes coding for the α-subunits of hemoglobin are located on chromosome 16 and include the zeta (ζ) gene, three pseudogenes, two alpha (α) genes, and a theta (θ) gene. In addition, an LCR region that regulates expression of the α-like genes is located 40 kb upstream of the zeta gene. The genes coding for the β-subunits of hemoglobin are located on chromosome 11 and include the epsilon (ε) gene, the G gamma (Gγ) and A gamma (Aγ) genes, a pseudogene, the delta (δ) gene, and the beta (β) gene. Located 6 kb upstream of the epsilon gene is the LCR region, which regulates expression of the β-like genes.*

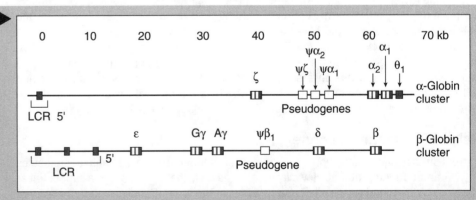

In the adult, the major hemoglobin present is HbA, a tetramer composed of two alpha chains and two beta chains. $HbA_2$ also is found in adults and makes up approximately 3% of the total hemoglobin present. Like HbA, $HbA_2$ contains two alpha chains, but in place of the beta chains, there are two delta chains. Fetal cells have fetal hemoglobin (HbF) that is composed of two alpha chains and two gamma chains, whereas embryonic cells contain three different hemoglobin molecules: Hb Gower 1, made up of two zeta and two epsilon chains; Hb Gower 2, made up of two alpha chains and two epsilon chains; and Hb Portland, made up of two zeta and two gamma chains (Table 8-3).

The presence of different forms of hemoglobin during development requires that different genes in the two clusters be turned on or off at precise times. This process, known as hemoglobin switching, allows the synthesis of the proper hemoglobin subunits at the proper state of development. Additionally, the site of synthesis of hemoglobin during development differs, as illustrated in Figure 8-8.

Early during development, Hb Gower 1, Hb Gower 2, and Hb Portland (see Table 8-3) are synthesized in the yolk sac. At approximately 8 weeks gestation, HbF and a small amount of HbA begin to be synthesized in the liver. By 18 weeks gestation, synthesis of hemoglobin begins in the bone marrow, with a switch in gene expression that progressively results in a decrease in the synthesis of the gamma chain and an increase in synthesis of the beta chain. By 12 months of age, HbF comprises less than 2% of the total hemoglobin, with the main hemoglobin present being HbA.

The mechanisms responsible for determining which globin genes are expressed during the embryonic, the fetal, and the adult stages of human development are determined by regulatory sequences within the promoter of each globin gene and a common regulatory sequence known as the locus control region (LCR; see Figure 8-7). The LCR, located 6–40 kb upstream of the structural genes, is required both for tissue-specific control of the globin genes and for determination of which genes within the locus are expressed at a particular time during development.

*Globin gene expression during different phases of development is controlled by regulatory sequences in the promoter of each globin gene and by sequences in a common regulatory locus known as the **locus control region**.*

Within the β-globin gene locus, a temporal expression of the genes occurs from the 5′ to the 3′ direction with respect to the position of the LCR. This directionality results in the epsilon gene, an embryonic gene, being expressed first in development; then the gamma genes, or the fetal genes, being expressed; and finally the beta gene, or the adult gene, being expressed. The order of the globin genes with respect to the position of the LCR appears to be an important aspect of control for their temporal expression.

The LCR for the β-globin gene locus, located 6 kb upstream of the epsilon gene, has multiple sites, or *cis*-acting elements, necessary for the binding of transcription factors. Two of these transcription factors are GATA-1 and NFE-2, which are expressed only in erythroid cells, thus providing tissue-specific regulation of the β-globin genes. Additionally, AP1 and Sp1, common transcription factors, also bind to sequences in the LCR. The binding of transcription factors at the LCR may alter the chromatin structure around the beta gene locus, allowing additional transcription factors access to the promoters of the individual genes in the cluster. The ability of a single controlling region like the LCR to regulate a cluster of genes rather than just a single gene provides a higher order of control over gene expression.

◀ **TABLE 8-3**
*Hemoglobin Subunit Composition at Different Stages of Human Development*

| Hemoblogin Subunit Composition | Designation | Developmental Period |
|---|---|---|
| $\zeta_2\epsilon_2$ | Hb Gower 1 | Embryonic |
| $\zeta_2\gamma_2$ | Hb Portland | Embryonic |
| $\alpha_2\epsilon_2$ | Hb Gower 2 | Embryonic |
| $\alpha_2\gamma_2$ | HbF | Fetal |
| $\alpha_2\beta_2$ | HbA | Adult, major form |
| $\alpha_2\delta_2$ | $HbA_2$ | Adult, minor form |

*Note.* Hb = hemoglobin; *HbA* = adult hemoglobin; *HbF* = fetal hemoglobin.

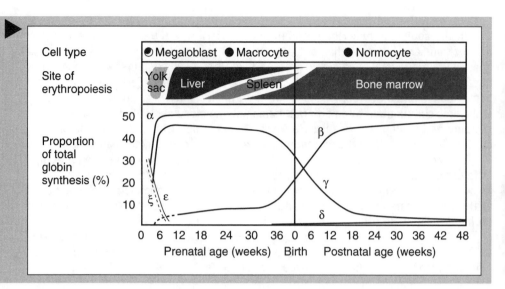

**FIGURE 8-8**

**Globin Synthesis during Embryonic and Fetal Development.** *Early during development, embryonic hemoglobin synthesis takes place in the yolk sac. At approximately 8 weeks of gestation, fetal hemoglobin, as well as a small amount of adult hemoglobin, begins to be synthesized in the liver. As development continues, synthesis of hemoglobin switches to the bone marrow, with the major form of hemoglobin synthesized being the adult forms, mainly hemoglobin A.* (Source: *Reprinted with permission from Weatherall DJ, Clegg JB, Higgs DR, et al: The hemoglobinopathies. In* The Metabolic and Molecular Bases of Inherited Diseases, *7th ed. Edited by Scriver C, Beaudet A, Sly W, et al: New York, NY: McGraw-Hill, 1995, p 3426.)*

*The switch in globin gene expression in different tissues depends on the presence of specific transcription factors in each tissue.*

The sequential expression of the globin genes during development may be a result of competition for the LCR function. One model would involve looping out of the DNA between the LCR and the gene to be expressed, which would allow only one gene in the cluster to be transcribed at a time. Binding of different transcription factors to various sequences within the LCR also may affect the interaction of the LCR sequence, with transcription factors bound at the different promoters of each gene in the cluster. This interaction could result in activation or silencing of the transcription of different genes.

The switch in tissue expression most likely depends on the presence of specific transcription factors in each tissue. For example, embryonic RBCs in the yolk sac have transcription factors that permit the LCR to activate the transcription of the epsilon gene. Switching hemoglobin synthesis from the yolk sac to liver cells is accompanied by switching the synthesis of the embryonic form of hemoglobin to the fetal form. This switch in transcribing the gamma gene and not the epsilon gene depends on the presence in hepatocytes of a repressor that binds to a silencer and represses transcription of the epsilon gene. In addition, progenitor fetal erythrocytes have specific transcription factors that recognize the promoter of the gamma genes and activate their transcription. This combined control by the LCR and specific transcription factors in different tissues results in an orderly switch in hemoglobin synthesis during development.

An LCR is also present 40 kb upstream of the zeta gene in the α-globin gene cluster. This LCR presumably functions like the one present in the β-globin gene cluster to control the expression of the zeta and alpha genes. Much is unknown about the molecular interactions of the LCR with transcription factors and with promoter sites at the individual genes within each globin locus and how this unusual control locus regulates gene expression. Control by LCRs is found in other gene clusters, such as the T-cell receptor genes. When these complex control systems are understood, they should provide additional insights into the complexity of gene control.

# ABERRANT TRANSCRIPTIONAL CONTROL AND CANCER

The preceding discussion focused on transcriptional regulation occurring during the growth and development of normal human cells. During the past 20 years, studies on the molecular basis of cancer have shown that interference with the functions of normal pathways used by cells to recognize and respond to growth factors, hormones, cytokines, and other regulatory molecules can result in the development of a tumor or a cancer cell. Two very important types of genes—oncogenes and tumor suppressor genes—play key roles in determining whether a cell remains under normal controls or whether it develops aberrant control signals and becomes a cancer cell.

## Oncogenes

An oncogene is defined as a viral or cellular gene that can induce neoplastic transformation or cancer when introduced into a normal cell [6]. Normal cellular genes, or proto-oncogenes, can be converted to oncogenes by various molecular changes that occur within the cell or by viral interference. In other words, all cellular and viral oncogenes originate by altering a normal cellular gene and thus altering the function of a normal gene product in the cell. Oncogenes arise at the single somatic cell level and act in a dominant fashion, but they are not inherited through the germ line.

***Mechanisms That Activate Proto-oncogenes.*** Activation of proto-oncogenes to oncogenes generally involves two types of alterations: a quantitative change leading to an increase in the amount of a normal gene product present in a cell or a qualitative change that involves a genetic mutation within a gene resulting in an abnormal gene product being produced. A proto-oncogene can be activated to become an oncogene by a number of different mechanisms, each of which is illustrated in Figure 8-9.

> **Oncogenes** *are a viral or cellular genes that can induce neoplastic transformation in normal cells.*

> **Activation of proto-oncogenes to oncogenes** *involves either an increase in the amount of a normal gene product or the production of abnormal gene product.*

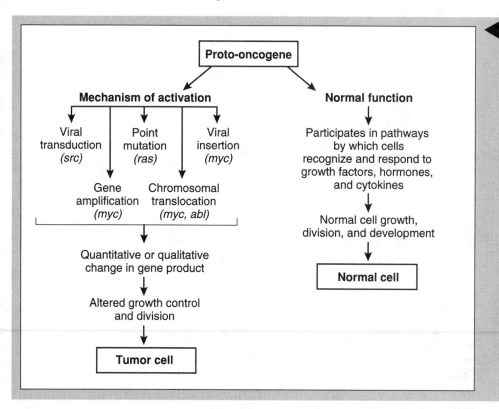

### FIGURE 8-9
***Mechanisms That Activate Proto-oncogenes.*** *Normal cellular genes or proto-oncogenes are activated to oncogenes by viral transduction, viral DNA insertion into the host DNA, point mutations within a host gene, amplification of host DNA sequences, and chromosomal translocations. These methods of altering a host gene result either in a quantitative change in the normal protein or a qualitative change in the normal protein to produce a mutant form. In both cases, the oncogenes produced result in altered growth control and cell division and lead to tumor production.*

Oncogenes were first discovered when it was shown that Rous sarcoma virus (RSV) carries a gene—the *src* gene (*src* for sarcoma)—that can transform a normal cell into a cancer cell. The finding by Varmus and Bishop that a homolog of the *src* gene is present in normal cells suggested that an alteration in a normal cellular gene might lead to cancer. When RSV and other viruses infect a host cell, the virus often picks up part or all of a host gene. This host gene, when present in the virus genome, may undergo modifications that result in the formation of an oncogene. When this oncogene is transduced back into a normal cell by viral infection, a cancerous cell is produced. This process of oncogene activation is known as viral transduction.

A second type of proto-oncogene activation resulting from viral infection is the insertion of viral DNA into the host DNA. If this insertion of viral DNA increases the efficiency of a promoter for a cellular gene or if it alters the cellular gene by causing a mutation within the gene, an increased amount of a normal product or an altered protein product can result.

Proto-oncogenes also can be activated to become oncogenes within a normal cell without infection by a virus. These mechanisms of activation include the following:

- Simple point mutations (missense mutations) within a gene result in the production of a gene product that has an altered function. These defects typically occur in genes that code for transcription factors or in genes that code for proteins involved in signal transduction pathways. The *ras* oncogene, which codes for a protein that differs from the normal gene product by just one amino acid, is a case in point [6].
- Gene amplification can occur in a cell in which a particular gene or gene sequence is amplified or increased hundreds of times in copy number. In such a cell, the amplified DNA is either located within a chromosome and appears as homogeneously staining regions (HSRs) upon G-banding, or it is located on extrachromosomal elements known as double minutes (see Chapter 5). The amplified gene sequences are capable of coding for very high levels (often more than 100-fold increase) of the normal gene product. This excessive increase in the amount of a normal gene product, particularly if it is a regulatory molecule, can have drastic effects on the normal growth regulation of a cell. The *myc* oncogene, which codes for a transcription factor, is found amplified in many neuroblastomas in which the highest degree of amplification often correlates with the more advanced stages of the cancer.
- Chromosomal translocations occur when a gene is translocated to a different position in the genome, resulting in an altered expression of the gene or sometimes the production of a fusion protein that has altered functions. Chromosomal translocations are associated with several types of cancer, including leukemias, Burkitt's lymphoma, and chronic myelogenous leukemia (CML). CML, which is associated with the Philadelphia (Ph) chromosome, occurs when a translocation between chromosome 9 and 22 results in the production of a fusion protein (Bcr-Abl) with protein kinase activity that is defective in its normal control. In Burkitt's lymphoma, a translocation of the normal *myc* gene from chromosome 8 to the site of the immunoglobulin heavy chain gene on chromosome 14 results in overexpression of the *myc* gene product.

In all of the five cases shown in Figure 8-9, the quantitative or qualitative change in a particular gene product results in altered growth control, altered cell cycle regulation, or altered transcriptional regulation of nuclear genes. When these normally controlled pathways are disturbed, the normal cell can develop into a cancerous cell.

*Oncoproteins and Their Functions.* More than 70 different oncogenes now have been identified. They code for oncoproteins that can be classified into several groups based on their biologic function. Examples of the four main classes of oncoproteins, listed in Table 8-4, include proteins that normally act as growth factors or that recognize external regulatory molecules, proteins that bind or modify these molecules, and proteins that regulate the genetic response to these molecules within the cell nucleus.

*Protein Kinases.* Tyrosine protein kinases phosphorylate proteins specifically on tyrosine residues and play a key role in signal transduction pathways. These proteins often are found either spanning the plasma membrane or are associated with the plasma membrane of cells. Tyrosine kinases can act as growth factor receptors, and they are important to the cell, enabling it to respond to growth factors and relaying the growth factor signal to the transcriptional machinery. The binding of a growth factor to a receptor stimulates the tyrosine kinase activity, often resulting in autophosphorylation and activation of the receptor. Many of the known oncogenes code for mutant forms of a normal tyrosine kinase in which the mutation alters the regulation of the kinase protein in such a way that it no longer requires a ligand (e.g., a growth factor) for activation. In these cases, the receptor is always "on" in the sense that it signals a message independent of the presence of the controlling ligand. Some of the tyrosine protein kinase oncogenes (e.g., *src* and *abl*) code for oncoproteins that function as membrane-associated kinases. Other tyrosine protein kinase oncogenes, such as *erb*B, *neu*, and *fms*, code for transmembrane kinases.

*Growth Factors.* Growth factors are the key ligands for cell cycle control, and they signal a cell to enter the G1 phase of the cell cycle from a resting phase or to bypass the restriction point in G1 (see Chapter 5). Important for the normal growth of a cell is the proper amount of a growth factor produced at the right time either in the growth cycle or in the differentiation of a cell. Two oncogenes that code for growth factors include the *sis* oncogene, which codes for the B chain of the platelet-derived growth factor, and the *int*-2

*Oncogenes can code for proteins that have protein kinase activity, that act as growth factors, that function in signal transduction pathways, or that function as nuclear factors (e.g., transcription factors).*

| General Class | Oncogene | Oncoprotein |
|---|---|---|
| **Protein kinases** | | |
| Membrane-associated proteins | *src, abl* | Tyrosine protein kinase |
| Transmembrane proteins | *erbB, neu, fms* | Tyrosine protein kinase |
| Cytoplasmic protein | *mos* | Serine/threonine protein kinase |
| **Growth factors** | *sis* | B-chain platelet-derived growth factor |
| | *int-2* | Fibroblast growth factor |
| **Signal transducers (GTP-binding proteins)** | H-*ras*, K-*ras*, N-*ras* | GTPases |
| **Nuclear factors** | *jun, fos* | Activator protein-1 |
| | *myc, myb* | DNA-binding proteins |
| | *bcl-1* | Cyclin D |

oncogene, which codes for a fibroblast-related growth factor. Tumor cells express and respond to these oncoproteins, resulting in uncontrolled growth, whereas their normal cell counterparts do not express the growth factor and thus remain quiescent until the growth factor is provided by the appropriate cell at the appropriate time.

*Signal Transducers.* An example of an oncogene that affects signal transduction is *ras*, the first cellular oncogene shown to be produced in response to a chemical carcinogen. This finding by Robert Weinberg in 1979 was the first indication that oncogenes could arise from normal cellular genes without viral intervention. Additional studies on *ras* have shown that it codes for a guanosine triphosphate (GTP)–binding protein with guanosine triphosphatase (GTPase) activity. Normal GTP-binding proteins, or G proteins, are important in signal transduction pathways in which the protein exists in a state of equilibrium between an active protein with GTP bound and an inactive protein with guanosine diphosphate (GDP) bound. In response to the appropriate signal, the inactive form of the G protein is converted to the active form by the exchange of GDP with GTP. The normal proto-oncogene is converted to the *ras* oncogene by a single base-pair substitution that results in the production of a missense mutant protein, which remains in the GTP-bound state and thus is always activated. Three *ras* family oncogenes are known—H-*ras*, K-*ras* and N-*ras*—each of which codes for a mutant protein of 21,000 daltons.

*Nuclear Factors.* A large number of oncogene products act in the nucleus as aberrant transcription factors. Oncogenes in this class include the *jun* and *fos* oncogenes. In a normal cell, the proto-oncogenes *fos* and *jun* code for subunits that make up the inducible transcription factor AP1. AP1, when activated in the presence of phorbol esters, binds to TRE sequences of genes and activates transcription initiation. Mutations in the normal *fos* or *jun* genes can lead to an oncogene that codes for protein subunits that no longer require activation by phorbol esters and can stimulate transcription in a constitutive manner. Other oncogenes, such as the *myc* and *myb* oncogenes, code for transcription factors that also have aberrant transcriptional regulatory properties. Included in this class of nuclear oncogenes are those that code for proteins that function in the normal progression of the cell cycle. An example is the oncogene *bcl-1*, which results from a mutation in the normal gene coding for cyclin D, a protein that is necessary for the activation of some cyclin-dependent kinases involved in the G1 phase of the cell cycle.

The understanding of oncogenes and their role in causing cancer has led to increased information about many key pathways that a cell uses in normal growth and development. It is very clear that cells have a fine line between a normal state and a cancerous state and that a single base-pair substitution in the DNA of any cell at any time in the life of an organism may result in the activation of a proto-oncogene and eventually lead to the development of a tumor.

## Tumor Suppressor Genes

The regulation of normal cell growth and division is subject to the action of positive activators that, in response to the proper stimuli, activate cytoplasmic products or the transcription of specific genes. Mutations in the genes that code for these positive activators often lead to the production of an altered gene product, an oncoprotein. Another important aspect of the regulation of normal cell growth is the role of gene products that function by negatively controlling cell proliferation. The normal growth of cells is a balance between regulation by positive activators, which stimulate growth, and regulation by negative activators, which inhibit growth. In either case, a mutation in the genes coding for these factors can lead to the loss of growth regulation and the development of cancer. When a mutation occurs in a gene that acts in a negative fashion and the product of the gene is either lost or altered, negative regulation is also lost, and uncontrollable cell growth occurs. These genes, whose loss or inactivation leads to uncontrollable cell proliferation, are known as tumor suppressor genes, and they code for products that normally suppress tumor formation [7, 8].

*Tumor suppressor genes code for proteins that suppress tumor formation by acting as negative growth regulators.*

## Tumor Suppressor Proteins and Their Functions

Products of tumor suppressor genes act in a number of different ways. They are important for regulating cell cycle progression, functioning to stop or start the growth of a cell at a particular phase in the cell cycle. They can act by regulating transcription of specific genes, either by activating transcription of some genes or by repressing transcription of other genes. Tumor suppressor proteins also can direct a cell to undergo differentiation, in which case the cell stops dividing. Finally, under the proper stimulatory signal, tumor suppressor proteins can direct a cell to apoptosis or a programmed death.

Mutations in tumor suppressor genes are, for the most part, loss-of-function mutations, in which the mutation in the gene results in the absence or loss of a functional gene product. These mutations are recessive; therefore, mutations in both alleles of the gene must occur before a mutant phenotype is observed. The recessive nature of tumor suppressor mutations allows the heterozygous state to be inherited through the germ line because a single defective allele does not interfere with the development of the fetus. Inheritance of a mutation in tumor suppressor genes is in contrast to oncogenes, which cannot be inherited because they are dominant and do not allow the development of a fetus. When one mutant allele of a tumor suppressor gene is inherited, the affected individual does not inherit cancer but inherits a predisposition to develop cancer. If a mutation in the second allele of the tumor suppressor gene occurs, it occurs at the somatic cell level, and that single cell becomes cancerous.

*Mutations in tumor suppressor genes are recessive and can be inherited through the germ line.*

In a normal individual, both alleles of a tumor suppressor gene are active and code for a functioning protein product. Therefore, two separate mutations must occur in a single cell to inactivate both alleles and produce a cancerous cell, as indicated in Figure 8-10. In an individual who has inherited one defective allele, every cell in the body carries that defect, and only one mutation at the somatic cell level is necessary to produce a totally defective or cancerous cell. Therefore, an individual who inherits one defective allele of a tumor suppressor gene has a greater chance of developing cancer than does a normal individual. Increasing numbers of tumor suppressor genes are currently being identified and associated with specific types of cancer. A list of some of the known tumor suppressor genes, their chromosomal location, and the type of tumor with which they are associated is given in Table 8-5.

The discovery of tumor suppressor genes in human cells came from several lines of experimentation. The first indication that normal cells contained a gene or genes that could suppress tumor formation came from studies with somatic cell hybrids (see Chapter 6). When a normal cell is fused to a tumor cell, the resulting cell hybrid is no longer tumorigenic, indicating that there is a positive factor in normal cells that can inhibit the tumor-producing capabilities of the tumor cell. A variation of this experiment using microcell fusion (see Chapter 6), in which microcells with a single human chromosome are fused with a tumor cell, led to the identification of particular chromosomes that can suppress tumorigenicity and suggested that a tumor suppressor gene was located on that chromosome. As indicated in Table 8-6, tumor suppressor genes

**Normal individual**

 Both alleles normal

 Alteration in one allele in a somatic cell

 Alteration in second allele in the same somatic cell
**(Tumor cell)**

**Individual with one defective allele inherited**

 Inherited mutation in one allele

 Alteration in second allele in a somatic cell
**(Tumor cell)**

**◄ FIGURE 8-10**

*Inheritance Pattern of a Defective Tumor Suppressor Gene. When an individual inherits a defective tumor suppressor gene, such as a defective retinoblastoma (Rb) gene, every cell in the body has one inactive copy of the gene. A second mutation occurring at the somatic cell level in one cell produces two defective alleles and the mutant phenotype, a tumor cell. In a normal individual who has both copies of the gene active in all cells, two individual mutations in a single cell must occur before the mutant phenotype, a tumor cell, is observed. Thus, the individual who inherits a defective tumor suppressor gene is said to have inherited a predisposition to develop cancer.*

are found throughout the human genome. Most likely, more will be identified in the near future.

Additional evidence for the presence of tumor suppressor genes comes from the studies of familial cancers such as retinoblastoma and the Li-Fraumeni syndrome. In these studies, inheritance of a predisposition to develop cancer can be followed through families, and the location of the responsible gene locus mapped by linkage analysis (see Chapters 7 and 11). Additionally, because some of the mutations in tumor suppressor genes are the result of deletions, the deletions can be detected by chromosome banding or hybridization studies, which aid in the localization and isolation of some of the tumor suppressor genes.

Often, when the second allele of a tumor suppressor gene is inactivated or lost in a somatic cell, not only are the DNA sequences within the tumor suppressor gene lost, but neighboring DNA sequences also are lost. This phenomenon, known as loss of heterozygosity (LOH) in which DNA sequences can be shown to be missing in an individual, is now used as a hallmark for the location and identification of tumor suppressor genes (see Chapter 11).

## TUMOR SUPPRESSORS AND HOW THEY FUNCTION

***Retinoblastoma (Rb) Gene.*** The *Rb* gene spans more than 200 kb of DNA and codes for a protein with a molecular weight of 105–110 kD. The Rb protein, which is expressed in the nucleus of all cells, regulates a number of important processes, including cell cycle progression, cell differentiation, and apoptosis, and is known to act at the level of gene transcription by forming protein–protein complexes with both upstream and basal transcription factors [9, 10]. This interaction can result in either positive or negative stimula-

**◄ TABLE 8-5**
*Human Tumor Suppressor Genes*

| Chromosome Location | Gene/Protein | Tumor/Disease |
|---|---|---|
| 3p26 | VHL (subunit of elongin) | Von Hippel-Landau disease |
| 5q21 | APC | Adenomatous polyposis coli |
| 9p21 | Cyclin-dependent kinase inhibitor, p16 | Familial melanoma |
| 11p13–15 | Transcription factor WT1 | Wilms' tumor |
| 11p15.1 | TSG101 | Breast cancer |
| 13q14 | Rb protein | Retinoblastoma |
| 13q12 | BRCA2 | Early-onset breast cancer |
| 17p13 | p53 | Li-Fraumeni syndrome |
| 17q21 | BRCA1 | Early-onset breast cancer |
| 17q11.2 | Neurofibromin (GTPase) | Neurofibromatosis 1 |
| 22q12 | Schwannomin | Neurofibromatosis 2 |

tion of expression of the genes coding for proteins that regulate progression through the cell cycle.

Most mutations that occur in the *Rb* gene lead to an alteration or a loss of the protein domain that interacts with other transcription factors. These mutations are the cause of retinoblastoma, a disease characterized by the development of eye tumors in young children. Two forms of retinoblastoma occur. One is the familial form, in which one defective allele of the *Rb* gene is passed through the germline with the second defective allele arising in somatic cells. The other form of retinoblastoma is the sporadic form, in which case both defective alleles arise at the somatic cell level. The disease is associated with a deletion that occurs on the long arm of chromosome 13 at 13q14, a finding that led to the identification and isolation of the *Rb* gene.

Approximately 40% of retinoblastoma patients have the familial form and a family history of the disease. In these individuals, tumors appear at an average age of 14 months and are usually bilateral; (i.e., tumors are present in both eyes). The tumors can be treated with radiation, cryosurgery, or surgical removal of the eye. An additional complication that occurs in many cases is the development of osteosarcomas in later life.

The remaining 60% of retinoblastoma cases occur sporadically with no history of the disease in the family. These individuals develop tumors usually in only one eye, and the tumors appear later (approximately 30 months of age). Both the familial form and the sporadic form of retinoblastoma result from the loss of function of both alleles of the *Rb* gene. However, in the case of the familial form, the patients already have one defective allele present in all cells; thus, they have a greater risk of a second mutation occurring in a somatic cell leading to the disease. In addition to retinoblastomas, mutations in the *Rb* gene can be found in a number of other types of cancers, including carcinomas of the bladder, the lung, the prostate gland, and the breast, as well as some forms of leukemia. Because children who inherit the mutant *Rb* allele do not generally develop carcinomas early in life, it is unclear what role the mutant *Rb* gene plays in these other cancers or what other factors are involved in the tumor development.

***Regulation and Functions of the Rb Protein.*** The functioning of the Rb protein is controlled by its phosphorylation state, which varies as a cell passes through the phases of the cell cycle (Figure 8-11). Cells in the G1 phase of the cell cycle contain the unphosphorylated form of Rb, which begins to be phosphorylated prior to the G1/S border. As the cells pass the G1/S border and enter into the S and G2 phases of the cell cycle, the Rb protein undergoes further phosphorylation. As cells leave M phase and proceed to a new G1 phase and the beginning of a new cell cycle, dephosphorylation of

The **Rb gene** codes for a tumor suppressor protein, the Rb protein, which is crucial in regulating normal progression through the cell cycle.

**FIGURE 8-11** ▶

***Modification of the Retinoblastoma (Rb) Protein by Phosphorylation in the Cell Cycle.*** *The functions of the Rb protein are regulated by its phosphorylation state, which varies as a cell moves through the cell cycle. In the G1 phase of the cell cycle, the Rb protein is in the unphosphorylated state and negatively controls cell cycle progression. This is partly caused by the binding of the transcription factor E2F, which prevents it from activating transcription. When the appropriate signal is given for a cell to move past the restriction point and to the G1/S border, Rb begins to become phosphorylated, causes E2F to be released and free to bind to DNA, and activates transcription of genes necessary for DNA replication. Rb protein continues to be phosphorylated through G2 and into M. As the cells leave M and enter again into G1, Rb is dephosphorylated and once again assumes its negative functions.*

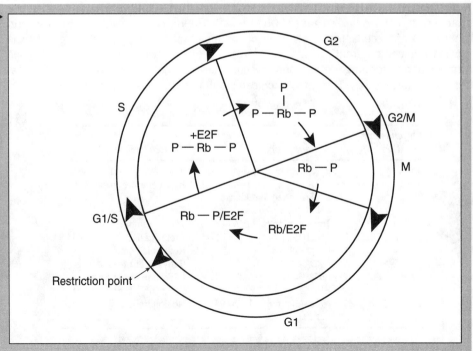

the Rb protein occurs. When Rb is in its unphosphorylated state (i.e., as in G1), it is thought to be in its growth-restricting state. Phosphorylation of the Rb protein, which leads to a loss of its negative functions, is carried out by specific cyclin-dependent kinases that function at various phases of the cell cycle (see Chapter 5).

Transcriptional regulation is the main function of the Rb protein, and it occurs by the interaction of the Rb protein with other transcription factors bound to core promoter and upstream promoter elements [10]. The Rb protein itself does not bind directly to DNA and does not contain a DNA-binding domain (Figure 8-12). A major portion of the protein structure—the A and B domains—is necessary for the interaction of the Rb protein with other transcription factors. Transcription factors that bind to the Rb protein include E2F, a protein regulator of genes involved in DNA replication, and Myc and Elf-1, which are sequence-specific binding proteins that regulate transcription initiation. The Rb protein also associates with the SWI/SNF complex of RNA polymerase II, the complex that is required for denaturing chromatin structure. Interaction of the Rb protein with other protein factors can result in either a repression or a stimulation of gene transcription. For example, binding of Rb protein to E2F represses transcription, whereas binding of Rb protein to Myc activates transcription.

The interaction of Rb protein with E2F is an important control process in regulating progression through the cell cycle (see Figure 8-11). When E2F is bound to the un-phosphorylated form of Rb protein, the complex is still capable of binding to DNA, but is no longer able to activate transcription and functions as a repressor that silences certain genes. When a cell is presented with growth-stimulating factors in G1, phosphorylation of Rb protein by cyclin D/cdk4 or cyclin E/cdk2, the cyclin-dependent protein kinases, occurs, releasing E2F from the Rb protein. With Rb protein no longer bound and not blocking the transactivation domain of E2F, E2F can now activate the transcription of genes necessary for the cell to pass the G1/S border and enter into S phase. Once the cells have passed the G1/S border, they are committed to cell division.

The Rb protein can regulate some genes by directly repressing their transcription. As indicated in Figure 8-13, the Rb protein can act in several ways to cause repression of transcription. In one case (see Figure 8-13A), Rb protein can block the binding of the transcription factor, TF, to the basal transcription complex, thus preventing activation of transcription. In another model (see Figure 8-13B), Rb protein can bind two upstream transcription factors, TF1 and TF2, and prevent their interaction with the basal transcription complex and their activation of the complex.

The Rb protein also can regulate transcription in a positive way by stimulating transcription mediated by a transcription factor such as SP1. The Rb protein can interact with TAFIID 250, a protein that is part of the basal transcription complex, and the transcription factor at the same time in such a way that the basal transcription complex is

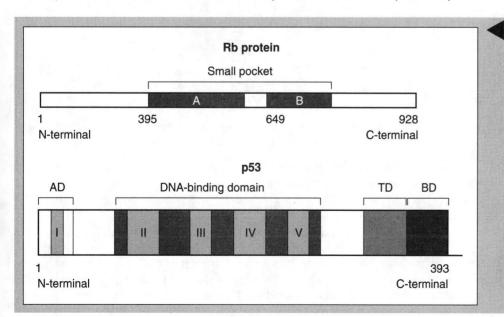

**FIGURE 8-12**

**Structure of the Retinoblastoma (Rb) Protein and p53.** The Rb protein is made up of 928 amino acids with a molecular weight of 110 kD. The central portion of the protein, the A and B domains, make up the region of the protein known as the small pocket. This region of the protein is necessary for Rb protein to interact with other regulatory proteins. The p53 protein contains only 393 amino acids and has a molecular weight of 53 kD. This protein contains four functional domains: the activating domain (AD), the DNA-binding domain, the tetramerization domain (TD), and a basic domain (BD). Five regions of the protein are highly conserved: I, II, III, IV, and V. Four of these occur in the DNA-binding domain and contain most of the mutations associated with loss of normal p53 activity.

*The **Rb protein** can regulate gene expression by activating transcription of some genes and repressing transcription of other genes.*

stabilized and activated to carry out transcription. The Rb protein may stimulate transcription by SP1 by its interaction with TAFIID 250 and some of the other TAFs, such as TAFIID 110, which is known to interact with SP1. This complex interaction between the basal transcription complex, the upstream transcription factors, and the Rb protein provides the cell with a tight connection between transcription initiation and cell cycle progression.

**FIGURE 8-13**
***Model for the Regulation of Transcription by the Retinoblastoma (Rb) Protein [10].*** *The tumor suppressor protein Rb can either repress the transcription of genes or activate the transcription of genes. Two models (A and B) present possible repressing action of the Rb protein. In A, Rb binds to both a transcription factor (TF) and to TFIID to prevent the interaction of the TF with TFIID. In B, Rb binds to two TFs, TF1 and TF2, and prevents their association with the basal transcription complex. In C, a possible model is presented for the activation activity of Rb. The Rb protein can stabilize the interaction between a TF and the basal transcription complex, facilitating transcription initiation. TATA = sequence in core promoter.*

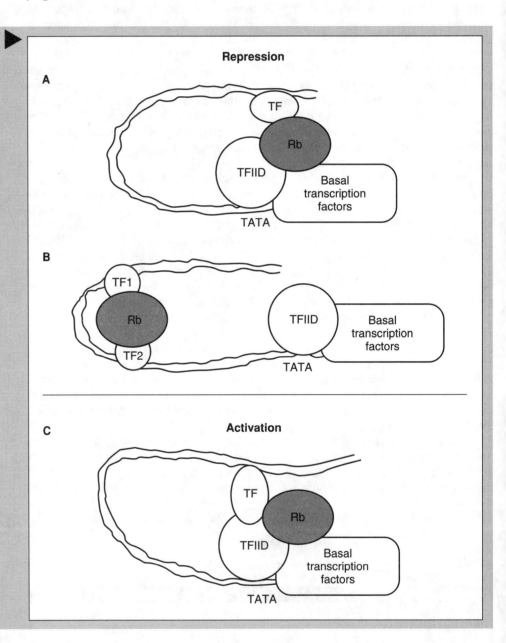

When mutations occur in both *Rb* alleles, the Rb protein is missing or has an altered function, and the regulation of transcription as well as cell cycle regulation is lost. The defective cell, no longer subject to normal controls, continues to progress through the cell cycle and is destined to become a tumor cell.

***The* p53 *Gene.*** Mutations in the tumor suppressor gene that codes for p53 are the most common mutations found in human cancers, occurring in more than 50% of all cancers. These mutations have been associated with carcinomas of the brain, breast, stomach, liver, lung, ovary, and prostate gland; osteosarcomas; and chronic myelogenous leukemia. Most of these mutations are sporadic, occurring at the somatic cell level; however, germ line transmission of a defective *p53* gene occurs in Li-Fraumeni syndrome, in which family members have a high rate of cancer development. A defect in *p53* is

*Mutations in the tumor suppressor gene p53 are the most common mutations found in human cancers.*

compatible with life because transgenic mice with two defective alleles have normal fetal development; however, they develop cancers at 3–6 months of age.

The *p53* gene spans 20 kb of DNA located on the short arm of chromosome 17 (17p13) and codes for a nuclear protein made up of 393 amino acids with a molecular weight of 53 kD. The p53 protein is a transcription factor that contains four functional domains: a DNA-binding domain, a transactivating domain, a tetramerization domain, and a basic domain (see Figure 8-12). Each of these domains is important for the various functions carried out by p53 in regulating cellular growth.

***Functions of the p53 Protein.*** The main function of p53 is to monitor cells for DNA damage during the cell cycle [11]. As discussed in Chapter 4, damage to the DNA is corrected by the DNA repair systems that are present in all cells. The role of p53 is to sense the presence of damaged DNA and to halt cells in the G1 phase of the cell cycle so the repair process can occur before the damaged DNA is replicated and passed on to daughter cells. The p53 protein carries out this important role by binding to DNA at specific p53 binding sites and activating the transcription of genes that are involved in DNA repair or in cell cycle control.

*The main function of the p53 protein is to monitor cells for DNA damage during the cell cycle.*

The transcription activating function of p53, which is necessary for transcription regulation, can be inhibited by a protein known as the Mdm-2 protein, which binds p53 and prevents it from activating the transcription of certain genes. The *mdm-2* gene, first identified as an oncogene present on mouse double minute chromosomes, is amplified in some types of tumors, resulting in increased levels of Mdm-2 protein. When present at high levels, the Mdm-2 protein obliterates the functions of p53. Under normal conditions, low levels of Mdm-2 protein regulate p53 activity, thus preventing p53 from stopping cells unnecessarily in the cell cycle or prematurely directing the cell to apoptosis.

The p53 protein serves multiple functions in transcriptional regulation, depending on its interaction with other transcription factors by way of its transactivation domain [12]. The p53 protein can interact with the TATA-binding protein as well as some of the TAFs in TFIID. It also can interact with TFIIH, the general transcription factor that contains subunits involved in DNA repair, and with the common transcription factor Sp1.

One important question is how p53 recognizes DNA damage and then is activated to regulate transcription of specific genes. When cells are damaged by ultraviolet light, single-strand breaks occur in the DNA. These breaks in the DNA appear to be the signal that results in an increase in the levels of p53. The increased levels of p53 are caused by a stabilization and an increased half-life of the protein, rather than by an increased synthesis of p53 mRNA. Additionally, p53 is subject to phosphorylation by cyclin-dependent kinases, which might be an important factor in its stabilization. When increased levels of p53 occur after DNA damage, transcription of a number of cellular genes is stimulated. These include genes that code for proteins necessary for cell-cycle regulation, such as p21, an inhibitor of cyclin-dependent protein kinases (see Chapter 5); GADD45, a protein important for DNA repair that is expressed after DNA damage and blocks the entry of cells into the S phase; and Bax, a protein involved in apoptosis. At the same time, other genes are repressed by p53. These include the genes *fos* and *jun*, which code for the subunits of the transcription factor AP1, and the *Rb* gene, which codes for the Rb protein.

Induction of p21 and GADD45 allows for cell cycle arrest to occur in the G1 phase so that the repair processes can function. In the event that DNA repair is not successful, the pathway to programmed cell death is activated. In either case, the damaged DNA is not perpetuated in future generations. However, cells with a mutation in p53 can no longer halt cells in the cell cycle in response to damaged DNA. These cells continue through the cell cycle, replicating the DNA damage and passing it on to daughter cells. Very often, this damaged DNA results in chromosome abnormalities such as translocations and gene amplification, which can activate proto-oncogenes to become oncogenes and lead to tumor formation.

***Other Tumor Suppressor Genes.*** As indicated in Table 8-6, other genes have been identified that code for tumor suppressor proteins. A number of these proteins appear to act as transcription factors, but their function in cells has yet to be elucidated. What is quite clear, however, is that a cell has multiple genes that function to prevent abnormal cell proliferation and maintain normal cell growth via a closely and accurately controlled regulation of transcription of its genome.

# COMPLEX TRANSCRIPTIONAL UNITS AND POST-TRANSCRIPTIONAL CONTROL

This chapter has discussed the transcription of genes that have a single promoter site, which determines the beginning of transcription and the sequences that are contained in the precursor mRNA molecule. The genes that contain a single polyadenylation signal, which is required for transcription termination by RNA polymerase II, were also discussed. However, some genes are more complex in that they have more than one promoter site or have more than one polyadenylation signal.

## Promoter Selection: The α-Amylase Gene

*The presence of more than one promoter within a gene can result in different amounts of the same gene product being produced in different tissues.*

Within the α-amylase gene there are two promoter sites that control the expression of this gene in a tissue-specific manner and provide a way for a cell to produce the α-amylase protein at high levels, low levels, or not at all. Figure 8-14 illustrates the α-amylase gene and its mechanism of transcription.

**FIGURE 8-14** ▶
*α-Amylase Gene Has Two Promoters. The α-amylase gene contains two 5' untranslated exons (gray boxes), and two translated exons (black boxes), which contain the coding information that determines the amino acid sequence of the final protein product. In the salivary gland, transcription factors recognize the first promoter site located at the arrow. This promoter site is a strong promoter site and results in a high level of mRNA being produced in the salivary gland. In liver cells, transcription factors recognize the second and weaker promoter located just 5' to the second untranslated exon indicated by the arrow. During the splicing process, the 5' untranslated exons in both cases are spliced to the first exon containing the amino acid sequence information, resulting in a different mRNA being produced with respect to the 5' untranslated region. The amino acid sequence of α-amylase is the same in both tissues; only the level of the protein varies.*

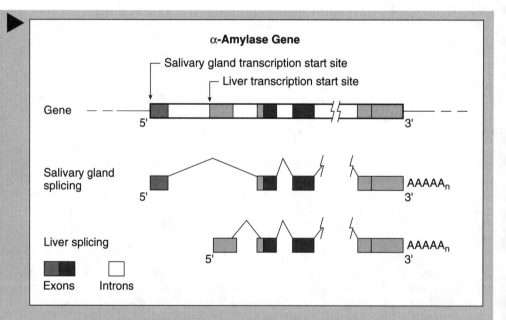

Cells in the salivary gland have very high levels of α-amylase, whereas liver cells have very low levels. This quantitative difference in enzyme amounts is controlled at the level of transcription of the α-amylase gene. In salivary gland cells, the first promoter site located just 5' to the first exon in the gene determines the start of transcription, as well as the rate of transcription of the α-amylase gene. This promoter is called a strong promoter because it has the ability to transcribe the gene at a high rate. The transcription of mRNA from this promoter site is controlled by specific transcription factors present only in cells of the salivary gland. When these transcription factors are present, only the first or strong promoter sequence in the gene is recognized by RNA polymerase II, and high levels of α-amylase result.

In liver cells, which express only low levels of α-amylase, the available transcription factors do not recognize the first strong promoter in the gene but instead direct RNA polymerase II to the second and weaker promoter site located just 5' to the second exon of the gene. The presence of liver-specific transcription factors and the binding of RNA polymerase II to the weaker promoter result in the same α-amylase protein being produced but at lower levels. When the precursor mRNA is spliced to the mature mRNA, the 5' untranslated exon in each cell type is spliced to the first exon containing the amino acid sequence information. The result is that the final mature mRNA in liver cells is different than the one found in the cells of the salivary gland with respect to the 5' untranslated sequences, although the amino acid coding regions are the same.

Other cells in the body do not produce any α-amylase protein, most likely because the transcription factors required to recognize either promoter sequence are missing in these cells. Thus, the interaction of specific transcription factors present in a cell with one or more promoters within a gene can regulate the amount of a specific protein present in a particular cell type.

## Alternative Polyadenylation Sites and Exon Selection: Immunoglobulin Heavy Chain Genes

Immunoglobulin M (IgM) is an antibody molecule that can occur either as a membrane-bound receptor for antigen or as a secreted form of the protein, depending on the state of development of B cells, or antibody-producing cells. This differential production of a membrane or secreted form of IgM depends on the structure of the heavy or mu chain that is part of the antibody molecule. The membrane form of IgM contains a mu heavy chain that has a carboxyl-terminal amino acid sequence rich in hydrophobic amino acids, which facilitates its interaction and binding to the cellular membrane. The secreted form of the antibody contains a mu heavy chain that is missing this carboxyl-terminal amino acid sequence and thus is no longer capable of binding to the membrane.

*Alternative polyadenylation sites and exon selection provide a cell with more flexibility in that a single gene can code for different polypeptides.*

During B-cell development, there is a mechanism that determines which form of the mu heavy chain mRNA is produced, which ultimately determines whether the membrane-bound or the secreted form of the molecule is synthesized. This mechanism operates at the level of polyadenylation of the mRNA by using alternative polyadenylation signals within the gene. Figure 8-15 illustrates alternative polyadenylation as it occurs in the mu heavy chain gene. The figure is a simplified schematic of the actual gene with most of the exons not depicted. When the mRNA coding for the membrane form of the mu heavy chain is produced, a polyadenylation signal present at the distal 3' end of the message determines the site of cleavage and polyadenylation of the mRNA. After polyadenylation of the mRNA occurs, splicing of all of the exons—including the 3' exon, which codes for the hydrophobic amino acid sequence found at the carboxyl-terminal end of the membrane-bound form of the mu heavy chain—occurs to produce the mature mRNA. Translation of this mRNA produces the form of the mu heavy chain, which has a hydrophobic tail and is found in membrane-bound IgM.

In cells in which the secreted form of the IgM molecule is produced, a second polyadenylation signal that is recognized by the polyadenylation system in mature B cells exists further upstream of the distal 3' polyadenylation signal. In these cells, cleavage and polyadenylation of the mRNA occur at this second site, and the exons located 3' to this site are no longer present in the mRNA produced. Following polyadenylation, the remaining exons are spliced together to form a mRNA that codes for a mu heavy chain deficient in the hydrophobic tail. Translation of this mRNA produces the mu heavy chain found in the secreted form of IgM.

***Exon Selection during Splicing: The Calcitonin Gene.*** Exon splicing during the processing of the precursor mRNA to a mature mRNA provides a cell with an additional level at which control of gene expression can occur. By determining which exons present in the precursor mRNA are conserved in the final mature mRNA, a cell gains the ability to produce more than one protein from the same gene. Calcitonin, a protein that is produced only in the thyroid gland, is important for calcium metabolism. The calcitonin gene-related peptide (CGRP) is a neuropeptide that is found in the neurons that innervate parts of the tongue, the esophagus, and the stomach and go into the spinal cord. The same gene codes for these proteins, although the proteins are quite different with respect to their amino acid sequence, their function, and their tissue location. The synthesis of these two very different proteins using the same genetic information occurs by a combi-

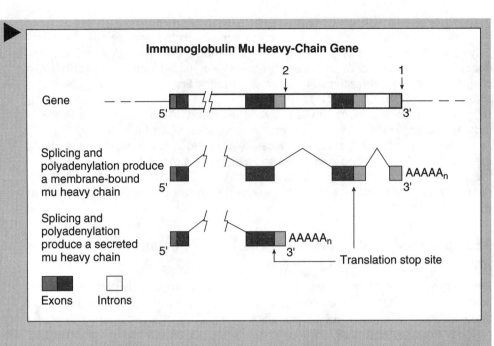

FIGURE 8-15 ▶

**Alternative Polyadenylation in the Immunoglobulin Mu Heavy-Chain Gene.** *This figure is a simplified diagram of the immunoglobulin mu heavy-chain gene with only some of the exons represented. In cells in which the membrane-bound form of the antibody is produced, a polyadenylation signal present at the distal 3' end of the mRNA, indicated by arrow 1, determines the site of cleavage and polyadenylation of the mRNA. During the splicing process of this mRNA, all of the exons, including the 3' exon coding for the hydrophobic amino acids found at the carboxyl-terminal end of the membrane-bound form of the mu heavy chain, are spliced together to produce the mature mRNA. In cells in which the secreted form of the antibody is produced, the upstream polyadenylation signal, arrow 2, is recognized and determines the site of cleavage and polyadenylation of the mRNA found in mature B cells. The mRNA produced after splicing is missing the exons located 3' to this polyadenylation signal, resulting in the production of a mu heavy chain that no longer contains a hydrophobic tail and is secreted.*

nation of alternative polyadenylation and differential exon selection. Figure 8-16 diagrams the processing of mRNA produced from the calcitonin gene.

The calcitonin gene contains two polyadenylation signals and 6 exons (1, 2, 3, 4, 5a, 5b). In the thyroid, the upstream polyadenylation signal is recognized by the polyadenylation machinery, resulting in the cleavage and polyadenylation of the mRNA at the 3' end of exon 4 to produce a precursor mRNA containing exons 1, 2, 3, and 4. These four exons are spliced together to form the mature mRNA, which codes for a calcitonin precursor peptide. Final protein processing of this peptide results in the calcitonin protein, which contains only amino acid sequence information from exon 4.

In neurons, the downstream polyadenylation signal is recognized by the polyadenylation machinery, at which point cleavage and polyadenylation of the mRNA occurs 3' to exon 5b to form a precursor mRNA containing exons 1, 2, 3, 4, 5a, and 5b. During the process of the splicing of this precursor mRNA, exon 4 is deleted, resulting in a mature mRNA that retains the exons 1, 2, 3, 5a, and 5b and codes for the precursor peptide to CGRP. Final processing of the precursor peptide results in a protein containing only the amino acid sequence information contained in exon 5a.

The use of alternative polyadenylation coupled with exon selection to produce two proteins that contain entirely different amino acid sequences, although they are coded for by the same gene, gives a cell additional flexibility in using a limited number of transcribed genes to carry out different functions. The tissue specificity of this process must rely on the presence of different regulatory proteins in the thyroid gland and in the neurons that can differentiate between the two polyadenylation signals.

Additionally, there must be a mechanism in cells that regulates the splicing process and directs exon selection. Insight as to how alternative splicing and exon selection are controlled in cells is just beginning to emerge with the identification of a family of proteins, the SR proteins, which contain a characteristic serine/arginine domain that may facilitate protein–protein interaction [13]. SR proteins also contain an RNA-binding domain, leading to the current speculation that they may act by binding specific RNA sequences to regulate splicing much like transcription factors bind specific DNA sequences to regulate transcription. Different SR proteins appear to be present in different cells and tissues, leading to further speculation that they could be instrumental in the tissue specificity of alternative splicing.

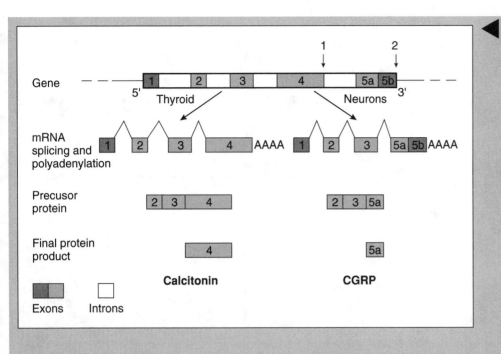

**FIGURE 8-16**
**Two Proteins Produced from the Same Gene by Exon Selection.** The calcitonin gene contains two polyadenylation signals and six exons (1, 2, 3, 4, 5a, 5b). In the thyroid, the upstream polyadenylation signal (arrow 1) is recognized, resulting in cleavage and polyadenylation of the mRNA at the 3' end of exon 4 to produce a precursor mRNA containing exons 1, 2, 3, and 4. These four exons are spliced together to form the mature mRNA, which codes for the calcitonin precursor peptide. The peptide is processed to produce calcitonin that contains amino acid sequence information only from exon 4. In neurons, the downstream polyadenylation signal (arrow 2) is recognized, resulting in cleavage and polyadenylation of the mRNA at the 3' end of exon 5b to form a precursor mRNA containing exons 1, 2, 3, 4, 5a, and 5b. During the splicing process, exon 4 is deleted, and the mature mRNA contains exons 1, 2, 3, 5a, and 5b, which code for the calcitonin gene-related peptide (CGRP). Final processing gives CGRP, which contains amino acid information found in exon 5a.

## Control of mRNA Stability and mRNA Translation

***Regulation of the Transferrin Receptor and Ferritin.*** In general, the amount of a protein present in a cell is directly related to the amount of mature mRNA present. This dependent relationship offers another level at which a cell can regulate protein concentrations. The steady-state level of a mRNA species is a balance between its rate of synthesis and its rate of decay or half-life. The half-life of mRNA molecules can vary in human cells from 30 minutes to hours and is controlled to some extent by the length of the polyadenylated sequence attached at the 3' end of the message. Control at the level of transcription determines the rate of synthesis of a mRNA molecule, whereas control at the level of stability determines the rate of decay. Thus, the level of a specific mRNA molecule can be increased or decreased either by changing its rate of synthesis or its rate of decay. An example in which the level of a protein is determined by altering the rate of decay of the mRNA is the synthesis of the transferrin receptor protein (Figure 8-17).

Free iron is carried in the blood by the carrier protein transferrin and is taken up by cells and internalized by the transferrin receptor protein. Located in the 3' untranslated region of the transferrin receptor mRNA is a sequence of nucleotides that can form a stem-loop structure by intrastrand hydrogen bonding. This stem-loop structure, known as the iron response element, acts as a binding site for a regulatory protein that responds to iron levels; that is, the iron regulatory protein (IRP). When intracellular levels of iron are depleted, the IRP binds to the iron response element and protects the transferrin receptor mRNA from degradation. This protection of the mRNA from degradation results in an increased level of transferrin receptor mRNA, leading to an increased synthesis of the transferrin receptor protein. In the presence of an increased level of the transferrin receptor protein, more iron can be internalized, which raises the intracellular iron concentration.

When the concentration of iron is in excess, it must be bound by the intracellular protein ferritin, or it will be toxic to the cells. In the presence of high concentrations of iron, the level of ferritin increases without a corresponding increase in the synthesis of ferritin mRNA. The active ferritin mRNA also has a stem-loop structure, an iron response element, located in the 5' untranslated region of the mRNA. When this stem-loop

Gene expression can be regulated at the level of mRNA stability and at the level of mRNA translation.

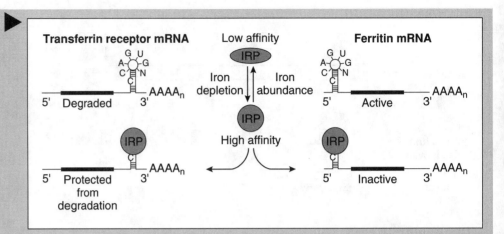

**FIGURE 8-17**

*Control of Protein Synthesis in the Liver by Iron. Located at the 3' end of the transferrin receptor mRNA is a stem-loop structure, the iron response element, which binds an iron regulatory protein (IRP) when the cell is depleted of iron. Binding of the IRP to the 3' end of the transferrin receptor mRNA protects the mRNA from degradation and results in an increase in the level of the transferrin receptor mRNA and a corresponding increase in the level of the transferrin receptor protein. At the 5' end of the ferritin mRNA molecule is a stem-loop structure that binds IRP when iron is depleted in the cell. Binding of the IRP to the 5' end of the ferritin mRNA blocks the translation of this mRNA and results in a decrease in the level of ferritin protein. When iron is in abundance, the ferritin mRNA no longer binds IRP and actively translates ferritin protein. At the same time, iron abundance prevents IRP from binding to the 3' end of the transferrin receptor mRNA, and the mRNA is degraded, which reduces the level of the transferrin receptor protein.*

structure is free or in its active state, the ferritin mRNA is actively translated to produce high levels of ferritin protein. However, when iron concentrations are depleted inside the cell, there is no longer a need for high levels of ferritin protein. In this case, the IRP binds to the 5' iron response element in the ferritin mRNA. When the IRP is bound to the 5' end of the ferritin mRNA, translation of the mRNA by the ribosomes is inhibited, and the levels of ferritin protein decrease.

In both cases, changes in the iron concentration can quickly regulate the synthesis of the necessary protein factor without the need for change in the rate of synthesis of mRNA. By using the same IRP, which can form alternative structures, depending on whether the concentration of iron is high or low, the cell can quickly respond to a change in the external milieu. The binding of the IRP to the iron response element occurs with both mRNA molecules when the levels of iron are low, but the ultimate effect on the synthesis of ferritin or the transferrin receptor is different. In one case, the binding of IRP to the mRNA increases transferrin receptor synthesis, and in the other case, the binding of IRP to the mRNA decreases the synthesis of ferritin. When the concentration of intracellular iron is high, it binds to the IRP and changes the structure so that it no longer recognizes the iron response element. Once again, this lack of binding has opposite effects. In the case of the transferrin receptor, the lack of binding of IRP to the mRNA results in degradation of the mRNA and a decrease in the amount of the transferrin receptor protein. In the case of the ferritin mRNA, the lack of binding of IRP to the mRNA frees the 5' untranslated region of the mRNA and allows the ribosomes to begin translation, and the level of ferritin increases.

# RESOLUTION OF CLINICAL CASE

After examining Jerry, the ophthalmologist suspected that the boy had retinoblastoma and sent him to the hospital for a more thorough examination of both eyes. The examination was performed under general anesthesia and revealed two tumors in his left eye—one about the size of a pea and the other at the tumor foci stage. In his right eye, there was also a tumor focus, indicating the beginning of a third tumor. Further tests were performed, including magnetic resonance imaging and ultrasonography, but they failed to reveal any other tumors in other tissues of his body.

After discussing the results with Marie and George, the oncologist recommended an initial course of radiation therapy. He also suggested that the parents have a genetic analysis performed so that they could be counseled as to the possible transmission of the disease to other offspring. Marie told the physician that no one in their families, including them, had eye tumors and that she does not understand how they could have the genetic defect that has caused Jerry's problem and still not show the symptoms.

The oncologist explained to Marie and George that sometimes a person may have the genetic alteration but not manifest the disease because of lack of penetrance. This is sometimes the case with retinoblastoma. Pulling out a sheet of paper, the oncologist drew a diagram to show Marie and George how retinoblastoma is inherited (see Figure 8-10). Approximately 40% of all cases of retinoblastoma are familial and inherited as an autosomal dominant trait; the remaining 60% of the cases arise as new mutations. When the disease is familial in origin, the patient develops the disease at an earlier age, and it is characterized by a bilateral distribution with tumors occurring in both eyes.

In the familial form of the disease, the affected individual inherits one defective allele, which is present in every cell in the body. However, this does not mean that every cell is a cancer cell or that cells will develop into cancerous cells. For a tumor to develop, there must be a second mutation that occurs in the somatic cells. With retinoblastoma, there is a high incidence of the second mutation occurring in cells of the eye. Having inherited one defective gene, the second mutation occurring in the other allele of the gene, known as the *Rb* gene, results in the loss of a protein, called a tumor suppressor protein. The function of this tumor suppressor protein is to prevent tumor development.

In Jerry's case, every cell in his body is probably defective in one allele of the *Rb* gene, and in the tumor cells, the second allele is defective as well. A person can inherit the defective allele but never develop a mutation in the second allele of the *Rb* gene. Thus, although one can carry the genetic defect, it is never expressed as a tumor in the eye. The oncologist told Marie and George that their carrier status and the molecular basis for Jerry's disease could be determined by a molecular test. The parents agreed to the test particularly when the oncologist explained that there is an increased risk for people who carry the defective *Rb* gene to develop osteosarcomas or bone tumors later in life. Jerry is at risk for these tumors, and he will need to undergo regular checkups to monitor the development of other tumors.

After undergoing radiation therapy, Jerry's prognosis was good. The large tumor in his left eye shrank significantly, and both tumor foci disappeared. Marie and George, as well as Jerry, provided blood samples for a molecular diagnosis of the disease. Marie did not tell the oncologist that George is not Jerry's biologic father. The results from the molecular analysis indicated that neither George nor Marie has a mutation in the *Rb* gene. However, chromosomal analysis showed that in all of Jerry's cells there is a deletion on one of his chromosomes at position 13q14, the position of the *Rb* gene. The oncologist suggested that the mutation most likely occurred in the father's sperm, and thus was present in all of Jerry's cells although neither parents are carriers. Marie said nothing, knowing that even if she revealed that George is not Jerry's biologic father, the outcome for Jerry would be the same. She took comfort in knowing that any children she and George might have would not be at any greater risk for developing retinoblastoma than the general population.

One year after the radiation therapy, Jerry remained tumor free. The oncologist recommended that the parents bring Jerry in on a regular basis to continue to monitor any further tumor development. At the age of 4 years, a new tumor was found in Jerry's left eye. Because the tumor was still small, the oncologist suggested that they try cryotherapy, a process in which the tumor is frozen at −80°F. The tumor was frozen for 30–60 seconds, and the process was repeated three times; microcrystals formed in the tumor and destroyed it.

The cryotherapy appeared successful, and Jerry has remained free of eye tumors for 3 years. He continues to have regular checkups and is routinely tested for any signs of osteosarcomas. Other than that, he lives a normal healthy life. Marie and George now have a 3-year-old daughter who has no signs of retinoblastoma.

# REVIEW QUESTIONS

**Directions:** The groups of questions below consist of lettered choices followed by several numbered items. For each numbered item, select the lettered option with which it is most closely associated. Each lettered option may be used once, more than once, or not at all.

**Questions 1–3**

Match each of the following definitions with the term it correctly describes.

(A) CAAT element
(B) TATA element
(C) Glucocorticoid response element
(D) Enhancer element
(E) Initiator element

1. This *cis*-acting element is the start site for transcription initiation by RNA polymerase II

2. This *cis*-acting element is a binding site for one of the general transcription factors

3. This *cis*-acting element can be located thousands of base pairs away from the gene it regulates

**Questions 4–6**

Match each of the following definitions with the term it correctly describes.

(A) TFIIF
(B) Sp1
(C) TFIIH
(D) TATA-binding protein
(E) Glucocorticoid receptor

4. This *trans*-acting factor binds to the glucocorticoid response element

5. This *trans*-acting factor contains subunits that function both in transcription and DNA repair

6. This *trans*-acting factor has zinc finger motifs that allow it to bind to GC sequences on the DNA and activate transcription

### Questions 7-9

Match each of the following definitions with the term it correctly describes.

**(A)** *src* gene

**(B)** *Rb* gene

**(C)** H-*ras* gene

**(D)** *p53* gene

**(E)** *jun* gene

7. This gene codes for a tumor suppressor protein that regulates transcription without binding to DNA

8. This gene codes for an protein that acts as a GTPase

9. This gene codes for a tumor suppressor gene that, in response to DNA damage, blocks cells in the G1 phase of the cell cycle

# ANSWERS AND EXPLANATIONS

**1-3. The answers are: 1-E, 2-B, 3-D.** The initiator (Inr) element is the site where RNA polymerase binds to the DNA and initiates synthesis of mRNA. It is known as the start site.

The TATA element is the site where the general transcription factor TFIID binds. The TATA-binding protein is one of the subunits of TFIID, and it binds first to the TATA element.

Enhancers are *cis*-acting sequences located large distances from the core promoter sites of the genes they regulate. They are also orientation-independent and can be located either 5' or 3' to the gene. Enhancers are binding sites for activator proteins, which are brought into proximity with the gene by DNA looping.

**4-6. The answers are: 4-E, 5-C, 6-B.** The glucocorticoid receptor, once bound to a glucocorticoid, enters the nucleus and binds to glucocorticoid response element (GRE) sequences in the DNA. All genes containing this GRE sequence are activated, and transcription occurs, which allows the cell to regulate a group of genes simultaneously in response to a single ligand.

The general transcription factor TFIIH contains subunits with DNA helicase activity, which are known to act as part of the nucleotide excision repair system.

Sp1 is a transcription factor with a DNA-binding domain composed of three zinc finger motifs. Sp1 binds to guanine–cytosine (GC) sequences on the DNA and stimulates transcription of genes that have this sequence proximal to the core promoter.

**7-9. The answers are: 7-B, 8-C, 9-D.** The retinoblastoma (*Rb*) gene codes for the Rb protein, a tumor suppressor protein that is important in regulating the progression of cells through the cell cycle. The Rb protein interacts with other transcription factors and exerts its effect without binding directly to DNA.

The Ras proteins are important proteins in signal transduction, and they exist in a GDP- or GTP-bound state. When the *ras* gene is mutated, it results in an oncogene that codes for a GTPase that is always in the active form.

The *p53* gene is known as the guardian of the genome, and it stops cell growth when a cell is exposed to DNA damage or to stress. The *p53* gene is a DNA-binding protein that can repress transcription of some genes and activate other genes, resulting in the stopping of a cell in the G1 phase of the cell cycle so that DNA repair can occur.

# REFERENCES

1. Orphanides G, Lagrange T, Reinberg D: The general transcription factors of RNA polymerase II. *Genes Dev* 10:2657–2683, 1996.
2. Nikolov DB, Burley S: RNA polymerase II transcription initiation: a structural view. *Proc Natl Acad Sci* 94:15–22, 1997.
3. Tjian R: Molecular machines that control genes. *Sci Am* 272:54–61, 1995.
4. Struhl K: Chromatin structure and RNA polymerase II connection: implications for transcription. *Cell* 84:179–182, 1996.
5. Weatherall DJ, Clegg JB, Higgs DR, et al: The hemoglobinopathies. In *The Metabolic and Molecular Bases of Inherited Diseases*, 7th ed. Edited by Scriver C, Beaudet A, Sly W, et al: New York, NY: McGraw-Hill, 1995, pp 3417–3484.
6. Park M: Oncogenes: genetic abnormalities of cell growth. In *The Metabolic and Molecular Bases of Inherited Diseases*, 7th ed. Edited by Scriver C, Beaudet A, Sly W, et al: New York, NY: McGraw-Hill, 1995, pp 589–611.
7. Levine A: The tumor suppressor genes. *Ann Rev Biochem* 62:623–651, 1993.
8. Weinberg R: Tumor suppressor genes. *Science* 254:1138–1146, 1991.
9. Newsham I, Hadjistilianou T, Cavenee W: Retinoblastoma. In *The Metabolic and Molecular Bases of Inherited Diseases*, 7th ed. Edited by Scriver C, Beaudet A, Sly W, et al: New York, NY: McGraw-Hill, 1995, pp 613–642.
10. Robbins PD, Horowitz J: Positive and negative transcriptional regulation by the retinoblastoma tumor suppressor protein. *Prog Nucleic Acid Res Mol Biol* (in press), 1997.
11. Levine A: p53, the cellular gatekeeper for growth and division. *Cell* 88:323–331, 1997.
12. Ko LJ, Prives C: p53: puzzle or paradigm. *Genes Dev* 10:1054–1072, 1996.
13. Manley J, Tacke R: SR proteins and splicing control. *Genes Dev* 10:1569–1579, 1996.

# 9

# PHENOTYPIC EXPRESSION OF CHROMOSOMAL ABNORMALITIES

Patricia A. Mowery-Rushton, Ph.D.
and Urvashi Surti, Ph.D.

## CHAPTER OUTLINE

## INTRODUCTION OF CLINICAL CASE

Mr. and Mrs. Smith and their 2½-year-old son, Mark, were referred to the medical genetics department for a genetic evaluation. Mark was born with multiple congenital anomalies and has moderate developmental delay and a severe delay in expressive language. Mrs. Smith was also concerned about the recurrence risk for her current pregnancy (12 weeks). The geneticist learned that Mrs. Smith had had two spontaneous abortions (13 and 18 weeks) since Mark was born. A review of Mark's medical history indicated that he was the product of a pregnancy complicated by intrauterine growth retardation (IUGR) and an active hepatitis B infection in a then 22-year-old prima gravida mother. He was delivered at 39 weeks by a scheduled cesarean section because of a breech presentation. He weighed 4 lbs, 5 oz, and his physical examination was unremarkable except for IUGR, bilateral undescended testes, and two pilonidal dimples. He required oxygen for mild respiratory distress and remained in the hospital for 3 weeks because of poor feeding, respiratory distress, and an eye infection. An ophthalmic examination revealed that he had ocular albinism. Assessment of his development indicated an abnormal brainstem auditory evoked response (BAER), abnormal automotor function and behavior, and abnormal hearing. Upon discharge from the hospital, the neonatologist recommended further follow-up and a genetic evaluation, which were not pursued.

The physical examination indicated that Mark was in good health except for recurrent bronchitis and otitis media. His height and weight were below the fifth percentile, and he demonstrated multiple anomalies, which included brachycephaly with a flat occiput, anteverted nose with a smooth philtrum and thin upper lip, epicanthal folds, down-turned mouth, high narrow palate, low-set, posteriorly rotated ears, bilateral clinodactyly of the fifth digits, and undescended testes. His psychomotor development was marked by early delay, but by the age of 2½ years, he was able to walk and climb but not run. He could pick up small objects and had learned to use simple signs appropriately.

The geneticist asked both Mr. and Mrs. Smith to fill out detailed family history forms, which contained questions concerning the health of their parents and siblings (and their children), as well as their ethnic backgrounds. Mr. Smith's family history was unremarkable. However, Mrs. Smith's family history revealed potentially important information concerning the cause of Mark's disorder. Mrs. Smith's mother had 11 recognized pregnancies that resulted in only three liveborn children and eight spontaneous abortions. Mrs. Smith was the oldest of the three children, followed by a 19-year-old sister and a 15-year-old brother. In addition, the family history also indicated that Mark's father was not the father of the current pregnancy. The geneticist asked for blood samples from Mrs. Smith and her son for cytogenetic analysis.

# CLINICAL IMPACT OF CHROMOSOMAL ABNORMALITIES

The occurrence of chromosomal abnormalities has a significant impact on both human fitness and reproduction. The effects of chromosomal abnormalities can be seen as a continuous spectrum from prenatally diagnosed congenital anomalies to dysmorphic and developmentally delayed infants, children, and adults to asymptomatic adults (Table 9-1).

**TABLE 9-1** ▶
*Indicators of Chromosomal Abnormalities*

| Prenatal | Postnatal | Adult |
|---|---|---|
| Intrauterine growth retardation | Dysmorphic features | Infertility |
| Congenital malformations | Developmental delay | Recurrent spontaneous abortions |
| Hydrops | Poor physical growth | Stillborn or liveborn dysmorphic infants |
| Abnormal maternal serum screen | | |

## Chromosomal Abnormalities in Spontaneous Abortions

It is believed that at least 15% of all recognized human conceptions are spontaneously aborted in the first trimester and that approximately 50% of these are chromosomally abnormal. The estimates would probably be much higher if unrecognized early embryonic losses (< 5 weeks) and stillborn infants (> 28 weeks) were included in these calculations [1]. The most common types of chromosomal abnormalities seen in spontaneous abortions are numerical, accounting for 95% of the cases. Structural rearrangements have been identified in approximately 4% of the cases (Table 9-2). Because most pregnancy losses with chromosomal abnormalities consist of placental material with or without identifiable embryonic or fetal parts, it is important for pathologists to be aware of the association between fetoplacental anomalies and specific cytogenetic abnormalities [2, 3].

**TABLE 9-2**
*Frequency of Chromosomal Abnormalities
in Spontaneous Abortions*

| Type | Frequency (%)[a] |
|---|---|
| Aneuploidy | |
|   45,X | 20 |
|   Autosomal monosomy | <1 |
|   Autosomal trisomy | |
|     Total | 52 |
|     Trisomy 16 | 16 |
|     Trisomy 18 | 3 |
|     Trisomy 21 | 5 |
|     Trisomy 22 | 5 |
|     Other trisomies | 23 |
| Triploidy | 16 |
| Tetraploidy | 6 |
| Structural rearrangements | 4 |

[a] Approximate percent.
*Source:* Reprinted with permission from Boue' A, Boue' J, Gropp A: Cytogenetics of pregnancy wastage. In *Advances in Human Genetics*, vol. 14. Edited by Harris H, Hirschhorn K. New York, NY: Plenum Press, 1985, p 10.

**Numerical Chromosomal Abnormalities in Spontaneous Abortions.** Numerical chromosomal abnormalities are divided into two types: aneuploidy and polyploidy. *Aneuploidy* involves the loss (monosomy) or gain (trisomy) of a single chromosome, resulting in either a 45- or 47-chromosome complement. Rarely, there may be two additional chromosomes present with a total of 48 chromosomes (double trisomy). *Polyploidy* involves the addition of one or more complete haploid (23) sets of chromosomes, giving rise to triploid (69) or tetraploid (92) chromosomal complements. Trisomy is the most common finding in chromosomally abnormal embryos, accounting for approximately 50% of the cases. Trisomies for all autosomes have been reported, including chromosome 1, which was recently reported in a clinically recognized pregnancy [4]. The most common trisomy reported in spontaneous abortions is trisomy 16.

The most common chromosomal abnormality seen in spontaneous abortions is 45,X or Turner's syndrome karyotype (20%). More than 95% of these conceptuses are lost during the first trimester of pregnancy and usually consist of hypoplastic chorionic villi with the presence of a growth-disorganized embryo. Fetal anomalies seen in later-gestation spontaneous abortions or on ultrasound may be very distinct and are discussed in more detail in Warburton [2] and in Sanders [5]. Phenotypic features of liveborn infants, children, and adults with Turner's syndrome are discussed later in this chapter.

Triploidy is the second most common cytogenetic abnormality seen in spontaneous abortions (16%). It occurs in approximately 1% of recognized conceptions, most of which end in spontaneous abortions. Very rarely a triploid fetus survives to term but dies either at or shortly after birth. Triploidy can arise by five mechanisms, which are summarized in Table 9-3. The parental origin of the extra set of chromosomes plays a role in the phenotype of the placenta. When the extra chromosomal complement is derived from the father, the placenta becomes enlarged and cystic with areas of hyperplasia. Often, the fetus suffers IUGR, most likely the result of placental insufficiency. The enlarged cystic placenta is referred to as a partial hydatidiform mole (PHM) and may cause additional complications (e.g., preeclampsia, hyperemesis, hyperthyroidism) to the mother if the triploid conceptus is not diagnosed until later in the pregnancy. When the extra chromosomal complement is derived from the mother, the placenta appears normal, and the fetus is usually better developed. The phenotype of the fetus is always abnormal and may include multiple congenital anomalies, such as hydrocephalus, holoprosencephaly, neural tube defects, facial dysplasia, syndactyly of the third and fourth digits, congenital heart disease, omphalocele, and renal anomalies.

**TABLE 9-3** ▶

*Mechanism of Origin of Triploid Conceptions*

| Mechanism | Ratio (Paternal:Maternal) | Frequency (%) |
|---|---|---|
| Dispermy: fertilization of a normal haploid egg by two individual sperm | 2:1 | 40 |
| Diandry I: fertilization of a normal haploid egg by a diploid sperm due to a meiosis I error | 2:1 | 15 |
| Diandry II: fertilization of a normal haploid egg by a diploid sperm due to a meiosis II error | 2:1 | N/A |
| Digyny I: a single haploid sperm fertilizing a diploid egg due to a meiosis I error | 1:2 | 7.5 |
| Digyny II: a single haploid sperm fertilizing a diploid egg due to a meiosis II error | 1:2 | 12.5 |
| Noninformative | ? | 25 |

*Note.* N/A = not available.

***Structural Chromosomal Abnormalities in Spontaneous Abortions.*** The most common structural abnormality is the reciprocal translocation of genetic material between two nonhomologous chromosomes. Balanced chromosomal translocations are present in approximately 1 in 500 newborns and do not usually have any phenotypic effect on the carrier because there has not been any loss or gain of genetic information. Despite their normal phenotype, translocation carriers have a significantly increased risk of having chromosomally unbalanced offspring or, in many cases, multiple spontaneous abortions or infertility. Inversions are a second type of chromosomal rearrangement that can also influence the reproductive fitness of an otherwise normal individual. Most carriers of balanced chromosomal rearrangements are usually not identified until multiple miscarriages or after the birth of an abnormal child with an unbalanced chromosomal complement. Unbalanced rearrangements usually involve duplications or deletions of regions of the chromosome, or both. The severity of the phenotype depends on the size of the segment and the specific genes located within that region. Chromosomal rearrangements, both balanced and unbalanced, occur spontaneously or are induced by environmental factors, such as ionizing radiation, viral infections, or certain chemicals.

In addition to whole chromosomal abnormalities and large structural abnormalities, there is increasing evidence that previously unidentified "cryptic" translocations may play a significant role in pregnancy loss and infant morbidity and mortality. Current technology using fluorescent in situ hybridization (FISH) only recently has allowed investigators to identify subtle translocations that may be associated with multiple spontaneous abortions, dysmorphic infants, and developmental delay.

## Genomic Imprinting and Uniparental Disomy (UPD)

The appearance of a normal diploid chromosomal complement in the product of spontaneous abortion or a dysmorphic or developmentally delayed infant or child may be deceiving. Although all of the appropriate genetic material may be present, there can be deficient or over expression of certain genes. This can occur when a conceptus inherits both copies of a homologous chromosome pair from only one parent (i.e., UPD), as opposed to inheriting one chromosome from each parent. The net imbalance is not in the amount of genetic material but rather in the expression of the genetic material. This phenomenon, known as genomic imprinting, involves the specific germline modification that produces functional differences in the expression of genetic material depending on its parental origin. Imprinting is a normal mechanism for gene regulation that involves only a small number of genes. Currently, there are only five chromosomal regions known to contain imprinted genes in humans: maternal 7q, paternal 11p, maternal 14q, maternal 15q, and paternal 15q. The imprinted genes appear to be arranged in clusters that have been evolutionarily conserved. Despite the relatively small number of imprinted genes, disruption of the proper expression patterns because of UPD, deletion, or relaxation of imprinting has been shown to have serious consequences in embryonic development, genetic disorders, and cancer.

The evidence for genomic imprinting was first inferred from experiments in mice and later confirmed in humans by the observations of specific diseases. Nuclear transplantation experiments in mice involving two male pronuclei (androgenesis) and two female pronuclei (gynogenesis) demonstrated that it is essential to have one male and one female pronuclei for normal development. Androgenesis results in better development of the trophoblast but poor fetal development, whereas gynogenesis results in better development of the fetus but poor trophoblast development. These abnormal conceptions in mice are very similar to complete hydatidiform moles (CHM) and ovarian teratomas in humans (see Chapter 5). The creation of mice with UPD for whole chromosomes or segments of chromosomes has aided in identifying specific chromosomal regions that contain imprinted genes and then subsequently the homologous imprinted genes in humans (Table 9-4).

**◀ TABLE 9-4**
*Homologous Imprinted Genes in Mice and Humans*

| Mouse Gene | Chromosome Location | Human Gene | Chromosome Location | Expression |
|---|---|---|---|---|
| Igf2 | 7 | IGF2 | 11p15.5 | Paternal |
| H19 | 7 | H19 | 11p15.5 | Maternal |
| p57$^{KIP2}$ | 7 | p57$^{KIP2}$ | 11p15.5 | Maternal |
| Snrpn | 7 | SNRPN | 15q11q13 | Paternal |
| Znf127 | 7 | ZNF127 | 15q11q13 | Paternal |

*Note.* p = short arm of chromosome; q = long arm of chromosome.

UPD for an imprinted gene produces distinct abnormal phenotypes that are associated with the parental origin of the disomic gene. In some cases, the phenotypes appear to represent opposite expression of genes involved in embryonic growth and development. An example of this phenomenon is the insulin-like growth factor 2 (*IGF2*) gene, which is repressed on the maternally derived chromosome. Maternal disomy involving the *IGF2* gene is characterized by a deficiency of IGF2 and leads to growth retardation in embryonic mice. On the other hand, paternal disomy for *IGF2* results in overexpression of the gene and subsequent overgrowth of embryonic mice. This gene is also involved in the human genetic disorder, Beckwith-Wiedemann syndrome, which is characterized by exomphalos, macroglossia, prenatal and postnatal overgrowth, characteristic facial features, ear lobe grooves or pits, visceromegaly, hemihypertrophy, neonatal hypoglycemia, and embryonal tumors. This disorder can arise by several mechanisms: (1) paternal UPD for chromosome 11 or partial disomy for 11p15.5, (2) duplication of part of the paternal chromosome 11, resulting in partial trisomy, or (3) disruption of the maternal imprint, resulting in expression of the maternal allele (normally silent). Several other imprinted genes are located within the same region of chromosome 11p and may be involved in the development of Beckwith-Wiedemann syndrome. The *H19* gene is closely linked to *IGF2* and is oppositely imprinted, meaning it is expressed from the maternal allele. The function of *H19* is not known, but it is widely expressed during embryonic development and can suppress tumor cell growth. A third imprinted gene, *p57$^{KIP2}$*, is a tumor-suppressor gene expressed only from the maternal allele. Lack of expression of this gene may play a role in the increased risk for embryonal tumors seen in Beckwith-Wiedemann patients [6].

The first case of UPD discovered in humans involved a 7-year-old girl with cystic fibrosis and short stature who was cytogenetically normal [7]. Molecular genetic studies indicated that she had inherited two copies of the maternal chromosome 7 with no paternal contribution. Unfortunately, the mother was a cystic fibrosis carrier, and her daughter had inherited two copies of the chromosome carrying the mutant gene (isodisomy). The occurrence of short stature in this patient, and in several others, was also associated with maternal UPD. Since then, UPD for chromosome 7 has been associated with Silver-Russell syndrome and primordial growth retardation [8].

UPD is primarily caused by meiotic nondisjunction followed by trisomy or monosomy "rescue" (Figure 9-1). The incidence is not known because only cases involving imprinted genes or the unmasking of autosomal recessive mutations have been ascertained in a clinical setting. There is an age-related increase in UPD, and in many cases it has been diagnosed prenatally because of the observance of trisomic mosaicism in chorionic villi. UPD for chromosomes 15 and 16 are the most frequent to be diagnosed prenatally following detection of trisomic cells. UPD for chromosome 15 can result in either Prader-Willi syndrome or Angelman's syndrome, which are the classic examples of genetic disorders influenced by genomic imprinting. Both can arise because of three distinct mechanisms: (1) a deletion of chromosome 15q, (2) UPD for chromosome 15, or (3) an imprinting error. However, the resulting phenotype depends on the parental origin of the chromosome involved. These syndromes are discussed in more detail in Chapter 5 and later in Chapter 9.

**FIGURE 9-1** ▶

**Mechanism for Uniparental Disomy.** This example illustrates a mechanism for uniparental disomy caused by a maternal meiosis I error involving chromosome 15. When the aberrant egg is fertilized by a normal sperm, the resulting zygote is trisomic for chromosome 15. For maternal uniparental disomy to arise, the paternal chromosome 15 must be lost during early embryonic development in cells destined to give rise to embryonic tissue.

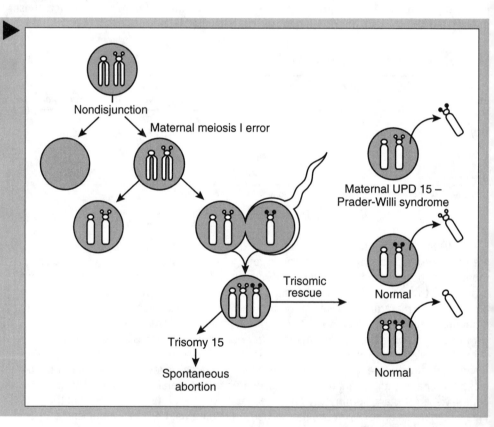

UPD for chromosome 16 is often associated with a poor pregnancy outcome, such as IUGR or fetal death. However, it is not known whether these effects are caused by UPD or by the presence of trisomic cells in the placenta. There appears to be a correlation between the level of trisomic cells and a poor pregnancy outcome. Currently there are no known imprinted genes located on human chromosome 16 or on any homologous sequences to mouse chromosomes that have been shown to have a significant phenotypic effect in uniparental animals. For additional information on UPD and genomic imprinting refer to Ledbetter and Engel [9] and Sapienza and Hall [10].

## Mosaicism

Chromosomal abnormalities are commonly thought to be present in all of the cells of an affected individual. However, in many cases, the abnormality is found only in a subset of cells, such that the individual is mosaic for two or more cell lines. The effects of mosaicism depend on several different factors: (1) the stage of development in which the second cell line arises, (2) the cell type or tissues involved, and (3) the type of chromosomal abnormality (i.e., numerical or structural). All of these factors play a role in the severity of the genetic disorder. Mosaicisms for numerical abnormalities are the most common and are usually caused by mitotic nondisjunction in early postzygotic development. If the original cell had a normal chromosomal complement, then the resulting cell lines would be monosomic and trisomic for the chromosome involved in the nondisjunction event. In some cases, the original cell line has an abnormal chromosomal complement (e.g., trisomy 21), with the nondisjunction event resulting in the creation of a "normal" cell line. This phenomenon, known as "trisomic rescue," is often seen in conceptions with nonviable trisomies (e.g., trisomy 15 or 16) in which the fetus would not normally have survived. Unfortunately, the karyotypically normal disomic cell line is not always genetically normal but instead has UPD for the original trisomic chromosome. UPD for chromosome 15 results in either Angelman's or Prader-Willi syndrome, and UPD for chromosome 16 results in IUGR or fetal demise.

There are many examples of somatic mosaicism cited in the literature that give rise to variable or incomplete phenotypic expression of known chromosomal abnormalities (e.g., mosaic trisomy 21, Turner's syndrome). When mosaicism is detected prenatally, it is very difficult to determine the phenotypic effects because only a small proportion of cells from a limited tissue type (e.g., chorionic villi or amniotic fluid) have been examined. In addition, the results of cytogenetic studies may not reflect the actual proportion of abnormal cells because cells from different tissues may have different percentages of abnormal cells. Another consideration is that the normal and abnormal cells may react differently to culture conditions, which can greatly influence the findings. One example is Pallister-Killian syndrome, which is a syndrome of multiple congenital anomalies and mental retardation, with tissue-specific mosaicism for an isochromosome for the short arm of chromosome 12, that is, i(12p). The cytogenetic diagnosis often is not made because of inadequate growth of the abnormal cells in peripheral blood leukocytes. When Pallister-Killian syndrome is suspected, it is usually necessary to obtain a skin biopsy or a bone marrow aspirate to confirm the diagnosis. It is important to remember that most cases of mosaicism are diagnosed as a result of a specific clinical problem, which is usually attributable to the presence of the mosaic cell line. However, it is not known how many undiagnosed cases of low-level mosaicism may be in the general population, because affected individuals may not display any abnormal features. Theoretically, every human is probably mosaic for one or more abnormal cell lines, which, because of the low number of abnormal cells, have no phenotypic effect. This has been documented in cryptic carriers of single gene disorders, such as hemophilia A, Duchenne's muscular dystrophy, Ehlers-Danlos syndrome, and osteogenesis imperfecta, in which an apparently unaffected parent is a low-level gonadal or somatic mosaic (or both) for a cell line containing the mutant gene.

## Autosomal Numerical Chromosomal Abnormalities in Liveborn Infants

Only three autosomal trisomies are known to allow a fetus to survive to term and are therefore seen in the population: trisomies 13, 18, and 21 (Table 9-5). Other autosomal trisomies occur very rarely and are usually present only in a mosaic form.

*Trisomy 21.* Down's syndrome (trisomy 21) is the most common chromosomal abnormality in newborns. It is seen in approximately 1 in 800 live births, which translates to 6000 children born with Down's syndrome in the United States each year. However, it occurs more frequently, with an incidence of 45 in 10,000 (1 in 222) recognized conceptions, most of which are lost as spontaneous abortions. The risk of having an infant with trisomy 21 increases with the age of the mother. The age-related risk of a woman younger than age 30 years is approximately 1 in 800 (population risk); however, this risk increases to 1 in 350 by the age of 35 years. The overall recurrence risk is 1%, partly due to rare families with germline mosaicism for trisomy 21. The actual recurrence risk would be far higher in these families, depending on the level of mosaicism.

Prenatal screening by ultrasound or analysis of biochemical markers in the maternal

**TABLE 9-5** ▶
*Frequency of Viable Trisomies in 100,000 Conceptions*

| Outcome | Conceptions | Spontaneous Abortion | Live Births |
|---------|-------------|----------------------|-------------|
| Trisomy 21 | 450 (~1/200) | 345 (77%) | 105 (~1/800) |
| Trisomy 18 | 200 (1/500) | 190 (95%) | 10 (1/8500) |
| Trisomy 13 | 195 (1/500) | 190 (97.5%) | 5 (1/17,000) |
| Total | 845 | 725 (85%) | 120 (~1/700) |

**Features of Down's Syndrome in a Neonate**
*Flat facial profile*
*Poor Moro reflex*
*Hypotonia*
*Upward-slanting palpebral fissures*
*Redundant loose neck skin*
*Hyperextensibility of joints*
*Dysplasia of pelvis*
*Simple, small round ears*
*Dysplasia of midphalanx of fifth finger*
*Single palmar crease*

serum can identify women under the age of 35 years who may have an increased risk for carrying a fetus with trisomy 21. Ultrasound indicators include nuchal thickening, cystic hygroma, duodenal stenosis or atresia (double-bubble sign), short femur length, and cardiovascular defects (e.g., endocardial cushion defect with atrial and ventricular septal defects, abnormal mitral and tricuspid valves). Measurement of the levels of certain biochemical markers in the maternal serum between 16 and 20 weeks' gestation can also identify an at-risk pregnancy. This type of test is referred to as a triple screen or maternal serum screen. In a pregnancy affected with Down's syndrome, there is usually a 25% reduction in the levels of α-fetoprotein (AFP) and unconjugated estradiol (uE3), and a twofold increase in human chorionic gonadotropin (HCG) hormone. These values are compared to the population mean values for the gestational age of the fetus. Triple-screen testing can detect approximately 60% of Down's syndrome pregnancies. There are three types of prenatal tests that can be used to confirm the diagnosis of Down's syndrome by examining the infant's chromosomes: amniocentesis, chorionic villus sampling (CVS), and percutaneous umbilical blood sampling (PUBS). Each of these tests can be performed at a particular gestational age, and each test carries some risk of miscarriage (see Chapter 13).

An infant with trisomy 21 usually presents in the newborn period with hypotonia and the characteristic facial appearance [11]. Approximately 85% of Down's syndrome infants survive to 1 year of age, and 50% can be expected to live longer than 50 years. The leading cause of death in infancy is congenital heart disease. However, with improvements in treatment, as many as 70% of infants with heart disease survive to 1 year of age. The degree of developmental delay varies greatly, and it is often impossible to predict the intellectual potential of young Down's syndrome children based on developmental milestones. In general, social and emotional development is near normal in the first year of life. However, delays in language and motor skills usually become apparent by 2 years of age. Early intervention programs, as well as physical and occupational therapy, may be very beneficial for proper individualized management (Figure 9-2). Most children demonstrate moderate-to-severe mental retardation with intelligence quotient (IQ) scores ranging from 20 to 85. This is contradicted by a higher-than-expected social development. Unfortunately, once those affected with Down's syndrome reach adulthood, there is a decline in their intellectual and cognitive abilities.

Children with Down's syndrome generally experience a reduction in their growth velocity, which ultimately leads to short stature, and most develop obesity in early childhood because of a lower-than-average metabolic rate. Other important features include hearing loss, ophthalmic disorders (e.g., cataracts, glaucoma), thyroid disorders, impaired cellular immunity, and leukemia. Many adults with Down's syndrome are capable of living in a supervised, group-home setting and can hold a steady job. By the age of 40 years, many begin to experience progressive Alzheimer-like dementia, and by 60 years of age, 75% exhibit signs and symptoms of Alzheimer's disease.

A clinical diagnosis of Down's syndrome should always be confirmed by cytogenetic analysis. The most common mechanism of origin for trisomy 21 is nondisjunction in the first meiotic division during formation of the oocyte. This accounts for approximately 90% of trisomy 21 cases. Only approximately 5% of cases are paternal in origin, with

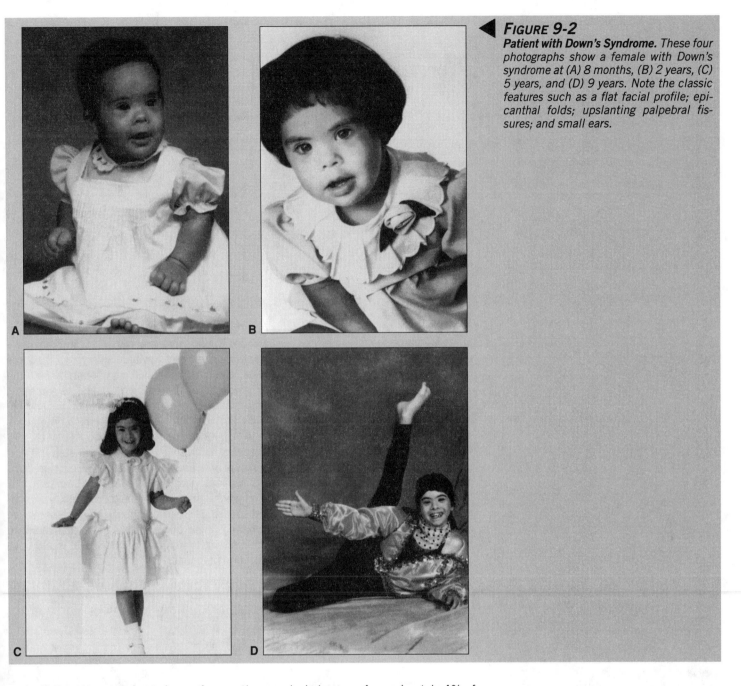

**FIGURE 9-2**
*Patient with Down's Syndrome. These four photographs show a female with Down's syndrome at (A) 8 months, (B) 2 years, (C) 5 years, and (D) 9 years. Note the classic features such as a flat facial profile; epicanthal folds; upslanting palpebral fissures; and small ears.*

meiosis II errors occurring twice as frequently as meiosis I errors. Approximately 4% of Down's syndrome cases are caused by Robertsonian translocations (Figure 9-3), which involve two acrocentric chromosomes (13, 14, 15, 21, 22) that fuse near the centromeric region with loss of the short arms. In those with translocation Down's syndrome, the most common Robertsonian translocations are t(14;21) and t(21;22). It is very important in cases of translocation Down's syndrome to perform chromosome analysis on both parents to rule out the possibility that either is a carrier of a Robertsonian translocation. Approximately 40% of affected individuals with a translocation between a D group chromosome (i.e., 13, 14, or 15; see Chapter 5) and chromosome 21, that is, t(D;21), inherit the aberrant chromosome from a carrier parent (the mother in 90% of cases). In contrast, only 7% of translocations between chromosome 21 and another G group chromosome, that is, t(G;21), are inherited, also usually from the mother. Carriers of balanced Robertsonian translocations are phenotypically normal but have an increased risk of chromosomally unbalanced offspring. A balanced Robertsonian translocation carrier has only 45 chromosomes, but has a derivative chromosome that replaces the two missing acrocentric chromosomes; that is, 45,XX,der(14;21)(q10;q10). The net

**FIGURE 9-3** ▶

**Robertsonian Translocation.** *The normal chromosomes 14 and 21 are shown on the left. The derivative (14;21) is shown on the right. The derivative chromosome may have either one of the centromeres, both centromeres, or a portion of both centromeres. The reciprocal derivative chromosome (p arms) is always lost. The Robertsonian translocation between chromosomes 14 and 21 is the most common.*

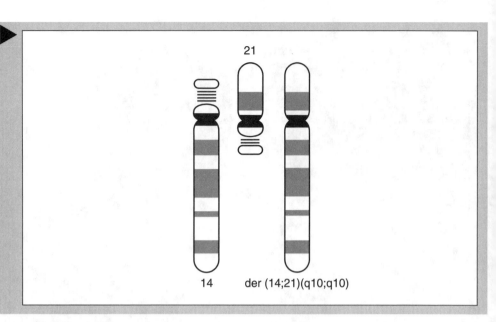

14        der (14;21)(q10;q10)

imbalance—the loss of the short arms from the two acrocentric chromosomes—does not have an adverse phenotypic effect. In a balanced Robertsonian translocation carrier there are six possible gametes (Figure 9-4), but only three give rise to a viable embryo. Of the three viable types, one is normal, one is balanced, and one is unbalanced, containing both the derivative chromosome and a normal chromosome 21. When the unbalanced gamete is fertilized by a normal gamete, the resulting imbalance produces translocation trisomy 14, which results in a spontaneous abortion or in translocation Down's syndrome. The corresponding karyotype for Down's syndrome is 46,XX,der(14;21),+21 or 46,XY,der(14;21), +21. If inherited from the mother, the observed recurrence risk is approximately 10%; if inherited from the father, the recurrence risk is only about 2%. The only exception is for carriers of a 21q;21q translocation. This derivative chromosome is thought to be caused by a failure in centromere division, producing an isochromosome for the long arms of chromosome 21. The potential progeny of a balanced 21q;21q carrier inherit either the derivative chromosome (leading to trisomy 21) or no chromosome 21 from the carrier parent (leading to monosomy 21). Monosomy 21 is embryonic lethal; therefore, this individual would have either miscarriages and or Down's syndrome term pregnancies.

**FIGURE 9-4** ▶

**Potential Gametes of a Robertsonian Translocation Carrier.** *During meiosis, six potential gametes can be formed. Three are nonviable and would give rise to aberrant embryos if fertilized by a normal sperm. The remaining three gametes would produce potentially viable embryos; one would be normal, one would be a balanced translocation carrier, and one would have Down's syndrome (trisomy 21).*

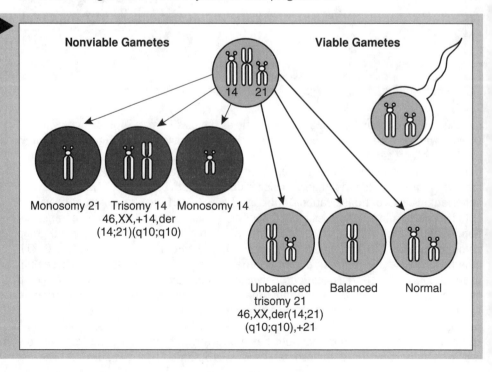

**Nonviable Gametes**                **Viable Gametes**

14    21

Monosomy 21    Trisomy 14    Monosomy 14
46,XX,+14,der
(14;21)(q10;q10)

Unbalanced    Balanced    Normal
trisomy 21
46,XX,der(14;21)
(q10;q10),+21

In a small number of cases, the patient is mosaic for the trisomic cell line and a normal cell line. Mosaicism can arise by two mechanisms: (1) a trisomic conceptus with loss of one chromosome 21 in an early progenitor cell, leading to a disomic cell line; or (2) a disomic cell undergoes nondisjunction during mitosis because of anaphase lag, giving rise to a trisomic cell line (see Chapter 5).

**Trisomy 18** is seen in approximately 1 in 8500 live births, but occurs in 1 of 500 conceptions. It is the second most common autosomal abnormality in humans. Approximately 95% of trisomy 18 conceptions are lost as spontaneous abortions. In the past decade, there has been a significant increase in the number of cases of trisomy 18 that are diagnosed prenatally in the second and third trimester. This is because of the increased use of ultrasound to detect major malformations in association with commonly diagnosed abnormalities.

Maternal serum biochemical marker screening can also detect as many as 80% of trisomy 18 fetuses. The serum levels of all three hormones are usually reduced below the population mean. Rapid confirmation of trisomy 18 by cytogenetic analysis is important so that appropriate management decisions can be made, including not only prenatal options but also delivery options and questions concerning immediate medical attention for severely affected infants. As many as 50% of trisomy 18 fetuses are delivered by cesarean section because of IUGR or fetal distress during labor; this option may not be indicated in prenatally diagnosed cases because of the extremely poor prognosis.

Only approximately 30% of the cases diagnosed prenatally survive to term, resulting in a survival rate lower than 5%. At birth, the sex ratio is skewed toward females, and excessive male mortality in the first several months of life contributes to a 3:1 sex ratio favoring survival of females with trisomy 18. The survival rate is based on limited information and suggests that as few as 5% of infants survive to 1 year. However, it is also important to note that there are documented cases of survival into the early 20s.

Infants with trisomy 18 are small for gestational age at the time of delivery and usually display severe hypotonia and impaired neurologic development. They have a typical facial appearance with a narrow head, open metopic sutures, small ears, a prominent occiput, and a receding jaw (Figure 9-5). Other features that are strongly suggestive of trisomy 18 include clenched fists, with the second and fifth digits overlapping the third and fourth; rocker-bottom feet with prominent calcanei; short sternum; and often cardiac and renal anomalies. The high mortality rate is usually attributed to the presence of cardiac malformations, the most common being ventricular septal defects. However, other factors also appear to play a role in the low survival (e.g., feeding difficulties and apnea caused by central nervous system [CNS] defects). Only a small number of infants with trisomy 18 survive the neonatal period; therefore, the information concerning developmental disability and future medical complications is severely limited. All affected children are severely or profoundly mentally retarded, displaying significant delay in both motor and speech development. Most children never acquire the ability to walk independently, and verbal communication in older children is usually limited to only a few words. Despite the severe limitations of trisomy 18, affected children do smile and laugh and are able to interact with their families. It should also be noted that there is no regression in their development once milestones are reached. In the subsequent medical management of infants and children with trisomy 18, there are certain problems that should be monitored, including gastroesophageal reflux, apnea, scoliosis, hearing loss, and Wilms' tumor.

The majority of trisomy 18 cases are caused by the presence of an extra chromosome 18; very few cases have been attributed to translocations or other chromosome 18 aberrations. In approximately 90% of the cases, the extra chromosome is maternal in origin, usually because of a meiosis II error. The rate of trisomy 18 increases with increasing maternal age, with the incidence detected at the time of amniocentesis rising from 1 in 2000 at the age of 35 years to 1 in 130 at the age of 43 years. In the small number of cases in which the extra chromosome is paternal in origin, most are caused by postfertilization mitotic errors. Mosaic trisomy 18 has also been reported. The clinical phenotype ranges from normal to full trisomy 18, depending on the level of mosaicism and the tissues involved. Because the level of mosaicism can vary greatly depending on

*Ultrasound Features of Trisomy 18*
*Severe, early-onset IUGR*
*Congenital heart lesions (VSD)*
*Diaphragmatic hernia*
*Polyhydramnios caused by esophageal atresia*
*Choroid plexus cyst*
*Omphalocele*
*Renal anomalies*
*Absence of corpus collosum*
*Clenched fists*

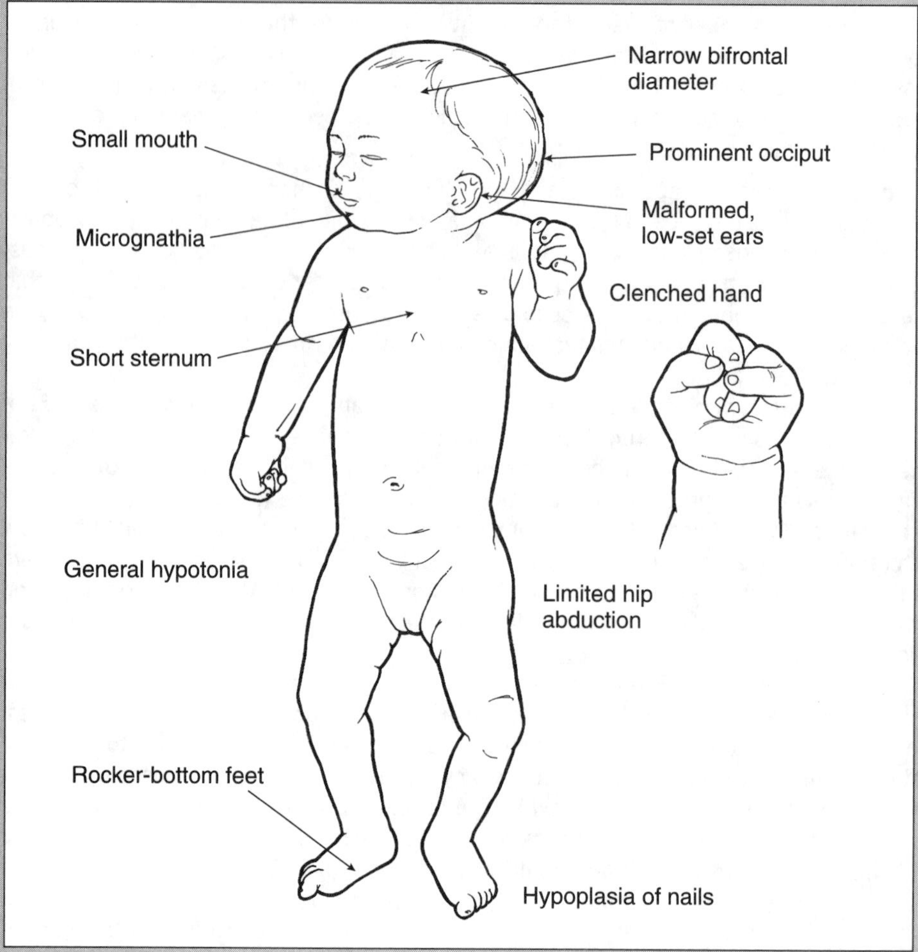

FIGURE 9-5 ▶

**Features of Trisomy 18.** *This figure illustrates the most common features seen in a trisomy 18 infant, which include typical facial features, prominent occiput, typical clenched hand, and rocker-bottom feet.*

the tissue type, it may be necessary to examine multiple cell types to confirm a diagnosis of mosaic trisomy 18.

Because trisomy 18 is very rare, little information is available about the recurrence risk. It is thought that some women may be predisposed to nondisjunctional aneuploidy; therefore, the risk for recurrence is usually adjusted to 1% above the age-related risk in older mothers and 1% below the age-related risk in younger mothers. If one of the parents is a balanced carrier of a structural rearrangement, then the risk may be substantially higher. The risk should be assessed based on the type of rearrangement and its pattern of segregation, based on family history.

**Trisomy 13** is extremely rare, occurring in only 1 in 17,000–25,000 live births. The incidence among recognized conceptions is not known. However, trisomy 13 accounts for 8% of all abortuses with autosomal trisomies, whereas trisomies 21 and 18 account for only 5% and 3%, respectively. The phenotype of a trisomy 13 fetus is usually very striking and easily diagnosed by ultrasound. The majority of fetuses display some degree of holoprosencephaly with incomplete development of the forebrain and olfactory and optic nerves, usually associated with facial clefting, anophthalmia, synophthalmia, or microphthalmia (Figure 9-6). Other associated abnormalities include congenital heart defects (e.g., atrial or ventricular septal defects, patent ductus arteriosis), renal abnormalities (e.g., hydronephrosis, polycystic kidney, hypoplasia), omphalocele, and polydactyly. Approximately 50% of the trisomy 13 cases are prenatally diagnosed by ultrasound. The diagnosis of trisomy 13 should be confirmed by cytogenetic analysis of fetal cells derived from either chorionic villi, amniotic fluid, or PUBS.

The prognosis for an infant born with trisomy 13 is extremely poor. Within the first month, 44% of trisomy 13 infants die; 70% die within 6 months. Only 15%–20% of infants survive the first year. Death is usually attributed to holoprosencephaly or congenital heart defects. There have been several reports of long-term survival of affected

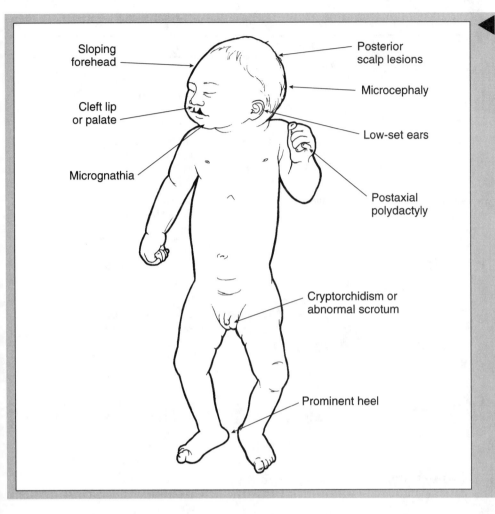

*Features of Trisomy 13.* This figure illustrates the most common features seen in a trisomy 13 infant, which include microcephaly with a sloping forehead, holoprosencephaly with cleft lip, cleft palate, or both, polydactyly, prominent heel, and cryptorchidism in males.

individuals, some even into adulthood. These individuals often suffer from seizures and may be deaf and blind. Mental and physical development is severely limited but does not appear to be as severe as in trisomy 18. Some older children with trisomy 13 have acquired the ability to walk unassisted, to understand and follow simple commands, and to interact with others.

## Autosomal Structural Chromosomal Abnormalities

The deletion or duplication of multiple unrelated genes that are located in proximity to each other is referred to as a *contiguous gene syndrome*. There are multiple mechanisms by which this can occur, but the ultimate effect is an alteration in the normal gene dosage. The resulting phenotype depends on the size of the region involved, as well as the genetic content. Most contiguous gene syndromes are characterized by some degree of mental retardation and one or more abnormal traits that are not usually associated with mental retardation. Each syndrome is characterized by a phenotypic spectrum of abnormal features, which may vary in intensity depending on the exact size and location of the deletion or duplication. In some cases, the actual structural change may be too small to be seen cytogenetically and may require molecular cytogenetic techniques to confirm. Although there are many contiguous gene syndromes (Table 9-6), only a few are discussed in this chapter. For additional information consult Jones [12] and Ledbetter and Ballabio [13], as well as Chapter 12.

### DELETION AND MICRODELETION SYNDROMES

*Prader-Willi Syndrome.* The incidence of Prader-Willi syndrome is 1 in 15,000 live births. In most cases, the condition is sporadic. Documented familial cases are few (familial cases of Angelman's syndrome are more common), although several families with multiple affected siblings with "apparently" normal karyotypes have been reported, and one case caused by a familial rearrangement is known. The recurrence risk is

**TABLE 9-6** ▶
*Autosomal Contiguous Gene Syndromes*

| Chromosome Location | Syndrome | Features |
|---|---|---|
| del(4)(p16.3) | Wolf-Hirschhorn syndrome | Dysmorphic features<br>Mental retardation<br>Growth retardation/microcephaly<br>Ocular hypertelorism<br>Cardiac defect |
| del(5)(p15.5) | Cri du chat syndrome | High-pitched, cat-like cry<br>Moderate-to-severe mental retardation<br>Hypotonia<br>Growth delay |
| del(7)(q11.2) | William's syndrome | Dysmorphic features<br>Failure to thrive<br>Developmental delay and overly friendly personality<br>Supravalvular aortic stenosis<br>Hypercalcemia/hypercalciuria<br>Musculoskeletal abnormalities |
| del(8)(q24.1) | Langer-Giedion syndrome/ trichorhinophalangeal | Multiple exostoses<br>Dysmorphic features<br>Variable intelligence |
| del(11)(p13) | WAGR syndrome | Wilms' tumor<br>Aniridia<br>Genitourinary dysplasia<br>Mental retardation |
| dup(11)(p15.5) paternal | Beckwith-Wiedemann syndrome | Exomphalos<br>Macroglossia<br>Macrosomia<br>Hypoglycemia<br>Hemihypertrophy/overgrowth |
| del(15)(q11q13) paternal | Prader-Willi syndrome | Mental retardation<br>Hypotonia<br>Obesity<br>Dysmorphic features<br>Hypopigmentation |
| del(15)(q11q13) maternal | Angelman's syndrome | Mental retardation<br>Seizures<br>Ataxic gait and hand flapping<br>Inappropriate laughter<br>Hypopigmentation |
| del(16)(p13.3) | Rubinstein-Taybi syndrome | Mental retardation<br>Dysmorphic features<br>Broad thumbs and toes |
| del(17)(p13.1–13.3) | Miller-Dieker syndrome | Lissencephaly (agyria, pachygyria)<br>Profound mental retardation<br>Dysmorphic features |
| del(17)(p11.2) | Smith-Magenis syndrome | Mental retardation<br>Dysmorphic features<br>Hyperactivity<br>Self-destructive behavior |
| dup(17)(p11.2) | Charcot-Marie-Tooth disease type 1A | Hypotonia<br>Decreased reflexes<br>Club foot |
| del(20)(p11.2p12) | Alagille syndrome | Chronic cholestasis<br>Dysmorphic features<br>Vertebral anomalies<br>Ocular embryotoxin |

| Chromosome Location | Syndrome | Features |
|---|---|---|
| del(22)(q11.2)<br>del(10)(p13) | DiGeorge syndrome and<br>  velocardiofacial syndrome | Absent or hypoplastic thymus and<br>  parathyroid<br>Cardiac malformations<br>Dysmorphic features<br>Mental retardation (variable)<br>Cleft palate |
| dup(22q) | Cat-eye syndrome | Coloboma of the iris<br>Anal atresia<br>Ear abnormalities<br>Cardiac defects |

*Note.* del = deletion; dup = duplication; p = short arm of chromosome; q = long arm of chromosome.

estimated to be lower than 1 in 1000. The most common mechanism giving rise to Prader-Willi syndrome is a deletion of the paternal chromosome 15 at bands q11 to q13 (70%). Maternal UPD appears to account for only 25% of the cases, and imprinting errors may account for as many as 5% of the cases.

Prenatal diagnosis of Prader-Willi syndrome is not common because the deletion is usually too small to be seen at routine band levels in amniotic fluid and chorionic villi (approximately 500 bands per haploid genome). There are no phenotypic abnormalities identifiable on ultrasound, and often the only indicator that is noted is a decrease in fetal movement. At birth, the affected infant is usually hypotonic and displays a poor suck reflex, often necessitating the use of gavage feedings. Diagnosis is usually not made until 1–2 years of age when developmental delay and behavioral problems begin to emerge. Other physical features may include obesity, hypogonadism, short stature, small hands and feet, hypopigmentation, and characteristic facial features (i.e., a narrow bifrontal diameter, almond-shaped eyes, small mouth with a thin upper lip, down-turned corners of the mouth) [Figure 9-7]. The behavioral problems are a hallmark for this syndrome and are often very problematic to the families and caregivers of individuals with Prader-Willi syndrome. The most marked behavioral problem is hyperphagia, which is always followed by obesity if not adequately controlled. The hyperphagic behavior includes food foraging, stealing, and garbage picking, and in some cases extends to spoiled food or nonfood items. Families with a child affected with Prader-Willi syndrome often have to lock up all food items and install internal security systems to prevent the child from leaving the house to obtain food. In addition, individuals with Prader-Willi syndrome are very argumentative and stubborn and may be prone to temper tantrums and violent outbursts. Children younger than 6 years display delayed motor, cognitive, and language development. The average individual has an IQ in the upper range of mild mental retardation (high 60s). Some children are able to attend school and participate in regular classroom activities; however, they tend to be disruptive and often require a specialized classroom environment. One behavioral feature that is often overlooked is compulsive skin picking, which can lead to chronic open sores and infection, sometimes requiring surgical intervention.

Analysis of genes located within the deletion region on chromosome 15 have identified six maternally imprinted genes, which are expressed exclusively from the paternal chromosome (Figure 9-8). The genes are *ZNF127, PW71, snRNP, PAR-5, IPW,* and *PAR-1.* The mechanism of imprinting is not understood but appears to involve parental-specific DNA methylation in regulating gene expression. The 5′ region of the *SNRPN* gene contains an imprinting center, which regulates in *cis* the chromatin structure, DNA methylation, and gene expression of this 2-Mb chromosomal domain. The imprinting center encodes for alternate *snRNP* transcripts, which appear to play a role in the parental imprint switch. A recent study by Dittrich and colleagues demonstrated that in families with Prader-Willi and Angelman's syndromes, microdeletions and point mutations within the imprinting center disrupt the expression of these transcripts, suggesting

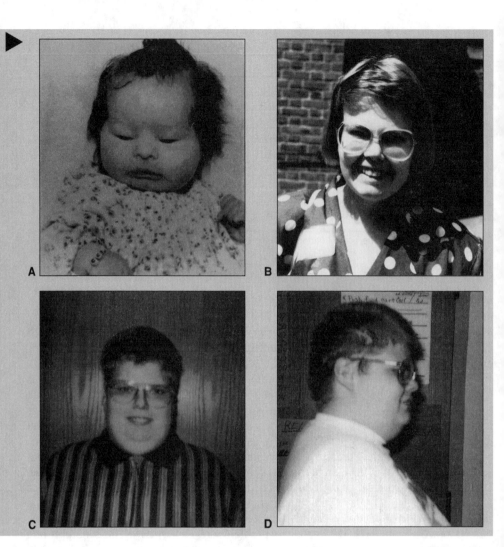

**FIGURE 9-7**
**Patients with Prader-Willi Syndrome.** *These photographs illustrate the wide range of phenotypic features seen in individuals with Prader-Willi syndrome. These features can change dramatically with age as is shown in Patient 1 as (A) a newborn and (B) a 26-year-old. Patient 2 (C) and (D) is a 19-year-old who displays many of the typical facial features such as narrow bifrontal diameter, almond-shaped eyes, small mouth with a thin upper lip, down-turned corners of the mouth, obesity and fair features.*

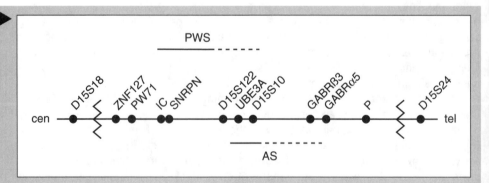

**FIGURE 9-8**
**Genetic Map of Chromosome 15q11-q13.** *This map illustrates the region on chromosome 15 that is responsible for Prader-Willi and Angelman's syndromes (PWS, AS). The solid line represents the critical deletion region for each syndrome. The dashed line represents regions thought to play a role in the cause of the syndromes.*

a possible role in the etiology of these syndromes, as well as a possible mechanism for the regulation of imprinted genes [14].

***Angelman's Syndrome.*** The incidence of Angelman's syndrome is approximately 1 in 25,000 live births. In most cases, the condition is sporadic, but a number of familial cases of Angelman's syndrome have been described. No recurrence risk has been established. Approximately 80% of the cases are caused by a deletion of the maternal chromosome 15 at bands q11 to q13. Most of the remaining 20% of cases appear to be caused by imprinting errors; fewer than 5% are caused by paternal UPD. Because of the more frequent occurrence of familial cases, it is thought that Angelman's syndrome may be the result of disruption of only a single gene. Mutations of the E6-AP ubiquitin–protein ligase (*UBE3A*) gene, which has been mapped to the critical deletion region, have been identified in several familial cases of Angelman's syndrome, suggesting that *UBE3A* may be the gene for Angelman's syndrome [15, 16].

A diagnosis of Angelman's syndrome usually is not made until after 1 year of age. The infants appear normal at birth with a normal head circumference and the absence of any major birth defects. The first indication of a problem may be the onset of seizures, which occurs in 30% of patients with Angelman's syndrome younger than 2 years. Electroencephalogram (EEG) studies show a characteristic pattern with large-amplitude, slow-spike waves facilitated by eye closure. The first developmental symptoms begin to emerge between 6 and 12 months, with a total lack of speech and motor delays complicated by movement and balance disorders. Most individuals with Angelman's syndrome have a characteristic wide-based gait and stiff, ataxic movements, which are often referred to as "puppet-like." The mental retardation is severe and is accompanied by unique behavioral features, which include frequent laughter and smiling, an apparently happy demeanor, and an easily excitable personality. Other physical features of Angelman's syndrome are microcephaly by age 2 years, midface hypoplasia, macrostomia, widely spaced teeth, protrusion of the tongue, and mandibular prognathism [17, 18].

Diagnostic tests should include standard and FISH chromosome analyses to identify a deletion, molecular analysis to identify UPD, and EEG studies to document neurologic abnormalities. Even in the absence of a microdeletion or paternal UPD the diagnosis of Angelman's syndrome cannot be ruled out because up to 20% of cases are caused by an imprinting error, most likely involving only a single locus.

**Smith-Magenis Syndrome** is a relatively common chromosome deletion syndrome with an incidence of 1 in 25,000 live births. All patients studied have a cytogenetically detectable deletion of chromosome 17p at band 11.2. The deletion appears to be relatively consistent and is approximately 5 megabases (Mb) in size. Two genes have been identified within this region, a small nuclear RNA U3 (*SNU3*), which is involved in pre-mRNA processing, and a microfibril-associated glycoprotein (*MFAP4*), but there is no evidence that deletion of these genes contributes to the phenotype of Smith-Magenis syndrome.

This syndrome is a clinically recognizable entity, consisting of multiple congenital anomalies, mental retardation, a characteristic phenotype, and behavioral features. The clinical phenotype includes minor craniofacial anomalies with brachycephaly; prominent forehead; synophrys; epicanthal folds; broad face and nasal bridge; ear anomalies; prognathism; and short, broad hands. People with Smith-Magenis syndrome also exhibit signs of peripheral neuropathy with decreased tendon reflexes and decreased sensitivity to pain, as well as hyperactivity and sleep disturbances, which may be related to a decrease in rapid eye movement (REM) sleep noted in 50% of the affected individuals studied. The most significant behavioral characteristic is self-destructive behavior, which often includes head banging, wrist-biting, and nail picking. Developmental delay is universal, with speech and language development more delayed than receptive language. All individuals with Smith-Magenis syndrome are mentally retarded, but there is considerable variation in IQ levels. Most individuals are mildly to moderately mentally retarded, but the range is from borderline normal to profoundly retarded. Other problems that may occur in patients with Smith-Magenis syndrome include valvular or structural heart defects (usually benign), mild thoracic scoliosis, and conductive hearing impairment [19]. A gene for autosomal recessive hereditary deafness has been mapped to the critical deletion region on chromosome 17. The diagnosis is usually made during cytogenetic evaluation for developmental delay or dysmorphic features. The phenotype becomes more striking with age and can often be diagnosed in adults based on clinical features alone. However, the diagnosis should always be confirmed cytogenetically.

**Miller-Dieker Syndrome.** Miller-Dieker syndrome is characterized by type I lissencephaly and a certain facial appearance. Lissencephaly is a severe brain malformation with absence of the typical convolutions (gyri and sulci) resulting in a smooth surface of the brain. The characteristic facial features include prominent forehead, bitemporal hollowing, a short nose with upturned nares, a protuberant upper lip with a thin vermilion border, and a small jaw. Miller-Dieker syndrome was originally thought to be an autosomal recessive disorder because of the presence of multiple affected siblings in several families but was found to be caused by a balanced rearrangement in one of the parents, leading to unbalanced products in the offspring. The deletion is on the terminal end of the short arm of chromosome 17 (17p13.1−13.3), which is more distal than the deletion responsible for Smith-Magenis syndrome. It can be detected cytogenetically in 50% of patients; in the remaining cases, the deletion is too small to be detected without the aid of FISH. Isolated lissencephaly, with normal facial features, has been found to be caused by a microdeletion or mutation of a single gene within the critical region for Miller-Dieker syndrome, *LIS1*. The function of this gene is not yet known, but it is expressed in all tissues, and sequence analysis indicates a possible role in signal transduction. This gene also plays a role in Miller-Dieker syndrome, but the tremendous phenotypic variability may be caused by the deletion of other genes within this region.

Infants born with Miller-Dieker syndrome are severely affected with profound mental retardation and development that ranges from achievement of no skills to showing some social response, rolling over, and gesturing. There is usually initial hypotonia followed by postnatal failure to thrive and death before 2 years of age.

***DiGeorge Syndrome and Velocardiofacial Syndrome (VCFS).*** These are related disorders involving deletions of either chromosome 22q11.2 or 10p13. Both disorders are characterized by conotruncal cardiac defects (e.g., interrupted aortic arch, truncus arteriosis, tetralogy of Fallot), cleft palate, and learning disabilities. Additional features may include absence or hypoplasia of the parathyroid and thymus gland (T-cell immunodeficiency) glands and facial dysmorphism (e.g., hypertelorism, prominent nose with a broad nasal root, narrow palpebral fissures, bifid uvula, retrognathia, low-set ears). The underlying cause is a developmental field defect of the third and fourth pharyngeal pouches during embryogenesis. The exact relationship between these two syndromes is currently undergoing re-evaluation because there is significant overlap between the phenotypes as well as the genetic etiology. VCFS may represent an autosomal dominantly inherited form of DiGeorge syndrome, or it may be caused by a specific molecular defect with a wide phenotypic spectrum that includes not only DiGeorge syndrome but also Robin sequence (early mandibular hypoplasia) and CHARGE association (i.e., coloboma, heart disease, atresia choanae, retarded growth and development, genital hypoplasia, ear anomalies).

These disorders are often characterized in infancy by failure to thrive, hypocalcemia, and cellular immunodeficiency with increased risk of infection. Death in infancy caused by cardiac failure, infection, or both is common. However, many cases are mild and go undiagnosed, in some cases until the birth of a severely affected child in the subsequent generation. The incidence is approximately 1 in 20,000 live births, but this is most likely an underestimation because of the underdiagnosis of milder cases. Only one-third of patients with DiGeorge syndrome have a cytogenetically detectable deletion of chromosome 22q11.2 (and less commonly 10p13−p14), whereas in patients with VCFS, deletions are detected in only a small number of cases. However, molecular methods have confirmed the presence of a deletion of 22q11.2 in almost 100% of both DiGeorge syndrome and VCFS cases. The number of cases with deletions of 10p is not yet known. The critical deletion region on chromosome 22 includes three common loci, and the severity of the disease does not appear to correlate with the size of the deletion. The deletions are often indistinguishable, but there may be considerable variation in the phenotypes of individuals, even within the same family. Almost all affected individuals with either syndrome have some degree of learning disability, with approximately 40% being mildly to moderately mentally retarded. Additional features may include microcephaly, short stature, inguinal and umbilical hernias, scoliosis, and eye abnormalities.

## PARTIAL TRISOMIES

***Trisomy 4p Syndrome.*** Trisomy for all or most of the short arm of chromosome 4 results in a characteristic clinical phenotype with prenatal onset of growth deficiency, hypertonia in infancy, seizures, severe mental retardation, and dysmorphic facial features, which include microcephaly, prominent forehead, glabella and supraorbital ridges, a bulbous nose with a depressed flat nasal bridge, synophrys, macroglossia, irregular teeth, small pointed mandible, enlarged misshapen ears, and an asymmetric crying mouth. Occasional internal anomalies such as cardiac defects, renal malformations and atresia, and absence of corpus collosum contribute to early infant mortality. In the absence of internal anomalies, the life span appears to be near normal. Individuals with trisomy for chromosome 4p continue to display features of postnatal growth deficiency, with the average adult height ranging from 4', 9" to approximately 4', 11". They also have a tendency to become obese with increasing age. Mental retardation is universal, with more severe delays in language skills than in social and fine motor skills.

***Partial Trisomy 10q Syndrome.*** Trisomy for the distal one-third of the long arm of chromosome 10 (10q24→qter) results in a multiple malformation syndrome with severe mental retardation and an infant mortality rate of up to 50%. There is prenatal onset of growth deficiency, cardiac and renal malformations, minor limb anomalies, kyphoscoliosis, cryptorchidism, and characteristic dysmorphic features including microcephaly; flat face with a high forehead and high, arched eyebrows; ptosis; short palpebral fissures; microphthalmia; broad and depressed nasal bridge; bow-shaped mouth with a prominent upper lip; and malformed, posteriorly rotated ears. The mental deficiency is severe, and most individuals who survive infancy are bedridden and unable to communicate.

***Trisomy 20p Syndrome.*** Trisomy 20p usually appears as a result of a familial structural rearrangement and in some cases has a very mild phenotype usually characterized by normal growth, abnormal narrowness of the palpebral fissures (blepharophimosis), large and poorly formed ears, and cubitus valgus. There may be mild-to-moderate mental retardation, hypotonia, and poor coordination but no major malformations. In some cases, mildly affected individuals are not diagnosed until the birth of a second affected sibling.

# SEX CHROMOSOMAL ABNORMALITIES

## Overview of the X Chromosome and X Inactivation

In 1949, Barr and Bertram discovered sex chromatin masses (i.e., Barr bodies) in the interphase nuclei of females but not males [20]. It was also observed that although most females were chromatin positive and most males were chromatin negative, females with Turner's syndrome (45,X) were Barr-body–negative, and males with Klinefelter's syndrome (47,XXY) were Barr-body–positive. We now know that the number of Barr bodies seen in an interphase cell is a reflection of the number of X chromosomes in the cell, such that it is always one less than the total number of X chromosomes. The Barr body represents the late-replicating, inactive X chromosome. The theory of X inactivation, the *Lyon hypothesis*, proposes that in the somatic cells of a normal female one of the two X chromosomes is inactivated, thus equalizing the expression of X-linked genes in the two sexes. In individuals with one or more extra X chromosomes, all but one of the X chromosomes is inactivated to maintain the correct dosage of genes located on the X chromosome. In this way, trisomy or tetrasomy for the X chromosome is less severe than trisomies for autosomes.

Inactivation occurs early in embryonic life, beginning in the morula stage approximately 3 days after fertilization is completed at the end of the first week of development. In the fetal cell, the inactivation process is random, and either the maternal or paternal X chromosome may be inactivated. Once the X chromosome is inactivated in a cell, all of the clonal descendants of that cell have the same inactivated X chromosome. As a result, females are mosaic with respect to their X-linked genes. However, preferential inactivation of the paternal X chromosome is a normal phenomenon seen in extraembryonic membranes during fetal development. The significance of selective expression of maternal alleles (imprinting) in this tissue is not known. The consequences of X chromosome inactivation are dosage compensation, variability of gene expression in heterozygous females, and mosaicism. Most of the genes located on the inactive X chromosome are not transcribed, with the exception of several segments that remain active, including the pseudoautosomal regions, the steroid sulfatase gene, and the *XIST* gene (Figure 9-9). The *XIST* gene is located at the X-inactivation center in the Xq13 region and is required for the X chromosome to become inactive. This gene is expressed only from the inactive X chromosome. The product is an RNA transcript that is not translated into a protein.

**FIGURE 9-9** ▶

***Idiogram of X Chromosome.*** *This idiogram shows the approximate band location of a few of the genes on the X chromosome. The pseudoautosomal regions are on the distal p and q arms, and the X-inactivation center (XIST) is located at Xq13. G6PD = glucose-6-phosphate dehydrogenase; HGPRT = hypoxanthine-guanine phosphoribosyltransferase.*

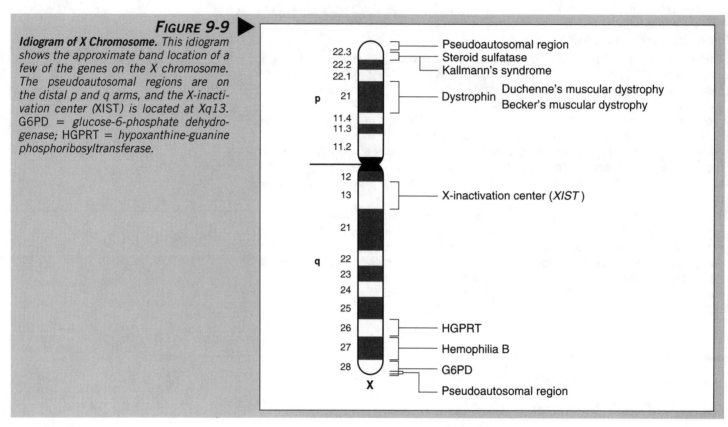

The clinical variation in expression of X-linked disorders in carrier females can be extreme, ranging from completely normal to full manifestation of the genetic defect. Manifesting heterozygotes have been described for many X-linked disorders, including Duchenne's muscular dystrophy (DMD), and represent an extreme example of unfavorable lyonization (X chromosome inactivation). In individuals with a structural abnormality of the X chromosome, the abnormal X chromosome is usually inactivated. Nonrandom X inactivation is also observed in cases of X;autosome translocations, in which the normal X chromosome is preferentially inactivated to prevent inactivation of an autosomal region.

## Overview of the Y Chromosome

The Y chromosome contains one or more genes that are responsible for male sexual development. The testis-determining factor or sex-determining region Y (*SRY*) gene is located on the short arm of the Y chromosome adjacent to the pseudoautosomal region (Yp11). It is the presence or absence of this gene that determines the sex of an individual. The *SRY* gene codes for a zinc finger protein that is most likely involved in regulation of gene expression by interaction with specific DNA sequences. In the presence of the *SRY* gene product, the primitive gonads develop into testicular tissue, which secretes androgens. The external male genitalia develop in response to the presence of androgens. In contrast, the absence of the *SRY* gene product results in the formation of ovaries or ovary-like structures and female external genitalia by default. Other genes located on the Y chromosome are summarized in Table 9-7.

## Numerical Sex Chromosomal Abnormalities

The incidence of numerical sex chromosomal abnormalities is not actually known. Individuals with additional copies of either the X or Y chromosome often have very mild or no phenotypic effect; therefore, it is believed that most cases are probably undiagnosed unless they present with a clinical problem, which may not be related to their abnormal

◀ **TABLE 9-7**
*Genes on the Y Chromosome*

| Genes | Location |
|---|---|
| *SRY* gene | Yp (pseudoautosomal region) |
| Colony-stimulating factor 2 receptor α-chain | Yp |
| *MIC2* gene | Yp |
| Zinc-finger Y gene (possible role in sex determination) | Yp |
| Ribosomal protein S4 gene | Yp |
| Histocompatibility-Y antigen gene (*H-Y* gene) | Yq11[a] |
| Major gene for stature | Yq11[a] |
| Azoospermia-related gene | Yq11[a] |
| Gonadoblastoma gene | Yq11[a] |

*Note.* SRY = sex-determining region Y.
[a] Tentative assignment.

chromosome complement. Similarly, individuals with monosomy for the X chromosome or Turner's syndrome often have such a mild phenotype that they remain undiagnosed until puberty.

***Turner's Syndrome.*** Turner's syndrome is associated with monosomy or partial monosomy for the X chromosome. Monosomy X (45,X) is the most common chromosomal abnormality seen in spontaneously aborted embryos, accounting for up to 20% of all spontaneous abortions. These data indicate that 45,X conceptions may account for as many as 2.25% of all recognized conceptions. However, more than 95% of Turner's syndrome conceptions are aborted, and it is seen in only approximately 1 in 2500 liveborn females. Despite the apparent reduced viability of these embryos, surviving infants, children, and adults may have relatively mild phenotypic abnormalities. Turner's syndrome is usually caused by either nondisjunction in females, resulting in exclusion of an X chromosome in one oocyte, or chromosome lag of the Y chromosome during meiosis in males. Molecular studies indicate that 66% of the cases retain the maternal X chromosome. There does not appear to be any maternal age effect, and there is no increased risk for future pregnancies.

Turner's syndrome should be considered in the differential diagnosis of any female with short stature, delayed or arrested puberty, or both. Recognition of the cardinal features of Turner's syndrome can provide affected individuals with treatment that can promote growth and the development of secondary sexual characteristics, as well as provide preventative testing for additional complications. There is considerable phenotypic variability among the common features of Turner's syndrome.

Prenatal diagnosis of Turner's syndrome is most commonly associated with phenotypic abnormalities that are detected during routine ultrasound examination. The most frequent abnormalities are cystic hygroma, fetalis hydrops, congenital heart disease (typically involving the left side of the heart), and renal abnormalities (i.e., renal agenesis, horseshoe kidney, pelvis kidney). Several of these features can be seen in infants, such as a short, webbed neck with a low posterior hairline; shield chest; edema of the hands and feet; prominent ears; and low birth weight. However, most infants and fetuses (75%–90%) with Turner's syndrome appear perfectly normal. The occurrence of low birth weight is noteworthy, in that it is the precursor to subsequent short stature, which is usually apparent by 3 years of age. Short stature is the hallmark feature of Turner's syndrome. It is a direct result of the chromosomal deficiency, most likely involving loss of the short arm of the X chromosome. During childhood, short stature is often the only visible feature of Turner's syndrome. Lack of secondary sexual characteristics or primary amenorrhea are common in adolescence. However, in some cases, individuals with Turner's syndrome develop sparse pubic hair or axillary hair and experience delayed breast development (5%–10%) and delayed menarche (1%) with menstrual irregularities. Therefore, it is important to consider Turner's syndrome in adult women with infertility, menstrual irregularities, or short stature. Almost all women with Turner's syndrome are infertile because of the degeneration of oocytes starting in midfetal life. Rare cases of pregnancy have been reported in the literature (21 cases reviewed in Kaneko and colleagues), but only 11 pregnancies resulted in a normal outcome; the remaining cases ended in miscarriages, stillbirths, an infant with Down's syndrome, and

**Features of Turner's Syndrome**
Short stature
Webbing of the neck
Low posterior hairline
Lack of secondary sexual characteristics
Primary or secondary amenorrhea
Gonadal dysgenesis

an infant with a partial cleft palate [21]. Therefore, it is appropriate to counsel women with Turner's syndrome that if they are able to conceive, they are at an increased risk for an adverse outcome (up to 50%), and they should be offered prenatal diagnosis. There have been several reports of women with Turner's syndrome conceiving by in vitro fertilization using a donor ovum [22].

Historically it was thought that Turner's syndrome was associated with mental retardation; however, intelligence is usually within normal limits, but quite often it is less than that of normal "non-Turner" siblings. Turner's syndrome has been associated with impairment in spatial and perceptual thinking with related cognitive impairment, resulting in difficulties with solving mathematical problems and orienting to left–right directions. Some Turner's syndrome individuals may experience delays in language development and visual–motor coordination, with reduced gross and fine motor skills. However, not all Turner's syndrome females have these problems; therefore, it is crucial that expectations do not become diminished because this may ultimately result in low self-esteem and subsequent poor performance.

Individuals with Turner's syndrome also are at risk for several other disorders that should be monitored throughout their lives. Autoimmune thyroiditis with or without hypothyroidism is seen more commonly in patients with Turner's syndrome than in the general population. As many as 50% of adolescents and young adults with Turner's syndrome have elevated levels of antithyroid antibodies. There is also an increased risk for hypertension and noninsulin-dependent diabetes in adults with Turner's syndrome, both of which are associated with obesity. In addition, abnormalities of the heart and kidneys should be ruled out. Approximately 35%–70% of children with Turner's syndrome have abnormalities of the kidneys or renovascular system (e.g., horseshoe kidney, unilateral pelvic kidneys, rotational abnormalities, duplications of the collecting system). These abnormalities usually have limited clinical significance but may contribute to hydronephritis, pyelonephritis, or hypertension. As many as one-third of Turner's syndrome children may also have cardiovascular abnormalities involving the left side of the heart, most commonly bicuspid aortic valves, coarctation of the aorta, or mitral valve prolapse.

Endocrine studies are often performed because of ovarian failure, and they usually provide a characteristic profile for Turner's syndrome. Ovarian failure in Turner's syndrome is not associated with defects in the hypothalamus or pituitary gland; therefore, after the age of 10 years, there are increased plasma levels of follicle-stimulating hormone (FSH) and luteinizing hormone (LH) associated with low estrogen levels. Elevated gonadotropin levels or an exaggerated response to gonadotropin-releasing hormones are highly suggestive of Turner's syndrome. However, it should be confirmed by cytogenetic studies for several reasons: (1) FSH and LH levels may be normal in midchildhood despite abnormal ovarian function, (2) some patients are still capable of secreting estrogen during childhood and early adolescence, and (3) other forms of gonadal dysgenesis resulting from the presence of an abnormal Y chromosome must be ruled out. In Turner's syndrome, nearly normal secondary sexual development can be achieved with estrogen treatment, demonstrating normal structure and function of all of the organs except for the ovaries. Estrogen replacement therapy has the added benefits of reducing the risk of heart disease and preventing osteoporosis.

Cytogenetic analyses of individuals with the Turner's syndrome phenotype have shown that it may be caused by a number of cytogenetic abnormalities, the most common being 45,X (50%). The remaining cases have either chromosomal mosaicism or structural abnormalities of the X or Y chromosomes. The use of FISH and polymerase chain reaction (PCR)–based assays suggests that the number of cases with mosaicism may be significantly higher, up to 74%. These cases involve mosaic karyotypes in which one of the cell lines is 45,X, whereas the other cell line may be either normal (46,XX or 46,XY) or abnormal, such as 46,X,i(X)(q10). The most common structural rearrangement of the X chromosome in Turner's syndrome is the formation of an isochromosome of the long arm; that is, i(Xq). This results in deletion of the short arm and duplication of the long arm of the X chromosome; however, the i(Xq) is always inactive. Females with this karyotype— 46,X,i(X)(q10)—can display the full phenotypic spectrum of Turner's syndrome. Similarly, deletions of all or part of the short arm of the X chromosome can also lead to Turner's syndrome. However, the severity of the phenotype often depends on the extent of the deletion. When the entire short arm is missing, it may result in the classic phenotype; whereas smaller deletions may manifest only as short stature, primary or secondary amenorrhea, or infertility. Deletions of the long arm have a wide range of phenotypic effects, but development of the Turner's syndrome phenotype is probably confined to a small region. Studies of women with small deletions of the long arm of the X chromosome suggest that genes responsible for normal gonadal development are located on Xq. Ring chromosomes 46,X,r(X) have also been seen in females with Turner's syndrome. The formation of a ring chromosome involves chromosomal breaks on both the long and short arms with fusion of the proximal ends. This results in loss of the terminal ends of both arms. These individuals can display typical features of Turner's syndrome and, in cases with smaller rings (i.e., loss of larger pieces of the X chromosome), it may be associated with mental retardation and other non-Turner features. This is most likely due to deletion of the XIST gene at Xq13, resulting in inappropriate overexpression of X-linked genes.

The presence of a marker chromosome in Turner's syndrome deserves special attention. Determination of the origin of the marker chromosome is especially important because the presence of Y DNA in these individuals may increase their risk for gonadal tumors, such as gonadoblastoma. The use of FISH with alphoid satellite probes specific for the Y chromosome is extremely useful for determining the origin of these marker chromosomes. If the Y chromosome is ruled out, then additional testing is usually not required. If the marker is identified as being derived from Y DNA, then it may be necessary for the affected female to have her ovaries removed to prevent formation of a gonadal tumor. The Yq region appears to be the critical region, and it is therefore thought that a gene responsible for gonadoblastoma is located within this region.

*Klinefelter's Syndrome.* Klinefelter's syndrome is the most common cause of hypogonadism in males and is characterized by a 47,XXY karyotype. The true incidence is not known, but it is thought that it occurs in at least 1 in 1000 liveborn males, with more than 3000 being born annually in the United States. The majority of 47,XXY conceptions survive the fetal period, and most are not diagnosed until adulthood when infertility or gynecomastia is detected. Approximately half of the cases are due to a maternal meiotic error (75% meiosis I), and the remaining half are due to paternal meiotic errors. There is an increased risk for Klinefelter's syndrome in older mothers, but there is no paternal age effect and no increase in the recurrence risk.

The classic features of Klinefelter's syndrome in an adult male include tall stature, small testes, and infertility, with or without gynecomastia. Infants with a 47,XXY karyotype have normal weight, length, and head circumference, as well as normal genitalia. They do not display any dysmorphic features or major congenital anomalies. They continue to appear physically normal until puberty, when they begin to experience lack of development of secondary sexual characteristics. Sexual development is usually normal until early adolescence, with normal initiation of pubertal changes and pituitary-gonadal function. There is an initial rise in serum testosterone, but this begins to decrease by the age of 15 years. There is a subsequent increase in the plasma levels of FSH, LH, and estradiol, which results in hypogonadism by midpuberty. The size of the penis is usually normal or slightly reduced, with the hallmark feature being testicular hypoplasia. The small testes are usually apparent on physical examination in teenage boys. Although the testes do not fully develop and do not produce sperm, adult males with Klinefelter's syndrome have normal sexual function and can maintain normal sexual relationships. Approximately 30% of young adult males with Klinefelter's syndrome develop gynecomastia, which cannot be diminished or prevented with hormone treatments. The only treatment in severe cases is plastic surgery, which is primarily for aesthetic reasons. Most men with Klinefelter's syndrome also have sparse growth of facial hair and usually do not develop male pattern baldness.

Individuals with Klinefelter's syndrome tend to have slightly lower intelligence than their siblings and score lower on standardized intelligence tests, with the average IQ ranging between 85 and 90. Approximately 70% of 47,XXY males have developmental and learning disabilities, compared with only 22% in the general population. These disabilities include speech delays, neurologic deficits, reading difficulties, and dyslexia. Many of these individuals also exhibit behavioral problems, but it is unclear whether this is caused by poor self-esteem and psychosocial development or a decreased ability to deal with stressful situations [23].

Hormone replacement therapy (HRT) can improve the self image and subsequently the behavior of many men with Klinefelter's syndrome. HRT increases sexual desire, as well as increases the size of the testes, and often makes patients feel better physically by providing more energy. One positive side effect is the prevention of osteoporosis, but it also has the negative side effect of causing prostatic hypertrophy, which should be monitored every year.

*47,XYY Syndrome.* The 47,XYY karyotype occurs in approximately 1 in 1000 liveborn males. It is not associated with any phenotypic abnormalities. Prenatal diagnosis is usually made only following routine prenatal diagnosis for advanced maternal age. Diagnosis after birth is usually coincidental to another unrelated clinical complaint. Males with a 47,XYY karyotype are usually taller than karyotypically normal individuals and may have an increased risk for behavioral problems. They do not, however, have a higher tendency toward criminal activity or violence, which has been previously reported in the literature. They have normal intelligence and are fertile. They do not appear to be at an increased risk for having chromosomally abnormal children.

*47,XXX Syndrome.* Trisomy X, or 47,XXX syndrome, occurs in approximately 1 in 1000 liveborn females, but most are never diagnosed. Molecular studies indicate that in nearly 100% of 47,XXX females, the extra X chromosome is derived from the mother, usually resulting from nondisjunction during maternal meiosis I. Affected females have manifestations similar to Klinefelter's syndrome, such as tall stature and mild developmental and learning disabilities. They do not have dysmorphic features, and their physical development is normal. The onset of puberty is usually normal, and most affected individuals are able to produce normal offspring. These patients are at an increased risk of developing psychiatric problems, specifically schizophrenia.

*Other Numerical Sex Chromosome Abnormalities.* Individuals with poly-Y karyotypes (e.g., 48,XYYY, 49,XYYYY) are extremely rare. There is little information about the prenatal diagnosis and early development of these individuals. Most cases are identified because of behavioral problems, developmental delay, hypogonadism, and radioulnar synostosis.

Other numerical abnormalities involving sex chromosomes tend to be more severe. Males with extra X chromosomes (48,XXXY or 49,XXXXY) have a phenotype that is consistent with severe Klinefelter's syndrome, but they are usually mentally retarded. Other features may include skeletal deformities, infantile appearance, and lack of sexual development. Males with multiple copies of both the X and Y chromosome (48,XXYY or 49,XXXYY) are also mentally retarded and may have behavioral problems.

The presence of multiple copies of the X chromosome in females (48,XXXX or 49,XXXXX) is associated with mental retardation with average IQs of 50 and 60, respectively. Phenotypically, the two disorders differ significantly. Females with a 48,XXXX karyotype are usually tall and have radioulnar synostosis, but they also may have epicanthal folds and hypertelorism. Approximately half of these women do not undergo complete development of secondary sexual characteristics, but the remaining 48,XXXX females appear to have normal menarche and menopause. In contrast, females with a 49,XXXXX karyotype are short and also have radioulnar synostosis. Their facial features are coarse, and they have congenital heart defects. Puberty is delayed, and secondary sexual characteristics are always incompletely developed.

## Structural Abnormalities of the X Chromosome

Deletions of the X chromosome in females can result in a variety of phenotypic abnormalities; the most common are features associated with Turner's syndrome (45,X), as discussed previously. Small deletions of both Xp and Xq have been associated with primary and secondary amenorrhea and infertility in phenotypically normal women. A deletion involving a small segment of the X chromosome is usually less deleterious to females because of preferential inactivation of the abnormal X. The opposite is true in balanced translocations involving the X chromosome and one of the 22 autosomes. In these cases, the normal intact X chromosome is preferentially inactivated rather than the translocated X chromosome. This prevents inactivation or functional monosomy for the region of the autosomal chromosome that is translocated to the *XIST*-bearing portion of the X chromosome. The presence of the *XIST* gene is very important in individuals with two or more X chromosomes. When the *XIST* gene is deleted or inactivated because of a structural rearrangement, such as a ring chromosome, the X chromosome does not undergo inactivation. The resulting phenotype includes severe mental retardation, which is most likely caused by the overexpression of X-linked genes.

In males, deletions of the X chromosome usually have a more severe phenotype. Many X-linked disorders have been shown to be caused by deletions involving one or more genes on the X chromosome, such as DMD, hemophilia A, and Kallmann's syndrome.

## Structural Abnormalities of the Y Chromosome

The Y chromosome shows considerable variation in its length within the male population; however, the length is a stably inherited trait and should be identical between fathers and sons. Therefore, it is very important to compare the paternal Y chromosome to the patient's Y chromosome when trying to determine whether there is a structural rearrangement or deletion.

Deletions, translocations, or structural rearrangements that lead to the deletion of the testis-determining factor (*SRY*) on the short arm of the Y chromosome always result in an individual with a male karyotype who has failed to undergo masculinization during embryogenesis. However, individuals with deletion of the long arm of the Y chromosome are phenotypically male but may have short stature, infertility because of arrested spermatogenesis, or hypogonadism. The phenotype depends on the size and location of the deletion. These findings suggest that genes responsible for stature and spermatogenesis are located within specific regions of the Y chromosome. Translocations between an X and Y chromosome during meiotic recombination occasionally result in the transfer of the *SRY* gene to the X chromosome. In such cases, offspring receiving either of the abnormal chromosomes may be 46,XX males or 46,XY females.

## Ambiguous Genitalia and Sex Reversal Phenotypes

Approximately 1 in 20,000 males has a 46,XX chromosomal complement. These individuals are considered to be *pseudohermaphrodites* whose condition may arise by several different mechanisms, all leading to some degree of gonadal dysgenesis. Most cases (75%) are due to the presence of the *SRY* gene translocated from the Y chromosome, most commonly to the X chromosome because of faulty meiotic recombination. These *SRY*-positive XX males have features similar to those of Klinefelter's syndrome except that they are of normal height and intelligence. The remaining cases are *SRY*-negative and are thought to arise because of inappropriate expression of a gene, which is normally switched on only after Y-induced male differentiation begins. These individuals often have ambiguous genitalia.

The occurrence of *XY females* represents *pure gonadal dysgenesis* in male conceptions and may be caused by several different defects. In most of the familial cases, the X-linked recessive form of the disorder is due to a mutation in a gene involved in a later event of the testis-determining pathway. These individuals have a normal Y chromosome with a functional copy of the *SRY* gene, but they are phenotypically female. The other familial cases are Y-linked recessive and are caused by a mutation of the *SRY* gene. Similarly, 10%–20% of the sporadic cases also have a mutation of the *SRY* gene. The phenotypes of XY females are nearly normal, except that these individuals do not undergo pubertal changes because they do not secrete the appropriate female hormones. Secondary sexual characteristics can be induced using HRT. These women are managed very similarly to Turner's syndrome women because they also have only one X chromosome. Prevention of gonadoblastoma is an important consideration because nearly 50% of these patients develop tumors, which can become malignant. Gonadectomy is recommended for the prevention of gonadoblastoma.

A second X-linked recessive disorder gives rise to another class of XY females who have testicular tissue that secretes androgens. Most cases (75%) are due to a defect in the androgen-receptor gene. This disorder was historically referred to as testicular feminization, but is now known as *complete androgen insensitivity syndrome*. These individuals appear to be phenotypically normal females with breast development and typically female contours. However, they have sparse axillary and pubic hair, a short vagina, and just remnants of a uterus and fallopian tubes. The primary complaint is usually amenorrhea and infertility. The incidence is approximately 1 in 20,000 females.

*True hermaphrodites* usually present with ambiguous genitalia at birth. The gonads are comprised of both testicular and ovarian tissue and may be in the form of two distinct organs or an ovotestis. The most common karyotype is 46,XX. Overall, this condition is relatively rare in the general population but is more prevalent among Bantu-speaking blacks in Southern Africa. Molecular analysis of individuals from this population has shown that they do not have the *SRY* gene and that they do not have UPD for the X chromosome. The exact mechanism is unknown but most likely involves activation of the testis-determining pathway. The clinical features vary widely, with some individuals having near-normal external male or female genitalia and others having ambiguous genitalia with both a phallus and a vagina. The development of female secondary sexual characteristics is incomplete, and amenorrhea is present in approximately 50% of XX true hermaphrodites. Although rare, pregnancies have been reported in some hermaphroditic females.

Another type of sex reversal phenotype involves the occurrence of ambiguous genitalia in conjunction with another abnormality, either physical, metabolic, or mental. Specific defects in enzymes of the adrenal cortex that are required for cortisol biosynthesis can result in overproduction of androgens, resulting in virilization of female infants. This group of disorders is called *congenital adrenal hypoplasia (CAH)*, and the most common defect is a deficiency of 21-hydroxylase, which occurs in 1 in 12,500 live births. The female infants are usually born with ambiguous external genitalia, which are surgically corrected, and normal internal genital structures. Males with this disorder are phenotypically normal. Diagnosis in infancy is extremely important because 75% of these infants have the more severe salt-wasting type of 21-hydroxylase deficiency, which may lead to neonatal death. If untreated, both sexes eventually experience rapid growth and accelerated skeletal maturation during childhood. The treatment is aimed at replacing the deficient hormones and reducing the levels of adrenocorticotropic hormone and adrenal androgens. Additional multiple malformation syndromes with sex reversal phenotypes are summarized in Table 9-8.

**TABLE 9-8** ▶

*Multiple Malformation Syndromes with Sex Reversal*

| Disorder | Phenotypic Abnormalities | Sexual Development |
|---|---|---|
| Congenital adrenal hypoplasia | Normal appearance at birth<br>Salt-wasting<br>Accelerated skeletal maturation<br>Rapid growth in childhood | Ambiguous genitalia in females<br>Normal in males |
| Smith-Lemli-Opitz syndrome | Microcephaly<br>Mental retardation<br>Hypotonia<br>Polydactyly<br>Small nose with anteverted nares<br>Ptosis | Ambiguous genitalia in males<br>Hypospadias<br>Complete failure to develop external male genitalia with mixed internal structures |
| Campomelic dysplasia | Dwarfism<br>Extreme bowing of leg bones<br>Small scapulae<br>Craniofacial changes<br>Defective tracheal bronchial cartilage→respiratory distress→neonatal death | 46,XY females with ovarian dysgenesis but normal uterus and fallopian tubes |
| Denys-Drash syndrome | Wilms' tumor<br>Aniridia<br>Severe urogenital aberrations | Ambiguous genitalia in males<br>Rudimentary uterus<br>Fimbriated fallopian tubes<br>Streak gonads |
| Meckel-Gruber syndrome | Encephalocele<br>Dysplasia of kidneys<br>Polydactyly<br>Neonatal death | Cryptorchidism<br>Incomplete development of external or internal genitalia |

# GENETIC COUNSELING FOR CHROMOSOMAL ABNORMALITIES

The consequences of a chromosomal abnormality can be devastating not only to the patient but also to the parents, and they often involve the entire extended family. Adequate genetic counseling is an important aspect of medical genetics and should be facilitated by either a trained genetic counselor or a board-certified medical geneticist. A genetic counselor provides the patient with information concerning the cause of the abnormality, the carrier status of the patient and the parents, the recurrence risk, and the availability of prenatal testing. Quite often, the patient is faced with a difficult decision concerning a current pregnancy, the care of an affected child, or future reproductive options, for which the counselor is called upon to provide balanced, nonbiased information on all available options. One of the basic principles of genetic counseling is to respect the autonomy of the patient and to be nondirective in counseling. Other important aspects of a counseling session are to reassure patients that this was not their fault and to provide them with an opportunity to ask questions. An excellent reference for genetic counseling for chromosomal abnormalities is provided in Gardner and Sutherland [24].

# CHROMOSOMAL ABNORMALITIES IN CANCER

As early as 1914, Theodore Boveri conceptualized the notion of chromosomal changes as the origin of cancer [25]. Detection of several specific cytogenetic and molecular changes in various tumors and isolation of oncogenes and tumor suppressor genes in the past 30 years have proven that Boveri's predictions were correct. It is currently believed that cancer is a genetic disease at a cellular level and that it results from an accumulation of inherited and somatic mutations. Cancer is a collection of diseases, with genetic and environmental components resulting in uncontrolled growth and cell cycle deregulation of different cell types (see Chapter 5). For example, cancer of the epithelial cells is termed carcinoma; cancer of connective tissues is a sarcoma; cancer of hematopoietic tissues is a leukemia. In this chapter, the phenotypic expression of constitutional chromosomal changes that are present in every cell of the body have been reviewed. When chromosomal changes associated with cancer are discussed, the focus is almost always on the acquired changes that are present only in the cancer cells resulting in uncontrolled cell growth.

## Chromosomal Changes Associated with Leukemia

In recent years, identification of consistent chromosomal abnormalities has resulted in molecular characterization of genes located at the break points. This has increased the understanding of the critical changes in gene structure and function leading to the formation and progression of various types of leukemia and lymphoma. Karyotypic abnormalities can be detected in most leukemias. Some of these abnormalities are listed in Table 9-9. The molecular consequences of a specific translocation in chronic myelogenous leukemia (CML) and a deletion in retinoblastoma are discussed here to illustrate the role of cytogenetic analysis of malignancies in clinical practice and research. The list of the genes thus identified is getting longer every day and is updated annually by the Committee on Chromosome Changes in Neoplasia [26].

*CML.* The first reported nonrandom abnormality was in the bone marrow of a patient with CML and was reported by Nowell and Hungerford (from Philadelphia) in 1960 [27]. This abnormality was termed the Philadelphia (Ph) chromosome and consisted of a small chromosome that appeared to be a deletion on chromosome 22. In 1973, Rowley reported that the correct abnormality consisted of a translocation between chromosomes 9 and 22 [28]. In 1982, molecular analysis revealed that t(9;22)(q34;q11) involves the transfer of the c-*ABL* (human homolog of the Abelson murine leukemia virus oncogene)

**TABLE 9-9** ▶

Examples of Structural Abnormalities in Leukemia and Lymphoma

| Disease | Abnormality | Genes Involved | |
|---|---|---|---|
| Chronic myelogenous leukemia | t(9;22)(q34;q11) | ABL | BCR |
| Acute myelogenous leukemia (M2) | t(8;21)(q22;q22) | ETO | AML1 |
| Acute lymphoblastic leukemia | t(10;14)(q24;q11) | HOX11 | TCRD |
| Burkitt's lymphoma | t(8;14)(q24;q32) | MYC | IGH |

*Note.* t = translocation; q = long arm of chromosome; p = short arm of chromosome.

gene from chromosome 9q34 to the break point cluster region (*BCR*) on 22q11. This abnormality is found in nearly 100% of patients with CML and is used clinically to confirm the diagnosis of CML. The normal c-*ABL* gene codes for two major transcripts, which are 6 kb and 7 kb in length, whereas the translocated *ABL* produces a novel 8.5-kb transcript. This hybrid *BCR-ABL* mRNA encodes a 210-kD polypeptide instead of a 145-kD *ABL* product. The fusion protein has increased *ABL* tyrosine kinase activity. Additional abnormalities are observed during the blast phase of the disease and include trisomies for chromosomes 8 and 19, an isochromosome for 17q, and additional copies of the Ph chromosome.

## Chromosomal Changes Associated with Solid Tumors

Although solid tumors are much more common than hematologic malignancies, technical difficulties in tissue dissociation, culture, and harvest procedures have hampered similar progress in the cytogenetic analyses of these tumors. Improved methodology in recent years has resulted in a dramatic increase in the number of solid tumors that are cytogenetically analyzed. Several recurrent, specific chromosomal abnormalities have been identified, and the cytogenetic analyses of selected solid tumors, especially some soft-tissue tumors, can provide important diagnostic and prognostic information. Unfortunately, many tumors display multiple chromosomal abnormalities, which makes differentiation of the critical changes more difficult. Many of these cases are processed for research purposes. In addition, tumor cells that are exposed to radiation or chemotherapy often show complex rearrangements. Many benign tumors also display specific chromosomal abnormalities and offer a unique opportunity to identify genes involved in growth disregulation. Some examples of specific chromosomal abnormalities found in solid tumors are listed in Table 9-10.

**TABLE 9-10** ▶

Examples of Chromosomal Abnormalities in Solid Tumors

| Tumor | Abnormality | Genes Involved | |
|---|---|---|---|
| **Malignant** | | | |
| Synovial sarcoma | t(X,18)(p11.2;q11.2) | SSX1,SSX2 | SYT |
| Myxoid liposarcoma | t(12;16)(q13;p11) | CHOP | FUS |
| Ewing's sarcoma | t(11;22)(q24;q12) | FLI1 | EWS |
| Retinoblastoma | del(13)(q14q14) | RB1 | |
| Wilms' tumor | del(11)(p13p13) | WT | |
| **Benign** | | | |
| Salivary gland tumor | t(1;12)(p22;q15) | HMGI-C | |
| Uterine leiomyoma | t(12;14)(q15;q24) | HMGI-C | |

*Note.* t = translocation; del = deletion; p = short arm of chromosome; q = long arm of chromosome.

***Retinoblastoma.*** Retinoblastoma is a rare childhood tumor of the retina that affects approximately 1 in 20,000 children. It presents clinically by the age of 5 years and is initiated when the retinal cells are still proliferating. Approximately 60% of the tumors are sporadic, and in 40% of the cases the mutation in the retinoblastoma (*RB1*) gene is inherited in a dominant fashion. A retinoblastoma with a cytogenetic deletion of the 13q14 region was one of the first microdeletion syndromes described. The *RB1* gene was also the first human tumor suppressor gene to be cloned.

Only 3% of cases have a cytogenetically visible deletion; most arise from point mutations or small deletions. However, some children with retinoblastoma tumors associated with other congenital anomalies and mental retardation were found to have a cytogenetically detectable constitutional deletion in the long arm of chromosome 13. The extent of the deletion varied, and the smallest region of the overlap was determined to be in the 13q14 region. Linkage of the *RB1* gene to the esterase-D (*ESD*) gene was established, and it was shown that in some deletion patients, one copy of the *ESD* gene was deleted constitutionally, whereas both copies were missing in patients with the tumor.

The retinoblastoma gene was first cloned in 1986 by Friend and colleagues [29]. Since that time, it has been the subject of a tremendous amount of research. The gene product is a 110-kD nuclear phosphoprotein consisting of 928 amino acids. Presence of the normal retinoblastoma gene helps to control the cell cycle and prevent uncontrolled cell proliferation. Knudson's two-hit theory of carcinogenesis states that both alleles of a tumor suppressor gene need to be inactive for a tumor to develop. Both copies of the *RB1* gene are inactivated in retinoblastoma patients. In familial cases, one mutation is inherited and is present in all the cells, followed by a second somatic mutation. Familial cases are usually bilateral and occur earlier than sporadic cases. Successful treatment of retinoblastoma has resulted in longer survival of the patients. Secondary osteosarcoma tumors develop with a high frequency, which calls for careful monitoring throughout the patient's life. Molecular testing is available, but because of the size of the gene and the large number of mutations that have been shown to inactivate *RB1*, it is expensive and not performed clinically at the present time.

## Chromosomal Changes Associated with Benign Tumors

Initially it was thought that chromosomal abnormalities were confined to malignant neoplasms, but recent data reveal that benign tumors also contain specific, recurrent abnormalities. Salivary gland adenomas, uterine leiomyomas (fibroids), lipomas, endometrial polyps, and other benign tumors all contain recurrent abnormalities. Uterine leiomyomas are common benign tumors of the smooth muscle of the uterus that occur in 30% of women of reproductive age. Forty percent of the leiomyomas have nonrandom chromosomal abnormalities. These include translocation t(12;14)(q13;q24), deletion of 7q22, and other abnormalities involving chromosomes 1, 3, 6, 10, and 13. Because 60% of tumors have a normal karyotype, the presence of the chromosomal abnormalities is not necessary for the formation of a leiomyoma but must represent a secondary change that might aid in the proliferation of the tumor cells. Recently, the gene involved in the translocation on chromosome 12 has been identified as *HMGI-C*, an architectural factor from the family of high-mobility group proteins. The gene on chromosome 7 has not yet been identified. *HMGI-C* is also altered in lipomas and endometrial polyps.

## Gene Amplification

Gene amplification is a phenomenon observed in several different types of tumors. It manifests in the form of either double minutes (DMs) or homogeneously staining regions (HSRs). DMs appear as small double dots and are made up of circular DNA molecules that usually lack a centromere and telomeres. They are seen in the metaphase cells of many different types of tumors and represent amplified copies of a small piece of the genome. Hundreds of extra copies may be present in a single tumor cell, resulting in increased gene product. The *N-myc* gene is amplified in late-stage neuroblastomas; in some types of breast cancer, the *ERBB2* and *myc* genes are both amplified. In general, the tumors with amplification represent more advanced stages and usually require more aggressive treatments. Several excellent references that describe chromosomal changes in cancer can be found in Heim and Mitelman [30], Mitelman [31], and Sandberg [32].

Appearance of either DMs or HSRs is also seen in cancer cells under experimental conditions. An example of this occurs during the development of resistance to methotrexate, which can be induced in cultured mammalian cancer cells by exposing them to strong doses of the drug. This effect is due to amplification of the dihydrofolate reductase gene, which is involved in the metabolism of methotrexate.

# RESOLUTION OF CLINICAL CASE

The Smiths returned for a follow-up visit 2 weeks later to discuss the results of the cytogenetic analyses. The results indicated that Mark had a partial deletion of the terminal short arm of chromosome 17 at band p13. The deletion was very subtle but could be seen on high-resolution banded chromosomes. This deletion is commonly associated with Miller-Dieker syndrome, which is a multiple malformation syndrome characterized by type I lissencephaly, seizures, severe mental retardation, and a characteristic facial appearance. The facial features include prominent forehead with vertical soft-tissue ridging and furrowing, bitemporal hollowing, a short nose with upturned nares, protuberant upper lip, thin vermilion border of the upper lip, and a small jaw. Most children born with Miller-Dieker syndrome have postnatal failure to thrive and die usually before the age of 2 years.

The geneticist found the cytogenetic results puzzling because Mark had only a few of the features associated with Miller-Dieker syndrome. He requested that the laboratory confirm the diagnosis using FISH analysis with a probe specific for the Miller-Dieker region on chromosome 17. Results of these studies indicated that Mark did not have deletion for this region of chromosome 17; his deletion was smaller and more terminal, which probably accounted for his more mild phenotype. Mrs. Smith's chromosomes appeared to be normal. Based on these results, the geneticist felt that Mark's deletion was most likely a new mutation and that Mrs. Smith's current pregnancy was not at increased risk for a similar genetic condition.

Six weeks later, the geneticist received a phone call from the cytogenetics laboratory. Mrs. Smith had an amniocentesis at 18 weeks because of an abnormal maternal serum AFP result, which indicated that she was at an increased risk for carrying a fetus with Down's syndrome. The results of the cytogenetic analysis on Mrs. Smith's amniotic fluid revealed that the fetus had a deletion of chromosome 17p13, similar to that seen in Mrs. Smith's son. The laboratory wanted to run an additional FISH probe corresponding to the subtelomeric region of chromosome 17 at p13.3 on the amniotic fluid sample, as well as on Mrs. Smith's and Mark's blood samples. The results of the FISH analysis would be available the following day. The geneticist approved the additional testing and scheduled the Smiths for a follow-up appointment the next day. The results of the FISH analysis indicated that both Mrs. Smith and the female fetus were carriers for a balanced translocation between the short arm of chromosome 17 and the long arm of chromosome 9. However, Mark carried only the derivative chromosome 17, which made him an unbalanced carrier, with partial trisomy for the terminal end of the long arm of chromosome 9 and partial monosomy for the terminal end of the short arm of chromosome 17. When Mr. and Mrs. Smith arrived, the doctor explained the results of the FISH analysis. He stressed that although the fetus appeared to have a deletion, that all of the genetic material was still present in the correct amount. Because Mrs. Smith had a normal phenotype, he expected that the fetus would also be normal. The reason the deletion was not originally detected in Mrs. Smith's blood sample was most likely because of lower band resolution.

The doctor reviewed Mrs. Smith's family history and explained that her two miscarriages were most likely due to the presence of an unbalanced chromosome complement in each of the pregnancies. However, because cytogenetics had not been performed on either conceptus, it would be impossible to confirm this. He also suggested that one of Mrs. Smith's parents might be a balanced translocation carrier, based on the history of multiple miscarriages. It would be important for her parents and her siblings to have blood drawn for cytogenetic analysis, so that a familial cryptic translocation could be ruled out. If Mrs. Smith's siblings were also carriers of the balanced translocation, then they would be at an increased risk for having children with chromosomal abnormalities, as well as multiple pregnancy losses. Mrs. Smith agreed to disclose this information to her family and told the doctor that she would have them contact him if they were interested in being tested. (This case was reprinted with permission from Estop AM, Mowery-Rushton, PA, Ciepy KM, et al: Identification of an unbalanced cryptic translocation t(9;17) (q34.3rp13.3) in a child with dysmorphic features. *J Med Genet* 32:819–822, 1995.)

# REVIEW QUESTIONS

**Directions:** For each of the following questions, choose the **one best** answer.

1. X chromosome inactivation is a mechanism for
   - **(A)** distinguishing between males and females
   - **(B)** maintaining polymorphisms in the population
   - **(C)** gene dosage compensation in mammals
   - **(D)** elimination of deleterious genes from the X chromosome

2. Genomic imprinting is best described as
   - **(A)** a phenomenon only seen in mice
   - **(B)** not involved in the etiology of any known human genetic disorder
   - **(C)** the preferential expression of specific genes that is dependent on the parental origin
   - **(D)** uniparental disomy
   - **(E)** responsible for most spontaneous abortions

3. Patients with Down's syndrome are best described by which one of the following statements?
   - **(A)** They always have a 47,XX,+21 or 47,XY,+21 karyotype
   - **(B)** They rarely live to adulthood
   - **(C)** They have a higher than average risk of impaired cellular immunity and leukemia
   - **(D)** They rarely have an intelligence quotient higher than 50
   - **(E)** They are usually confined to a wheelchair

4. Klinefelter's syndrome is best described by which one of the following statements?
   - **(A)** It occurs in a phenotypic female with a 46,XY karyotype
   - **(B)** It is characterized by ambiguous genitalia at birth
   - **(C)** It is caused by a mutation involved in cortisol biosynthesis
   - **(D)** It is characterized by a 47,XXY karyotype with hypogonadism after puberty
   - **(E)** It is seen at a higher frequency in males in prison than in the general population

5. Characteristics of women with Turner's syndrome include which of the following?
   - **(A)** They almost always display short stature
   - **(B)** They usually undergo normal pubertal changes
   - **(C)** They are mentally retarded
   - **(D)** They have a higher than normal risk of developing breast cancer

6. Triploidy is best described by which one of the following statements?
   - **(A)** It is a conception with one extra chromosome
   - **(B)** It is a conception with one extra haploid set of chromosomes
   - **(C)** It is usually not diagnosed in an infant until after the first year of life
   - **(D)** It can arise by five distinct mechanisms involving errors in the development of the egg
   - **(E)** It is the most common chromosomal abnormality seen in spontaneous abortions

7. The Philadelphia (Ph) chromosome is best described as

(A) a deleted chromosome 22

(B) a derivative chromosome formed by a translocation between chromosomes 9 and 22

(C) most commonly associated with acute myelogenous leukemia

(D) a cause of mental retardation when inherited constitutionally

(E) commonly seen in spontaneous abortions

8. The development of a retinoblastoma is most accurately described as being

(A) commonly associated with mental retardation and multiple congenital anomalies

(B) caused by a translocation between chromosomes 12 and 14

(C) caused by a deletion of the esterase D gene

(D) caused by mutations or deletion of both copies of the RB1 gene

(E) always associated with osteosarcoma

**Directions:** The group of questions below consists of lettered choices followed by several numbered items. For each numbered item, select the appropriate lettered option with which it is most closely associated. Each lettered option may be used once, more than once, or not at all.

**Questions 9–14**

Match each of the following phenotypes with the appropriate karyotype.

(A) 46,XX,dup(11)(p15.5)pat

(B) 47,XY,+i(12p)/46,XY

(C) 45,XX,der(14;21)(q10;q10)

(D) 46,XY,del(17)(p13.1)

(E) 46,XX,del(22)(q11.2)

(F) 46,XX,der(14;21)(q10;q10),+21

9. Down's syndrome

10. DiGeorge syndrome

11. Normal female

12. Pallister-Killian syndrome

13. Beckwith-Wiedemann syndrome

14. Miller-Dieker syndrome

# ANSWERS AND EXPLANATIONS

**1. The answer is C.** X chromosome inactivation is seen in any individual with two or more X chromosomes and is responsible for maintaining dosage compensation for genes on the X chromosome.

**2. The answer is C.** The classic definition of genomic imprinting is preferential expression of specific genes that is dependent on the parental origin. Imprinting was first studied in mice, but it is also present in humans, as well as other species. It is known to be involved in at least six disorders but is not likely to play a major role in pregnancy loss. Uniparental disomy, a separate phenomenon from genomic imprinting, only causes a clinical problem when the chromosome or chromosomal region contains an imprinted gene.

**3. The answer is C.** Individuals with Down's syndrome have impaired cellular immunity and are at a 15-fold increased risk of developing acute leukemia. The most common karyotypes are 47,XX,+21 or 47,XY,+21 (95%), but approximately 4% of cases arise because of Robertsonian translocations. Approximately 50% of Down's syndrome infants survive to 50 years of age with intelligence quotients ranging from 25 to 85. Most affected individuals are not wheelchair-bound.

**4. The answer is D.** Individuals with Klinefelter's syndrome usually appear normal with only tall stature until puberty at which time secondary sexual characteristics fail to develop, and hypogonadism becomes apparent. The genitalia are not ambiguous. Defects in cortisol biosynthesis are related to virilization of female infants in congenital adrenal hypoplasia; affected male infants appear normal. There is no evidence that individuals with Klinefelter's syndrome are more prone to criminality than the general male population.

**5. The answer is A.** Short stature is a universal feature of women with Turner's syndrome. They are at increased risk of developing autoimmune thyroiditis, but they are not at an increased risk for breast cancer. They usually do not undergo complete development of their secondary sexual characteristics. They are not mentally retarded, but they may experience certain learning disabilities.

**6. The answer is B.** Triploidy is a conception with one extra haploid set of chromosomes, giving rise to a 69,XXX (XXY or XYY) chromosomal complement, whereas trisomy is a conception with only one extra chromosome. Most of these conceptions do not survive to term, and those that survive usually die within several hours of birth. The five mechanisms giving rise to triploidy can involve either the egg or the sperm. It is the second most common cytogenetic abnormality seen in spontaneous abortions, accounting for 16% of chromosomally abnormal cases.

**7. The answer is B.** A translocation between chromosomes 9q34 and 22q11 yields a small derivative chromosome referred to as the Philadelphia (Ph) chromosome, which is most commonly associated with chronic myelogenous leukemia (although it is seen in acute myelogenous leukemia). There is no association between the Ph chromosome and mental retardation or spontaneous abortions.

**8. The answer is D.** The retinoblastoma gene is a tumor-suppressor gene; therefore, for development of a tumor, both copies must be inactivated because of either mutations or deletions. The majority of retinoblastoma patients have normal karyotypes, but microdeletions of chromosome 13q14 (which is also associated with mental retardation and multiple congenital anomalies) or translocations involving this region of chromosome 13 have been seen in a minority of patients. Osteosarcomas are often seen as secondary tumors in familial retinoblastoma patients, but this usually occurs later in life.

**9–14. The answers are: 9-F, 10-E, 11-C, 12-B, 13-A, 14-D.** The carrier of a Robertsonian translocation who carries more than one extra copy of the chromosomes involved is trisomic for the extra chromosome. This individual has two normal chromosomes 21 in addition to the der(14;21) chromosome, resulting in trisomy 21 or Down's syndrome.

DiGeorge syndrome is caused by an interstitial deletion of the long arm of chromosome 22.

A balanced Robertsonian translocation carrier does not usually have an abnormal phenotype. The only genetic material that is lost is the short arms of the two acrocentric chromosomes, which contain nucleolar organizing regions that are found in multiple copies within a cell.

Pallister-Killian syndrome is a multiple congenital anomalies/mental retardation syndrome that is characterized by the presence of an isochromosome for the short arm of chromosome 12. This marker is always mosaic and is not usually seen in metaphase cells prepared from cultured peripheral blood.

Beckwith-Wiedemann syndrome can arise from several different genetic aberrations, all of which involve chromosome 11p15.5: duplication of the paternal chromosome 11p15.5, uniparental disomy for the paternal chromosome 11p15.5, or a translocation of the maternal chromosome 11p15.5, which results in disruption of the maternal imprint.

Miller-Dieker syndrome is caused by a deletion of the distal short arm of chromosome 17.

# REFERENCES

1. Hook EB: Chromosome abnormalities: prevalence, risks and recurrence. In *Prenatal Diagnosis and Screening*. Edited by Brock DJH, Rodeck CH, Ferguson-Smith MA. Edinburgh, UK: Churchill Livingstone, 1992, pp 351–392.
2. Warburton D, Byrne J, Canki N: *Chromosomal Abnormalities and Prenatal Development: An Atlas*. Oxford Monograph on Medical Genetics No. 21. Oxford, UK: Oxford University Press, 1991.
3. Genest DR, Roberts D, Boyd T, et al: Fetoplacental histology as a predictor of karyotype: a controlled study of spontaneous first trimester abortions. *Hum Pathol* 26:201, 1995.
4. Hanna JS, Shires P, Matile G: Trisomy 1 in a clinically recognized pregnancy. *Am J Med Genet* 68:98, 1997.
5. Sanders RC, Blackmon LR, Hogge WA, et al: *Structural Fetal Abnormalities: The Total Picture*. St. Louis, MO: Mosby-Yearbook, 1996.
6. Hatada I, Ohashi H, Fukushima Y, et al: An imprinted gene *p57^KIP2* is mutated in Beckwith-Wiedemann syndrome. *Nat Genet* 14:171–173, 1996.
7. Spence JE, Periaccante RG, Greig GM, et al: Uniparental disomy as a mechanism for human genetic disease. *Am J Hum Genet* 42:217–226, 1988.
8. Kotzot D, Schmitt S, Bernasconi F, et al: Uniparental disomy 7 in Silver-Russell syndrome and primordial growth retardation. *Hum Mol Genet* 4:583–587, 1995.
9. Ledbetter DH, Engel E: Uniparental disomy in humans: development of an imprinting map and its implications for prenatal diagnosis. *Hum Mol Genet* 4:1757–1764, 1995.
10. Sapienza C, Hall JG: Genetic imprinting in human disease. In *The Metabolic and Molecular Bases of Inherited Disease*, 7th ed. Edited by Scriver CR, Beaudet AL, Sly WS, et al. New York, NY: McGraw-Hill, 1995, pp 437–458.
11. Hall B: Mongolism in newborn infants. *Clin Pediatr* (Phila) 5:4–12, 1966.
12. Jones KL: *Smith's Recognizable Patterns of Human Malformation*, 4th ed. Philadelphia, PA: W. B. Saunders, 1988.
13. Ledbetter DH, Ballabio A: Molecular cytogenetics of contiguous gene syndromes: mechanisms and consequences of gene dosage imbalance. In *The Metabolic and*

*Molecular Bases of Inherited Disease*, 7th ed. Edited by Scriver CR, Beaudet AL, Sly WS, et al. New York, NY: McGraw-Hill, 1995, pp 811–839.

14. Dittrich B, Buiting K, Korn B, et al: Imprint switching on human chromosome 15 may involve alternative transcripts of the *SNRPN* gene. *Nat Genet* 14:163–170, 1996.

15. Kishino T, Lalande M, Wagstaff J: *UBE3A/E6-AP* mutations cause Angelman's syndrome. *Nat Genet* 15:770–773, 1997.

16. Matsuura T, Sutcliffe JS, Fang P, et al: De novo truncating mutations in *E6-AP* ubiquitin protein ligase gene (*UBE3A*) in Angelman's syndrome. *Nat Genet* 15:74–77, 1997.

17. Williams CA, Angelman H, Clayton-Smith J, et al: Angelman's syndrome: consensus diagnostic criteria. *Am J Med Genet* 56:237–238, 1995.

18. Buntinx IM, Hennekam RCM, Brouwer OF, et al: Clinical profile of Angelman's syndrome at different ages. *Am J Med Genet* 56:176–183, 1995.

19. Greenberg F, Lewis RA, Potocki L, et al: Multidisciplinary clinical study of Smith-Magenis syndrome (deletion 17p11.2). *Am J Med Genet* 62:247–254, 1996.

20. Barr ML, Bertram EG: A morphological distinction between neurons of the male and female, and the behavior of the nuclear satellite during accelerated nucleoprotein synthesis. *Nature* 163:676–677, 1949.

21. Kaneko N, Kawagoe S, Hiroi M: Turner's syndrome: review of the literature with reference to a successful pregnancy outcome. *Gynecol Obstet Invest* 29:81–87, 1990.

22. Saenger P: *Turner syndrome. N Engl J Med* 335:1749–1754, 1996.

23. Netley CT: Summary overview of behavioral development in individuals with neonatally identified X and Y aneuploidy. In *Prospective Studies on Children with Sex Chromosome Aneuploidy*. Edited by Ratcliffe SG, Paul N. New York, NY: Alan R. Liss, 1986, pp 293–306.

24. Gardner RJ, Sutherland GR: *Chromosome Abnormalities and Genetic Counseling*, 2nd ed. New York, NY: Oxford University Press, 1996.

25. Boveri T: *Zur Frage der Entstehung maligner Tumoren*. Jena: Gustav Fischer, 1914.

26. Mitelman F, Kaneko Y, Berger R: Report of the committee on chromosomal changes in neoplasia. In *Human Gene Mapping 9 Compendium*. Baltimore, MD: John Hopkins University Press, 1995, pp 1332–1350.

27. Nowell PC, Hungerford DA: A minute chromosome in human granulocytic leukemia. *Science* 132:1497, 1960.

28. Rowley JD: The Philadelphia chromosome translocation. A paradigm for understanding leukemia. *Cancer* 65:276–293, 1973.

29. Friend SH, Bernards R, Rogelj S, et al: A human DNA segment with properties of the gene that predisposes to retinoblastoma and osteosarcoma. *Nature* 323:643–646, 1986.

30. Heim S, Mitelman F: *Cancer Cytogenetics*, 2nd ed. New York, NY: Wiley-Liss, 1995.

31. Mitelman F: *Catalog of Chromosome Aberrations in Cancer*, 5th ed. New York, NY: Wiley-Liss, 1995.

32. Sandberg AA: *The Chromosomes in Human Cancer and Leukemia*, 2nd ed. New York, NY: Elsevier, 1990.

# 10 INHERITANCE PATTERNS IN HUMAN POPULATIONS

Mary L. Marazita, Ph.D.

## CHAPTER OUTLINE

## INTRODUCTION OF CLINICAL CASE

Mr. and Mrs. Green, both Caucasian, presented for cystic fibrosis (CF) genetic counseling because Mrs. Green has a positive family history. Mr. Green was adopted and thus has no knowledge of his birth family. They wanted to know their chances of having an infant with CF.

CF is one of the most common genetic disorders in North American Caucasians, affecting approximately 1 in 2500 Caucasian newborns. It was clinically identified in 1938 and was originally called "cystic fibrosis of the pancreas" because of the fibrotic lesions that develop in the pancreas. The pancreas is unable to secrete digestive enzymes in most CF patients, which eventually leads to malnutrition. The sweat glands are abnormal, resulting in high levels of chloride in the sweat. Approximately 10% of newborns with CF have meconium ileus. More than 95% of males with CF are sterile because of abnormalities in the vas deferens. The most serious complication of CF is the obstruction of the lungs by thick mucus that cannot be cleared effectively, which greatly increases the risk of lung infections.

# POPULATION GENETICS

## Introduction

Population genetics is the study of the action of genes at a population-wide level. Population genetics strives to estimate allelic and genotypic frequencies within a population, using probability and statistical analysis models based on mendelian predictions. Population genetics also investigates factors (e.g., mutation, selection) that can change genetic frequencies through time. The results of population genetics are important for recurrence risk estimation, for determining probability of paternity, for forensic applications, and for studying the composition and migration patterns of populations.

## Basic Probability Concepts

Calculations for population genetics and the other topics in this chapter rely on several basic probability concepts. The probability (P) of an event is the chance that the event will occur. Probabilities can range in value from 0 to 1 and are sometimes expressed as percentages.

*The **probability** of event A is denoted as P(A).*

The first basic concept is that the probabilities of a set of mutually exclusive and exhaustive events will sum to 1. For example, when tossing a coin, there are two possible "events," heads or tails. Therefore,

$$P(heads) + P(tails) = 1.0$$

Note that this could be rewritten as:

$$P(heads) = 1.0 - P(tails)$$

There are two important rules for combining probabilities of independent events: the *multiplication rule* and the *addition rule*. The multiplication rule is used to determine the combined probability of multiple independent events.

$$P(A \text{ and } B \text{ and } C) = P(A) \times P(B) \times P(C)$$

That is, the combined probability of independent events A, B, and C occurring is the product of the individual probabilities. For example, if a coin is tossed three times, and the result is "heads" each time, the probability would be:

$$P(heads \text{ and } heads \text{ and } heads) = P(heads) \times P(heads) \times P(heads)$$

For a fair coin, the combined probability would be:

$$P(heads \text{ and } heads \text{ and } heads) = (0.5) \times (0.5) \times (0.5) = 0.125$$

The addition rule, on the other hand, is used to calculate the probability of any one of a list of multiple independent events occurring.

$$P(A \text{ or } B \text{ or } C) = P(A) + P(B) + P(C)$$

That is, the probability of A or B or C occurring is the sum of the individual probabilities.

## Genetic Loci in Populations

Humans have two types of genetic loci. *Autosomal loci* are located on one of the 22 autosomes, and *sex-linked loci* are located on the X chromosome. Humans are diploid organisms, so each person has two alleles at each autosomal locus. Females also have two alleles at each sex-linked locus, but males have only one allele for sex-linked loci because males have only one X chromosome.

Genetic loci vary in their number of possible alleles. Some loci have only one possible allele (so all individuals are genetically identical at such loci), and other loci have scores of possible alleles, such as the human leukocyte antigen (HLA) locus. Loci with two or more possible alleles are said to be *polymorphic*.

Within a particular individual, the two alleles at a locus may be the same or different. If the alleles are the same, the individual is said to be *homozygous* at that locus; if the alleles are different, the individual is *heterozygous*. Consider an autosomal locus with two alleles, *A* and *B*. There are three possible genotypes—two homozygous (*AA* and *BB*) and one heterozygous (*AB*).

> *For a locus with "n" total alleles, there are n(n + 1)/2 total genotypes; "n" of them are homozygous, and the remainder are heterozygous.*

## Parameters of Population Genetics

A fundamental aim of population genetics is to estimate three frequencies for genetic loci: *allelic frequencies* (sometimes termed *gene frequencies*), *genotypic frequencies*, and *phenotypic frequencies*. Each of these frequencies is derived as population-based probabilities. For example, the allelic frequencies are the probabilities of the alleles within a population.

***Estimating Allelic and Genotypic Frequencies.*** Consider an autosomal locus with two alleles *A* and *B* (hence three genotypes). Assume that the population is randomly mating with respect to this locus and that there is no mutation or selection operating. A population that meets these assumptions is said to be in *Hardy-Weinberg equilibrium*, after the two statisticians who developed the concept.

> *Random mating is mating without regard to the genotype or phenotype of one's partner. The opposite of random mating is **assortative mating**. After one round of random mating (one generation), a population will be in Hardy-Weinberg equilibrium.*

The probability of allele *A* in the population, P(A), also termed the allelic frequency or gene frequency, can be conceptualized as the proportion of *A* alleles in the population. For a two-allele system, the two allelic frequencies are usually denoted with "p" and "q," and their sum must equal 1.0. Therefore,

$$P(A) = p \quad \text{and} \quad P(B) = q = 1 - p$$

The corresponding genotypic frequencies are conceptualized as the proportion of each genotype in the population and are then P(genotype *AA*), P(*AB*), and P(*BB*). Because the three genotypes are mutually exclusive and exhaustive, the sum of the three probabilities must equal 1.0:

$$P(AA) + P(AB) + P(BB) = 1.0$$

If the population is in Hardy-Weinberg equilibrium, then the genotypic frequencies can be readily calculated from the allelic frequencies. The genotypic frequencies under Hardy-Weinberg equilibrium are sometimes called the *Hardy-Weinberg proportions*.

$$P(AA) = p^2$$
$$P(AB) = 2pq = 2p(1 - p)$$
$$P(BB) = q^2 = (1 - p)$$

As for any set of genotypic frequencies, the sum of these Hardy-Weinberg frequencies equals 1.0.

$$p^2 + 2pq + q^2 = 1.0$$

For a sex-linked locus with two alleles, *A* and *B*, the proportions are calculated similarly but with sex-specific allelic frequencies.

$$P(A \text{ in males}) = p_m \quad P(A \text{ in females}) = p_f$$
$$P(B \text{ in males}) = q_m \quad P(B \text{ in females}) = q_f$$

For calculating the genotypic frequencies for a sex-linked locus, recall that males are hemizygous and receive their X chromosome from their mothers. Therefore, the genotypic frequencies for males are equal to the allelic frequencies for females.

$$P(\text{hemizygous genotype } A \text{ in males}) = p_f$$
$$P(\text{hemizygous genotype } B \text{ in males}) = q_f$$

Females, on the other hand, get one X chromosome from their mothers and another from their fathers.

$$P(\text{genotype } AA \text{ in females}) = p_m p_f$$
$$P(\text{genotype } AB \text{ in females}) = p_m q_f + p_f q_m$$
$$P(\text{genotype } BB \text{ in females}) = q_m q_f$$

Note how the multiplication and addition rules were used to derive, for example, the P(AB in females). A woman could be AB in two ways: either A came from her mother and B from her father, or A came from her father and B from her mother. Therefore,

$$P(\text{genotype } AB \text{ in females}) =$$
$$P(A \text{ from father and } B \text{ from mother or } B \text{ from father and } A \text{ from mother}) =$$
$$P(A \text{ from father and } B \text{ from mother}) + P(B \text{ from father and } A \text{ from mother}) =$$
*(by addition rule)*
$$P(A \text{ father}) \times P(B \text{ mother}) + P(B \text{ father}) \times P(A \text{ mother}) =$$
$$p_m q_f + p_f q_m$$
*(by multiplication rule)*

***Examples of Allelic and Genotypic Frequencies.*** Probabilities are always estimated from data. When collecting data about genetic loci, the basic observations are the genotypes of the population members. Therefore, the observed genotypic proportions are used to calculate the underlying allelic frequencies. The Hardy-Weinberg equilibrium proportions can be used to calculate allelic frequencies from genotypic proportions, and they can also be used to test whether a population is in equilibrium (i.e., randomly mating) by comparing the observed genotypic proportions to those expected under Hardy-Weinberg equilibrium.

Table 10-1 gives the observed genotypic proportions for two different populations with respect to an autosomal locus with two alleles, A and B.

**TABLE 10-1** ▶

*Observed Genotypic Frequencies for Two Populations*

|  | Proportions of Each Genotype | | |
|---|---|---|---|
|  | **AA** | **AB** | **BB** |
| Population I | 0.64 | 0.32 | 0.04 |
| Population II | 0.36 | 0.28 | 0.36 |

When reviewing population genotypic data, the sum of the three genotypic frequencies should be verified as equaling 1.0.

For population I:    $0.64 + 0.32 + 0.04 = 1.0$
For population II:   $0.36 + 0.28 + 0.36 = 1.0$

Next, the frequencies should be checked for consistency with Hardy-Weinberg equilibrium. There are many ways to do this; one way is described. Under Hardy-Weinberg equilibrium, $P(AA) = p^2$, $P(BB) = q^2$, and the square roots would be p and q, respectively. (Recall that $p + q = 1.0$.)

Therefore, if the sum of the square roots of P(AA) and of P(BB) equals 1.0, the data are consistent with Hardy-Weinberg equilibrium.

| | $\sqrt{P(AA)}$ | $\sqrt{P(BB)}$ |
|---|---|---|
| Population I: | 0.8 | 0.2 |
| Population II: | 0.6 | 0.6 |

Summing $\sqrt{P(AA)} + \sqrt{P(BB)}$ for each population reveals that population I is consistent with Hardy-Weinberg equilibrium (i.e., the sum equals 1.0), but population II is not. Therefore, p = 0.8 and q = 0.2 in population I.

Because population II is not consistent with Hardy-Weinberg equilibrium, p and q must be estimated in another way. The most straightforward way to estimate p and q in this situation is known as *gene (or allele) counting*. Basically, the numbers of each allele in a population are counted. In population II (see Table 10-1), 36/100 individuals are *AA*, 28/100 are *AB*, and 36/100 are *BB*. The 36 *AA* individuals have a total of 72 *A* alleles, the 28 *AB* individuals have 28 *A* alleles and 28 *B* alleles, and the 36 *BB* individuals have 72 *B* alleles, for a total of 72 + 28 = 100 *A* alleles and 28 + 72 = 100 *B* alleles (100 individuals will, of course, have a total of 200 alleles). Therefore, the P(*A*) in population II is 100/200 = 0.50, and the P(*B*) is 100/200 = 0.50.

***Estimating Phenotypic Frequencies.*** If it is possible to distinguish among the genotypes at a genetic locus, then the locus is said to be *codominant*, and the phenotypic frequencies equal the genotypic frequencies. For many of the genetic traits of medical interest, however, there are *dominance relationships* between the genotypes such that some genotypes are indistinguishable phenotypically.

Consider a two-allele, autosomal genetic locus for a disorder with two phenotypes, "affected" and "unaffected." Table 10-2 lists the possible genotypes and genotype/phenotype relationships under two dominance relationships: "*A*" allele dominant versus "*A*" allele recessive.

> **Dominant Allele**
> The phenotype of the heterozygote is the same as the phenotype of the allele's homozygote; that is, only one allele is necessary to express the phenotype.

> **Recessive Allele**
> The phenotype of the heterozygote is the same as the phenotype of the other allele's homozygote; that is, two alleles are necessary to express the phenotype.

◀ **TABLE 10-2**
Genotype/Phenotype Relationships for an Autosomal Locus (Alleles A and B) with Dominance

| | Phenotype if: | |
|---|---|---|
| Genotype | A Allele Dominant/ B Allele Recessive | A Allele Recessive/ B Allele Dominant |
| AA | Affected | Affected |
| AB | Affected | Unaffected |
| BB | Unaffected | Unaffected |

As more is learned about the molecular bases of human diseases, the concept of dominance is an oversimplification in some cases. However, "dominance" is useful for population genetics, for describing diseases that are not yet understood at a molecular level, and for genetic counseling.

The phenotypic frequencies for a locus with two phenotypes are then P(affected) and P(unaffected). To calculate these frequencies, one must know the *penetrance* of each genotype, which is the probability that an individual with a particular genotype will be affected.

$$f_i = P(\text{person with genotype } "i" \text{ is affected})$$

For a locus with alleles *A* and *B*, there are three genotypes and three penetrances ($f_{AA}$, $f_{AB}$, and $f_{BB}$). If $f_{AA} = f_{AB} = 1.0$, and $f_{BB} = 0.0$, then allele *A* is dominant. If $f_{AA} = 1.0$, and $f_{AB} = f_{BB} = 0.0$, then allele *A* is recessive. If the penetrances all equal 1 or 0, then the locus is said to be *fully penetrant*. If some of the penetrances do not equal 1 or 0, then the locus is *incompletely penetrant*. To determine the phenotypic frequencies, the penetrances and the genotypic frequencies are combined.

> **Incompletely Penetrant Locus**
> Some individuals carrying a disease allele do not express it. This can occur for a variety of reasons. For example, there may be advanced age of onset for the disorder, or the disease locus may actually be a susceptibility locus to an environmental trigger. If the environmental exposure is not present, then the phenotype will not be expressed.

$$P(\text{affected}) = P(AA)f_{AA} + P(AB)f_{AB} + P(BB)f_{BB}$$
$$P(\text{unaffected}) = 1 - P(\text{affected})$$

Under Hardy-Weinberg equilibrium, this equals:

$$P(\text{affected}) = p^2 f_{AA} + 2pq f_{AB} + q^2 f_{BB}$$

Therefore, if *A* is dominant and fully penetrant (i.e., $f_{AA} = f_{AB} = 1.0$, and $f_{BB} = 0.0$):

$$P(\text{affected}) = p^2 + 2pq$$
$$P(\text{unaffected}) = q^2$$

***Examples of Phenotypic Frequencies.*** For most genetic disorders, the data collected within a population are stated in terms of the phenotype (i.e., affected, unaffected). Population geneticists then use the phenotypic frequencies to infer the genotypic and

> **Phenocopies** are clinically identical phenotypes that are not caused by segregation of alleles at the disease locus under investigation. Thus, there will be a nonzero probability of expressing the phenotype even if an individual does not carry a disease gene.

allelic frequencies. For example, consider a population with 0.01% of individuals affected with an autosomal recessive disorder. Assume the population is in Hardy-Weinberg equilibrium and that the alleles are fully penetrant. Then,

$$P(\text{affected}) = 0.0001 = q^2 f_{BB} = q^2(1.0) = q^2$$
$$q = 0.01, \quad p = 0.99$$

## Factors Influencing Genotypic Frequencies

For many years, population geneticists have realized that genotypic frequencies at most genetic loci differ among populations. Much effort has gone into defining and quantitating the factors that lead to such population differences. In this section, some of the major factors are briefly summarized.

*Inbreeding.* Inbreeding is a factor that can alter genotypic frequencies on a population level. In particular, inbreeding can increase the amount of homozygosity in a population. To determine if homozygosity is caused by inbreeding, *identity by descent (IBD)* and *identity by state (IBS)* are distinguished.

If an individual is homozygous at a particular locus, both alleles are the same. If these two identical alleles are copies of the same stretch of DNA from a single ancestor, then the alleles are said to be IBD. Figure 10-1 depicts how an allele may be passed down to become IBD in a descendent, such as individual IV-1 in Figure 10-1.

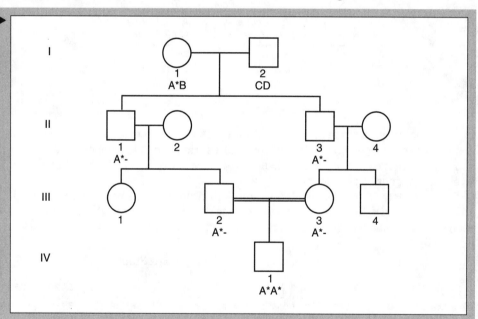

**FIGURE 10-1**
*Family Illustrating Identity by Descent (IBD). Individual IV-1 has the genotype A\*A\*. The A\* alleles are IBD because the A\* allele originated in individual I-1, the grandmother of IV-1. Note that the parents of IV-1 are first cousins, and each received the A\* allele from their common ancestor.*

**Consanguineous mating** *is mating involving two individuals with at least one ancestor in common; generally, no more remote than a great-grandparent.*

**Inbreeding** *is mating between two consanguineous individuals; the offspring of such a mating are said to be inbred.*

**Incest** *describes consanguineous marriages that are forbidden by law. Note, some cultures encourage inbreeding (see Figure 9-3).*

If two identical alleles did not come from a single ancestor, then they are merely IBS (Figure 10-2). Having two alleles IBD implies that an individual is the product of a consanguineous mating.

IBD and IBS are also used to compare the alleles of two individuals. Individuals III-2 and III-3 in Figure 10-1 share one allele IBD, allele *A\**. Individuals I-1 and I-2 in Figure 10-2 have identical alleles (*A*), but they are merely IBS, and they share no alleles IBD. For any particular locus, two individuals can share zero, one, or two alleles IBD, and the expected probabilities of sharing alleles IBD are used to quantitate inbreeding. In Figure 10-2, II-1 and II-2 share no alleles IBD although they are siblings.

The *coefficient of inbreeding* (F) equals P(an individual has both alleles IBD). Therefore, F gives a probability for a particular way to be homozygous at a locus (i.e., being homozygous as a result of inbreeding). A similar coefficient, the *coefficient of relatedness* (R) equals P(two individuals share an allele IBD). Therefore, R gives an estimate of how closely related two people are, and F estimates how closely related an individual's parents are (note: R = 2F). Table 10-3 tabulates F and R for different degrees of relationship.

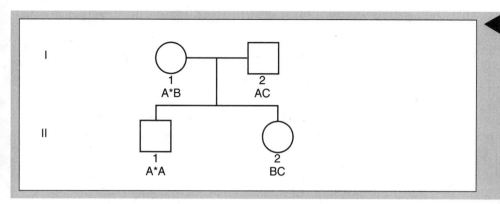

**FIGURE 10-2**
**Family Illustrating Identity by State (IBS).**
*Individual II-1 is homozygous for allele A, receiving one allele A from his mother (I-1, allele A\*) and the other from his father (I-2, allele A). Since these two A alleles did not originate in a single individual, II-1's homozygosity is IBS.*

**TABLE 10-3**
*Coefficients of Inbreeding (F) and Relatedness (R) for Various Degrees of Relationship (R = 2F)*

| Relatives | R | F |
|---|---|---|
| Siblings | ½ | ¼ |
| Uncle or aunt; niece or nephew | ¼ | ⅛ |
| First cousins | ⅛ | 1/16 |
| Second cousins | 1/32 | 1/64 |

The amount of inbreeding in a population can be quantitated by calculating a weighted average inbreeding coefficient ($\mu_F$), which can range from 0 (no inbreeding) to 1 (completely self-fertilizing). Table 10-4 gives the average inbreeding coefficient for several human populations. Then, the magnitude of the effect of inbreeding on a population can be estimated. Inbreeding increases homozygosity, consequently decreasing heterozygosity. If $H_0$ is the amount of heterozygosity before inbreeding, and $H_F$ is the amount of heterozygosity after inbreeding, then:

$$H_F = (1 - \mu_F)H_0$$

**TABLE 10-4**
*Average Inbreeding Coefficients ($\mu_F$) for Human Populations*

| Population | Time Period | $\mu_F$ |
|---|---|---|
| Belgium [2] | 1918–1959 | 0.0005 |
| Canada [3] | 1959 | 0.0004 |
| India (Andra-Pradesh) [4] | 1957–1968 | 0.0320 |
| Japan [5] | 1900s | 0.0046 |
| U.S. (Roman Catholic) [2] | 1959–1960 | 0.0001 |

For the Indian (Andra-Pradesh) population (see Table 10-4), $\mu_F = 0.032$; therefore,

$$H_F = (1 - \mu_F)H_0 = (0.978)H_0$$

That is, with $\mu_F = 0.032$, heterozygosity is decreased to 97.8% of what it would be without inbreeding. The amount of homozygosity after inbreeding is then:

$$1 - H_F = 1 - [(1 - \mu_F)H_0] = 1 - (0.978)H_0$$

The increase in homozygosity could be either good or bad, depending on whether homozygosity at a particular locus has deleterious effects or not. Figure 10-3 depicts Cleopatra's family tree, which has a high degree of inbreeding. Cleopatra, therefore, had a high degree of homozygosity and yet was a high-functioning leader in her society.

**Genetic Isolates.** *Genetic isolates* are groups whose members tend to mate within the group. Isolates may be formed because of geographic constraints (e.g., a group living on a

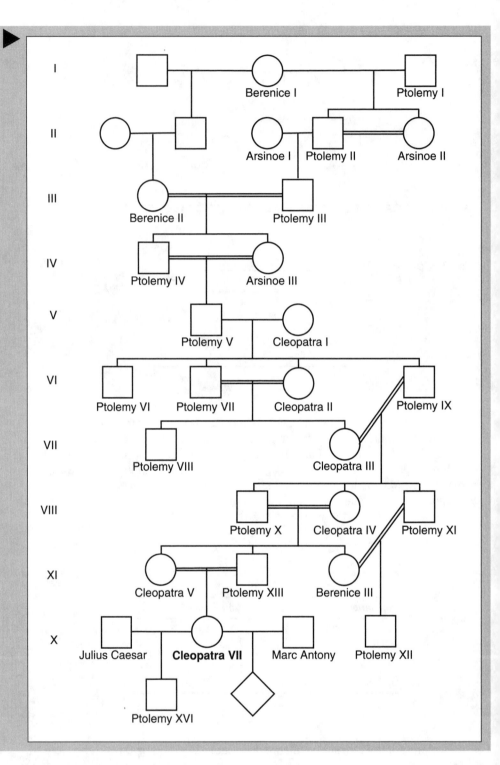

**FIGURE 10-3** ▶
***Cleopatra's Pedigree.*** *Among the ruling classes of ancient Egypt, the preferred mating was between brother and sister. Thus, a high degree of inbreeding is evident in this family history.*

small, isolated island) or by social constraints (e.g., religious beliefs, linguistic commonalities).

Genetic isolation can lead to changes in genotypic frequencies for a number of reasons: (1) *Founder effect* is caused by isolates that are typically initiated by a limited group of original settlers, the "founders" of the isolate. The genotypes of the descendants of the founders reflect the genotypes of the founders, and founders with many descendants make a disproportionate contribution to the genes in the descendant population. If one or more founders were carriers of a genetic disorder, the disorder is proportionately more common in the isolate than in other populations. (2) *Random genetic drift* is caused by chance events that may remove certain genotypes from a genetic isolate. For example, if one of the founders dies before reproducing, his or her genotype is lost to descendants.

(3) *Inbreeding* is sometimes increased in genetic isolates with a concomitant increase in homozygosity as described earlier.

***Mutation and Selection.*** Genetic mutations, or DNA changes in gametes, can add new alleles to a locus and potentially new phenotypes, can change an existing allele without changing the phenotype (so-called null mutations), or can change an existing allele in such a way as to be nonviable. Selection pressures operate on new mutations to determine whether they will survive. The frequency of a particular allele in a population results as a balance between the mutation rate for a particular locus and the effect of selection for or against the allele.

Selection may operate in complicated ways, resulting in persistence of mutations that initially appear quite deleterious. The classic example is sickle cell anemia in which heterozygotes for the abnormal sickle hemoglobin allele are resistant to malaria and therefore more "fit" than normal hemoglobin homozygotes in areas where malaria is endemic. Therefore, the frequency of the abnormal sickle hemoglobin allele has remained relatively high in such areas, despite the deleterious effect of being homozygous for sickle hemoglobin.

# MENDELIAN INHERITANCE PATTERNS

## Introduction

In the late 1800s, Gregor Mendel outlined rules for the inheritance of alleles at single genetic loci. Because Mendel's rules are now known to be accurate, such inheritance patterns are referred to as mendelian inheritance. Recognizing mendelian inheritance patterns is a major focus of human genetics. Human geneticists study the pattern of phenotypes in families, which then can be interpreted in light of the predictions of mendelian inheritance.

## Pedigree Notation

A number of designs have been used to depict phenotypes in families, with pedigree diagrams the most widespread. Pedigree diagrams graphically depict a large amount of information in an easily grasped format. The information summarized in a pedigree diagram comes from a careful family history to depict relationships accurately, physical examinations of family members to determine phenotype/diagnosis, and medical records or other information sources to confirm diagnoses (or to determine diagnoses on deceased or unavailable family members). Figure 10-4 summarizes many of the currently accepted pedigree symbols, definitions, and abbreviations.

## Patterns of Mendelian Inheritance in Pedigrees

There are four patterns of mendelian inheritance for monogenic traits, depending on the chromosomal location of the genetic locus and the dominance of the affected allele: autosomal dominant, autosomal recessive, sex-linked dominant, and sex-linked recessive. Each pattern has unique characteristics that can be observed in pedigree diagrams.

***Segregation of Alleles and Dominance.*** Mendel deduced several important principles from his research into phenotypes, such as height and seed shape expressed in garden peas. The first important principle is that of segregation of alleles, which Mendel called *units* because at the time of Mendel's experiments there was no knowledge of chromosomes, meiosis, genetic loci, or genes. This principle stated that each parent in sexually reproducing organisms possesses units (alleles/genes) in pairs and that only one member of each pair is transmitted to the offspring; that is, the alleles at a genetic locus segregate to the offspring. This was a revolutionary notion in Mendel's time because the prevailing dogma was that parental characteristics were blended in the offspring. Segregation of alleles means that parental alleles are not blended but rather remain intact from parents to children, then to grandchildren, and so on, thereby allowing inheritance to be traced from generation to generation.

*Modes of Inheritance*
- Single locus or monogenic: *Phenotype is due to mendelian inheritance of alleles at a single genetic locus.*
- Oligogenic: *Phenotype is due to interaction of a few loci.*
- Polygenic or multilocus: *Phenotype is due to many loci.*
- Multifactorial: *Phenotype is due to many loci and environmental factors.*
- Mixed: *Phenotype is due to a single locus plus multifactorial modification.*

*Segregation Analysis*
*This is a statistical technique used to determine the most likely mode of inheritance from family data.*

FIGURE 10-4 ▶
*Common Pedigree Symbols, Definitions, and Abbreviations.* (Source: *Adapted with permission from Bennett RL, Steinhaus KA, Uhrich SB, et al: Recommendations for standardized human pedigree nomenclature. Am J Hum Genet 56:745–752, 1995.*)

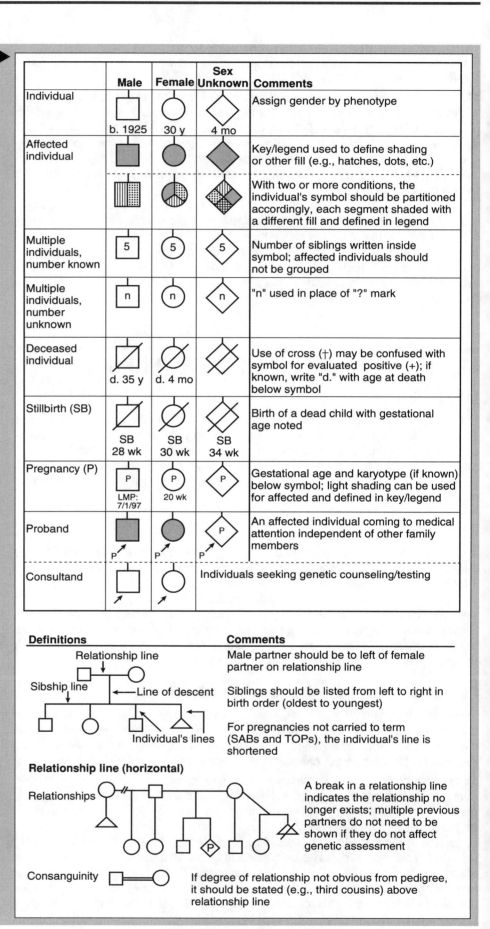

*Punnett squares* are used to illustrate the segregation of alleles into gametes and to predict the relative proportions of the resulting offspring genotypes. Table 10-5 is a Punnett square of a mating between an *AB* parent and a *CD* parent.

| | Gametes | Parent One (genotype AB) | |
|---|---|---|---|
| | | A | B |
| Parent Two | C | AC | BC |
| (genotype CD) | D | AD | BD |

Note. Shaded area signifies the possible offspring from the mating of an *AB* parent and a *CD* parent.

**TABLE 10-5**
General Punnett Square

The *AB* parent has *A* gametes and *B* gametes; the *CD* parent has *C* gametes and *D* gametes. There are four possible offspring from such a mating in equal proportions: ¼ *AC*, ¼ *AD*, ¼ *BC*, and ¼ *BD*.

A second major principle from Mendel's work is that of *independent assortment* of alleles at more than one locus. In particular, alleles at different loci are transmitted independently. This principle is discussed in more detail later in this chapter in the section on genetic linkage. Table 10-6 illustrates the independent assortment of alleles at two loci.

| | Gametes | Parent One (genotype BC;12) | | | |
|---|---|---|---|---|---|
| | | B,1 | B,2 | C,1 | C,2 |
| Parent Two | A,1 | AB;11 | AB;12 | AC;11 | AC;12 |
| (genotype AA;12) | A,2 | AB;12 | AB;22 | AC;12 | AC;22 |

Note. Shaded area signifies all possible offspring from a mating of (AA;12) × (BC;12) parents, illustrating independent assortment of the alleles at the two loci.

**TABLE 10-6**
Punnett Square for Two Loci (locus I has alleles A, B, and C; locus II has alleles 1, 2, and 3)

A third observation from Mendel's work is that the effects of one allele may mask those of another; that is, dominance relationships may exist between alleles of a genetic locus. Table 10-2 depicts the genotype/phenotype relationships for dominant and recessive alleles at a locus with two alleles.

Family histories as summarized in standardized pedigree diagrams can be interpreted in light of the patterns predicted by Mendel's laws. In this section, the major characteristics of each of the four patterns of mendelian inheritance are summarized. Each of the four patterns has a different set of characteristics, which can be used to infer the mode of inheritance of a particular trait given a pedigree diagram.

***Autosomal Dominant Inheritance.*** Figure 10-5 is a typical autosomal dominant pedigree, illustrating four main characteristics:

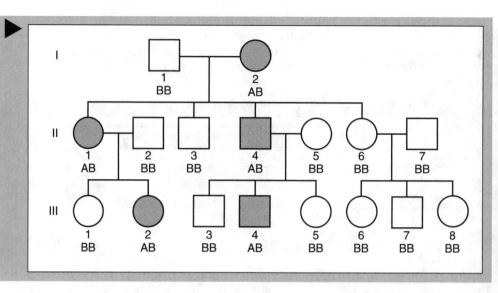

**FIGURE 10-5** ▶
**Typical Autosomal Dominant Pedigree.** *Disease locus with alleles* A *and* B*;* A *is the affected allele. Pedigree symbols are defined in Figure 10-4.*

1. *Any child of an affected parent has a 50% chance of inheriting the trait.* Assume that the disease locus has two alleles, *A* and *B*, with *A* being the affected allele. Because the affected allele is dominant, *AA* and *AB* individuals are affected. However, genetic disease allele frequencies are generally very low, so virtually all affected individuals are heterozygotes. Table 10-7 depicts the expected offspring of an *AB* × *BB* mating: half of the offspring will have genotype *AB* (affected), and half will be *BB* (unaffected).

2. *In an autosomal dominant pedigree, the phenotype appears in every generation.* Because only one allele is required for a person to be affected, as the affected allele segregates through a pedigree from generation to generation, so does the affected phenotype (see Figure 10-5).

3. *Unaffected individuals in a pedigree do not transmit the phenotype.* For a fully penetrant autosomal dominant trait, unaffected individuals do not carry an affected allele and therefore cannot have affected children.

4. *Males and females are equally likely to be affected.* This characteristic is true of any autosomal trait, as opposed to sex-linked traits, in which males and females are not equally likely to be affected (see sex-linked locus characteristics below).

**TABLE 10-7** ▶
*Expected Offspring of a Mating Between Affected and Unaffected, Autosomal Dominant Transmission*

| | | Affected Parent (AB) | |
|---|---|---|---|
| | **Gametes** | *A* | *B* |
| **Unaffected Parent** | *B* | *AB* | *BB* |
| **(BB)** | *B* | *AB* | *BB* |

*Note. Shaded area* signifies affected offspring genotypes.

**Autosomal Recessive Inheritance.** Figure 10-6 depicts a typical autosomal recessive pedigree, illustrating four main characteristics, which result from the fact that two alleles are required to be affected:

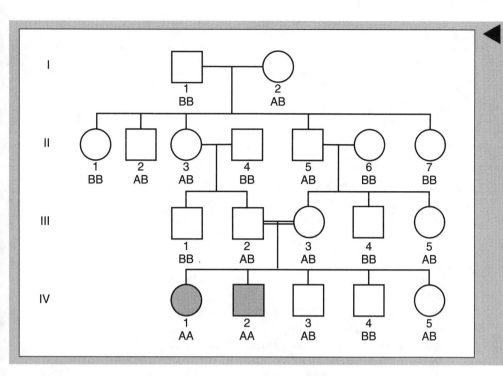

◀ **FIGURE 10-6**
***Typical Autosomal Recessive Pedigree.***
*Disease locus with alleles A and B; A is the affected allele. Pedigree symbols are defined in Figure 10-4.*

1. *On average, the parents of an affected individual will usually be unaffected.* To be affected, an individual must have received an affected allele from each parent. The allele frequencies at human disease loci are very low. Therefore, it is most likely that the parents of an affected individual are unaffected heterozygotes (see Figure 10-6).

   Table 10-8 shows the expected offspring genotypes for the mating between heterozygotes (*AB* × *AB*, *A* the affected allele). There will be 25% *AA*, 50% *AB*, 25% *BB* offspring; or 25% affected (*AA*) and 75% unaffected (*AB*, *BB*).

   If the phenotype is very common and not lethal (i.e., if affected individuals can reproduce), then there may be some matings between unaffected and affected individuals (*AB* × *AA*) with affected children. For an *AB* × *AA* mating, 50% of the offspring will be *AB* (unaffected) and 50% *AA* (affected). Because these proportions of affected and unaffected offspring mimic the proportions seen with autosomal dominant inheritance, this situation is termed *pseudodominance*. It is essential to examine multiple generations to distinguish between pseudodominance and dominance.

2. *In a pedigree, vertical transmission is observed; that is, the phenotype usually occurs in siblings, or sometimes cousins (rather than parents or offspring), of affected individuals.* In contrast to autosomal dominant pedigrees, affected individuals do not occur in each generation (see Figure 10-6).

3. *Consanguinity is increased in the parents of affected individuals.* Both parents of an affected individual must carry the affected allele. It is more likely that two parents are both heterozygous if they are related to each other and share the affected allele IBD.

4. *Males and females are equally likely to be affected.* This characteristic is true of any autosomal trait, as opposed to sex-linked traits, in which males and females are not equally likely to be affected.

**TABLE 10-8 ▶**
*Expected Offspring, Unaffected × Unaffected Mating Type, Autosomal Recessive Transmission*

| | | Unaffected Parent (AB) | |
|---|---|---|---|
| | Gametes | A | B |
| Unaffected | A | AA | AB |
| Parent (AB) | B | AB | BB |

Note. Shaded area signifies affected offspring genotypes.

**Sex-linked Recessive Inheritance.** The characteristics of the inheritance patterns of sex-linked loci are quite different than those of autosomal loci because males have only one X chromosome. Figure 10-7 is a typical sex-linked recessive pedigree. The four main characteristics of sex-linked recessive inheritance are as follows:

1. *There is no father–son transmission of sex-linked recessive traits.* Men always transmit their Y chromosome to their sons and their X chromosome to their daughters. Therefore, because sex-linked loci are located on the X chromosome, affected men can transmit the affected allele only to their daughters, never to their sons.

2. *For sex-linked recessive traits, males are more likely to be affected than females.* Females homozygous for the affected allele are affected, whereas males hemizygous for the affected allele are affected. Therefore, to be affected, females must receive an affected allele from each parent (i.e., a female's father must be affected, and her mother must be a carrier).

   The probability of a female having the affected genotype (homozygous) is:

   $$P(\text{female } AA) = p_m p_f = p^2 \text{ (assuming male and female allele frequencies are the same)}$$

   The probability of a male having the affected genotype (hemizygous $A$) is:

   $$P(\text{male } A) = p_f = p \text{ (assuming male and female allele frequencies are the same)}$$

   Therefore, for sex-linked recessive traits, males are more likely to be affected than females. For example, if the frequency of the affected allele (p) is 0.001, the probability of a male being affected is 0.001 and of a female being affected is $(0.001)^2 = 0.000001$.

3. *Affected males in a kindred are related to each other through carrier females.* See Figure 10-7 for examples of this characteristic, which is a consequence of the lack of male-to-male transmission of sex-linked traits.

4. *All daughters of an affected man are carriers; half of the sons of carrier females are affected.* Because a man has only one X chromosome to transmit to his daughters, if

*Heterozygotes are sometimes called **carriers** because they carry an affected allele.*

**FIGURE 10-7 ▶**
**Typical Sex-linked Recessive Pedigree.** *Disease locus with alleles A and B; A is the affected allele. Females can have genotype AA, AB, or BB; males can have hemizygous genotypes A or B. The female symbol with a dot represents carriers. Other pedigree symbols are defined in Figure 10-4.*

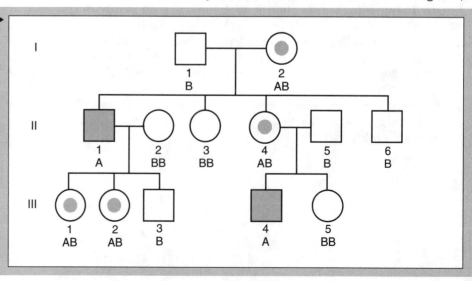

he is affected, then all his daughters receive an X chromosome carrying the affected allele. Carrier females have one X chromosome with an affected allele and one with an unaffected allele. Therefore, the sons of a carrier female have a 50% chance of receiving the X chromosome with the affected allele. Similarly, the daughters of a carrier female have a 50% chance of receiving the affected allele (i.e., of also being a carrier).

***Sex-linked Dominant Inheritance.*** Of the known sex-linked disease loci, most exhibit recessive inheritance. Figure 10-8 is a typical pedigree pattern for a sex-linked dominant trait, illustrating the following characteristics:

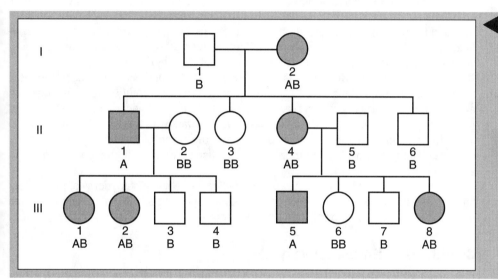

### FIGURE 10-8
***Typical Sex-linked Dominant Pedigree.*** *Disease locus with alleles A and B; A is the affected allele. Females can have genotypes AA, AB, or BB; males can have hemizygous genotypes A or B. Pedigree symbols are defined in Figure 10-4.*

1. *There is no father–son transmission of sex-linked dominant traits.* See the discussion above for sex-linked recessive inheritance.

2. *For sex-linked dominant traits, females are approximately twice as likely to be affected as males.* As for sex-linked recessive traits, the probability of a male being affected is:

$$P(\text{hemizygous } A) = p$$

The probability of a female being affected is:

$$P(\text{homozygous } AA \text{ or heterozygous } AB) =$$
$$P(\text{homozygous } AA) + P(\text{heterozygous } AB) =$$
$$p^2 + 2pq = p^2 + 2p(1 - p) = p^2 + 2p - 2p^2 = 2p - p^2$$

Because the allele frequencies are small, $p^2$ is very small (close to zero); thus, the probability of a female being affected is approximately $2p$, which is twice the probability of a male being affected.

3. *All daughters of an affected male are affected, half the offspring of a heterozygous (affected) female are affected, and all offspring of a homozygous (affected) female are affected.* These characteristics follow from the transmission and dominance patterns; see Figure 10-8 for examples. These characteristics are very similar to autosomal dominant inheritance and thus make it very difficult to distinguish between autosomal dominant and sex-linked dominant conditions. If any father–son transmission is seen, then a trait cannot be sex-linked dominant. A preponderance of females among the affected is also indicative of a sex-linked versus an autosomal dominant trait. However, a large number of families is necessary to demonstrate such a preponderance with statistical confidence. Because sex-linked dominant traits are so uncommon, a dominant pattern in a pedigree is more likely to be autosomal.

# RECURRENCE RISK ESTIMATION

## Introduction

A very important aspect of medical genetics and genetic counseling is to assess the probability of becoming affected with particular genetic disorders. To do so, medical geneticists must start with an accurate diagnosis of the disorder in a particular family. To arrive at an accurate diagnosis, a complete family and pregnancy history must be obtained, affected family members must have a dysmorphology examination, and additional tests may be necessary (e.g., cytogenetic analyses, biochemical assays). Based on such findings, a medical geneticist arrives at a diagnosis that includes whether there is an isolated anomaly in the family versus multiple anomalies, whether multiple anomalies fit into a known syndromic pattern, and whether the family pattern or syndromic classification has a clear etiology (e.g., mendelian segregation of alleles at a single locus, versus teratogenic exposure, versus a multifactorial etiology).

Once an accurate diagnosis is reached, medical geneticists then use statistical methods to arrive at risk figures. Risk is another word for probability, or the chance of an event occurring. A very common risk calculation that medical geneticists must derive is *recurrence risk*. For example, a couple may have an infant with a birth defect and wish to know the chance that their next child will have the same birth defect. An individual related to someone with a genetic disorder may wish to know their chance of having the same disorder.

If the cause of a disorder is known, the medical geneticist uses axioms of probability and predicted probabilities based on mendelian segregation to calculate recurrence risks. If the cause is unknown, then empiric recurrence risks are used.

**Recurrence risk** *is the chance that a condition will recur in a relative of an affected person.*

## Conditions of Known Etiology

If a genetic disorder is diagnosed, and the mode of inheritance is clear-cut, then recurrence risks are straightforward. For example, if the disorder is a fully penetrant, autosomal dominant condition, recurrence risks are based on the expected mendelian inheritance patterns and the dominance relationships between alleles. For an affected person, the risk of having an affected child is 50% (assuming the allele frequency is rare so that affected individuals are heterozygous). For an unaffected person, there is no risk of having an affected child.

For many human genetic disorders, however, the mode of inheritance is not completely clear-cut. Many dominant conditions have reduced penetrance, advanced age of onset, or variable expressivity. Some recessive conditions have the same complications; in addition, it is difficult to determine the genotype of unaffected individuals when calculating recurrence risks. Furthermore, there is sometimes additional information, such as the results of a biochemical test, to be incorporated into the risk calculation. To modify risk calculations for factors complicating inheritance patterns or to incorporate additional information, a framework termed Bayesian probability is used.

***Reduced Penetrance***
*An individual with at-risk genotype does not express the expected phenotype.*

***Advanced Age of Onset***
*A particular phenotype is not expressed at birth but manifests later in life.*

***Variable Expressivity***
*Individuals with the same genotype do not always express the same phenotype. Some individuals may express only a mild form of a disorder, whereas others are more severely affected.*

***Bayesian Probability Calculations.*** These calculations are used to determine the relative probabilities of two mutually exclusive alternatives. For each alternative, four components are calculated:

1. *Prior probability*, the initial probability without regard for complications or for additional information
2. *Conditional probability*, the probability of the additional information under each alternative
3. *Joint probability*, the product of the prior and conditional probabilities
4. *Posterior probability*, the joint probability of each alternative divided by the sum of the two joint probabilities

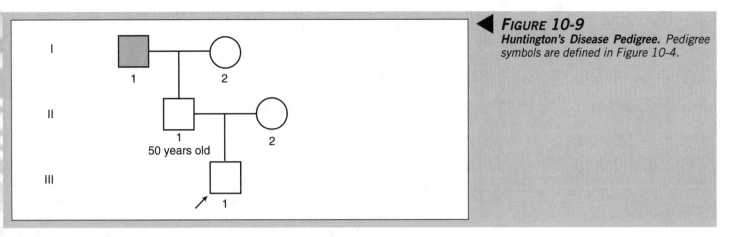

**FIGURE 10-9**
*Huntington's Disease Pedigree. Pedigree
symbols are defined in Figure 10-4.*

***Example of Bayesian Calculation.*** Consider the family depicted in Figure 10-9. Individual III-1 came for genetic counseling because his grandfather (I-1) was affected with Huntington's disease (HD), an autosomal dominant degenerative neurologic disorder with an age of onset typically of 40 years or later. Individual III-1 wished to know the risk that he will become affected with HD, given that his father (II-1) was unaffected at age 50 years.

With just the information that III-1 has an affected grandfather, the risk that III-1 would become affected is:

P(II-1 is a carrier) x P(transmitting the HD allele to III-1) = 1/2 x 1/2 = 1/4

There is additional information that can be incorporated; that is, that II-1 was unaffected at age 50 years. Based on the histories of many affected individuals, 80% of HD allele carriers are affected by age 50 years. This information can be used to modify the risk that III-1 will become affected.

To do this calculation in the Bayesian framework, the relevant alternatives are determined first. In this case, III-1 can become affected only if his father is a carrier of the HD allele. Therefore, the two alternatives are "II-1 is a carrier" versus "II-2 is not a carrier." Table 10-9 summarizes the calculation. For the *prior probabilities*, one assumes that I-1 was a heterozygote because I-1 was affected, and HD is rare. Therefore, the prior probability that II-1 is a carrier (i.e., that II-1 received a HD allele from I-1) is ½. Similarly, the prior probability that II-1 is not a carrier (i.e., received the unaffected allele from I-1) is also ½. As a check, note that the sum of the two prior probabilities must equal 1.

| Probability | Carrier | Not a Carrier |
|---|---|---|
| Prior | ½ | ½ |
| Conditional | P(not affected by 50 years) = ⅕ | 1 |
| Joint | ½ × ⅕ = ⅒ | ½ × 1 = ½ |
| Posterior | (⅒) ÷ (⅒ + ½) = ⅙ | (½) ÷ (⅒ + ½) = ⅚ |

*Note.* Calculated are the relative probabilities that individual II-1 in Figure 10-9 is either a carrier for Huntington's disease or not a carrier.

**TABLE 10-9**
*Summary of Bayesian Calculation for Figure 10-9*

The *conditional probabilities* are then determined. For the first alternative, the conditional probability is P(II-1 is not affected by age 50 years, assuming that II-1 is a carrier). The probability of a carrier being affected by age 50 years is 80%, so the probability of the converse (i.e., a carrier not being affected by age 50 years) is:

1 - 0.8 = 0.2, or 1/5

For the second alternative, the conditional probability is P(II-1 is not affected by age 50 years, assuming that II-1 is not a carrier). This probability is 100%; if II-1 is not a carrier, there is no chance that he will develop HD. Note that the sum of the two conditional probabilities does not equal 1.

The *joint probabilities*, the product of the prior and conditional probabilities for each alternative, are then determined. The sum of these values does not equal 1.

For the *posterior probabilities*, each joint probability is divided by the sum of the two joint probabilities. The sum of the two posterior probabilities must equal 1.

As shown in Table 10-9, the final probability that II-1 is a carrier is $\frac{1}{6}$, or $\frac{5}{6}$ that he is not a carrier. The probability that III-1 has an HD allele is:

P(II-1 is a carrier) x P(II-1 transmitted the HD allele to III-1) = 1/6 x 1/2 = 1/12.

Adding the information about II-1's age reduced the risk estimate for III-1 from $\frac{1}{4}$ to $\frac{1}{12}$.

This very simple example illustrates the concepts and method of applying a Bayesian probability framework to a recurrence risk estimation. Bayesian calculations are also used in very complex risk calculations, incorporating, for example, information about reduced penetrance, other phenotypes within the family, or results of diagnostic tests.

**Sibling Carrier Calculations.** In autosomal recessive conditions, recurrence risk calculations are complicated by the difficulty in determining the genotype for unaffected individuals in a pedigree. Consider Figure 10-10.

Individuals II-2 and II-3 wonder what the risk is that their children would be affected with the same autosomal recessive disorder as II-2's brother, II-1. Because II-1 is affected, I-1 and I-2 must each be heterozygous for the disease allele. Table 10-8 depicts the expected offspring from a mating between two unaffected individuals: $\frac{1}{4}$ affected (*AA* genotype) and $\frac{3}{4}$ unaffected ($\frac{1}{2}$ *AB*, $\frac{1}{4}$ *BB*). In the group of unaffected offspring (*unshaded* portion of Table 10-8), $\frac{2}{3}$ are *AB* and $\frac{1}{3}$ *BB*. Therefore, because II-2 is unaffected, the probability that she is heterozygous is $\frac{2}{3}$.

To have affected children, II-3 must also be a carrier. With no family history, the risk that II-2 is a carrier is the population heterozygote frequency (2pq under Hardy-Weinberg equilibrium). Therefore,

P(II-2 is a carrier *and* II-3 is a carrier) =
P(II-2 is a carrier) x P(II-3 is a carrier) = 2/3 x 2pq = 4pq/3
P(II-2 *and* II-3 carriers, *and* they have an affected child) =
P(both carriers) x P(affected offspring) = 4pq/3 x 1/4 = pq/3

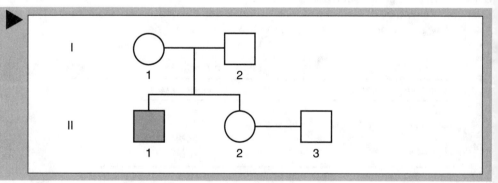

***FIGURE 10-10***
***Autosomal Recessive Pedigree for Sibling Carrier Calculation.*** *Pedigree symbols are defined in Figure 10-4.*

## Conditions of Unknown Etiology

There are many conditions for which the cause is unknown or perhaps multifactorial in nature. In these cases, it is necessary to use *empiric recurrence risks*. Empiric risk figures are tabulated for many disorders of unknown etiology and are simply counts of recurrence for a population-based sample of affected individuals and their relatives. Table 10-10 is an example of empiric recurrence risks for relatives of schizophrenic probands. For example, the oldest child in a family of four children has recently been diagnosed with schizophrenia. The chance that any of the other siblings will also develop schizophrenia is 7.3%, which is increased over the population incidence of schizophrenia of approximately 1.0%.

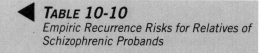

| Relationship | Recurrence Risk (%) |
|---|---|
| Monozygotic twin | 44.3 |
| Dizygotic twin | 12.1 |
| Offspring | 9.4 |
| Sibling | 7.3 |
| Niece/nephew | 2.7 |
| Grandchild | 2.8 |
| First cousin | 1.6 |
| Spouse | 1.0 |

*Source:* Adapted from McGue M, Gottesman II, Rao DC: The analysis of schizophrenia family data. *Behav Genet* 16:75–88, 1986.

**TABLE 10-10**
*Empiric Recurrence Risks for Relatives of Schizophrenic Probands*

# LINKAGE ANALYSIS

## Introduction

So far in this chapter, the focus has been on patterns of inheritance at single genetic loci. The co-segregation of alleles at two or more loci is the focus of this section.

Genetic loci on the same chromosome are said to be *syntenic*. Syntenic loci that are close to each other are *linked* and represent an exception to Mendel's law of independent assortment in that the alleles of linked loci are inherited together more often than expected. *Linkage analysis* is the term for statistical techniques that exploit this exception to Mendel's law, so the distance between linked loci can be determined. Linkage analysis is thus used to *map* loci to their chromosomal locations. In this section, the basic concepts of linkage analysis are reviewed; for more details, refer to Ott [1].

## Linkage Phase

As diploid organisms, humans have a pair of each type of chromosome, one from each parent. The members of a pair are known as *homologous chromosomes* or *homologues*. The alleles at linked loci, which are received from one parent (i.e., are on one homologue), make up a *haplotype*. For example, consider two linked loci, each with two alleles. Locus I has alleles *A* and *B*; locus II has alleles *1* and *2*. The possible haplotypes are *A1*, *B1*, *A2*, and *B2*. The segregation of haplotypes in a family can be followed in the same way as the segregation of alleles. Figure 10-11 shows the segregation of these haplotypes in a family.

*Linkage phase* describes which alleles at linked loci are on the same homologue in an individual. Alleles are in *coupling phase* if they are on the same homologue. They are in *repulsion* if the alleles are on opposite homologues. Individual II-1 in Figure 10-11 has genotype *AB* at locus I and genotype *12* at locus II. Given the genotypes of II-1's parents, the linkage phases of these alleles can be determined. Alleles *A* and *1* are in coupling, as are *B* and *2*; alleles *A* and *2* are in repulsion, as are *B* and *1*.

If linkage phase is known, then linkage analysis is much more efficient because it is then easier to determine which individuals represent recombinational events (see next section). If linkage phase cannot be determined, then statistical methods are used to incorporate the relative weights for each possible phase.

## Recombination

Consider individual II-1 in Figure 10-11, with genotypes *AB* and *12*. The possible haplotypes that this individual can transmit to his offspring are *A1*, *B2*, *A2*, and *B1*. Because the linkage phase is known for II-1's alleles (i.e., *A1*, *B2*), *A2* and *B1* represent a *recombination* of II-1's haplotypes; *A2* and *B1* are also called *recombinants*. A1 and B2 are therefore *nonrecombinants*.

**FIGURE 10-11**

**Genotypes and Phase at Two Linked Loci.**
*Locus I has alleles* A *and* B. *Locus II has alleles* 1 *and* 2. *Pedigree symbols are defined in Figure 10-4. Individuals III-1, III-2, and III-4 are nonrecombinants, with the same phases as found in their father (II-1). Individual III-3 is a recombinant, with a crossover occurring between II-1's homologous chromosomes.*

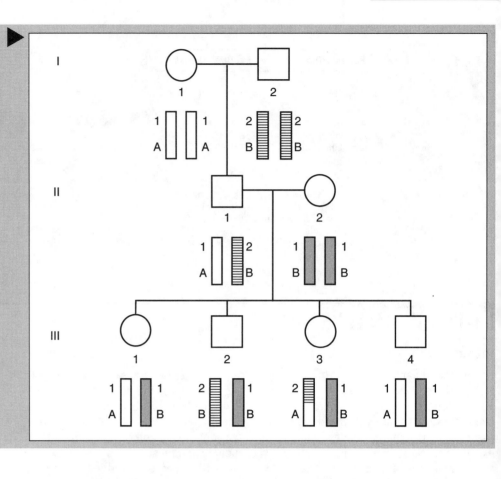

The **recombination fraction (θ)** between two loci is the probability of a crossover occurring between the loci. This probability depends on the distance between the two loci. **Centimorgans (cM)** are the units of distance between loci based on the amount of recombination: 1 cM = 1% recombination.

The physical basis of recombination is *crossing over* between homologous chromosomes during meiosis. To recognize recombinant offspring, at least one parent must be heterozygous at each locus, as is individual II-1. The other parent (II-2) is homozygous at each locus, so there is only one haplotype that that individual can transmit to her offspring. Recognition of recombination is the foundation of linkage analysis; thus, for a nuclear family to be informative, at least one parent must be doubly heterozygous (i.e., heterozygous at both loci).

The proportion of recombinant offspring is termed the *recombination fraction* and is denoted by the Greek letter theta, θ. The recombination fraction between two loci is a measure of how closely linked the loci are: the more recombination, the further apart the loci. Conceptually, θ is the number of recombinant offspring divided by the total number of offspring.

The recombination fraction between two loci can range from 0% to 50%. A value of 50% implies no linkage between the loci. Consider the doubly heterozygous individual II-1, with genotypes *AB* and *12*. The possible offspring haplotypes are summarized in Table 10-11. If the loci are not linked, there are equal proportions (25%) of each offspring haplotype. Two of the haplotypes are nonrecombinants (total probability 50%), and two are recombinants (total probability 50%). If the loci are linked with recombination fraction θ, then the total proportion of recombinants is θ, and the proportion of each of the two possible recombinant haplotypes is θ/2. The total proportion of nonrecombinants is then (1 − θ), and the proportion of each of the two possible nonrecombinant haplotypes is (1 − θ)/2.

If the linkage phases of the offspring of II-1 can be determined, then their recombinant status also can be determined. Because II-2 is doubly homozygous, there is only one possible haplotype that II-2 could transmit to her offspring, that is, *B1*. Then, the other haplotype (received from II-1) can be easily determined for each child (see Figure 10-11). Offspring III-1, III-2, and III-4 are nonrecombinants, and III-3 is a recombinant. An estimate of θ in this family is the number of recombinants divided by the total number of offspring, or ¼. Because the recombination fraction is less than ½ (the expected value

| | Probabilities under: | |
|---|---|---|
| **Possible Offspring Haplotyes** | **No linkage** | **Linkage at Recombination $\theta$** |
| A1  Nonrecombinant | ¼ | $(1 - \theta)/2$ |
| B2  Nonrecombinant | ¼ | $(1 - \theta)/2$ |
| A2  Recombinant | ¼ | $\theta/2$ |
| B1  Recombinant | ¼ | $\theta/2$ |

**TABLE 10-11**
Possible Recombinant and Nonrecombinant Haplotypes in the Gametes of a Doubly Heterozygous Individual

if the loci are not linked), one might suspect that the loci are indeed linked. To test whether loci are linked, a variety of statistical tests have been proposed over the years. Currently, the most common is the method of LOD scores.

## LOD Scores

"LOD" is an abbreviation for "log of the odds," in this case, the logarithm of the odds for and against linkage. LOD scores are usually calculated at a range of values for $\theta$; the value of $\theta$ yielding the largest LOD score is the best estimate of $\theta$ for the data.

The LOD score at $\theta$ is:

$$Z(\theta) = \log_{10} [L(\theta)/L(\theta = 1/2)],$$

where $L(\theta)$ is the *likelihood* of a family, assuming $\theta$ (i.e., linkage) versus $L(\theta = \frac{1}{2})$, or no linkage. Within a nuclear family, the likelihood expression is based on the binomial probability distribution:

$$Z(\theta) = \log_{10} \frac{\theta^R (1 - \theta)^{NR}}{0.5^{NR+R}}$$

where R is the number of recombinants, and NR is the number of nonrecombinants.

LOD scores for a family can be calculated for any arbitrary value of the recombination fraction, or LOD scores can be maximized to yield the best estimate of the recombination fraction for a family. When reporting linkage analysis results, the convention is to report the LOD scores for several arbitrary values of $\theta$, plus the maximum LOD score and the corresponding estimate of $\theta$. Reporting the LOD scores at arbitrary values of $\theta$ allows other investigators to readily compare their results and also to combine results across studies. LOD scores calculated in different study samples but at the same values of $\theta$ can be added together to result in combined LOD scores.

Table 10-12 is a LOD score table at arbitrary values of $\theta$ for the family depicted in Figure 10-11 (one recombinant, three nonrecombinants).

> The **likelihood** of an event is closely related to the probability of an event. Within a particular set of data, the likelihood equals a constant times the probability. For linkage analysis within a nuclear family, the likelihoods are calculated based on the binomial probability distribution.

> Likelihoods, such as those used in LOD scores, can be maximized with respect to the parameters of the function (the recombination fraction in the case of LOD scores). To maximize the likelihood is simply to find the value of the parameters that result in the largest value of the likelihood. The parameter values at that point are called the **maximum likelihood estimates** of the parameters, and they represent the best estimate of the parameters in the data.

| Recombination fraction ($\theta$) | LOD Scores |
|---|---|
| 0.01 | −0.808 |
| 0.05 | −0.164 |
| 0.10 | 0.067 |
| 0.20 | 0.214 |
| 0.25 | 0.227 |
| 0.50 | 0.0 |

**TABLE 10-12**
LOD Scores Calculated at Arbitrary Values of $\theta$ for the Family Depicted in Figure 10-11

The equation for the calculations is:

$$Z(\theta) = \log_{10} \frac{\theta^R (1 - \theta)^{NR}}{0.5^{NR+R}} = \log_{10} \frac{\theta^1 (1 - \theta)^3}{0.5^{3+1}}$$

For example, for $\theta = 0.10$, the calculation is:

$$Z(0.10) = \log_{10} \frac{0.10^1 (0.90)^3}{0.5^4} = 0.067$$

A LOD score greater than 0 is evidence in favor of linkage, whereas a LOD score less than 0 is evidence against linkage. By convention, a LOD score of 3.0 or greater (i.e., 1000:1 odds in favor of linkage) is *statistically significant* evidence of linkage, and a LOD score of −2.0 or less is statistically significant evidence against linkage. Note that the LOD score at $\theta = 0.5$ will always equal 0 (the $\log_{10} [1.0]$).

As can be seen in Table 10-12, the maximum LOD score occurred at $\theta = 0.25$, therefore 0.25 is the best estimate of $\theta$ from these data. However, there are no statistically significant LOD scores obtained from this family. Data from additional families are needed to be able to draw conclusions about whether locus I and locus II are linked.

# RESOLUTION OF CLINICAL CASE

Figure 10-12 shows Mr. and Mrs. Green, including Mrs. Green's family (recall that Mr. Green was adopted and has no knowledge of his birth family). This family pattern exhibits the characteristics of an autosomal recessive disorder.

CF is an autosomal recessive disorder present in approximately 1/2500 Caucasian newborns. Therefore, the frequency of the CF allele is the square root of 1/2500 = 1/50, and the frequency of the normal allele is 1 − (1/50) = (49/50), or approximately 1.0. From the Hardy-Weinberg equilibrium expectations, the carrier (heterozygote) frequency is 2 × (1/50) × 1.0 = 1/25. That is, approximately 1/25 Caucasians are carriers of cystic fibrosis.

**FIGURE 10-12** ▶
*Pedigree Diagram for Mr. and Mrs. Green. Mr. and Mrs. Green are individuals IV-2 and IV-3. Pedigree symbols are defined in Figure 10-4. Affected individuals have cystic fibrosis.*

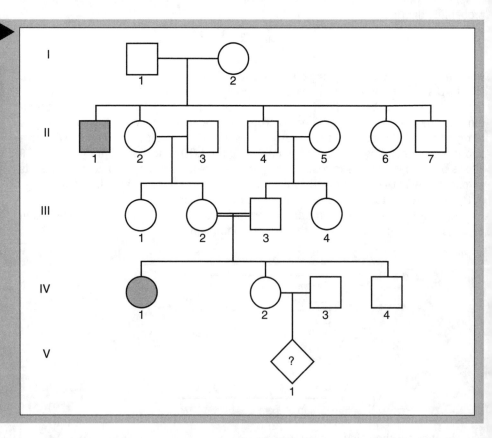

If no information was available about Mr. and Mrs. Green's family history, the chance that both of them were heterozygotes would be 1/25 × 1/25 = 1/625. Then, the chance of having an affected child would be 1/4 (the probability that two heterozygotes have an affected child) × 1/625 = 1/2500 (the population incidence of CF).

Mrs. Green is the sister of a woman with CF; therefore, her chance of being a carrier is 2/3. Mr. Green has no known family history, so his chance of being a carrier is approximately 1/25. Therefore, the chance that both Mr. and Mrs. Green are carriers is

$2/3 \times 1/25 = 2/75$. Usually, 1/4 of the offspring of two heterozygotes will be affected; the risk that the Greens will have an offspring with CF is $1/4 \times 2/75 = 1/150$, which is an increase over the general population risk (incidence) of 1/2500 births.

Using linkage analysis between CF and polymorphic DNA markers, CF was mapped to chromosome 7 in 1985 by investigators in London, Toronto, and Salt Lake City. The gene was cloned in 1989 by investigators in Michigan and Toronto and found to code for a protein termed cystic fibrosis transmembrane regulator (CFTR). Much effort has gone into characterizing the CFTR locus, its mutations, and its protein product. The locus is large (approximately 250 kb) and includes 27 exons. CFTR is involved in the transport of $Cl^-$ ions across the membranes of specialized epithelial cells. To date, DNA sequencing has discovered more than 300 possible mutations (alleles) at the CFTR locus. One allele, labelled $\Delta$F508 (for deletion of phenylalanine at position 508 of the protein product) is the most common, accounting for approximately 70% of mutations among Caucasians.

Mrs. Green's DNA was screened for the most common CFTR variants using PCR-based methods. She was found to be a carrier for the $\Delta$F508 mutation. Given the results for Mrs. Green, Mr. Green was also screened for $\Delta$F508 and found not to be a carrier. Therefore, Mrs. Green is heterozygous, and Mr. Green is homozygous unaffected, so there is essentially no risk that they will have a child affected with CF. The availability of the tests for the CFTR alleles reduces their risk from 1/150 (based on the family history and population incidence of CF) to approximately 0.

# REVIEW QUESTIONS

**Directions:** For each of the following questions, choose the **one best** answer.

1. The ABO blood type locus has three alleles: *A, B*, and *O*. In a particular population, the frequency of the *A* allele is 0.3, and the frequency of the *O* allele is 0.6. What is the frequency of the *B* allele?

    **(A)** 0.1

    **(B)** 0.2

    **(C)** 0.3

    **(D)** 0.4

    **(E)** 0.5

## Questions 2–4

Consider the following four populations, with the specified genotypic frequencies for the MN blood group system (an autosomal codominant blood group locus with two alleles, *M* and *N*):

| | *Genotypes* | | |
| --- | --- | --- | --- |
| *Population* | **MM** | **MN** | **NN** |
| 1 | 0.81 | 0.18 | 0.01 |
| 2 | 0.22 | 0.45 | 0.33 |
| 3 | 0.0 | 1.0 | 0.0 |
| 4 | 0.32 | 0.64 | 0.04 |

2. Which one of the populations is in Hardy-Weinberg equilibrium?

    **(A)** 1

    **(B)** 2

    **(C)** 3

    **(D)** 4

3. What is the frequency of the *M* allele in population 4?

    **(A)** 0.04

    **(B)** 0.32

    **(C)** 0.50

    **(D)** 0.64

    **(E)** 1.0

4. What would be the frequency of the *MM* homozygote in population 3 after one round of random mating?

    **(A)** 0.0

    **(B)** 0.25

    **(C)** 0.50

    **(D)** 0.75

    **(E)** 1.00

**5.** What is the probability that two siblings will share an allele identical by descent at the ABO locus?

(A) 0.0

(B) 0.25

(C) 0.50

(D) 0.75

(E) 1.00

## Questions 6 and 7

Consider the following fully penetrant genetic disorders, in a population in Hardy-Weinberg equillibrium:

| Disorder | Mode of Inheritance | Population Incidence |
|---|---|---|
| Phenylketonuria | Autosomal recessive | 1 in 10,000 |
| Hemophilia | Sex-linked recessive | 1 in 10,000 males |

**6.** What is the carrier frequency for phenylketonuria?

(A) 0.0

(B) 0.01

(C) 0.0001

(D) 0.0198

(E) 0.0019

**7.** What is the frequency of the hemophilia allele?

(A) 0.0

(B) 0.1

(C) 0.01

(D) 0.001

(E) 0.0001

## Questions 8–10

Refer to the following pedigree diagram, which depicts the family pattern of a fully penetrant, rare, genetic disorder.

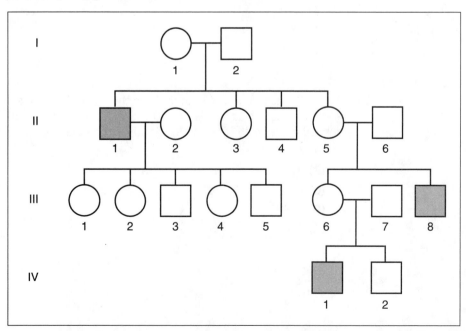

8. What is the probability that individual III-1 shares an allele identical by descent with individual III-7?

   **(A)** 0.0

   **(B)** 0.125

   **(C)** 0.25

   **(D)** 0.5

   **(E)** 0.625

9. What is the most likely mode of inheritance of this trait?

   **(A)** Autosomal recessive

   **(B)** Autosomal dominant

   **(C)** Sex-linked recessive

   **(D)** Sex-linked dominant

10. Individual III-6 is pregnant. What is the chance that her infant will be affected?

    **(A)** 0.0

    **(B)** 0.25

    **(C)** 0.50

    **(D)** 0.75

    **(E)** 1.00

11. Pyloric stenosis, which is severe projectile vomiting in early infancy, is much more common in boys. Refer to the following table of empiric recurrence risks.

| Sex of Proband | Approximate Risk to Relatives of Probands | | | |
| --- | --- | --- | --- | --- |
| | Brothers | Sisters | Sons | Daughters |
| Male | 2% | 2% | 6% | 3% |
| Female | 11% | 9% | 23% | 12% |

A woman has a son who suffered from pyloric stenosis as an infant. She now has a newborn daughter. What is the chance that the daughter will also exhibit the disorder?

    **(A)** 2%

    **(B)** 3%

    **(C)** 6%

    **(D)** 12%

    **(E)** 23%

12. Mr. Smith's father and Mrs. Smith's brother each have the same rare, autosomal recessive genetic disorder. Mr. and Mrs. Smith are both unaffected. What is the prior probability that both Mr. and Mrs. Smith are carriers of the affected allele?

    **(A)** 4/9

    **(B)** 1/3

    **(C)** 1/2

    **(D)** 2/3

    **(E)** 1/4

**13.** What is the recurrence risk of a rare, autosomal recessive disorder for the siblings of an affected individual?

    **(A)** 0.0

    **(B)** 0.10

    **(C)** 0.25

    **(D)** 0.33

    **(E)** 0.50

**14.** A recombination fraction of 50% between two loci implies that the loci are

    **(A)** in Hardy-Weinberg equilibrium

    **(B)** both autosomal dominant

    **(C)** not linked

    **(D)** sex-linked

**Questions 15 and 16**

Refer to the following family segregating a very rare autosomal dominant disease with alleles $A$ (affected) and $B$ (unaffected). Alleles $1$ and $2$ are at a codominant marker locus. Assume that alleles $A$ and $2$ are in the coupling phase in individual I-1.

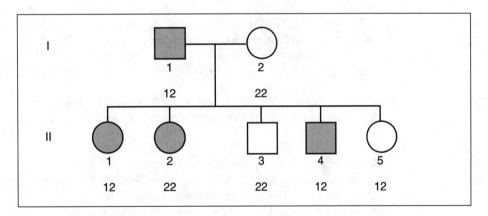

**15.** What is the estimated recombination fraction between the disease and marker loci in this family?

    **(A)** 0.8

    **(B)** 0.6

    **(C)** 0.4

    **(D)** 0.2

    **(E)** 0.0

**16.** What is the LOD score in this family calculated at a recombination fraction of 0.5?

    **(A)** 0.0

    **(B)** 0.2

    **(C)** 0.3

    **(D)** 0.4

    **(E)** 0.5

# ANSWERS AND EXPLANATIONS

**1. The answer is A.** The sum of the allele frequencies for a locus must be 1.0, so the $B$ allele frequency is obtained by subtraction:

$$1.0 - 0.3 - 0.6 = 0.1$$

**2. The answer is A.** If a population is in Hardy-Weinberg equilibrium, the frequency of $MM$ equals the frequency of $M$ squared. The square root of $0.81 = 0.9$. The frequency of $N = 1 - 0.9 = 0.1$, or the frequency of $NN$ should equal $(0.1)^2 = 0.01$, which it does. Therefore, population 1 is in Hardy-Weinberg equilibrium. None of the other populations match these predictions, so none of the other three populations are in Hardy-Weinberg equilibrium.

**3. The answer is D.** Because population 4 is not in Hardy-Weinberg equilibrium, allele counting must be used to calculate the allele frequencies. In a population of 100 individuals, there are 200 total alleles: 64 $M$ alleles from $MM$ homozygotes + 64 from heterozygotes, 8 $N$ alleles from $NN$ homozygotes + 64 from heterozygotes, or 128/200 $M$ alleles = 0.64, and 72/200 $N$ alleles = 0.36.

**4. The answer is B.** Currently, the population is not in Hardy-Weinberg equilibrium. Using allele counting, the frequency of the $M$ allele is 0.5, and the frequency of the $N$ allele is also 0.5. After one round of random mating, the population would be in Hardy-Weinberg equilibrium, so the frequency of $MM$ would be $(0.5)^2 = 0.25$.

**5. The answer is C.** The coefficient of relatedness (R) between siblings is 0.50, and R is defined as the probability that two relatives will share an allele identical by descent at any locus.

**6. The answer is D.** Because the population is in Hardy-Weinberg equilibrium, the incidence of phenylketonuria (PKU) equals the allele frequency squared. Therefore, the allele frequency of PKU is the square root of $0.0001 = 0.01$; the allele frequency of the unaffected allele is $1 - 0.01 = 0.99$. The carrier frequency is $2pq = 2(.01)(.99) = 0.0198$.

**7. The answer is E.** Because males are hemizygous, the incidence in males equals the allele frequency.

**8. The answer is B.** Individuals III-1 and III-7 are first cousins; therefore, their coefficient of relatedness (R) is 1/8.

**9. The answer is C.** Autosomal dominant inheritance can be ruled out because there are affected children from matings between unaffected individuals. A sex-linked dominant trait is ruled out for the same reason and is also unlikely because there are no affected females. Autosomal recessive inheritance is unlikely because there are no affected females. The most likely mode of inheritance is sex-linked recessive because only males are affected, there is no father–son transmission, and affected males are related through female relatives.

**10. The answer is B.** Because III-6 has one affected son, she is a known carrier of the sex-linked disorder. Therefore, none of her daughters will be affected, but half of her sons will be affected. The probability needed is P(infant is male *and* receives the affected allele from III-6) = P(male) × P(affected allele) = ½ × ½ = ¼.

**11. The answer is A.** The newborn daughter is the sister of a male proband, so the empiric recurrence risk is 2%.

**12. The answer is D.** To solve this probability problem, one must seek P(Mr. Smith carrier *and* Mrs. Smith carrier) = P(Mr. Smith carrier) × P(Mrs. Smith carrier). Because Mr. Smith's father is affected, and the disorder is recessive, the father must be homozygous affected. Therefore, Mr. Smith must be a carrier (probability = 1.0). Mrs. Smith is an unaffected sibling of an affected, so her probability is ⅔. The total probability is 1.0 × ⅔ = ⅔.

**13. The answer is C.** If the disorder is rare, then an affected person is most likely to be the product of a mating between two phenotypically unaffected individuals (i.e., a heterozygous × heterozygous genotypic mating type). The offspring of this mating type will be 25% homozygous unaffected, 50% heterozygous, and 25% homozygous affected; therefore, the recurrence risk for siblings will be 25%.

**14. The answer is C.** If two loci are not linked, the relative proportion of recombinant to nonrecombinant offspring is 50:50.

**15. The answer is B.** If the disorder is very rare, the most likely genotype at the disease locus for I-1 is *AB*. I-2 can only give haplotype *B2* to her offspring. I-1 transmits nonrecombinant haplotypes *A2* and *B1*, or recombinant haplotypes *A1* and *B2*. Therefore, individuals II-1, II-3, and II-4 are all recombinants. The recombination fraction is the number of recombinants divided by the total number of offspring, which is ⅗, or 0.6, in this family.

**16. The answer is A.** The LOD score at a recombination fraction of 0.5 is always 0.0, or $\log_{10}(1.0)$.

# REFERENCES

1. Ott J: *Analysis of Human Genetic Linkage* (revised edition). Baltimore, MD: Johns Hopkins University Press, 1991.
2. Twisselmann F: De l'evolution du taux de cansanguinité en Belgique entre les annés 1918–1959. *Proc Second Int Congress of Hum Genet* 1:142–150, 1961.
3. Freier-Maia N: Inbreeding levels in American and Canadian populations: a comparison with Latin America. *Eugenics Quarterly.* 15:22–27, 1968.
4. Sanghvi LD: Inbreeding in India. *Eugenics Quarterly.* 13(4):291–301, 1966.
5. Schull WJ, Neel JV: *The Effects of Inbreeding on Japanese Children.* New York, NY: Harper and Row, 1965.

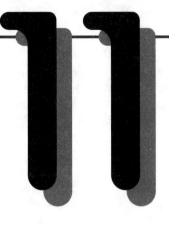

# MONOGENIC INHERITED DISORDERS

## INTRODUCTION OF CLINICAL CASE

Lisa, a 28-year-old woman, gravida 1 (18 weeks gestational age), visited her obstetrician for a checkup. The obstetrician noted the presence of café au lait spots on her left leg during a routine prenatal examination and recommended that she see her physician for further analysis. Two weeks later, Lisa was examined by her physician. He saw no neurofibromas on a limited examination of the skin, although a full body examination was not performed. No scoliosis was clinically obvious, although Lisa did seem to have a large head. She is heavily freckled. Lisa could not recall any family members with medical problems, but thought her 22-year-old brother had similar spots on his skin, and an 8-year-old nephew has freckles and learning difficulties. The physician sent Lisa to an ophthalmologist for a retinal examination and notified the ophthalmologist that the patient had a possible diagnosis of neurofibromatosis.

## PHENOTYPIC CONSEQUENCES OF MUTATION

Inherited genetic diseases are a result of mutations that occur in the DNA base sequence of specific genes. These mutations, which were discussed in detail in Chapter 4 and listed in Table 4-1, can alter the products coded for by genes in a number of different ways. The way in which a gene product is altered ultimately determines the clinical phenotype of a disease, as well as the type of inheritance pattern that the disease follows. In this chapter, the major types of inheritance patterns associated with inherited monogenic disorders are discussed, and specific examples for each type are examined.

## Autosomal Recessive Inheritance

Recessive inheritance is generally attributed to loss-of-function mutations, in which both copies of the patient's gene are dysfunctional. For most genes, one functional copy is sufficient to preclude clinical symptoms, although individuals with one inactive allele of a gene are carriers of a recessive disease. When both parents are carriers (i.e., each parent has one active and one inactive allele), their risk of having an affected child with both alleles inactive is 25%. Any mutation that disables the gene, the RNA, or the protein such that little or no functional product is produced can lead to a loss-of-function mutation. Chapters 4 and 8 presented the many ways in which a gene can be disabled. These include deletions; insertions; disruption of the gene by chromosomal breaks; defects in transcription, RNA splicing, or translation; missense or nonsense codons in the middle of the coding regions of a gene; and mutations that effect the stability of the RNA or the protein product.

Because there are multiple ways of inactivating a gene or disabling a gene product, loss-of-function mutations are quite common and account for many of the inherited human diseases we see. Examples of some recessively inherited human diseases are listed in Table 11-1. Many of these diseases result from mutations occurring in genes that code for proteins involved in cellular metabolism. These include the enzymes involved in the metabolism of carbohydrates, nucleotides, amino acids, lipids, and organic acids. In addition, proteins involved in maintaining the structure of the cell, homeostasis, transport, cell–cell communication, and growth control often are inactivated by loss-of-function mutations.

*Loss-of-function mutations are generally inherited as autosomal recessive disorders.*

**TABLE 11-1** ▶

Examples of Autosomal Recessive Inherited Disorders

| Disorder | Protein Altered | Frequency |
|---|---|---|
| Sickle cell anemia | β-globin | 1/650 Blacks (in United States) |
| Cystic fibrosis | Transmembrane conductance regulator | 1/2500 Whites |
| Tay-Sachs disease | Hexosaminidase A | 1/3000 Ashkenazi Jews |
| Gaucher's disease | β-glucosidase | 1/6000 (1/600 Ashkenazi Jews) |
| Phenylketonuria | Phenylalanine hydroxylase | 1/12,000 |
| Severe combined immunodeficiency disease | Adenosine deaminase | Extremely rare |

## Autosomal Dominant Inheritance

Dominant inheritance usually is caused by a change-of-function mutation in which clinical symptoms are associated with a defect in only one allele of a gene. In these disorders, even if only one parent is affected with the disorder, the risk of having an affected child is 50%. Change-of-function mutations are caused by a mutation in the DNA that results in the production of an abnormal gene product. These mutations are much more restricted when compared to loss-of-function mutations because they require that a protein or RNA be produced and remain stable, although it is functionally abnormal. Many change-of-function mutations are caused by a missense mutation occurring in the coding region of a gene, resulting in the production of a mutant protein that has an altered function. In some instances, the presence of this mutant protein can result in a dominant-negative phenotype, in which the abnormal protein inactivates or inhibits the gene product of the normal allele. Dominant-negative effects usually involve proteins

*Gain or change-of-function mutations often are inherited as dominant disorders.*

that form oligomers. In this instance, the mutant subunit forms a complex with the normal subunits, and the resulting oligomer is inactive or altered in its function.

Another type of change-of-function mutation that shows dominant inheritance is represented by the expansion disorders, such as Huntington's disease and myotonic dystrophy. As is found with missense mutations, expansions of the trinucleotide repeats within a gene can result in a protein or RNA being produced, but the protein or RNA functions in an abnormal way.

## X-Linked Inheritance

Genes located on the X chromosome show different inheritance patterns for the two sexes. A woman has two copies of the X chromosome, one copy from her mother and one copy from her father. Therefore, a woman can be either heterozygous or homozygous for a defective gene located on the X chromosome. A man, however, has only one copy of the X chromosome, which always comes from the mother. Therefore, a man is always homozygous for all genes located on the X chromosome. For X-linked recessive inherited disorders, the disease is expressed in males who receive the disease gene from the mother, but it is expressed in females only if they receive a disease gene from both the mother and the father and are homozygous for the disease gene. For this reason, most X-linked recessive diseases are expressed in males and rarely in females.

> *X-linked disorders* are predominately expressed in males and rarely in females.

When an X-linked disease is expressed in the female heterozygous state, it is referred to as an X-linked dominant inherited trait. With an X-linked dominant disorder, affected males cannot transmit the disease to their sons, but they do transmit the disease to all of their daughters. When a female is affected with an X-linked dominant disorder, there is a 50% chance that any male or female child will inherit the disorder.

## Haploinsufficiency

Examples of loss-of-function mutations that can result in a dominant inheritance pattern are the haploinsufficiency diseases. These diseases occur when the amount of product produced from a single normal allele is not sufficient to prevent a clinical phenotype (i.e., half is not enough). In haploinsufficiency diseases, individuals who are heterozygous for the defective allele do express the clinical phenotype. Generally, the disease is lethal in the homozygous state.

## Dominant Inheritance of a Predisposition to Develop Cancer

A special type of a dominantly inherited disease involves inactivation of a single copy of a tumor suppressor gene, resulting in the inheritance by an individual of a predisposition to develop cancer. Tumor suppressor genes code for proteins that are crucial for the prevention of abnormal cell growth that might lead to tumor formation. A single abnormal tumor suppressor gene is passed from a carrier parent, who has one allele inactivated, to 50% of the children. Inheritance of a single defective allele does not produce cancer; however, when the second allele is inactivated in a somatic cell, the cell has a propensity to develop a tumor. Tumor suppressor genes and their function are discussed in more detail in Chapter 8.

> *Mutations in tumor suppressor genes* result in the inheritance of a predisposition to develop cancer.

# MONOGENIC DISORDERS WITH DIFFERENT INHERITANCE PATTERNS

## Autosomal Recessive Inheritance Attributed to Loss-of-Function Mutations

### SICKLE CELL ANEMIA

Genetic disorders involving mutant hemoglobin (Hb) are numerous, as discussed in Chapter 2. One of the most commonly inherited disorders of the β-globin gene is sickle cell anemia, which results from a single amino acid change of a glutamic acid to a valine at position 6 of the protein. Individuals homozygous for this mutation are affected with sickle cell disease, which is often fatal. The disease is most common in the black population of the United States, with a frequency of 1 in 650 live births.

In Chapter 3, the clinical case describes the phenotype of a child with sickle cell anemia. The clinical phenotype is not obvious at birth but begins to become apparent within the first 2 years of life. The Hb molecule in an individual with the disease is made up of two α chains and two mutant β-subunits. Although the mutant Hb molecules can still bind oxygen, when they are present in deoxygenated blood, they become less soluble than normal Hb and tend to form aggregates of rod-shaped polymers. The presence of these aggregates in the red blood cells (RBCs) causes a distortion and a sickling of the cells. Once distorted, the RBC can no longer enter the capillaries, resulting in a blockage and a lack of blood flow. The RBCs, which are unable to move into the capillaries, are destroyed, resulting in the hemolytic anemia associated with this disease.

Individuals who are heterozygous for the hemoglobin S (HbS) mutation are not often affected with the disease. They do, however, show a greater resistance to malarial infections, and, if they develop a malarial infection, it tends to be a much milder form. This phenotype gives the heterozygote a selective advantage in a geographic area where malaria is a common problem and is no doubt related to the high frequency of sickle cell anemia in parts of Africa and Asia.

As discussed in Chapter 3, the sickle cell mutation can be identified in the heterozygous and homozygous state by looking for a change in the electrophoretic pattern of the Hb protein or by looking for a specific alteration in a restriction endonuclease cleavage site in the β-globin gene. The mutation associated with sickle cell anemia is specifically caused by the replacement of an adenine base by a thymine base in a DNA sequence recognized by the restriction enzyme *Mst* II. DNA from the normal allele is cut by *Mst* II to give a 1.15-kb fragment, whereas the mutant DNA, which is missing this *Mst* II site, gives a 1.35-kb DNA fragment. The presence of a 1.35-kb fragment in a Southern blot analysis is diagnostic for the sickle cell mutation.

## CYSTIC FIBROSIS

*Sickle cell anemia, cystic fibrosis, Tay-Sachs disease and Gaucher's disease are inherited as autosomal recessive disorders.*

Cystic fibrosis, the most common potentially fatal autosomal recessive disorder found in whites, occurs with a frequency of 1 in 2500 live births. The disease generally affects children and young adults who suffer from mucus accumulation obstructing their airways. Infections with *Pseudomonas* and *Staphylococcus* can often lead to respiratory failure. The mucus accumulation also can obstruct the pancreatic ducts, as well as cause problems with the intestines and the liver. Males with the disease tend to be infertile. Patients with cystic fibrosis have excessive salt loss in their sweat and can be diagnosed for the disease by the sweat test. Patients undergoing treatment can live to adult life with a median age of 29 years, although improved treatment of the disease results in patients now surviving until the age of 40 years.

Numerous biochemical studies on cystic fibrosis were consistent with the disease being associated with a defect in the normal efflux of chloride ions ($Cl^-$) across epithelial cell membranes, but the exact molecular basis for the disease remained unknown until 1989. Attempts to map the cystic fibrosis gene were hampered by the lack of known chromosomal deletions or translocations associated with the disease. However, a large number of families were available for linkage analysis, providing data that allowed the gene to be mapped to chromosome 7q. In addition, recombination analysis and a study of linkage disequilibrium were instrumental in the final identification and isolation of the gene. Linkage disequilibrium results from a mutation occurring many years ago in one gene of one person. That person's gene has a specific set of polymorphisms associated with it. The polymorphisms within 30,000 bp of the mutation undergo no recombination with the disease mutation and remain associated with that mutation forever. Individuals with cystic fibrosis have certain alleles at nearby polymorphic loci at a much higher frequency than found in the general population. This information was extremely useful in mapping the cystic fibrosis gene close to specific DNA markers on chromosome 7 and its localization at 7q31.2.

In 1989, a 250-kb sequence of DNA was isolated from this region of chromosome 7 and found to be closely linked to the cystic fibrosis gene [1, 2]. The DNA fragments were tested for the presence of exon sequences, as discussed in Chapter 7, and finally a candidate gene was identified and found to be the cystic fibrosis gene. The cystic fibrosis gene is made up of 27 exons spanning 230 kb of DNA and codes for a 6.5-kb mRNA that contains information for a protein made up of 1480 amino acids. This protein is known as the cystic fibrosis transmembrane conductance regulator (CFTR) [Figure 11-1].

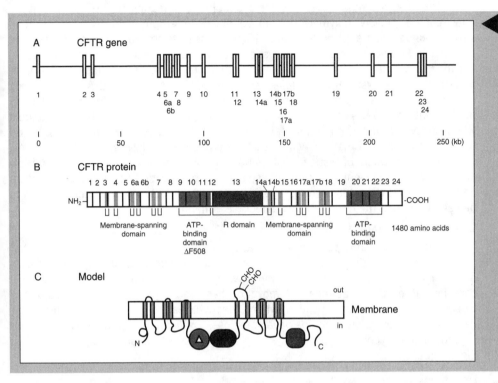

◀ **FIGURE 11-1**
**Cystic Fibrosis (CF) Gene and Its Protein Product.** *The CF gene (A) is made up of 27 exons (open boxes) that span 230 kb of DNA. The gene codes for the CF transmembrane conductance regulator (CFTR) protein (B), which is made up of 1480 amino acids. Corresponding exons are shown as numbers above the protein structure. The CFTR has several domains that are important for its function: the membrane-spanning domains, the adenosine triphosphate (ATP)–binding sites, and the regulatory (R) domain. The structure is similar to other transport proteins that have been sequenced. The ΔF508 mutation is in exon 10 within one of the ATP-binding domains. The model (C) proposes a possible structure for the CFTR protein and its interaction with the cellular membrane. (Source: Reprinted with permission from Tsui LC, Buchwald M: Biochemical and molecular genetics of cystic fibrosis. In* Advances in Human Genetics, *vol 20. Edited by Harris H, Hirschhorn K. New York, NY: Plenum Press, 1991, p 208.)*

The Δ**F508 mutation**, which is associated with 70% of the mutant alleles in the cystic fibrosis gene, results in a defective CFTR protein.

Comparison of complementary DNA (cDNA) base sequences taken from normal individuals and cystic fibrosis patients revealed a specific 3-bp deletion in exon 10 that leads to the loss of a phenylalanine residue at position 508 in the protein (ΔF508). This particular mutation is associated with 70% of the mutant alleles in cystic fibrosis. The high frequency of the ΔF508 mutation is an example of a founder effect, in which a common mutation is derived from a common ancestor and maintained within the population for many decades, perhaps forever. Once the gene was cloned, further studies identified more than 350 additional mutations within the gene that are associated with the remaining 30% of the mutant alleles of cystic fibrosis.

The CFTR is a glycosylated protein with a molecular mass of 170,000 daltons (D). Amino acid sequence analysis indicates that CFTR has homology to a family of transport proteins known as traffic adenosine triphosphatases (ATPases) or the adenosine triphosphate (ATP)–binding cassette family. Similar to these proteins, the CFTR has two hydrophobic transmembrane domains and two nucleotide binding folds. In addition, the protein has a regulatory (R) domain containing serine residues that are the target of protein kinase A phosphorylation. The wild-type protein functions as a cyclic adenosine monophosphate (cAMP)–regulated Cl⁻ channel, with the transmembrane domains spanning the epithelial cell membrane as part of the channel pore (see Figure 11-1).

Mutations in the cystic fibrosis gene can result in the following:

- A complete loss of the protein
- A defect in protein processing that occurs with the ΔF508 mutation
- A defective regulation of the channel
- A defective conduction through the channel

All of these types of mutations can lead to the observed phenotype and result in a defect in the transport of Cl⁻. In addition to serving as the Cl⁻ channel, CFTR also appears to regulate the activity of several other ion channels. These include an outward-rectifying $Cl^-$ channel distinct from CFTR; a sodium ion ($Na^+$) channel; and a calcium ($Ca^{2+}$)-activated $Cl^-$ channel. Understanding how this regulation occurs is important in eventually understanding the pathophysiology of cystic fibrosis.

Diagnosis of cystic fibrosis is based primarily on clinical presentation and the sweat test. Molecular diagnostic methods are available for detecting the ΔF508 mutation. These generally involve polymerase chain reaction (PCR) amplification followed by allele-specific oligonucleotide hybridization (ASOH). These tests are most important in genetic counseling. Population screening for cystic fibrosis is being considered, although screening for the ΔF508 mutation alone detects only 70% of carriers. Some 10–30 other mutations in the cystic fibrosis gene can be tested for by PCR and ASOH, allowing approximately 90% of the mutations associated with cystic fibrosis to be detected. In families with a 25% risk of having an affected child, fetuses can be tested by chorionic villus biopsy followed by mutation analysis of the DNA samples. Newborns can be tested by obtaining a cheek brushing, which is used for molecular analysis.

Current treatment of cystic fibrosis patients involves percussion of the chest, antibiotic treatment to prevent infection, and sometimes pancreatic enzyme replacement. In addition, various approaches are being explored that would allow somatic gene therapy to be used to correct the cystic fibrosis defect (see Chapter 14).

## LYSOSOMAL STORAGE DISEASES

Intracellular cytoplasmic particles known as lysosomes contain numerous enzymes that are required for the digestion of many types of macromolecules. Among these macromolecules are glycoproteins, glycosaminoglycans, and glycolipids, all of which have complex carbohydrates attached. Removal of these carbohydrate residues is catalyzed by specific sugar hydrolases, which, when defective, lead to the accumulation of their respective substrates. Glycolipids have galactose or glucose linked to ceramide; when degraded, the glycolipids release the sugar moieties and ceramide as illustrated in Figure 11-2.

### FIGURE 11-2 ▶

*Glycolipid Degradation in the Lysosomes. The glycolipids are degraded in the lysosomes by enzymes that sequentially remove carbohydrate residues to produce free sugars and ceramide. A defect in any one of these enzymes leads to accumulation of the respective substrate, resulting in a clinical phenotype. 1 = N-acetylhexosaminidase A (defective in Tay-Sachs disease); 2 = galactosidase A (defective in Fabry's disease); 3 = β-glucosidase (defective in Gaucher's disease); 4 = sphingomyelinase (defective in Niemann-Pick disease).*

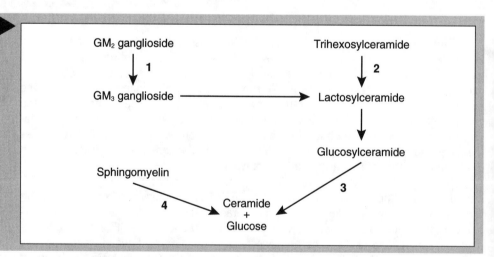

Mutations that result in a defect in the enzymes involved in the individual degradation steps lead to the accumulation of intermediates that are inhibitory to cells. As shown in Figure 11-2, a defect in enzyme 1, hexosamidase A, results in Tay-Sachs disease; a defect in enzyme 2, galactosidase A, results in Fabry's disease; a defect in enzyme 3, β-glucosidase, results in Gaucher's disease; and a defect in enzyme 4, sphingomyelinase, results in Niemann-Pick disease. Tay-Sachs, Gaucher's, and Niemann-Pick diseases are inherited in an autosomal recessive manner, whereas Fabry's disease is an X-linked disorder.

*Tay Sachs disease and Gaucher's disease, disorders of glycolipid metabolism, are most common among individuals of Ashkenazi Jewish descent.*

Disorders of glycolipid metabolism are most common among individuals of Ashkenazi Jewish descent. The Ashkenazi (i.e., "German" in Hebrew) Jewish population originated in Israel. In the 9th century, they migrated to the Rhineland, and in the 14th century moved to Eastern Europe and Russia. Today, most of the Ashkenazi Jewish population lives in the United States, the former Soviet Union, and Israel. In this population, 3%–4% are carriers of Tay-Sachs disease, 1% are carriers of Niemann-Pick disease, and 4%–6% are carriers of Gaucher's disease. Among non-Jewish individuals, the incidence of the carrier state is estimated to be at least ten times lower.

***Tay-Sachs Disease.*** Gangliosides (glycosphingolipids made up of ceramide and a hydrophilic polysaccharide chain with $N$-acetylneuraminic acid attached) such as $GM_2$ and $GM_3$ are broken down in a stepwise manner with the sugars being removed sequentially to produce the ceramide core. Ganglioside $GM_2$ is degraded by hexosaminidase A to produce $GM_3$. In individuals with Tay-Sachs disease, the hexosaminidase A is defective, leading to the accumulation of $GM_2$ [3].

Hexosaminidase A is made of two subunits, $\alpha$ and $\beta$. The gene *HexA* maps to chromosome 15q23-24 and codes for the $\alpha$-subunit, which is mutated in Tay-Sachs disease. The *HexA* gene, isolated by functional cloning in the mid-1980s, contains 14 exons spanning 35 kb of DNA. Mutations within the *HexA* gene that are associated with the disease include deletions, insertions, and base substitutions. These mutations generally lead either to a complete loss of function or to a drastic reduction in the enzyme activity. In cases in which partial enzyme activity is retained, the disease has a later onset and a slower progression.

Mutations that result in a complete lack of enzyme activity lead to the infantile form of Tay-Sachs disease, which is a particularly devastating form of the disease. Children born with infantile Tay-Sachs are generally asymptomatic before 3 months of age, at which time they begin to develop mild motor weakness followed by a development of progressive neurologic degeneration. Death occurs at 2–4 years of age. A characteristic indicative of infantile Tay-Sachs disease is the appearance of a cherry-red spot in the eye of the infant. Additionally, there is an accumulation and precipitation of $GM_2$ ganglioside in the brain.

Two specific mutations account for the severe infantile form of the disease that occurs in individuals of Ashkenazi Jewish descent (Table 11-2). One mutation, found in 79% of carriers, is the result of a 4-bp insertion in exon 11, which leads to a frameshift mutation and a downstream stop codon. The second mutation, which is found in 15% of carriers, involves the donor splice junction in intron 12 and results in the production of aberrant mRNA molecules that are improperly spliced. A third mutation, found in 3% of carriers, is a glycine to serine missense mutation in exon 7. This mutation gives a late-onset adult form of the disease in which the mutant protein is produced but is unstable, resulting in the amount of enzyme activity being reduced sufficiently to result in a clinical phenotype.

In other populations, the 4-bp insertion in exon 11 is seen in 20% of non-Jewish carriers with the exon 7 missense mutation present in 5% of non-Jewish carriers. Approximately 15% of non-Jewish carriers have a splice site mutation in intron 9, which is different than the splice-site mutation common in Ashkenazi Jewish populations. The high percentage of these common mutations leading to Tay-Sachs disease occurring in both the Jewish and non-Jewish carrier populations indicate the possibility of a founder effect or some selective advantage for the heterozygous state.

In the diagnosis of Tay-Sachs disease, assays for hexosaminidase A are carried out on synthetic substrates or on $GM_2$ ganglioside, using tissue samples, serum samples, or leukocytes obtained from patients. With the identification of common mutations causing Tay-Sachs disease, molecular diagnosis can be carried out using PCR amplification and ASOH.

Treatment for Tay-Sachs disease is limited to management of the patient's problems involving infections and nutrition. There is no cure for the disease, but efforts are underway to try enzyme replacement therapy and somatic cell gene therapy. Inexpensive and accurate methods are available for heterozygote screening. In the Jewish population with a high incidence of the disease, genetic counseling and prenatal diagnosis are available.

***Gaucher's Disease.*** Gaucher's disease, the most common lysosomal storage disease, is the result of a defect in $\beta$-glucosidase, the enzyme that is responsible for the degradation of glucocerebrosides in the lysosomes. The accumulation of glucocerebrosides leads to enlargement of the liver and spleen, as well as the production of lesions in the bone.

***TABLE 11-2*** ▶

*Common Mutations Associated with Some Autosomal Recessive Inherited Disorders*

| Disorder | Protein Altered | Mutation |
|---|---|---|
| Sickle cell anemia | β-globin | Glu-to-Val missense mutation |
| Cystic fibrosis | CFTR | ΔF508 3-bp deletion mutation |
| Tay Sachs disease | Hexosaminidase A | 4-bp insert in exon 1 (a frameshift mutation)<br>Splicing mutation in intron 12<br>Gly-to-Ser missense mutation in exon 7 |
| Gaucher's disease | β-glucosidase | L444P in exon 10 (a missense mutation)<br>IVS2+1 (a splicing mutation)<br>84GG (a frameshift mutation)<br>N370S in exon 9 (a missense mutation) |
| Phenylketonuria | Phenylalanine hydroxylase | IVS12nt1 (a deletion of exon 12 and stop mutation)<br>R158Q (a missense mutation)<br>R261Q (a missense mutation)<br>R408W (a missense mutation) |

*Note.* CFTR = transmembrane conductance regulator; Glu = glutamine; Gly = glycine; Ser = serine; Val = valine.

Gaucher's disease, which can vary from a mild to a severe form, is classified into types I, II, and III [4, 5].

Type I Gaucher's disease, the most common form, is non-neuronopathic and shows no primary involvement of the central nervous system. This form of the disease is characterized by enlargement of the spleen and liver and skeletal disease. There is a broad spectrum with regard to the severity of type I Gaucher's disease, from mildly affected adults to severely affected children who may die within the first 2 decades of life. The children suffer from abnormalities of the liver and spleen and have skeletal defects. Type I Gaucher's disease is found at a very high frequency in the Ashkenazi Jewish population, with 1 in 400 to 1 in 600 individuals being affected.

Type II Gaucher's disease, the most severe form of the disease, is characterized by extensive neurologic defects and leads to death of affected infants in the first 2 years of life.

Type III Gaucher's disease shows an intermediate severity with involvement of the visceral organs as well as some neurologic symptoms. However, the symptoms appear later than type II and are much less severe, with patients often surviving to young adulthood.

β-glucosidase, the enzyme defective in Gaucher's disease, is required for the breakdown of glucosylceramide to glucose and ceramide (see Figure 11-1). The monomeric protein is a membrane-associated glycoprotein that is 497 amino acids in length with a molecular weight of 55,575–65,000 D, depending on its state of glycosylation.

The β-glucosidase gene, mapping to chromosome 1q21, was cloned by functional cloning with the use of an expression DNA library and an antibody formed to the purified protein. The gene contains 11 exons spanning 32 kb of DNA and codes for 2.5-kb mRNA. Located 16 kb downstream of the functional gene lies a pseudogene. The pseudogene (5769 bp) is smaller than the normal gene (7604 bp), contains several mutations, including those involved in the normal splicing of the messenger RNA (mRNA), and does not code for a functional protein (Figure 11-3) [4].

More than 30 different mutations in the β-glucosidase gene that cause Gaucher's disease have been identified. These loss-of-function mutations include missense, nonsense, insertions, deletions, and splice-site mutations. Only four of these, however, occur with any frequency among individuals with Gaucher's disease (Table 11-2). A missense mutation in exon 10 (L444P), in which a leucine is changed to a proline residue at amino acid 444, is found most often in type III Gaucher's disease, which occurs in the Swedish population. Among the Ashkenazi Jewish population, 70% of individuals with type I Gaucher's disease have a missense mutation in exon 9 (N370S), in which an asparagine is changed to a serine residue at amino acid 370. Approximately 12% have a single base insertion (G) [84GG] to give a frameshift mutation at position 1035 that leads to a severe

β-glucosidase, which degrades glucocerebrosides in the lysosomes, is defective in individuals with **Gaucher's disease**.

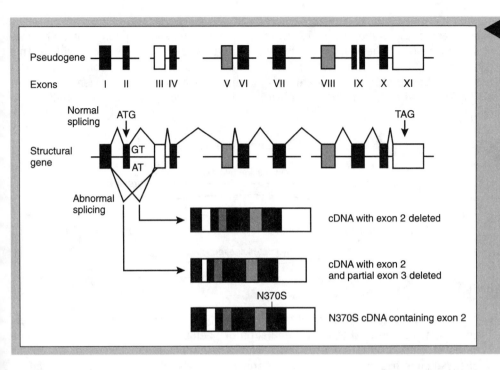

**FIGURE 11-3**
*β-Glucosidase Normal Gene and Pseudogene.* The β-glucosidase pseudogene is diagramed at the top, and the structural gene is shown at the bottom of the figure. The 11 exons of each gene are represented as boxes. The guanine-to-adenine mutation (IVS2+1) marked at the exon 2–intron 2 junction of the structural gene results in abnormal splicing and the production of the two cDNA products, indicated with a deletion of exon 2 or a deletion of exon 2 and some of exon 3. Additionally, the N370S missense mutation is noted. (Source: *Reprinted with permission from Grabowski GA: Gaucher's disease. In* Advances in Human Genetics, vol 21. *Edited by Harris H, Hirschhorn K. New York, NY: Plenum Press, 1993, p 404.)*

type II form of Gaucher's disease. Approximately 4.5% have the L444P mutation, and 1%–2% have a base substitution at position 1067 (IVS2+1), which results in aberrant splicing and type II Gaucher's disease. The mutation L444P is also associated with types I, II, and III Gaucher's disease; however, individuals with this mutation who have the severe form of Gaucher's disease may also have secondary mutations within the gene that may account for the severity of the disease.

Molecular analyses indicate that the carrier frequencies for Gaucher's disease alleles in the Ashkenazi Jewish population could be as high as 1 in 10–12 people, leading to a disease frequency of 1 in 400–600. The severity of the disease may be related to the basic mutation, as well as to whether any residual enzyme activity is present in the lysosomes.

Gaucher's disease can be diagnosed by an enzymatic assay for β-glucosidase that is carried out using peripheral blood cells, water-soluble substrates, and an antibody to the protein. The assay is not able to differentiate types II and III Gaucher's disease from type I. However, DNA technology is available to detect the four common mutations that account for nearly 95% of the alleles in Ashkenazi Jewish individuals with Gaucher's disease. In non-Jewish populations, mutations L444P and N370S account for 75% of the disease alleles. For an at-risk family, there is an excellent chance of detecting the mutant alleles using PCR and ASOH. Once an allele is detected, it is often possible to predict the prognosis or the disease. Homozygotes for the L444P mutation will probably have a severe visceral and neurologic disease. Individuals with the N370S mutant enzyme present generally are not at risk for neurologic involvement.

Bone marrow transplantation is available for treating the disease, but generally is restricted to the most severe cases. Enzyme replacement therapy has been used to treat patients with type I Gaucher's disease with some success, but the procedure is very expensive. Current research is underway to develop somatic cell gene therapy protocols using retroviral vectors. Beta-glucosidase can be measured in amniotic fluid cells and chorionic villus samples, making prenatal diagnosis a valuable tool, particularly for those families at high risk.

## PHENYLKETONURIA (PKU)

Classic PKU represents one of the first known inherited metabolic diseases. It was described in 1934 by Folling as being associated with an inherited form of severe mental retardation. Since that time, PKU has been extensively studied both at the biochemical and molecular levels [6, 7]. The basic biochemical defect in classic PKU is the loss of function of the enzyme phenylalanine hydroxylase, which catalyzes the conversion of phenylalanine to tyrosine (Figure 11-4). In addition to the need for phenylalanine hydroxylase, conversion of phenylalanine to tyrosine requires a cofactor, a reduced pteridine. This cofactor, 5,6,7,8-tetrahydrobiopterin ($BH_4$), is oxidized to quinoid dihydrobiopterin ($qBH_2$) and is maintained at the appropriate level in the cell by its subsequent regeneration from $qBH_2$. The regeneration of $BH_4$ from $qBH_2$ is catalyzed by the enzyme dihydropteridine reductase (DHPR).

**FIGURE 11-4**

*Conversion of Phenylalanine to Tyrosine. Phenylalanine is converted to tyrosine by the enzyme phenylalanine hydroxylase. The reaction requires the cofactor tetrahydrobiopterin ($BH_4$), which is regenerated from quinoid dihydrobiopterin ($qBH_2$) by the enzyme dihydropteridine reductase. Classic phenylketonuria is caused by a mutation that leads to a loss of phenylalanine hydroxylase activity.*

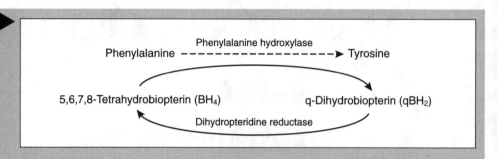

Because the conversion of phenylalanine to tyrosine requires two enzymes, different mutations can lead to an accumulation of phenylalanine in cells. However, most cases of classic PKU are associated with a mutation in the gene that codes for phenylalanine hydroxylase, with only approximately 1%–2% being caused by a defect in a deficiency of $BH_4$.

Phenotypic expression of PKU varies in its intensity, leading to the classification of the disease into three groups. Group 1, the classic form of PKU, is characterized by serum phenylalanine levels exceeding 1200 µM, a complete loss of phenylalanine hydroxylase activity, and severe mental retardation if not treated very early in life. Group 2, a milder form of PKU in which the serum phenylalanine levels range from 800–1200 µM, is associated with decreased expression of phenylalanine hydroxylase activity. Although individuals with this form of PKU require diet control, they can tolerate some phenylalanine in their diets. Group 3, a form of PKU in which the serum phenylalanine levels range from 250–800 µM, occurs in individuals who have no clinical symptoms and usually do not require diet control.

The phenylalanine hydroxylase gene was cloned by functional cloning using antibodies to the purified enzyme as a method to isolate the cDNA sequence. The isolation and analysis of the complete cDNA sequence revealed an open reading frame that coded for a protein with 452 amino acids and a molecular weight of 51,800 D. The gene for phenylalanine hydroxylase, which maps to 12q22–q24 and spans 90 kb of DNA, contains 13 exons and codes for a mature mRNA of 2.4 kb as diagramed in Figure 11-5.

**FIGURE 11-5**

*Phenylalanine Hydroxylase Gene. The phenylalanine hydroxylase (PAH) gene is made up of 13 exons (noted as thin, dark lines) and spans 90 kb of DNA. Four of the most common mutations that lead to phenylketonuria are indicated: R158Q, a missense mutation in exon 5; R261Q, a missense mutation in exon 7; R408W, a missense mutation in exon 12; and IVS12nt1, a splice junction mutation that leads to a deletion of exon 12. (Source: Adapted with permission from Scriver C, Kaufmann S, Eisensmith R, et al: The hyperphenylalaninemias. In The Metabolic and Molecular Bases of Inherited Diseases. Edited by Scriver CR, Beaudet AL, Sly WS, et al. New York, NY: McGraw-Hill, 1995, p 1028.)*

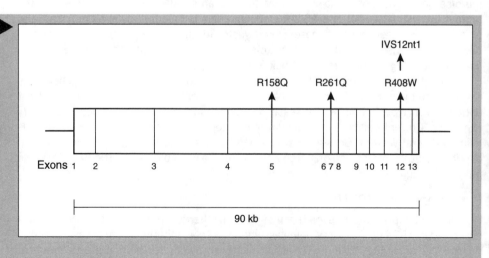

Nearly 200 different mutations have been identified in the gene. These include the types of mutations that are predicted to lead to loss of function, such as missense, nonsense, frameshift, and splicing mutations as well as deletions and insertions. The classic or severe form of PKU is found in individuals with very low or undetectable levels of phenylalanine hydroxylase. Milder forms of the disease result from missense mutations in which a protein is produced but has reduced activity.

Several specific mutant alleles are found in European individuals with PKU (see Figure 11-5 and Table 11-2). A splicing mutation in intron 12, IVS12nt1, which is associated with Scandinavians, results in a loss of exon 12 and the generation of a stop codon. Premature termination of protein synthesis leads to a lack of phenylalanine hydroxylase protein and a severe form of PKU. Mutation R408W, which accounts for 50% of the mutations that occur in Eastern Europeans, results in an arginine residue being converted to a tryptophan residue at amino acid position 408. This leads to the production of a defective protein and a severe form of the disease. Mutation R158Q has a wide distribution but is found most frequently in Dutch and Belgian populations. This severe form of the disease is a result of an arginine residue being converted to a glutamine residue at amino acid 158, which produces a defective phenylalanine hydroxylase protein. Mutation R261Q, associated mainly with Swiss and Turkish populations, is caused by an arginine residue being converted to a glutamine at amino acid 261. This missense mutation results in a phenylalanine hydroxylase protein with approximately 30% of the normal activity and leads to a milder form of PKU.

Treatment of PKU requires early intervention within a few months of birth because seizures begin at 6–12 months of age. Newborns are routinely screened for PKU by a bacterial assay known as the Guthrie test, which is discussed in Chapter 13. This test is carried out on a blood spot and measures the amount of phenylalanine in the blood. Infants with high levels of phenylalanine are placed on a low-phenylalanine diet by 4 weeks of age and are maintained on this diet, sometimes for life. It is particularly important for women who have PKU to maintain a restrictive diet during pregnancy to prevent damage to the fetus.

*Newborns are routinely screened for PKU by the Guthrie test because early intervention (within a few months) is essential to prevent the clinical symptoms.*

With the cloning of the phenylalanine hydroxylase gene, there is potential for using somatic cell gene therapy for this disease. Ideally, vectors carrying the cDNA coding for the enzyme should be delivered to hepatocytes. Retroviruses and adenoviruses currently are being used as potential vectors. The advantages and disadvantages of using these vectors are discussed in Chapter 14.

## Severe Combined Immunodeficiency Disease (SCID)

A rare, autosomal recessively inherited form of SCID is caused by a deficiency of the enzyme adenosine deaminase (ADA). ADA catalyzes the conversion of adenosine and deoxyadenosine to inosine and deoxyinosine, respectively. Lack of ADA accounts for approximately 20% of SCID cases and is characterized by a loss of both B and T lymphocytes. The disease becomes apparent during the first 3 months of life, and the patient usually dies within 2 years of massive infections. A characteristic of this form of SCID is the accumulation of high levels of deoxyadenosine triphosphate (dATP) in the patient's erythrocytes, which results from the absence of ADA and the conversion of deoxyadenosine to dATP. High levels of dATP also accumulate in lymphocytes and block DNA replication of the dividing T cells. The block in DNA replication in the lymphocytes results from the inhibition of the enzyme ribonucleotide reductase by the high levels of dATP. Ribonucleotide reductase is the key enzyme in the conversion of ribonucleotides to deoxyribonucleotides. When the activity of ribonucleotide reductase is inhibited by dATP, the cell is depleted of the deoxyribonucleotides required for DNA synthesis.

Several treatments can be used for children with ADA deficiency. These include protective isolation and bone marrow transplantation from a human leukocyte antigen (HLA)–identical sibling. Bone marrow transplants are available for only a minority of patients. Enzyme replacement therapy using ADA-containing erythrocytes or bovine ADA conjugated to polyethylene glycol has resulted in a life-saving treatment for some of these children. More recently, ADA deficiency has been treated by somatic cell gene therapy using retrovirus vectors carrying the cDNA for the ADA gene. The use of somatic cell gene therapy for the treatment of ADA deficiency is discussed in Chapter 14.

## Autosomal Dominant Inheritance Attributed to Change-of-Function Mutations

### DISEASES RESULTING FROM TRINUCLEOTIDE EXPANSION

Human diseases that result from the expansion of trinucleotide repeats were discussed in Chapter 4 (see Table 4-2). In these diseases, repeats of trinucleotides that occur at the beginning or end of a gene undergo expansion and alter the protein or the RNA coded for by the gene. This expansion can occur at the 5' or 3' noncoding regions of the gene or within the first exon or first intron of the gene. The site of the expansion determines the effect of the mutation on the RNA or protein product. Two diseases, Huntington's disease and myotonic dystrophy, both result from trinucleotide expansion, and both are inherited as an autosomal dominant trait. In these two diseases, the trinucleotide expansion has altered the gene in such a way that it codes for an aberrant protein or RNA molecule, which has a dominant effect over the normal gene product. This type of change-of-function mutation leads to autosomal dominant inheritance, in which individuals with only one mutant allele are affected with the disease and pass it on to 50% of their offspring.

*Huntington's Disease.* Huntington's disease, a late-onset, progressive, and fatal disease named after George Huntington, who described the disease in 1872, has been identified as one of the trinucleotide expansion disorders. The trinucleotide, CAG, is expanded in the coding region of the first exon of the gene. This codon (CAG) codes for glutamine, resulting in the huntingtin protein containing multiple glutamines at its amino-terminal end and having an altered function.

Characteristics of Huntington's disease include a degeneration of the basal ganglia leading to uncontrollable muscle movements, the loss of cognitive functions, and the development of psychiatric problems. The disease can be divided into four main stages that span from the age of onset of approximately 36 years to clear expression of the phenotype by age 65 years. In the initial stage of the disease, the patient has small losses of muscular coordination, forgetfulness, and some personality changes. During the second stage of the disease, uncontrollable movements begin and are accompanied by speech disorders. By the third stage of the disease, the patients lose control of bodily functions and are no longer able to care for themselves. The fourth and final stage finds the patient in a vegetative state during which death occurs. Because of the late age of onset, generally past the reproductive years, the transmission of the disease to offspring often takes place before the disease has been diagnosed.

> *Huntington's disease, inherited as an autosomal dominant disorder, is caused by the expansion of CAG repeats at the 5' end of the Huntington's disease gene and the presence of excess glutamines at the N-terminal end of the protein, huntingtin.*

The gene for Huntington's disease was isolated by classic positional cloning and involved extensive linkage analyses followed by molecular analysis of the DNA at the suspected region on chromosome 4. No chromosomal rearrangements were associated with the disease; therefore, all of the potential genes in the region had to be isolated and tested for their identity to the candidate gene. The gene, finally identified and isolated in 1993 by The Huntington's Disease Collaborative Research Group, maps to chromosome 4p16.3 [8]. Initially known as *IT15*, the gene spans 210 kb of DNA and encodes the huntingtin protein. The function of huntingtin remains unknown, although the protein, which contains 3144 amino acids, appears to be a DNA-binding protein. In the first exon, at the 5' end of the normal gene, there is a sequence of 21 copies of CAG that results in 21 glutamines being present in the N-terminal portion of the protein (Figure 11-6). In the disease gene, the stretch of CAG repeats is expanded to 40–100 copies, resulting in the production of an altered protein that contains 40–100 glutamine residues. The degree of expansion of the CAG repeat and the number of glutamines in the protein correlate with the severity of the disease. For example, a child with 100 copies of the CAG repeat may show an early onset of the disease at age 2 years, whereas his father, who has fewer copies of the repeat sequence, will have a later onset of the disease at approximately age 40 years.

The biochemical defect of Huntington's disease is still unknown, consequently there is no cure or treatment for the disease. Recent studies have focused on the potential effects that extra glutamines at the N-terminal end of a protein could have in a cell. Other diseases have been identified with the same type of defect, notably dentatorubral-pallidoluysian atrophy, spinobulbar muscular atrophy, and spinocerebellar ataxia (types

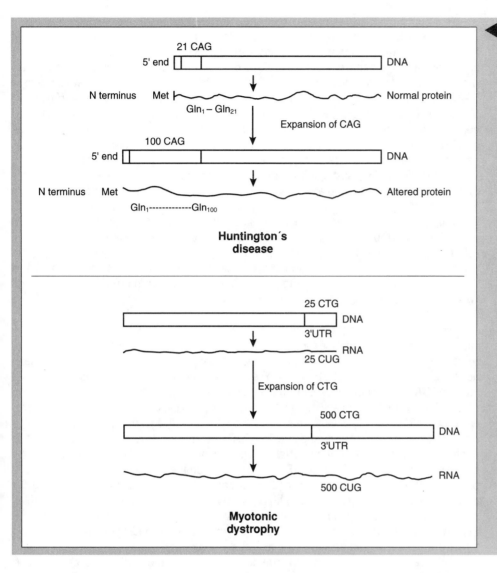

**FIGURE 11-6**

**Position of Trinucleotide Repeats in the Expansion Disorders, Huntington's Disease, and Myotonic Dystrophy.** *The Huntington gene contains a repeated CAG sequence at the 5' end of the gene within the first exon. The CAG codon codes for glutamine, resulting in a stretch of approximately 21 glutamines in the N-terminal end of the normal protein. In the disease state, the CAG repeat is expanded, resulting in up to 100 CAG sequences in the first exon of the gene and the presence of 100 glutamines in the mutant protein. The presence of the additional glutamines changes the function of the protein, leading to the disease. The myotonic dystrophy gene contains a repeated CTG sequence in the 3' untranslated region (UTR) of the normal gene. This sequence is transcribed into RNA but is not translated into protein. The CTG sequence undergoes expansion from approximately 25 in the normal gene to 500 or more repeats in the mutant gene. However, because the expansion takes place in the 3' UTR of the gene, the effect of the expansion is reflected only in the structure of the 3' end of the RNA and not in the protein. The presence of the expanded sequence in the RNA molecule appears to affect the polyadenylation of the myotonic dystrophy RNA as well as interfere with normal processing of other RNA molecules.*

1 and 3). All of these diseases show genetic anticipation, in which the repeat length is unstable and has a significant probability of increasing in length with each generation. Attempts to identify other proteins that might interact with the excessive glutamine residues in these disease proteins have tentatively suggested that the enzyme glyceraldehyde-3-phosphate dehydrogenase (G3PD) may do just that. Furthermore, the role that G3PD plays in glucose metabolism suggests a possible basis for the phenotypic expression of these diseases in the brain. Energy production in the brain relies on glycolysis; therefore, any disturbance in the dehydrogenase activity could lead to decreased energy in the brain and adversely affect the neurons. The role that G3PD plays in the disease phenotype is still speculative and under intense investigation.

***Myotonic Dystrophy.*** Myotonic (repetitive muscle firing) dystrophy (weakness and wasting) is an autosomally dominant inherited disease associated with CTG repeats in the noncoding 3' end of the last exon of the gene coding for a cAMP-dependent kinase (i.e., dystrophia myotonica [DM] kinase) [see Figure 11-6]. The symptoms of myotonic dystrophy are quite variable and are related to the age at which the patient expresses the phenotype. When expressed in newborns, the disease is characterized by respiratory problems, which lead to death if the infant is not given ventilation assistance. In childhood, the expressed phenotype is characterized by mental retardation. Adults may demonstrate some myotonia and muscle weakness, and some older adults may develop only cataracts. All four phenotypes can be expressed in different members of the same family. The usual presentation of the disease is a development of progressive muscular dystrophy that is often induced by cold and associated with an inability to relax the muscles. The severity of the disease is directly related to the size of the trinucleotide repeat [9].

*Myotonic dystrophy, inherited as an autosomal dominant disorder, is caused by the expansion of CTG repeats in the 3' untranslated end of the myotonic dystrophy gene and the production of a RNA transcript with an excess of CUG repeats.*

The gene for myotonic dystrophy maps to chromosome 19q13.3 and codes for a DM kinase that is expressed in numerous tissues (e.g., skeletal muscle, heart, brain). There are 15 exons in the gene, which spans 13 kb of genomic DNA. In the final 3' exon within the untranslated region, there is a run of CTG repeats. Normal individuals have a range of 5–30 CTG repeats. Individuals who show minimal effects of the disease have repeats in the range of 50–80 copies, which is increased in adult patients to a range of 100–500 copies. Early onset of the disease is associated with an expansion of the CTG repeats ranging from 500–2000 copies. The neonatal form of myotonic dystrophy shows maternal transmission in which length of the expansion increases from mother to child.

In the other diseases resulting from trinucleotide expansion (e.g., Huntington's disease), which are inherited as autosomal dominant disorders, the expanded repeats are located at the 5' end of the gene and affect the structure of the protein product. Therefore, the inheritance pattern is consistent with a change-of-function mutation in which the mutant protein has an altered structure leading to an altered function in the cell. In the case of myotonic dystrophy, in which the expansion of the trinucleotide repeat occurs in the 3'-untranslated region of the gene, the repeat sequence is not translated into protein and does not lead to the production of a mutant protein (see Figure 11-6). What then accounts for the autosomal dominant inheritance pattern of myotonic dystrophy?

Studies on RNA levels in muscle biopsies from patients with myotonic dystrophy suggest a possible explanation for this enigma [10]. In the total pool of RNA from muscle, there is a small decrease of DM kinase RNA in the diseased tissue, but this decrease is not specific for myotonic dystrophy. However, measurements of DM kinase polyadenylated mRNA in the nuclei reveal dramatic disease-specific decreases in the level of both mutant and normal DM kinase mRNA. These results suggest that a RNA molecule with an expanded repeat at the 3' end may have a defect in polyadenylation. Furthermore, a RNA-binding protein that is specific for CUG repeats appears to accumulate in the nucleus of myotonic dystrophy patients. These results suggest a model in which the poorly polyadenylated DM kinase mRNA precipitates in the nucleus and sequesters CUG-binding proteins. This action disrupts the normal processing of RNA transcripts present in the nucleus, affecting normal DM kinase mRNA function and possibly other mRNAs, resulting in the clinical variability seen in myotonic dystrophy. The pleiotropic effects of the disease could be related to the ability of the mutant RNA species to interact with and affect different RNA molecules in the nucleus.

The presence of the expanded trinucleotide repeats and PCR-based analysis can determine if individuals are carriers of the disease and can be used as a tool for prenatal diagnosis using chorion biopsy samples. Currently there is no therapy available to alter the progression of the disease. Patients are advised to avoid anesthesia during surgery because many of these patients have a high sensitivity to anesthetics and may undergo heart block and sudden death.

## OSTEOGENESIS IMPERFECTA: A DISORDER OF COLLAGEN BIOSYNTHESIS

Collagen is coded for by at least 28 different genes and is found in the body in many forms, designated from type I to type XIV.

*Type I.* Type I collagen, which is found in skin, bone, tendons, and ligaments, is altered in individuals with the autosomal dominant disorder *osteogenesis imperfecta*. Type I collagen is composed of three $\alpha$ chains (two $\alpha_1$ chains and one $\alpha_2$ chain). Each $\alpha$ chain forms a left-handed helix that is made up of three amino acids per turn. Three of these left-handed helices come together to form a right-handed triple helical structure (Figure 11-7) [11, 12]. There are three amino acids per turn in the left-handed helix, and every third amino acid is glycine, giving a repeat of Gly-X-Y, where X and Y represent other amino acid residues. Approximately 10% of the X residues are proline, and approximately 10% of the Y residues are hydroxyproline, accounting for the rigidity of the collagen molecule. Analysis of the sequence of the exons that code for collagen show that the sequence coding for Gly-X-Y is repeated numerous times and is found in all of the exons. A single exon can be made up of 54 nucleotides, with the codons for the Gly-X-Y sequence repeated 6 times. The small size of the glycine residue in the structure allows the amino acid to fit into the central core of the helix. This is extremely important as

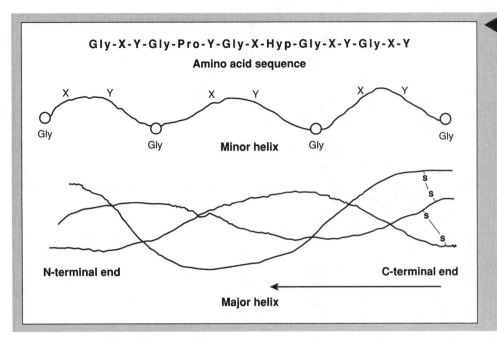

**FIGURE 11-7**
**Structure of Type I Collagen.** *The amino acid sequence of collagen, shown at the top, is represented by a glycine (Gly) at every third residue and X and Y, representing other amino acid residues. The minor helix shown in the middle of the figure represents the folding of each individual chain. At the bottom of the figure is the major or triple helix, made up of two $\alpha_1$ chains and one $\alpha_2$ chain. The chains are held together by disulfide bonds. Formation of the helical structure begins at the C-terminal end of the triple helix and proceeds to the N-terminal end. (Source: Adapted with permission from Nicholls AC: Collagen disorders. In* The Inherited Metabolic Diseases. *Edited by Holton J. London, UK: Churchill Livingston, 1994, p 381.)*

demonstrated by the finding that many missense mutations that result in a substitution of the glycine residue by another amino acid result in the production of a mutant collagen.

The final triple helical molecule, made up of two $\alpha_1$ chains and one $\alpha_2$ chain, is held together at the C-terminal end of the protein by interchain disulfide bonds. The formation of the triple helix begins at the C-terminal end and moves toward the amino-terminal end of the molecule. This directionality to the helix formation appears to be directly related to the effect that different missense mutations have on the severity of the clinical phenotype of osteogenesis imperfecta.

In a normal individual, two $\alpha_1$ helixes and one $\alpha_2$ helix come together to form the right-handed triple helix or type I collagen. In an individual with a splicing mutation in one of the $\alpha_1$ chain alleles, a normal molecule of type I collagen is made but at one-half the normal amount because of the reduced level of $\alpha_1$ chain (Figure 11-8). Individuals with splicing or null mutations are affected with the mild type I form of osteogenesis imperfecta characterized by blue sclera and brittle bones that can lead to as many as 50 fractures prior to puberty. Hearing loss is present in approximately one-half of the patients, but there is no skeletal deformity. The frequency of type I osteogenesis imperfecta is 1 in 10,000.

The molecular basis for the decrease of $\alpha_1$ protein can vary from one patient to another, but one common form of the diseases is known to be due to a 5-bp deletion near the 3' end of the gene, *COL1A1*. This 5-bp deletion results in a frameshift mutation, which extends the length of the protein a predicted additional 84 amino acids. The increased size of the protein appears to make it unstable, accounting for the deceased level of type I collagen. In addition, any mutation that results in improper splicing often gives rise to type I disease.

In the other three types of osteogenesis imperfecta (types II, III, IV), the molecular defects are usually missense mutations that result in a substitution of a crucial glycine residue, with another amino acid leading to a disruption of the helix. The severity of osteogenesis imperfecta is a reflection of where the substitution occurs and which amino acid is placed in the mutant $\alpha_1$ chain.

*Type II.* A lethal form of the disease in infants, type II presents with severe deformity and death within 1 month. This form of the disease has a frequency of 1 in 20,000 and is inherited as an autosomal dominant disease. The missense mutations that result in a substitution of a glycine residue in the triple-helical domain of the protein generally occur in the C-terminal end of the gene and result in an altered protein that interferes

*A defect in collagen biosynthesis can result in four types of* **osteogenesis imperfecta***: types I, II, III, and IV.*

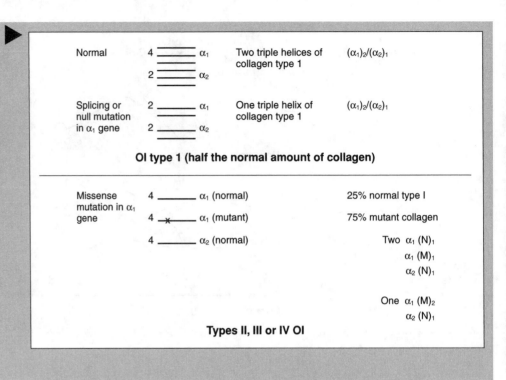

**FIGURE 11-8**

*Dominant Negative Effect of Mutated $\alpha_1$ Molecules.* Normal type I collagen, shown at the top, is formed from two normal $\alpha_1$ chains and one $\alpha_2$ chain to give the triple helix. A splicing or null mutation that reduces the amount of $\alpha_1$ chain by 50% (i.e., that produced from the normal allele) results in 50% the normal amount of type I collagen. An individual with only half the amount of type I collagen presents with type I osteogenesis imperfecta (OI). Below the line is a schematic of what happens when a missense mutation arises in the $\alpha_1$ gene and results in a defective $\alpha_1$ chain being produced. The mutant $\alpha_1$ chain is incorporated into the final collagen molecule and destroys its function. Therefore, the individual makes only 25% of the normal amount of type I collagen and 75% mutant or defective collagen. These individuals present with types II, III, or IV OI. The severity of the disease is related to the position of the missense mutation within the gene and the amino acid substitution that occurs. Generally, the closer the mutation is to the C-terminal end of the protein the more severe the disease.

with normal assembly. The normal allele is still producing a normal $\alpha_1$ molecule; therefore, 25% of the triple helical molecules formed are normal. However, since the composition of the triple helix is made up of two $\alpha_1$ chains, the presence of the abnormal molecule results in the formation of three abnormal triple helices for each normal molecule (see Figure 11-8). Therefore, an individual is better off with a null mutation that results in 50% of the normal amount of collagen than a mutation that results in an altered protein and a dominant negative effect on the normal protein.

Infants who present with the lethal form type II osteogenesis imperfecta appear to have new mutations rather than an inherited mutation from one of the parents. The exception to this is when parental mosaicism occurs. Type II osteogenesis imperfecta is lethal, with death usually resulting from pulmonary complications. The disease can be diagnosed by ultrasound screening during 14 and 18 weeks of gestation, which reveals abnormal limb length and morphology. Abnormal collagen can also be detected by analysis of the chorionic villi.

*Types III and IV.* Type III disease is characterized by a progressive deformation of the bone, limited growth, and hearing loss. Type III disease also is caused by the production of an $\alpha_1$ missense protein, in which substitutions of glycine residues in the triple-helical domain interfere with normal collagen formation. Type IV osteogenesis imperfecta causes moderate deformity, fractures, and hearing loss and is due to missense mutations in which a glycine residue is substituted by another amino acid. The decreased severity of types III and IV osteogenesis imperfecta appears to be related to which glycine in the $\alpha_1$ chain is changed and which other amino acid is substituted. In general, the closer the substitution is to the amino-terminal end of the molecule, the less severe the clinical phenotype, although the type of substitution is also important in determining the clinical severity.

Although mutations in the $\alpha_2$ chain also arise, they tend to cause less severe forms of the disease than mutations in the $\alpha_1$ chain. This is probably a reflection of the presence of only one $\alpha_2$ chain per triple helix so that these individuals are still capable of producing 50% normal molecules.

## HAPLOINSUFFICIENCY AS A DOMINANTLY INHERITED DISORDER

Genes localized to autosomal chromosomes are present in two copies in humans. If one of the alleles is rendered defective (i.e., the heterozygous state), the individual generally is unaffected with the particular disease associated with that defective allele. In these cases, the heterozygote is a carrier and is usually clinically asymptomatic even with only 50% of the normal gene product. Exceptions to this are the autosomal dominant disorders described above (e.g., myotonic dystrophy, type II osteogenesis imperfecta), in which the abnormal gene product interferes with the function of the normal gene product. In those cases, the mutant phenotype is dominant to the normal phenotype, resulting in clinical disease even in the heterozygous state.

Another exception occurs when 50% of the normal gene product is not enough for normal functioning, as found in type I osteogenesis imperfecta. Although type I osteogenesis imperfecta is generally a mild form of the disease, there is an obvious clinical phenotype. Diseases in which 50% of the normal amount of protein is not enough for a normal phenotype and show clinical phenotypes in the heterozygous state are called haploinsufficiency diseases. One of the most important diseases that shows haploinsufficiency and is inherited as an autosomal dominant trait is familial hypercholesterolemia.

*Familial hypercholesterolemia* is an example of a haploinsufficiency disorder in which half of the normal gene product is not enough to prevent clinical symptoms.

### Familial Hypercholesterolemia: A Defect in the Low-Density Lipoprotein (LDL) Receptor.

Familial hypercholesterolemia in the heterozygous state has an incidence of 1 in 500 individuals and accounts for 5% of the coronary artery disease seen in the Western world. Individuals with familial hypercholesterolemia have myocardial infarctions before the age of 55 years and have cholesterol levels twice that of a normal individual. The disease in the homozygous state has a frequency of 1 in 1 million. Homozygous individuals suffer myocardial infarction before the age of 20 years and have cholesterol levels 3–4 times the normal level.

Cholesterol is carried by LDL and is taken into cells by the attachment of the LDL–cholesterol complex to the LDL receptors located in the coated pits on the surface of cells. In patients with familial hypercholesterolemia, the number of LDL receptors is reduced, resulting in an increase in the levels of cholesterol in the plasma. More than 300 mutations in the LDL receptor gene have been identified as typical loss-of-function mutations (i.e., missense, nonsense, insertions, deletions). The position of the mutation within the gene alters the receptor protein differently, depending on which of the LDL receptor protein functions are impaired (Figure 11-9) [13, 14].

The LDL receptor gene maps to chromosome 19p13 and is made up of 18 exons spanning 45 kb of genomic DNA. The mature mRNA is 5.3 kb in length and codes for a protein of 839 amino acids. As indicated in Figure 11-9, various exons within the gene code for different domains in the protein. Mutations that occur in the specific exons alter specific functions of the LDL receptor protein and have been classified into five groups. Each of these classes represents a defect in different stages involved in the synthesis or function of the receptor protein.

The LDL receptor is synthesized in the endoplasmic reticulum (ER) and then transported first to the Golgi complex and then to the cell surface. Exons 7–14 contain the information for the movement of the LDL receptor from the ER, exons 16–17 contain information for membrane spanning of the protein from inside to outside the cell, and exons 17–18 contain information for localization of the protein to the coated pits on the cell surface. Once on the cell surface, the LDL receptor binds the protein moiety of the LDL–cholesterol complex, and this complex is brought into the cell by endosome invagination. The endosomes carry the LDL–cholesterol complex to the liposomes, with the release and recycling of the LDL receptor to the surface of the cell where it then binds more LDL–cholesterol complexes. Exons 2–6 contain the information for the binding of the receptor to LDL. (Exon 1 is required for the signal sequence necessary for processing of the receptor protein, and exon 15 contains the sequences necessary for post-translational modification by glycosylation.)

The five classes of mutations resulting in familial hypercholesterolemia are related to the effect the mutation has on the LDL receptor protein (Figure 11-10) [14]. *Class 1* mutations are null mutations (nonsense, frameshift, or splicing mutations) in which no protein is made by the mutant allele; an individual heterozygous for the mutation has 50% of LDL receptors. *Class 2* mutations occur in exons 7–14. Individuals with these mutations synthesize normal amounts of a defective LDL receptor that stays in the ER and is degraded there. Mutations of *Class 3* occur in exons 2–6, and these individuals make a defective LDL receptor that cannot bind LDL. *Class 4* mutations occur in exon 18. These individuals produce a LDL receptor protein with a defective cytoplasmic tail that

**FIGURE 11-9**

*Exons in the Low-Density Lipoprotein (LDL) Receptor Gene Code for Domains of the LDL Receptor Protein.* The LDL receptor gene is made up of 18 exons and spans 45 kb of DNA. The domains of the LDL receptor protein and their function are indicated, as well as the corresponding exons that code for each domain. Exon 1, which codes for the 21 amino acid signal sequence, is not indicated. (Source: Reprinted with permission from Hobbs HH, Russell D, Brown M, et al: The LDL receptor locus and familial hypercholesterolemia. Ann Rev Genet 24:136, 1990.)

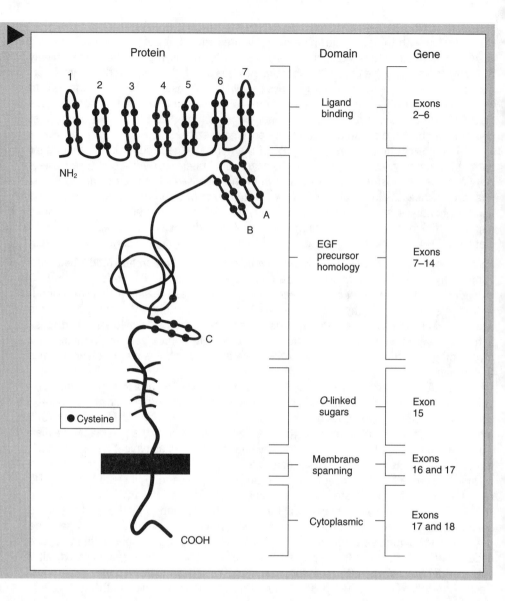

*Mutations in the LDL receptor generally fall into five classes that reflect the effect the mutation has on the synthesis or function of the receptor protein.*

interferes with the ability to internalize the bound LDL. *Class 5* mutations interfere with the dissociation of the receptor from the LDL and block the recycling of the receptor back to the surface of the cell. The basic effect of all of these mutations is a reduction by one half of the amount of the normal LDL receptors. Half is not enough, and heterozygous individuals are affected with familial hypercholesterolemia.

The elucidation of the function of the LDL receptor and an identification of the mutations and their effects on this protein have increased the understanding of this form of coronary disease. Many patients now can be identified by a family history of early heart attacks and elevated cholesterol levels. Treatment involves a change in diet to lower the level of cholesterol, as well as the use of medications that interfere with the synthesis of cholesterol in the liver.

### AUTOSOMAL DOMINANT DISORDERS WITH A PREDISPOSITION TO DEVELOP CANCER

Considerable evidence accumulated during the past 10 years indicates that families can be affected with rare autosomal dominantly inherited genetic predispositions to develop cancer (Table 11-3). These rare autosomal dominant disorders include: retinoblastoma (RB), Li-Fraumeni syndrome, early-onset breast and ovarian cancer, familial adenomatous polyposis (FAP), hereditary nonpolyposis colon cancer (HNPCC), and neurofibromatosis 1 and 2 (NF-1, NF-2). All of these disorders, with the exception of HNPCC, are caused by mutations that lead to a loss of function of one allele of a gene that codes for a tumor suppressor protein. The actual cancer that develops results from a somatic cell mutation that inactivates the second allele of the tumor suppressor gene, making the

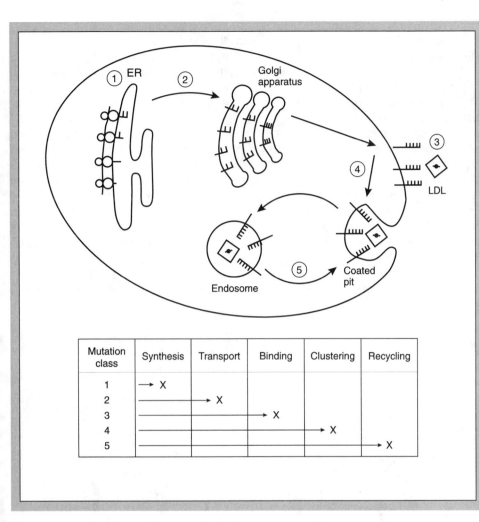

◀ **FIGURE 11-10**
**Classes of Low-Density Lipoprotein (LDL) Receptor Mutants.** *Each mutant class (1– 5) is correlated with the abnormal function of the mutant LDL receptor protein. Mutations can disrupt synthesis of the LDL receptor in the endoplasmic reticulum (ER) [class 1], transport of the protein to the Golgi apparatus (class 2), binding of ligands to the receptor (class 3), clustering of the receptor in the coated pits (class 4), or recycling of the receptor in the endosomes (class 5).* (Source: Reprinted with permission from Hobbs HH, Russell D, Brown M, et al: The LDL receptor locus and familial hypercholesterolemia. Ann Rev Genet 24:141, 1990.)

◀ **TABLE 11-3**

*Examples of Dominantly Inherited Disorders*

| Disorder | Protein Altered | Frequency |
|---|---|---|
| **Change of function** | | |
| Huntington's disease | Huntingtin | 1/10,000 |
| Myotonic dystrophy | Protein kinase | 1/10,000 |
| Osteogenesis imperfecta | Collagen | 1/10,000 |
| **Haploinsufficiency** | | |
| Familial hypercholesterolemia | Low-density lipoprotein | 1/500 |
| **Predisposition to developing cancer** | | |
| Retinoblastoma | pRb | 1/15,000 |
| Li-Fraumeni syndrome | p53 | Rare |
| Early-onset breast, ovarian, prostate, colon cancer | BRCA1 | 1/200 |
| Early-onset breast cancer | BRCA2 | 1/200 |
| Colon cancer (FAP) | APC | 1/7000 |
| Colorectal cancer (HNPCC) | MSH-2, MLH-1, PMS-1, PMS-2, and GTBP | 1/500 |
| Neurofibromatosis 1 | Neurofibromin (GTPase) | 1/4000 |
| Neurofibromatosis 2 | Merlin/schwannomin | 1/40,000 |

*Note.* FAP = familial adenomatous polyposis; HNPCC = hereditary nonpolyposis colon cancer; GTPase = guanosine triphosphatase.

cell more likely to lose growth control and develop into a malignant cell. HNPCC, discussed in Chapter 4, is not caused directly by the loss of a tumor suppressor protein but results from mutations that occur in genes required for DNA repair. In this case, the loss of normal DNA repair functions leads to an overall increase in mutation rate and an increased probability that a mutation will occur in a tumor suppressor gene and inactivate the gene.

The identification and crucial role that tumor suppressor genes play in the regulation of cell growth came from studies on RB, an inherited disorder expressed in children as a development of tumors of the retina. Studies on the inheritance patterns of RB by Knudson led to "the two-hit hypothesis" to explain the inherited nature of RB (Figure 11-11) [15]. According to this hypothesis, a single mutation occurs in the germ line and inactivates one allele of a crucial gene. This mutation is passed on to the offspring of an affected parent. A second mutation occurs in a somatic cell of the offspring, inactivating the other allele to give a homozygous state and a predisposition of the cell to develop cancer.

**FIGURE 11-11** ▶

*Two-Hit Hypothesis of Knudson. A model developed by Knudson to explain the inheritance pattern of retinoblastoma (RB) proposed that a single allele of the Rb gene is inactivated in the germ cell and transmitted to the offspring. This single hit is recessive to the normal allele. However, somatic cells carrying one defective allele have a greater chance of having a second hit, in which the remaining normal allele is inactivated or lost, leading to loss of heterozygosity. If these mutations occur in tumor suppressor genes, the cell is predisposed to develop cancer. The second hit can occur by multiple events, including the loss of the normal chromosome, the loss of the normal chromosome and duplication of the mutant chromosome, recombination in the somatic cell leading to loss of the normal allele, methylation of the promoter and silencing of the normal gene, and deletions or point mutations that inactivate the normal allele. (Source: Adapted with permission from Lasko D, Cavenee W, Nordenskjold M: Loss of constitutional heterozygosity in human cancer. Ann Rev Genet 25:286, 1991.)*

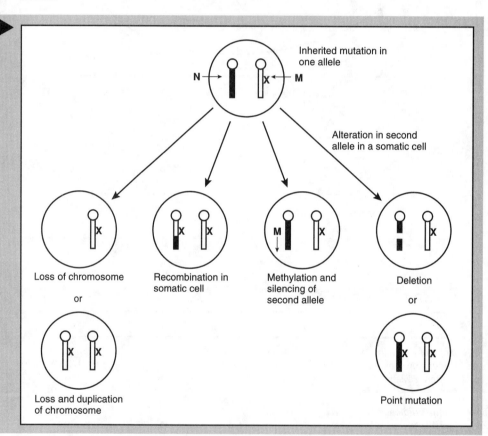

The **two-hit hypothesis** proposes that a mutation in a single allele of tumor suppressor gene is inherited as a dominant mutation and that a second mutation in the other allele occurs by a variety of mechanisms in a somatic cell to give the homozygous defective state.

This hypothesis proved correct for RB, in which it was shown that a single gene at 13q14 was inactivated in families that have the inherited form of RB. This loss-of-function mutation would be expected to be inherited as an autosomal recessive trait; in fact, this initial mutation does not in itself lead to the development of cancer and is recessive to the normal allele. However, cells with this single allele inactivated have a greater probability of developing cancer if another mutation inactivates the second allele to give a homozygous state. In a normal cell with two functional alleles, two somatic cell mutations occurring in each of the alleles are necessary to produce the homozygous state and place that cell at the same risk for developing cancer. The probability of two mutations occurring in a single cell is much lower than the probability of a single mutation occurring in the cell. In some cancers, more than two mutations may be required for cancer development. However, in all cases, a family with an initial mutation in one of the tumor suppressor gene alleles has a greater risk of developing cancers than those without the inherited mutation.

Once a defective tumor suppressor gene has been inherited, there are a number of ways in which the second allele can be inactivated, as illustrated in Figure 11-11. This process, termed *loss of heterozygosity* (LOH), can occur by the loss of the chromosome with the normal allele or the loss of the chromosome with the normal allele and a duplication of the chromosome with the defective allele. Recombination can occur in the somatic cell by sister chromatid exchange, resulting in the loss of the normal allele and the presence of two chromosomes, each with the defective allele. Deletions or point mutations can occur within the normal allele, leading to a loss of function and a homozygous defective cell. Additionally, methylation of the promoter sequences of the normal gene can result in a lack of transcription and silencing of the second allele. All of these events occur with a certain frequency in all cells, but if the loss of the second allele of a tumor suppressor gene gives a cell a selective growth advantage, that cell will outgrow the normal cells and ultimately may lead to a tumor.

The presence of tumor suppressor genes in normal cells also is supported by somatic cell hybridization studies, in which the transformed phenotype of a cell can be suppressed by fusing the transformed cell with a normal cell. These studies provide evidence that a gene or a protein product in a normal cell can suppress the malignant phenotype and suggest that the transformed cell is deficient in that gene or gene product. Many examples of tumor suppressor genes are now known.

*Retinoblastoma.* The gene responsible for RB is mapped to 13q14 and codes for the protein pRb, which has a molecular weight of 110,000 D. Protein pRb plays a crucial role both in cell cycle control and transcriptional regulation (see Chapters 5 and 8). Phosphorylation and dephosphorylation of pRb, which is controlled by cyclin-dependent protein kinases during cell cycle progression, determines the ability of pRb to initiate or suppress transcription. Protein pRb does not bind DNA directly, but it interacts with other transcription factors involved in regulating cell cycle genes. In this way, pRb can function to regulate progression of the cell cycle or, under certain conditions, block the progression. Inactivation or modification of pRb potentially disrupts the normal cell cycle controls, leading to uncontrolled growth and a potentially cancerous cell.

*Li-Fraumeni Syndrome.* A study of children with rhabdomyosarcomas carried out in 1969 by Li and Fraumeni identified four families in which there was an increased incidence of soft tissue tumors, breast cancer, brain tumors, osteosarcomas, and other neoplasms. This familial cancer syndrome is now known as the autosomal dominantly inherited Li-Fraumeni syndrome. This syndrome is the result of a germline mutation in the gene coding for the tumor suppressor protein, p53. Although Li-Fraumeni syndrome and the inheritance of a defective p53 gene is quite rare, the presence of a mutated p53 in human cancers is quite common. In fact, more than 50% of all human cancers have a mutation in the gene coding for p53, which plays a central role in normal growth control. Loss or mutation of p53 can have multiple effects on growth regulation because p53 functions both as a transcription factor and as a controlling factor to sense DNA damage in the genome. When DNA damage occurs, p53 signals the cell to stop the cell cycle in G1 to allow DNA repair to correct the damage. In the event that the DNA is not corrected, p53 may direct the cell to programmed death (i.e., apoptosis) [see Chapters 4 and 8]. In the absence of a normal p53, a damaged cell that is not repaired continues to grow and transfers the damage to daughter cells, allowing the defect to be perpetuated in subsequent generations.

*Hereditary Breast and Ovarian Cancer.* The autosomal dominantly inherited form of breast and ovarian cancer accounts for approximately 5%–10% of the total breast and ovarian cancers found in United States. During the past few years, mutations have been identified in two genes (*BRCA1* and *BRCA2*) that appear to account for nearly two-thirds of the hereditary form of breast cancer [16]. Women who have a *BRCA1* mutation have a lifetime risk of 80% for developing breast cancer and a 50% risk of developing ovarian cancer. A loss of heterozygosity is associated with the disease, suggesting that the *BRCA* genes code for tumor suppressor proteins. *BRCA* mutations are rare in the general population, and those that have been identified are found spread throughout both genes. However, women in the Ashkenazi Jewish population have four specific mutations in the *BRCA* genes that appear to occur at a high frequency (Figure 11-12) [17].

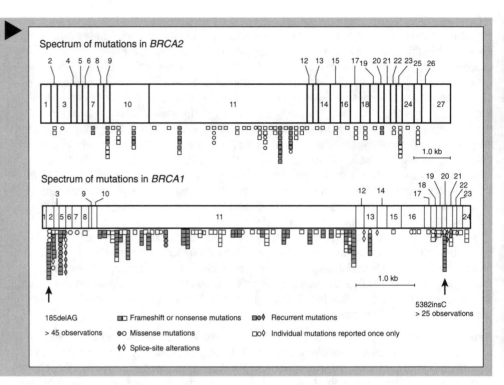

**FIGURE 11-12** ▶

**BRCA2 and BRCA1 Genes.** *The BRCA2 gene is made up of 27 exons and codes for a protein containing 3418 amino acids and also may be a tumor suppressor protein. More than 100 different mutations have been identified within the gene. The mutation, 6174delT, present at a high frequency in the Ashkenazi Jewish population, is located in exon 11 of BRCA2. It is a frameshift mutation that leads to premature termination of protein synthesis. The BRCA1 gene is made up of 24 exons spanning more than 100 kb of DNA. The gene codes for a protein of 1863 amino acids, which appears to function as a tumor suppressor protein. More than 200 different mutations have been identified scattered throughout the gene. Two common mutations, 185-delAG and 5382insC, arising at a high frequency in the Ashkenazi Jewish population are indicated. (Source: Reprinted with permission from Kahn P: Coming to grips with genes and risk. Science 274:496, 1996.)*

Two mutations, one in BRCA1 *and one in* BRCA2, *occur at a high frequency in individuals of the Ashkenazi Jewish population.*

The *BRCA1* gene contains 24 exons, spans 100 kb of DNA, and codes for a protein of 1863 amino acids. The function of the protein is unknown; however, in the C-terminal end of the protein coded for by *BRCA1*, there is a transcriptional transactivation domain that is mutated in families predisposed to breast and ovarian cancer. Consistent with this transcriptional regulatory function is the recent observation that transfecting breast and ovarian cancer cells in vitro with wild-type *BRCA1* inhibits the growth of the tumor cells. All of these characteristics support the role of *BRCA1* as a tumor suppressor gene. However, an additional function of the *BRCA1* protein has been described in which the central portion of the protein contains a granin protein consensus sequence. Granins are proteins found in the trans-Golgi network that function in the regulation of protein secretion. This finding is consistent with BRCA1 protein having both a membranous and a secretory form. The *BRCA2* gene contains 27 exons and codes for a protein of 3418 amino acids, which has an unknown function. However, the BRCA2 protein also contains a granin protein consensus sequence.

Two specific mutations (185delAG in *BRCA1*, 6174delT in *BRCA2*), one in each of the *BRCA* genes, are found in 1% of the Ashkenazi Jewish population. Female carriers of *BRCA1* mutations have an 80%–90% risk of developing breast cancer and a 40%–50% risk of developing ovarian cancer. The *BRCA1* mutation, 185delAG, has been found in 20%; and the *BRCA2* mutation, 6174delT, has been found in 8% of Ashkenazi Jewish women with early-onset breast cancer. The mutation 185delAG, which occurs in exon 2, results in a frameshift mutation and the production of a truncated protein. The mutation 6174delT occurs in exon 11 of *BRCA2* and also is a frameshift mutation that leads to premature protein termination. Two other mutations that are common in the Ashkenazi Jewish population are the frameshift mutation 188del11, found in exon 2 of *BRCA1*, and the frameshift mutation 5382insC, found in exon 20 of *BRCA1*. The presence of these common mutations in Jewish women may provide a means for testing and screening this high-risk population.

***Familial Adenomatous Polyposis (FAP).*** FAP is inherited as an autosomal dominant disorder and is characterized by the development of multiple benign colorectal tumors, some of which undergo additional mutations and eventually lead to the development of carcinomas. In the germline, the adenomatous polyposis coli (*APC*) gene located on chromosome 5q21 undergoes a mutation in one allele. The APC protein of 310,000 molecular weight (MW) is made up of 2840 amino acids and is believed to be a tumor suppressor protein, although its exact function is still not known. The most common mutations seen in FAP involve the production of a truncated protein caused by nonsense or frameshift mutations within the gene. Loss of the APC function leads to an increase in the rate of additional mutations, eventually resulting in the alteration of K-*ras* and p53 functions. As in other disorders that result in mutations in tumor suppressor genes, only the initial mutation is inherited, with the development of cancerous cells arising from additional mutations in somatic cells.

*Hereditary Nonpolyposis Colorectal Cancer (HNPCC).* A second and more common form of hereditary colon cancer is HNPCC, which accounts for nearly 10%–15% of all colon cancers. Development of HNPCC is caused by mutations in the genes involved in the mismatch repair system described in Chapter 4. The mismatch repair system removes improperly paired bases from the DNA, thus preventing mutations from occurring at the next replication cycle. Five human genes have been identified that function to keep mismatches from persisting in the DNA. These genes include: *MSH-2*, which codes for a DNA-binding protein and is mutated in 60% of HNPCC patients; *MLH-1*, which functions in the excision of the mismatch base from the DNA and is mutated in 30% of HNPCC patients; and *PMS-1* and *PMS-2*, which are homologs to *MLH-1* and *GTBP*, a DNA-binding protein that accounts for the remaining 10% of HNPCC patients. Although mutations in these genes result in an inherited predisposition to develop cancer, the molecular basis for cancer development differs from that seen with the classic tumor suppressor mutations. In patients with HNPCC, the cells have a defective mismatch repair system, which leads to a mutator phenotype (i.e., a cell with a dramatic increase of 100–1000-fold in the overall mutation rate). In such a cell, there is an increased risk that a mutation will occur in a tumor suppressor gene or other crucial genes and will lead to the development of cancer. The growth properties of HNPCC cells are not initially altered, but the increased mutation rate increases the likelihood that mutations in other crucial genes will occur that will lead to the loss of growth control. One hallmark of HNPCC is microsatellite instability in which the trinucleotide and tetranucleotide repeat sequences vary in number. This microsatellite instability appears to be caused by mistakes by DNA polymerase when replicating these repeat sequences. Normal cells with a functional mismatch repair system correct the mistakes; however, cells with a defect in the mismatch repair system are unable to correct the defect, and the mistake becomes a permanent part of the DNA sequence.

> **HNPCC** results from mutations in the genes involved in DNA mismatch repair, which lead to a mutator phenotype.

As with some other types of inherited cancers, several mutations appear to be found more often than others. In patients with HNPCC, a deletion of exon 5 of the gene *MSH-2*, a deletion of exon 16 of the *MLH-1* gene, and a deletion of exon 6 of the *MLH-2* gene are most common. This pattern is suggestive of a founder effect, in which a common mutation occurs in an individual and perpetuates through many generations.

*Neurofibromatosis Types 1 and 2 (NF1 and NF2).* NF1 affects approximately 1 in 4000 individuals and is inherited as an autosomal dominant disorder. This common disorder is characterized by the appearance of café au lait spots (at least six spots larger than 15 mm in size) and at least two Lisch nodules on the iris. Additionally, the development of benign neurofibromas, auxiliary freckling, and optic gliomas occur. The disease, initially known as von Recklinghausen disease, can vary in its severity, but generally no significant decrease in life expectancy occurs [18]. The disease shows a high mutation rate, and many cases (30%–50%) are caused by new mutations. Individuals with NF1 are predisposed to some cancers at a low risk (twice the overall risk) including neurofibrosarcomas, malignant schwannomas, and gliomas.

> **Neurofibromatosis 1** varies in its severity, but most individuals with the disease have no significant decrease in life expectancy and often only express minor clinical symptoms such as café au lait spots and Lisch nodules.

Linkage analysis mapped the *NF1* gene to 17q, and with the finding of two NF1 patients with translocations at 17q11.2, the map position of the gene was finally pinpointed. The *NF1* gene, approximately 400 kb in size, was cloned by positional cloning and shown to code for a 2800-amino acid–containing protein known as neurofibromin. The protein shows the presence of a guanosine triphosphatase (GTPase)–activating protein (GAP) domain, which is homologous to two Ras inhibitor proteins found in yeast. Ras is a protein that binds guanosine triphosphate (GTP) and is involved in the signal transduction pathway. The function of neurofibromin is to interact with GTP-bound Ras, thereby participating in the hydrolysis of GTP to guanosine diphosphate (GDP) with subsequent inactivation of Ras. If Ras has undergone mutation to its oncogene form, it is no longer subject to inactivation by neurofibromin and can result in an abnormal or cancerous cell. When mutations occur in the *NF1* gene, the defective neurofibromin can no longer inactivate Ras, resulting in increased levels of GTP-bound Ras and an alteration in signals that control cell growth and division. These results are consistent with the normal function of neurofibromin as a tumor suppressor protein, which functions by inhibiting Ras.

NF2 is also inherited as an autosomal dominant disorder but at a much lower frequency (1 in 40,000) than NF1. The disease is a result of mutations in the *NF2* gene located on chromosome 22q12 and is a separate disease from NF1. The primary difference is the development of bilateral vestibular schwannomas (acoustic neuroma) that occur in adolescents or young adults, leading to the development of deafness and balance disorders. In addition, individuals with NF2 tend to develop tumors of the brain, the spinal cord, and peripheral nerves.

The *NF2* gene was cloned by positional cloning and shown to code for a protein that has homology to a family of cytoskeleton-associated proteins. The protein (named merlin) is also referred to as schwannomin for its involvement in schwannomas. The function of this protein is to serve as a link between the cytoskeleton and the cell membrane. How a mutation in merlin/schwannomin, which appears to act like a tumor suppressor protein, results in altered growth control is currently under investigation.

## X-Linked Inheritance

X-linked inheritance has specific patterns caused by the presence of two X chromosomes in the mother and only one X chromosome in the father. A father never passes on a copy of his X chromosome to his son; thus, male-to-male transmission of X-linked disorders is never seen. Transmission from father to daughter, however, is 100% because the daughter must receive one of her X chromosomes from her father. The mother is a heterozygous carrier if one allele on one of her X chromosomes is affected, and she will transmit the affected allele to 50% of her sons and to 50% of her daughters. For most X-linked disorders, the female is generally a heterozygous carrier, and there is a much greater chance that males will be affected with the disorder than females. Most of the commonly inherited X-linked disorders are inherited as X-linked recessive disorders (Table 11-4).

*Duchenne's Muscular Dystrophy (DMD).* DMD, discussed previously in Chapters 4 and 7, is the most severe and most common form of muscular dystrophy and affects 1 in 3500 males. The disease is particularly devastating because the boys are born asymptomatic and only begin to show clinical symptoms at the age of 4 years. The disease is generally fatal and currently has no treatment. Because of the excessively large size of the DMD gene (discussed in Chapter 7), nearly 2.5 million bp, many cases of the disease are caused by new mutations without a previous family history of the disease. Current efforts are underway to develop somatic cell gene therapies for DMD (see Chapter 14).

*Fragile X Syndrome.* One of the most common inherited forms of mental retardation seen in males results from fragile X syndrome. This disorder, which affects 1 in 1500 males, is characterized cytogenetically by the presence at the end of the long arm of the X chromosome of a nonstaining region (Figure 11-13). This cytogenetic abnormality is seen when cells from individuals with fragile X syndrome are cultured in a medium deficient in folic acid and thymidine. Metaphase chromosomes prepared from cells grown under these culture conditions show a lack of staining and apparent breakage on the X chromosome at Xq27.3.

**TABLE 11-4** ▶
*Examples of X-linked Inherited Disorders*

| Disorder | Protein Altered | Frequency |
|---|---|---|
| Duchenne's muscular dystrophy | Dystrophin | 1/3500 males |
| Fragile X syndrome | FMR1 protein | 1/1500 males<br>1/3000 females |

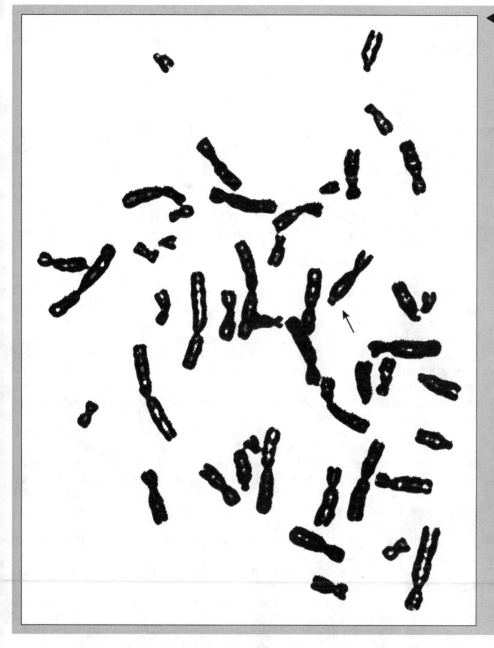

Individuals with fragile X syndrome are mildly to severely mentally retarded, usually have long and narrow facies with enlarged ears, and males have increased testicular volume (macroorchidism). Fewer females present with clinical symptoms, with an incidence of 1 in 3000. Fragile X syndrome is one of the newly described expansion disorders discussed in Chapter 4 and previously in this chapter. With the cloning of the gene responsible for fragile X syndrome (*FMR1*), the molecular basis for the disease has become much clearer. The *FMR1* gene spans 38 kb and consists of 17 exons. In the noncoding 5′ end of the gene's first exon is a trinucleotide repeat sequence, CGG, which is present in normal individuals in a range of 6–50 copies. Expansion of this trinucleotide repeat sequence from 60 to 1000 copies is associated with the disease.

The expanded CGG repeats and nearby CG sequences are heavily methylated in affected individuals, resulting in a transcriptionally inactive *FMR1* gene and an absence of *FMR1* mRNA. In a normal individual, the gene is unmethylated and produces normal amounts of mRNA, with the highest expression occurring in the brain and testes. The gene undergoes alternative splicing, and multiple protein isoforms are found in the cells. The FMR1 proteins are found in the cytoplasm and contain RNA-binding domains, suggesting a possible role for the proteins in RNA metabolism. FMR1 proteins appear to be associated with the 60S ribosomal subunit, so the normal protein may have a role in

*The **fragile X syndrome**, inherited as an X-linked trinucleotide expansion disorder, is the most common inherited form of mental retardation seen in males.*

translation. The absence of the proteins in an individual with fragile X syndrome may potentially interfere with normal translation, especially in the brain and testes. Additionally, heavy methylation or the large expansion region may alter DNA replication timing or chromatin structure in such a way that it affects the condensing of the chromatin at mitosis, which is reflected in the cytogenetic fragile X structure.

# MITOCHONDRIAL DISORDERS

Crucial to the normal life of a cell is the production of energy in the form of ATP, a process carried out in the mitochondria of the cell. These organelles, present in about 100 copies per cell, carry out oxidative phosphorylation using enzyme complexes located in their inner membranes to provide cells with their major source of ATP. The five complexes involved in the respiratory chain are made up of multiple subunits; the nuclear genome codes for some and the mitochondrial DNA codes for others.

The mitochondrial DNA is 16.5 kb in size and codes for 22 transfer RNA (tRNA) molecules, 2 ribosomal RNA (rRNA) molecules, and 13 protein subunits that are part of the complexes involved in the respiratory chain. Unlike nuclear DNA, mitochondrial DNA has no introns, is not complexed to histones, and appears not to be subject to the various repair processes that take place in the nucleus. These unique characteristics of the mitochondrial DNA no doubt contribute to the nearly ten-fold increase in its mutation rate as compared to nuclear DNA. The absence of intron sequences leads to a greater chance that substitution mutations will occur in coding regions and in crucial parts of a gene, thus increasing the chance that a single base change will lead to a mutant phenotype. In addition, the absence of DNA repair processes allows for any mutation that occurs to persist and replicate.

> Mitochondria have their own DNA genome, which codes for 22 tRNA molecules, 2 rRNAs, and 13 protein subunits.

Mitochondria are believed to have originated from the engulfment of bacteria-like particles during evolution, accounting for the similarity of many of their biosynthetic processes to prokaryotes rather than to eukaryotes. Thus, the ribosomes in mitochondria resemble those of bacterial cells, and their DNA is tightly organized and free of introns. Additionally, the genetic code varies from that of the nuclear DNA, with AGA and AGG serving as termination codons, UGA (a termination codon in the nuclear DNA) coding for tryptophan in the mitochondria, and AUA coding for methionine in the mitochondria. The presence of bacteria-like ribosomes in mitochondria is significant in a certain form of deafness (see below) found in humans.

The five complexes involved in oxidative phosphorylation in the mitochondria are listed in Table 11-5. Complex I, which involves the transfer of electrons from reduced nicotinamide adenine dinucleotide (NADH) to coenzyme Q, is the largest of the complexes and contains approximately 34 subunits, seven of which are coded for by the mitochondrial genes *ND 1–6*. Complex II, the succinate–coenzyme Q reductase, has subunits coded for only by nuclear genes. Complex III, which is responsible for the transfer of electrons from coenzyme Q to cytochrome *c*, is made up of 11 subunits, one of which (cytochrome *b*) is coded for by the mitochondrial DNA. Complex IV, which catalyzes the transfer of electrons from cytochrome *c* to oxygen, is made up of 13 subunits; the mitochondrial genes *CO I–III* code for three of these. The final complex, V, is ATP synthetase, which catalyzes the phosphorylation of adenosine diphosphate (ADP) to form the high-energy compound ATP. ATP synthetase is made up of 14 subunits; mitochondrial genes code for two of these (ATPase 6 and ATPase 8). Mutations in mitochondrial DNA that alter the structure of these subunits or that alter the structure of the tRNAs or rRNAs necessary to synthesize these subunits can lead to mitochondrial disorders.

Mitochondrial disorders are predominately maternally inherited [19, 20]. The human egg cell contains more than 200,000 mitochondrial DNA molecules, which are passed to offspring. Sperm contain mitochondria but contribute an exceedingly small percentage (0.1%) of mitochondria to the children. Every cell in the body contains hundreds of mitochondria, with each mitochondrion containing 2–10 molecules of DNA. Therefore, each cell can have thousands of copies of mitochondrial genomes. Within a single cell, and even within a single mitochondrion, there can exist DNA molecules that differ from one another in base pair sequence.

| *Complex* | *No. Mitochondrial Subunits* | *Mitochondrial Genes* |
|---|---|---|
| I. NADH → coenzyme Q | 7 | *ND1–6*[a] |
| II. Succinate coenzyme A reductase | 0 | — |
| III. Coenzyme Q → cytochrome *c* | 1 | *Cytochrome b* |
| IV. Cytochrome *c* → oxygen | 3 | *CO I–III* |
| V. ATP synthetase (ADP + $P_i$ → ATP) | 2 | *ATPase 6; ATPase 8* |

*Note.* ADP = adenosine diphosphate; ATP = adenosine triphosphate; NADH = reduced nicotinamide adenine dinucleotide; ATPase = adenosine triphosphatase; $P_i$ = inorganic phosphate.
[a] ND4 codes for two subunits.

***Table 11-5***
*Complexes Involved in Oxidative Phosphorylation*

Two terms are especially important when referring to the content of mitochondria in a single cell. The term *homoplasmy* refers to a cell that contains all normal or all mutant mitochondrial DNA. The term *heteroplasmy* refers to the presence of both mutant and normal mitochondrial DNAs in the same cell. Often, the ratio of mutant to normal mitochondrial DNA in a cell can determine the severity of the clinical phenotype associated with the mutant DNA. Furthermore, some cells in the body require more energy than others (e.g., muscle, eye), resulting in thresholds at which a certain level of mutant DNA must be present before a cellular dysfunction is observed. This may be the reason many of the mitochondrial disorders are chronic, late-onset diseases that involve environmental factors.

*__Homoplasmy__ and __heteroplasmy__ are terms used to refer to the content of mitochondria in a cell. Homoplasmy is the presence of all normal or all mutant mitochondrial DNA molecules, whereas heteroplasmy is the presence of both mutant and normal mitochondrial DNA molecules in the same cell.*

## Deletion Mutations That Lead to Mitochondrial Disorders

A 5-kb region of the mitochondria DNA located between *ND 5* and ATPase 8 is a region referred to as the *common deletion area* (Figure 11-14).

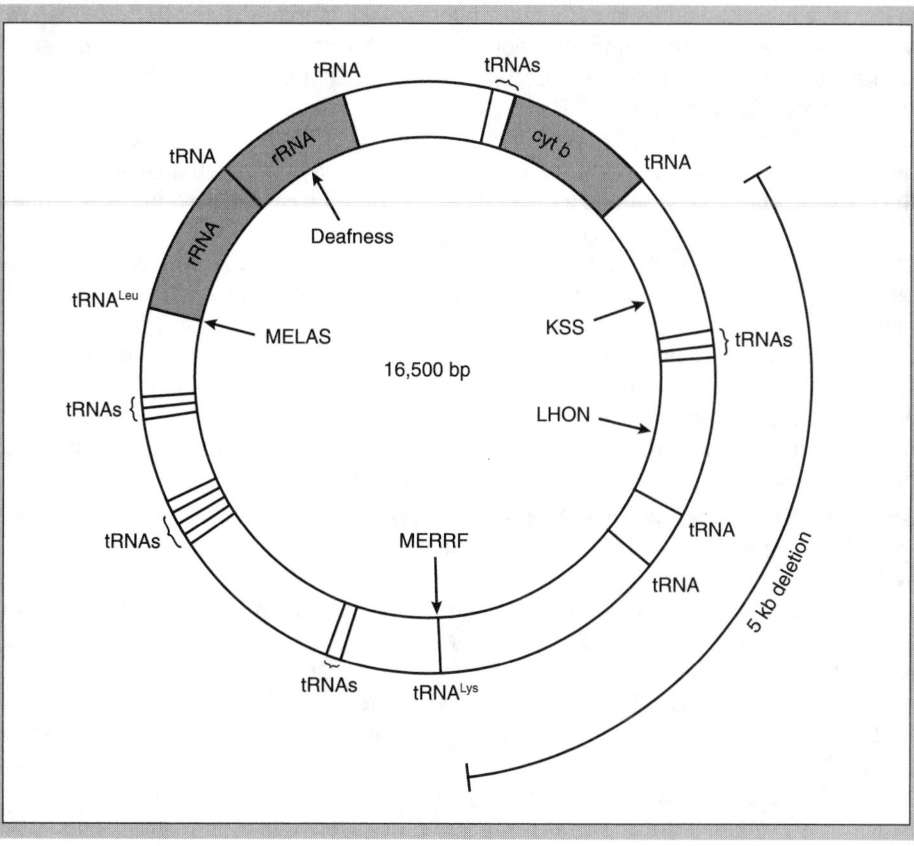


***Figure 11-14***
***Site of Mitochondrial Mutations Leading to Common Mitochondrial Disorders.*** *The mitochondrial DNA codes for 22 transfer RNAs (tRNAs), 2 ribosomal RNAs (rRNAs), and 13 proteins. A common 5-kb deletion is indicated as the site of one of the common deletion mutations leading to Kearns-Sayre syndrome (KSS). Specific point mutations leading to mitochondrial disorders include a missense mutation in* ND4*, resulting in Leber's hereditary optic neuropathy (LHON); a base substitution in tRNA-Leu, resulting in mitochondrial myopathy, encephalopathy, lactic acidosis, and stroke-like episodes (MELAS); a base substitution in tRNA-Lys, resulting in myoclonic epilepsy with ragged red fibers (MERRF); and a base substitution in the 12s rRNA gene resulting in deafness. (Source: Adapted with permission from Johns D: The other human genome: mitochondrial DNA and disease.* Nat Med *2:1066, 1996.)*

Many of the deletions that result in the syndrome known as Kearns-Sayre syndrome (KSS) occur in this region and result in defects in complexes I, III, and IV, all of which have subunits for which this region of the mitochondrial DNA codes. KSS is characterized by progressive external ophthalmoplegia, pigmentary degeneration of the retina, and heart block, or cerebellar syndrome. The onset of symptoms occurs before 20 years of age. This disease is heteroplasmic, and most cases appear to be caused by spontaneous deletions occurring in the mitochondrial DNA during embryogenesis rather than by maternal transmission.

## Point Mutations That Lead to Mitochondrial Disorders

The entire base sequence of the mitochondrial DNA is known, allowing the identification of specific point mutations or substitutions that are associated with specific disease phenotypes. Some of the most common of these are listed in Table 11-6 and shown in Figure 11-14 [21].

**TABLE 11-6** ▶
Point Mutations in Mitochondrial DNA Leading to Mitochondrial Disorders

| Disease | Common Mutations Associated with the Disease |
| --- | --- |
| LHON | Missense mutation in ND4 (LHON 11778); defects in complex I |
| MELAS | Base substitution mutation in tRNA-leucine (MELAS 3243) |
| MERRF | Base substitution mutation in tRNA-lysine (MERRF 8344) |
| Deafness | Base substitution mutation in 12s rRNA (12srRNA 1555) |

*Note.* LHON = Leber's hereditary optic neuropathy; MELAS = mitochondrial myopathy, encephalopathy, lactic acidosis, and stroke-like episodes; MERRF = myoclonic epilepsy with ragged red fibers.

***Leber's Hereditary Optic Neuropathy (LHON).*** LHON is one of the most common inherited causes of blindness in young males younger than the age of 25 years. The disease is maternally inherited and shows increased penetrance in males (three to four times that seen in females with the same mutation). The blindness, which is of sudden onset and with a rapid progression, is due to degeneration of the optic nerve. Several mutations in mitochondrial DNA that lead to LHON have been identified. The most common mutation (LHON 11778) is a missense mutation at base pair 11778 in a gene coding for the protein ND4, where the substitution of a guanine residue by an adenine residue leads to the conversion of an arginine to a histidine at position 340 in the protein. ND4 is one of the subunits found in complex I, which catalyzes the transfer of electrons from NADH to coenzyme Q. This complex, the largest of the five complexes involved in oxidative phosphorylation, is made up of more than 30 subunits and may have its normal activity modulated by the presence of the missense ND4 protein. Many individuals with this LHON mutation are homoplasmic; however, heteroplasmic individuals also exist. LHON appears to be subject not only to sex-related differences but also to different responses to environmental factors (e.g., smoking). The variability of the disease may be related to heteroplasmy in different individuals. Recovery of vision occurs in fewer than 5% of individuals whose blindness is due to the LHON 11778 mutation.

***Mitochondrial Myopathy, Encephalopathy, Lactic Acidosis, and Stroke-like Episodes (MELAS).*** MELAS is one of the mitochondrial encephalopathies that affect the nervous system, the eye, and some somatic organs. This is a progressive disease, with the first symptoms occurring between 5 and 15 years of age. Symptoms include ataxial tremors and myoclonic jerks. MELAS is distinguished from other similar diseases by the occurrence of stroke-like episodes, although the basis for the stroke symptoms is unknown. The disease may lead to cortical blindness and often dementia. Nearly 80% of individuals with MELAS are heteroplasmic, with the mutant mitochondrial DNA having a specific adenine-to-guanine substitution at base pair 3243 in the gene that codes for mitochondrial transfer RNA–leucine (tRNA-Leu) [see Figure 11-14]. The mutated tRNA-Leu may result in defective protein synthesis within the mitochondria, leading to a deficit in the necessary mitochondrial-coded protein subunits that carry out oxidative phosphorylation.

*Myoclonic Epilepsy with Ragged Red Fibers (MERRF).* Individuals with MERRF present with muscle fibers that have too many mitochondria. This abnormal accumulation of both normal and mutant mitochondria beneath the muscle cell membrane gives it a ragged edge. Staining of the muscle tissue using the Gomori trichrome method produces a red color. Symptoms of MERRF appear between 5 and 12 years of age and are similar to those seen with MELAS. However, stroke-like symptoms are not found in patients with MERRF. The disease has variable severity, is usually heteroplasmic, and is characterized by progressive myoclonic epilepsy. Patients commonly have seizures and develop dementia, hearing loss, and optic atrophy.

The most common mutation, present in 80%–90% of the cases, is due to a guanine-to-adenine substitution at base pair 8344 in the gene coding for transfer RNA–lysine (tRNA-Lys). This single base pair change in tRNA-Lys results in defects in complexes I and IV, although defects in complex III also occur. Often, individuals with tRNA-Lys–mutant mitochondria do not develop symptoms until childhood. Once symptoms develop, they become progressively worse with age. The severity of the symptoms is related to the ratio of mutant to normal mitochondria as well as to the organ effected.

*Nonsyndromic Deafness.* Nonsyndromic deafness shows a maternal inheritance pattern suggesting that the disease may be caused by mitochondrial mutations. This disease is characterized by an irreversible hearing loss that is associated with the use of aminoglycoside antibiotics (e.g., streptomycin, gentamicin, kanamycin). In the mitochondria, there are two ribosomal RNA (rRNA) genes that code for rRNA that is used in the synthesis of proteins in the mitochondria. The mitochondrial ribosomes are similar to bacterial ribosomes. The 70S bacterial ribosome, in the presence of aminoglycoside antibiotics, misreads mRNA molecules during translation. In addition, the aminoglycoside antibiotics bind to bacterial rRNA and interfere with translation. The similarity of the 12S rRNA made in the mitochondria to bacterial rRNA suggests that aminoglycoside antibiotics might have a similar effect on mitochondrial rRNA in cases in which a mutation is present in the rRNA gene. Indeed, a substitution of adenine to guanine at base pair 1555 in the 12S rRNA gene was shown to be maternally inherited and associated with nonsyndromic deafness. This mutation appears to be homoplasmic, with family members either affected or not affected. The position of the mutation in the rRNA gene occurs in a highly conserved region, which is a region also conserved in bacterial rRNA and known to bind aminoglycosides. The administration of aminoglycoside antibiotics and their accumulation in the ear may result in the binding of the antibiotic to the mutant rRNA, thus interfering with translation within the mitochondria and leading to a loss of mitochondrial function.

Other disorders are being identified that are maternally inherited and most likely caused by mutations occurring in the mitochondrial DNA. Knowledge of the complete base sequence of the mitochondrial DNA allows not only the association of a specific mutation with a disorder but also provides information to develop specific methods for their detection and ultimately may lead to ways of developing therapy for these disorders [21].

# RESOLUTION OF CLINICAL CASE

Lisa returned to her physician for a follow-up visit. The ophthalmologist's report stated that Lisa has multiple Lisch nodules but gave no specific diagnosis. There were no presenile posterior lenticular opacities (i.e., cataracts). The physician informed Lisa that she does indeed have NF1. Lisa immediately panicked and asked about the implications of the diagnosis to her and to her unborn child.

The physician assured Lisa that most cases of NF1 are relatively benign in nature, just as hers is. One of the most difficult problems she might face would be if the diagnosis of NF1 had been included on her ophthalmologist's report, which could have future implications with respect to obtaining insurance and even employment. Lisa asked if there are further tests that could confirm the diagnosis. The physician informed her that no other tests were needed because she met the minimum of two criteria: six café au lait spots and the presence of at least two Lisch nodules.

He further reassured Lisa that, most likely, her infant will be only mildly affected, but the recurrence risk to the infant is 50% because of the dominant inheritance pattern of the disease. The disease also shows a high mutation rate, making it difficult to use genetic testing for diagnosis because many of the mutations that occur are new ones and not easily detected by current methods. Furthermore, even if the child is definitively diagnosed as having NF1, there is no way to predict the clinical severity of the disease. Possible conditions that might be seen in the infant include the café au lait spots and the Lisch nodules, just as Lisa has. Additionally, changes in vision and hearing or some scoliosis might develop, as well as benign growths that occur along the nerves (neuro-fibromas). Unfortunately, there is no way to determine if the child will be more severely affected, less severely affected, or affected the same as Lisa.

Six months later Lisa returned to the physician with her infant. After carefully examining the infant girl, the physician found no café au lait spots or any other signs that might indicate the child had inherited NF1 from the mother. After hearing the good news, Lisa was overjoyed. However, the physician repeated what he had told her previously. NF1 is a dominantly inherited disease, and each child she has in the future will have a 50% chance of inheriting the condition.

# REVIEW QUESTIONS

**Directions:** For each of the following questions, choose the **one best** answer.

1. The biochemical causes of both myoclonic epilepsy with ragged red fibers (MERRF) and mitochondrial myopathy, encephalopathy, lactic acidosis, and stroke-like episodes (MELAS) are most often which one of the following?

  **(A)** Defects in a single nuclear gene/protein, resulting in abnormal mitochondrial function

  **(B)** Deletions of mitochondrial DNA, resulting in defects in multiple mitochondrial protein complexes

  **(C)** Mutations in a mitochondrial transfer RNA gene, resulting in defects in multiple mitochondrial protein complexes

  **(D)** Mutations in a single mitochondrial structural gene affecting a single mitochondrial protein complex

  **(E)** Mutations in a mitochondrial ribosomal RNA gene, resulting in defects of multiple mitochondrial protein complexes

**Directions:** The groups of questions below consist of lettered choices followed by several numbered items. For each numbered item, select the appropriate lettered option with which it is most closely associated. Each lettered option may be used once, more than once, or not at all.

**Questions 2–5**

Match each of the following definitions to the term it describes.

  **(A)** Autosomal recessive inheritance

  **(B)** Autosomal dominant inheritance

  **(C)** X-linked inheritance

  **(D)** Mitochondrial inheritance (maternally inherited)

2. This term describes the inheritance pattern for cystic fibrosis

3. This term describes the inheritance pattern for Duchenne's muscular dystrophy

4. This term describes the inheritance pattern for Leber's hereditary optic neuropathy

5. This term describes the inheritance pattern for Huntington's disease

**Questions 6–9**

Match each of the following conditions with the biochemical abnormality that characterizes it.

(A) Change or gain of function

(B) Loss of function

(C) Haploinsufficiency

(D) Dominant negative

(E) Single cell somatic loss of function

6. Retinoblastoma

7. The dominantly inherited form of familial hypercholesterolemia

8. Type II osteogenesis imperfecta

9. Tay-Sachs disease

# ANSWERS AND EXPLANATIONS

**1. The answer is C.** Both myoclonic epilepsy with ragged red fibers (MERRF) and mitochondrial myopathy, encephalopathy, lactic acidosis, and stroke-like episodes (MELAS) are caused by mutations in transfer RNA (tRNA) genes. MERRF is caused by a mutation in the gene for tRNA-lysine, and MELAS is caused by a mutation in the gene for tRNA-leucine. Other known mitochondrial diseases are Leber's hereditary optic neuropathy, which is caused by a mutation in the *ND4* structural gene that leads to a defect in oxidative phosphorylation complex I; Kearns-Sayre syndrome, which is caused by a deletion mutation in the mitochondrial DNA; and deafness, which can be caused by a mutation in the 12S RNA gene.

**2–5. The answers are: 2-A, 3-C, 4-D, 5-B.** Cystic fibrosis results from a defect in the cystic fibrosis transmembrane conductance regulator protein and is inherited as an autosomal recessive disorder in which the homozygous patient receives one defective gene from the mother and one defective gene from the father.

The gene for Duchenne's muscular dystrophy is located on the X chromosome and is expressed predominantly in males and is expressed only rarely in females. A female with one defective gene is a carrier of the disease and passes it on to 50% of her children. Any male who receives the defective allele will have the disease.

Leber's hereditary optic neuropathy is the result of a missense mutation in a mitochondrial DNA gene that codes for a protein subunit. Because children receive their mitochondria almost exclusively from their mother, the disease is maternally inherited.

Huntington's disease is one of the trinucleotide expansion disorders and is inherited as an autosomal dominant disorder.

**6–9. The answers are: 6-E, 7-C, 8-D, 9-B.** Retinoblastoma results from a defect in pRB, a tumor suppressor protein involved in transcriptional regulation. A single defective allele is inherited, but the disease develops only after a second mutation occurs in the other allele in a somatic cell.

Familial hypercholesterolemia, a defect in the low-density lipoprotein (LDL) receptor, is a haploinsufficiency disease expressed in the heterozygote. Heterozygotes express only one-half of the amount of the LDL receptor, and half is not enough to prevent clinical symptoms.

Type II osteogenesis imperfecta results from the production of a missense $\alpha_1$ chain that has a dominant negative effect on the collagen molecule, which is made up of two $\alpha_1$

chains and one $\alpha_2$ chain. The presence of a mutant $\alpha_1$ chain in the molecule destroys the function of the oligomer and thus is dominant to the normal.

Tay-Sachs disease is an autosomally inherited disorder that is caused by the loss of function of both alleles of the gene that codes for hexosaminidase A.

# REFERENCES

1. Welsh MJ, Tsui L-C, Boat TF, et al: Cystic fibrosis. In *The Metabolic and Molecular Bases of Inherited Disease*, 7th ed. Edited by Scriver CR, Beaudet AL, Sly WS, et al. New York, NY: McGraw-Hill, 1995, pp 3799–3876.

2. Tsui L-C, Buchwald M: Biochemical and molecular genetics of cystic fibrosis. In *Advances in Human Genetics, vol 20*. Edited by Harris H, Hirschhorn K. New York, NY: Plenum Press, 1991, pp 153–266.

3. Gravel RA, Clarke J, Kaback MM, et al: The GM$_2$ gangliosidoses. In *The Metabolic and Molecular Bases of Inherited Disease*, 7th ed. Edited by Scriver CR, Beaudet AL, Sly WS, et al. New York, NY: McGraw-Hill, 1995, pp 2839–2879.

4. Grabowski GA: Gaucher disease: enzymology, genetics and treatment. In *Advances in Human Genetics*, vol 21. Edited by Harris H. Hirschhorn K. New York, NY: Plenum Press, 1992, pp 377–441.

5. Beutler E: Gaucher disease. In *Advances in Genetics*, vol 32. Edited by Hall JC, Dunlap JC. New York, NY: Academic Press, 1995, pp 17–49.

6. Scriver CR, Kaufman S, Eisensmith R, et al. The hyperphenylalaninemias. In *The Metabolic and Molecular Bases of Inherited Disease*, 7th ed. Edited by Scriver CR, Beaudet AL, Sly WS, et al. New York, NY: McGraw-Hill, 1995, pp 1015–1075.

7. Eisensmith RC, Woo S: Molecular genetics of phenylketonuria. In *Advances in Genetics*, vol 32. Edited by Hall JC, Dunlap JC. New York, NY: Academic Press, 1995, pp 199–271.

8. The Huntington's Disease Collaborative Research Group: A novel gene containing a trinucleotide repeat that is expanded and unstable on Huntington's disease chromosomes. *Cell* 72:971–983, 1993.

9. Harper P: Myotonic dystrophy and other autosomal muscular dystrophies. In *The Metabolic and Molecular Bases of Inherited Disease*, 7th ed. Edited by Scriver CR, Beaudet AL, Sly WS, et al. New York, NY: McGraw-Hill, 1995, pp 4227–4251.

10. Wang J, Pegoraro E, Menegazzo E, et al: Myotonic dystrophy: evidence for a possible dominant-negative RNA mutation. *Hum Molec Gen* 4:599–606, 1995.

11. Nicholls AC: Collagen disorders. In *The Inherited Metabolic Diseases*. Edited by Holton J. London, UK: Churchill Livingstone, 1994, pp 379–420.

12. Byers PH: Disorders of collagen biosynthesis and structure. In *The Metabolic and Molecular Bases of Inherited Disease*, 7th ed. Edited by Scriver CR, Beaudet AL, Sly WS, et al. New York, NY: McGraw-Hill, 1995, pp 4029–4077.

13. Goldstein J, Hobbs H, Brown M: Familial hypercholesterolemia. In *The Metabolic and Molecular Bases of Inherited Disease*, 7th ed. Edited by Scriver CR, Beaudet AL, Sly WS, et al. New York, NY: McGraw-Hill, 1995, pp 1981–2030.

14. Hobbs HH, Russell DW, Brown M, et al: The LDL receptor locus and familial hypercholesterolemia. *Ann Rev Genetics*, 24:133–170, 1990.

15. Knudson AG: Mutation and cancer: a statistical study of retinoblastoma. *Proc Natl Acad Sci* 68:820–823, 1971.

16. Kahn P: Coming to grips with genes and risk. *Science* 274:496–498, 1996.

17. Roa B, Boyd A, Volcik D, et al. Ashkenazi Jewish population frequencies for common mutations in BRCA1 and BRCA2. *Nat Genet* 14:185–187, 1996.

18. Wallace M, Collins FS: Molecular genetics of von Recklinghausen neurofibromatosis. In *Advances in Human Genetics*, vol 20. Edited by Harris H, Hirschhorn K. New York, NY: Plenum Press, 1991, pp 267–307.

19. Shoffner J, Wallace D: Oxidative phosphorylation diseases. In *The Metabolic and Molecular Bases of Inherited Disease*, 7th ed. Edited by Scriver CR, Beaudet AL, Sly WS, et al. New York, NY: McGraw-Hill, 1995, pp 1535–1609.

20. Bindoff L, Turnbull DM: Defects of the mitochondrial respiratory chain. In *The Inherited Metabolic Diseases*. Edited by Holton J. London, UK: Churchill Livingstone, 1994, pp 265–295.
21. Johns D: The other human genome: mitochondrial DNA and disease. *Nat Med* 2:1065–1068, 1996.

# 12 SYNDROMES AND DYSMORPHOLOGY

Kenneth L. Garver, M.D., Ph.D.

## CHAPTER OUTLINE

## INTRODUCTION OF CLINICAL CASE

While in college, Mary O'Malley met a young man, Edmund Casey. They began dating and decided to marry after Edmund's graduation. During their premarital examination, their family physician obtained a medical history and a family history as part of his evaluation. Because of the good health of both Edmund and Mary and because of a completely negative history, their family physician assured them that they would have the slight risk that everyone has of having a child with a serious birth defect, which is about 3%.

After 2 years of marriage, Mary became pregnant; the obstetrician verified the pregnancy at 10 weeks and reported that Mary seemed to be in good health. After an uneventful pregnancy, the couple had a healthy infant girl. Eighteen months later, Mary was again pregnant, and she made a routine appointment with her obstetrician at about 8 weeks. The obstetrician indicated that everything appeared to be normal and asked Mary if she was interested in any prenatal diagnostic tests, particularly a triple screen. The obstetrician indicated that the triple screen would help to determine whether she had an increased risk of having an infant with several birth defects, including the open neural tube defects (ONTDs) [i.e., anencephaly, spina bifida cystica] and Down's syndrome. Because of religious beliefs, Mary decided against prenatal screening. Furthermore, with all other factors being normal, she believed that her chances of having an infant with a birth defect were slim. Her obstetrician concurred with this. However, at 4 months, a routine ultrasound to determine the growth of the fetus revealed a suspicious finding in the lumbar area. A diagnostic sonogram was then performed, which revealed a small cystic mass. The couple was told that their infant had spina bifida cystica.

The obstetrician described spina bifida cystica as being part of an ONTD, a common but serious birth defect. In most areas of the United States, the incidence of ONTDs (i.e., anencephaly, spina bifida cystica) is a little over 2 per 1000 per year, with a birth prevalence for anencephaly of 1 in 1000 and for spina bifida cystica, a little lower than 1 in 1000. The obstetrician explained that during embryogenesis, the neural groove completes its closure on the 25th day of gestation. If the head (cephalic) end of the embryo fails to close, the infant is born with anencephaly (Figure 12-1). If the lower end of the neural groove fails to close, the infant is born with spina bifida cystica, which can involve the cervical, thoracic, or lumbar areas (Figures 12-2 and 12-3).

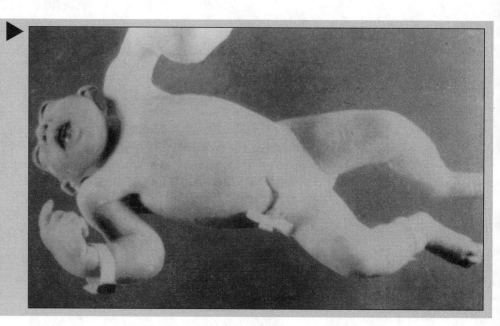

**FIGURE 12-1**

**Newborn Infant with Anencephaly.** *This infant was born alive, although the entire bony top of the head was missing. The infant's brain had been exposed to the amniotic fluid and to the pressure of the uterine wall, so that it was very atrophied. The ears are low placed and very prominent. The infant died within 1 hour of birth. (Courtesy of Dr. Leland A. Albright, Director of Neurosurgery, Children's Hospital of Pittsburgh and The University of Pittsburgh, Pittsburgh, Pennsylvania.)*

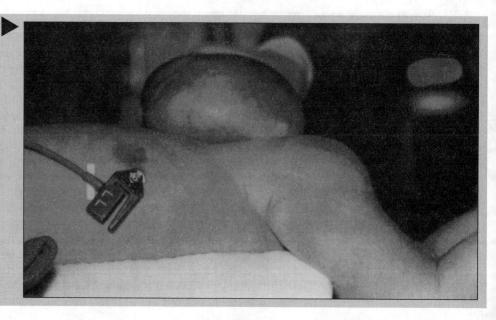

**FIGURE 12-2**

**Newborn Infant with a Closed Spina Bifida Cystica (Meningocele or Myelomeningocele).** *This infant was born with no paralyses. The infant could move its arms and legs; as it grew, it had control of both bowel and bladder. Infants who are born with a covered spina bifida cystica usually have less damage to their spinal cord. (Courtesy of Dr. Leland A. Albright, Director of Neurosurgery, Children's Hospital of Pittsburgh and The University of Pittsburgh, Pittsburgh, Pennsylvania.)*

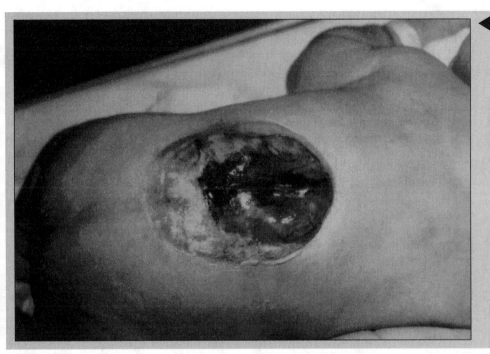

**FIGURE 12-3**
**Newborn Infant with an Open Spina Bifida Cystica.** *An infant who is born with an open spina bifida cystica has a greater chance of damage to the spinal cord and to the spinal nerves and also a much greater chance of having a serious infection of the meninges. This infant did have some paralysis of the legs but otherwise had an uncomplicated postpartum period. (Courtesy of Dr. Leland A. Albright, Director of Neurosurgery, Children's Hospital of Pittsburgh and The University of Pittsburgh, Pittsburgh, Pennsylvania.)*

# SYNDROMES AND DYSMORPHOLOGY

## History of Congenital Malformations

From the earliest recorded history, people of all countries and cultures have been interested in the cause and significance of congenital malformations in infants, children, and adults. One of the first recorded instances of congenital malformations was at least 2500 years ago in the ancient city-state of Sparta. Shortly after birth, newborn children who exhibited any weakness or congenital malformation were brought before the elders of the city, who decided whether the child would live or be killed. If the child was defective or weak, it was taken to the mountains and either left exposed to the elements and animals or thrown into a deep ravine to be killed instantly. This is an early example of negative eugenics.

The most ancient sculpture depicting a birth defect (i.e., a double-headed twin goddess) was discovered in southern Turkey and dates from approximately 6500 B.C. [1, 2]. In Egypt, mummies have been discovered with achondroplasia, clubfeet, and cleft palate. Ancient Egyptian statuettes show figures with achondroplastic proportions and facial features and also achondroplastic proportions with normal facial features. Ancient Peruvian pottery depicts individuals with cleft lip and limb reduction defects. Statuettes showing double-headed human beings have been found in pre-Columbian graves and tombs in Mexico and Central America that date from 500 B.C. to 800 A.D.

In 1963, Dr. R. B. Greenblat described what he believes to be the first written description of congenital adrenal hyperplasia, which is described in the Book of Genesis in the Old Testament of the Bible [3], that is, the story of the twins, Esau and Jacob, who were born to Isaac. Because Esau was the first born, he inherited the wealth of his father. Esau is described as having red skin that was hairy and as being a very muscular man and a great hunter. He also had episodes of severe faintness and weakness, so much so that he was afraid that he might die during one of these spells. Dr. Greenblat pointed out that the physical features are typical of a child with adrenal hyperplasia and that it has been recognized in the 20th century that hypoglycemia is an important part of this genetic disease. It is reasonable to speculate that Esau renounced his birthright for a bowl of porridge because of his genetically induced hypoglycemia.

## Principles of Dysmorphology

Dysmorphology is a word coined by Dr. D. W. Smith in 1966 to describe the study of human congenital defects—that is, abnormalities of body structure that originate before birth [4]. Malformations are morphologic defects of an organ, part of an organ, or a larger region of the body that result from an intrinsically abnormal developmental process. Congenital malformations are anatomic or structural abnormalities that are present at birth, although they may not be diagnosed until later. A syndrome is a pattern of multiple anomalies thought to be pathogenetically related and not known to represent a single sequence or a polytopic field defect.

> *Malformations* are morphologic defects of an organ, part of an organ, or a larger region of the body resulting from an intrinsically abnormal developmental process.

Malformations (dysmorphic anomalies) can occur in any part of the body, and most arise during the first 3 months of intrauterine life. Some of these anomalies are negligible or have only cosmetic significance, but approximately 3% of all children are born with a serious structural defect that interferes with normal body function and can lead to a life-long handicap or even early death. There are many different types of birth defects, and each is individually rare. However, together they account for a disproportionately large fraction of childhood morbidity and mortality. Several studies have indicated that approximately one-third of pediatric inpatients are hospitalized because of congenital abnormalities.

> A *major malformation* is one that may be lethal or of serious surgical, medical, or cosmetic importance to the patient.

Congenital malformations are usually divided into major and minor. A *major malformation* is one that may be lethal or of serious surgical, medical, or cosmetic importance to the patient. Examples of major malformations are bilateral cleft lip and palate, anencephaly, spina bifida cystica, and transposition of the great vessels. A *minor malformation* (more commonly called minor anomaly) is a minor physical defect that occurs in fewer than 4% of patients. Examples of minor malformations are single palmar crease, hypoplastic nails, inner epicanthal folds, bifid thumbs (Figure 12-4), edema of the extremities (Figure 12-5), low-set ears, and syndactyly of the second and third toes. If the malformation occurs in more than 4% of patients, the term "phenotypic variant" is used. Anyone who examines infants, children, or adults should be aware of the difference between minor malformations and major malformations. Minor malformations (minor anomalies) may serve as indicators of altered morphogenesis in a general sense and may aid in the diagnosis of a specific syndrome. The suspicions of altered morphogenesis may lead to a search for occult major anomalies that may be ameliorated by treatment. Furthermore, minor anomalies may provide valuable information regarding the time of onset of a more serious problem. Fewer than 3% of all newborns have more than three minor anomalies, and depending upon the series, 20%–90% of these also have a major malformation.

> A *minor malformation* (more commonly called minor anomaly) is a minor physical defect that occurs in fewer than 4% of patients.

> The suspicions of altered morphogenesis may lead to a search for occult major anomalies that may be ameliorated by treatment.

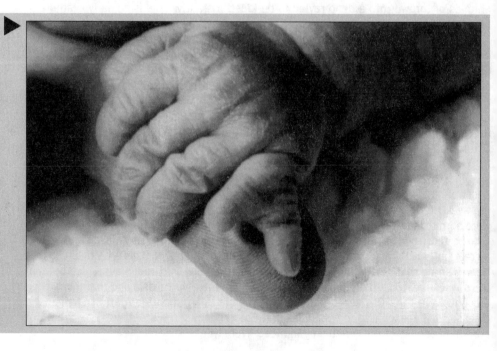

**FIGURE 12-4**
**Hand of an Infant Showing Polydactyly due to a Bifid Thumb.** *Polydactyly and bifid fingers are two of several hundred minor malformations that can occur in a child with multiple malformations. This child's major and minor malformations were caused by trisomy 13.*

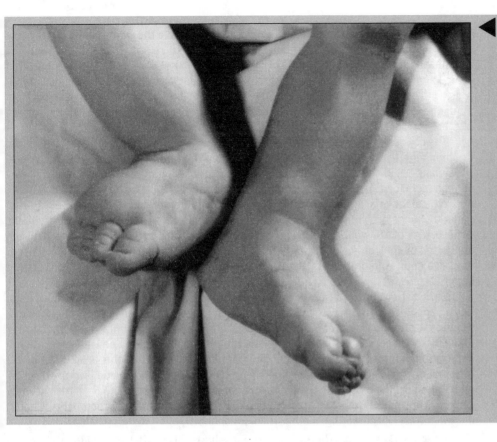

**FIGURE 12-5**
*Feet of an Infant with Marked Edema. There are many causes for edema in infants. The cause of this infant's minor malformation, the edema, was having only one X chromosome (a 45,X karyotype).*

To understand dysmorphology and syndromes better, other concepts are important. A *disruption* is a morphologic defect of an organ, part of an organ, or a larger region of the body that results from the extrinsic breakdown of, or an interference with, an originally normal developmental process. A *deformation* is an abnormal form, shape, or position of part of the body caused by mechanical forces. A *dysplasia* is an abnormal organization of cells into tissues and the morphologic results. Thus, it involves a process of histiogenesis. A *polytropic field defect* is a pattern of anomalies derived from the disturbance of a single developmental field [5, 6].

## Counseling for Syndromes and Congenital Malformations

*Human genetics* is the scientific study of human variation and heredity, *medical genetics* is the study of the hereditary nature of human disease, and *clinical genetics* involves the care, diagnosis, and counseling of patients with congenital malformations or genetic diseases. During the past 40 years, there has been an explosion of new information in medical genetics, which has been applied to clinical genetics in various ways. In the 1950s, population and quantitative genetics primarily involved public health physicians. After a 1956 publication described an easily reproducible technique for examining human chromosomes, there was much new information delineated concerning the cause of birth defects and mental retardation [7]. In the 1970s, cytogenetics became more sophisticated because of the introduction of banding techniques. Somatic cell and biochemical genetics were also developed at this time. During the 1980s, the introduction of molecular probes made the science of cytogenetics even more exact. In the 1990s and into the 21st century, the use of molecular genetic techniques to aid in diagnosing and treating patients will continue.

An important part of clinical genetics is speaking to patients regarding the genetic information that is available. This has been best defined by the Committee on Genetic Counseling of The American Society of Human Genetics, which wrote, "Genetic counseling is a communication process that deals with the human problems associated with the occurrence, or the risk of occurrence, of a genetic disorder in a family. This process involves an attempt by one or more appropriately trained persons to help the individual or family to: (1) comprehend the medical facts, including the diagnosis, probable course of

*Human genetics is the scientific study of human variation and heredity.*
*Medical genetics is the study of the hereditary nature of human disease.*
*Clinical genetics deals with the care, diagnosis, and counseling of patients with congenital malformations or genetic diseases.*

*An important part of clinical genetics is speaking to patients regarding the genetic information that is available.*

*It is important that the primary care physician has available an accurate diagnosis and has an understanding of the medical management, the risk determination, the risk options, the reproductive options, and the professional and lay support services available in the community.*

*The patients involved in genetic counseling may be presented with very serious moral or ethical questions, and the decision should be based on the patient's moral and ethical beliefs, not those of the counselor.*

*Retrospective genetic counseling is given after an individual develops a genetic disease or after a couple has a child with a particular birth defect or genetic disease.*

*Prospective genetic counseling is anticipatory in that an individual or couple is identified as being at increased risk for developing a genetic disease or birth defect, or of having a child with a similar problem, before its occurrence.*

the disorder, and the available management; (2) appreciate the way heredity contributes to the disorder and the risk of recurrence in specified relatives; (3) understand the alternatives for dealing with the risk of recurrence; (4) choose a course of action that seems appropriate in view of the risk, family goals, and the ethical and religious standards, and to act in accordance with that decision; and (5) make the best possible adjustment to the disorder in an affected family member and/or to the risk of recurrence of that disorder" [8].

The counseling of patients who are at risk for a syndrome, congenital malformation, or other genetic disease is radically different from the traditional counseling by general physicians [9]. The patient involved in genetic counseling may be presented with very serious moral or ethical questions, and decisions should be based on the patient's moral and ethical beliefs, not those of the counselor. Therefore, most clinical geneticists, genetic counselors, and primary physicians who counsel for clinical genetic problems believe strongly in *nondirective counseling*, in which the patient makes the decision supported by the counselor, his or her family, or a religious person. This is in contrast to directive counseling, in which the physician tells the patient what his or her decision should be.

Each patient has a different moral, ethical, and ethnic background, as does each physician. Each patient has a different need to have his or her own children, and it requires a strong commitment to care for a child with a congenital malformation or another type of genetic disease. Physicians should not interject their own biases into the decision-making of their patient. Genetic information given to patients is only part of the information they need for decision-making. In many cases, the decision is a very difficult one. Patients should be given time to talk privately with each other and, in some cases, to seek outside help. This may involve speaking to respected family members, friends, clergy, or possibly obtaining another medical opinion.

*Retrospective Counseling.* Retrospective counseling is given after an individual develops a genetic disease or after a couple has a child with a particular birth defect or genetic disease. Until recently, most genetic counseling was retrospective. Currently, more at-risk situations can be identified before conception, and the individual or couple can be given prospective counseling.

*Prospective Counseling.* Prospective genetic counseling is anticipatory in that an individual or couple is identified as being at increased risk for developing a genetic disease or birth defect or for having a child with a similar problem. Examples of prospective genetic counseling include counseling a couple in which the father or the mother is older or in which one member of the couple is at risk for having a particular genetic disease that is more prevalent in their ethnic or racial groups (e.g., whites, cystic fibrosis; blacks, sickle cell disease; Ashkenazi Jews, Tay-Sachs disease; Mediterraneans, β-thalassemia; Asians, α-thalassemia) [9].

Other examples suitable for prospective genetic counseling include a woman who is planning a pregnancy and is taking a medication that is potentially teratogenic, a woman who had been successfully treated for phenylketonuria (PKU) but who has reverted to a normal diet and now has high blood phenylalanine levels, a woman who has insulin-dependent diabetes mellitus, or a woman who has no history of rubella and has a nonimmune rubella titer. In all of the above cases, judicious advice to the patient and prospective treatment could increase greatly the chance of the couple having a normal child.

# ETIOLOGY OF SYNDROMES AND CONGENITAL MALFORMATIONS

When determining the cause of a syndrome or congenital malformation, it is important to take a complete history. This should include an extensive family history to see if any other member on either side of the family has problems similar to the patient and also to evaluate carefully the pregnancy history of the mother of the patient. Many syndromes and birth defects superficially resemble each other. To give adequate counseling and risk assessment, not only to the patient but also to members of his or her family, it is necessary to make an accurate diagnosis. In addition to a thorough history and physical examination, many other techniques such as a cytogenetic study (possibly using molecular probes), biochemical studies, biopsy, and, in some cases, radiographs, computerized axial tomography (CAT) scans, and sonographs can be helpful. The etiology of syndromes and congenital malformations can be considered under five headings: environmental, chromosomal, mendelian, multifactorial, and sporadic.

## Environmental Causes

Table 12-1 outlines some environmental conditions that cause birth defects and syndromes when the fetus is exposed in utero to these agents. Environmental factors can be divided into the following categories: medications, drugs, infectious agents, deformations, maternal disease, and other factors, which include radiation, maternal hyperthermia, and disruptions (early amnion rupture sequence).

> *Many syndromes and birth defects superficially resemble each other. To give adequate counseling and risk assessment, not only to the patient but also to members of his or her family, it is necessary to make an accurate diagnosis.*

> **Etiology of Syndromes and Congenital Malformations**
> *Environmental*
> *Chromosomal*
> *Mendelian*
> *Multifactorial*
> *Sporadic*

◀ **TABLE 12-1**
*Examples of Environmental Causes of Birth Defects*

| Medications | Drugs | Infectious Agents | Deformations | Maternal Disease | Other |
|---|---|---|---|---|---|
| Dilantin | Alcohol | Rubella | Equinovarus | Phenylketonuria | Radiation |
| Retinoic acid | | Toxoplasmosis | Facial asymmetry | | Maternal hyperthermia |
| Warfarin | | Cytomegalovirus | | | Amniotic band sequence (early amnion rupture sequence) |
| | | Syphilis | | | |

***Medications.*** Fetal Dilantin syndrome is caused by the ingestion of *phenytoin* (*Dilantin*) during the first trimester of pregnancy. Physical features in newborns include mild-to-moderate prenatal growth deficiency, mild mental retardation, ocular hypertelorism, depressed nasal bridge, hypoplasia of the distal phalanges with small nails, low-set hairline, and abnormal palmar crease.

*Retinoic acid* (Accutane) is a very effective treatment of acne. However, when used by a pregnant woman, retinoic acid can cause serious congenital defects in the exposed fetus, including small or missing ears, narrow sloping forehead, various forms of congenital heart disease, hydrocephalus, microcephaly, and thymic abnormalities.

Fetal warfarin syndrome is seen in about one-third of infants whose mothers took *Coumadin* (*warfarin sodium*) any time during pregnancy. The marked features of this syndrome are severe nail hypoplasia and stippled epiphyses, particularly of the axial skeletal, the proximal femora, and the calcanei. There is occasionally severe mental retardation.

***Drugs.*** Fetal alcohol syndrome is probably the major cause of mental retardation in children, and it is completely preventable by prospective counseling of women before they become pregnant. Other findings in this syndrome include prenatal and postnatal growth retardation; microcephaly; short palpebral fissures; short nose; smooth philtrum with thin, smooth upper lip; occasional congenital heart disease; and a small fifth finger and nail. The harmful effects of maternal alcohol ingestion extends throughout the entire pregnancy.

***Infectious Agents.*** Fetal rubella syndrome is caused when the mother develops rubella during the first trimester of pregnancy and the fetus is exposed to the virus. The defects include deafness, cataracts, glaucoma, congenital heart defects, microcephaly, and mental retardation. When the infection occurs later in pregnancy, the defects include purpura, hepatitis, splenomegaly, myocarditis, interstitial pneumonitis, and encephalitis. The fetal rubella syndrome is also preventable by immunizing any woman before

> *The **fetal alcohol syndrome** is probably the major cause of mental retardation in children and is completely preventable by prospective counseling of women before they become pregnant not to ingest alcohol.*

pregnancy who does not have a definite history of rubella or who has a nonimmune titer when tested.

Congenital toxoplasmosis results from transplacental transmission of *Toxoplasma gondii* during pregnancy. The manifestations in the fetus include hydrocephalus or microcephaly, mental retardation, convulsions, chorioretinitis, deafness, psychomotor disturbances, and cerebral calcifications that are diffused throughout the cortex.

The infection in infants by cytomegalovirus is termed cytomegalovirus inclusion disease (CID) and is characterized by hepatosplenomegaly, hyperbilirubinemia, thrombocytopenia with petecchiae or purpura, and variable involvement of the central nervous system (CNS), including cerebral calcifications, microcephaly, chorioretinitis, deafness, and psychomotor retardation.

Syphilis in a pregnant woman can affect the unborn child and can cause many manifestations in any organ or tissue of the fetus. Characteristic later manifestations include saddle nose, Hutchinson teeth, mulberry molars, saber shin, and various CNS abnormalities.

***Deformations.*** The most common cause of equinovarus (clubfoot) is an extrinsic constraint, such as a lack of or deficiency of amniotic fluid during pregnancy. This results in the walls of the uterus exerting increased pressure on the fetus, which causes transient malformations. Equinovarus can also be a manifestation of neuromuscular disorders, limb deficiency, or a skeletal dysplasia.

With a face and brow deformation, the face is the compressed presenting part. The growth of the mandible and nose is deficient and, as a consequence, the growth of the mandible and nose may be restrained. Although these children have marked facial features at birth, the parents can be reassured that the infant is normal and that the face will grow into normal proportions.

***Maternal Disease.*** Women who were successfully treated during infancy for phenylketonuria (PKU) usually revert to a normal diet during childhood. This causes them to have elevated blood phenylalanine levels, which are then transmitted to their fetuses in utero and cause the infants to be born with growth deficiency, mental retardation, microcephaly, epicanthal folds, and strabismus. It is important to instruct women who have been successfully treated for PKU either to continue their low-phenylalanine diet or to revert to a low-phenylalanine diet prior to pregnancy.

> *It is important for females who have been successfully treated for **PKU** either to continue their low-phenylalanine diet into adulthood or to revert back to low-phenylalanine diets prior to pregnancy.*

***Radiation.*** Radiation exposure can be categorized as diagnostic, therapeutic, or occupational. Possible effects from radiation prior to and during pregnancy should be considered when evaluating the cause of congenital malformations.

***Maternal Hyperthermia.*** There is growing evidence from both animal and human studies that hyperthermia during the first trimester not only causes an increased risk of congenital malformations in the fetus but also leads to an increase in spontaneous abortion, stillbirth, and prematurity.

***Amniotic Band Sequence (Early Amnion Rupture Sequence).*** The malformations vary depending on the timing of the rupture of the amnion. Early in pregnancy, anencephaly and facial clefting may be seen. Later in pregnancy, limb reduction, polydactyly, and syndactyly may be seen, usually before 12 weeks (Figure 12-6). This is a sporadic event, and the family can be counseled that it is extremely unlikely to recur.

## Chromosomal Causes

Some of the more common chromosomal abnormalities that result in specific syndromes can be subdivided under autosomal, X-chromosome, and microdeletion syndromes (Table 12-2).

### AUTOSOMAL CHROMOSOME SYNDROMES

***Down's Syndrome (Trisomy 21; 47,XX or XY,+21).*** Approximately 92% of persons with Down's syndrome have the primary trisomy type, and approximately 4% have a translocation type in which the extra chromosome 21 is joined to another chromosome. The characteristics that are frequently found in affected infants and children include hypotonia, flat facial profile, depressed nasal bridge, upward-slanting palpebral fissures, epicanthal folds, prominent tongue, extra skin at the nape of the neck, single transverse palmar crease, and increased distance between the first and second toes (Figures 12-7

TABLE 12-2
Examples of Chromosomal Causes of Birth Defects

| Chromosomal Cause | Associated Birth Defects |
|---|---|
| **Autosomal chromosome syndromes** | |
| Down's syndrome (trisomy 21) | Hypotonia; flat face; upward-slanting palpebral fissures; Brushfield's spots; bilateral epicanthal folds; flat nasal bridge; extra skin at nape of neck; broad, short hands; mental retardation; and frequently malformations of all major organs. |
| Patau's syndrome (trisomy 13) | Usually more serious and obvious defects including cleft lip or palate, micrognathia, microthalmia, coloboma, simian crease, flexion of fingers with or without overlapping, polydactyly of hands and sometimes feet, and serious malformations of internal organs (particularly heart and CNS). Many infants born with trisomy 13 die within hours or days of birth, and 80% die within the first month of life. The few infants who survive to 6 months have severe mental and physical defects. |
| Cri-du-chat syndrome (5p−) | The cat-like cry is almost diagnostic. Defects include microcephaly; hypertelorism; round face; wide, flat nasal bridge; micrognathia; hypotonia; and mental retardation. |
| **X-chromosome syndromes** | |
| Turner's syndrome (45,X) | Short stature, lymphedema over dorsum of fingers and toes, broad chest with widely spaced nipples, prominent ears, webbed neck with low posterior hairline, cubitus valgus of elbow, and sometimes coarctation of aorta. These children lack breast development, have amenorrhea and mild mental retardation, and are sterile. |
| Klinefelter's syndrome (47,XXY) | Normal appearance as an infant; long limbs; at adolescence, a small penis and testicles; gynecomastia in 40% during adolescence; and usually sterile. Fifteen percent have an IQ lower than 80. |
| **Microdeletion syndromes** | |
| Prader-Willi syndrome, del(15q11–q13) | Hypotonia in infancy, obesity, polyphagia, small hands and feet, short stature, almond-shaped eyes, hypogonadism, and mild mental retardation. |
| Angelman's syndrome, del(15q11–q13) | Severe mental retardation, hyperactivity, seizures, ataxia, hand flapping, absence of speech, and paroxysms of inappropriate laughter. |
| DiGeorge syndrome, del(22q11.2) | Hypertelorism, short philtrum, downward-slanting palpebral fissures, aortic arch anomalies, hypoplasia-to-aplasia of thymus (allowing severe infectious disease), and hypoplasia-to-absence of parathyroid gland (results in severe hypocalcemia and seizures in early infancy). |
| Miller-Dieker syndrome, del(17p13.3) | Microcephaly with bitemporal narrowing, variable high forehead, vertical ridging, small nose with anteverted nostrils, upward-slanting palpebral fissures, and protuberant upper lip. The most marked feature is incomplete development of the brain, often with a smooth surface (lissencephaly). |

FIGURE 12-6 ▶

**Lower Extremities of a Stillborn Infant with the Amniotic Band Sequence.** *The amniotic band sequence is usually a sporadic event that can mimic other birth defects, depending on the location of the amniotic bands. The minor defects in this infant were mostly in the extremities, although some children with the amniotic band sequence can have cleft lip and other clefting abnormalities of the face.*

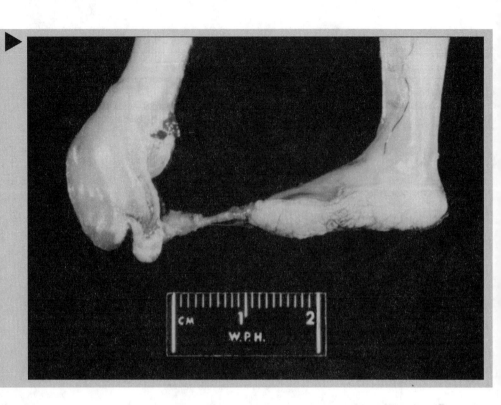

FIGURE 12-7 ▶

**Frontal View of an Infant's Face with Down's Syndrome.** *There are more than 100 major and minor malformations that occur more frequently in infants with Down's syndrome than the general population. This infant has an increased distance between the eyes, which are slanting upward and outward. The tongue seems to be more prominent than usual.*

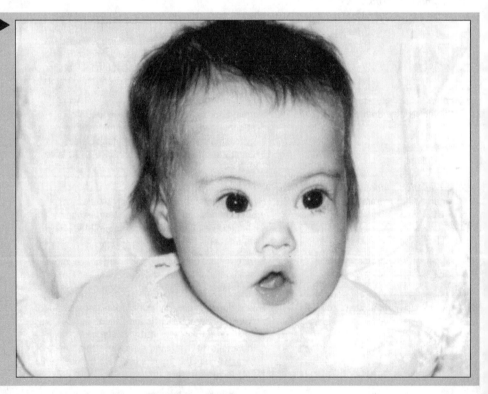

and 12-8). Approximately 40% of children with Down's syndrome have congenital heart disease, and all affected children are, to some extent, mentally retarded.

**Patau's Syndrome (Trisomy 13).** Approximately 90% of children born with this syndrome die within the first year, and most affected children die by the end of the second year. These children have severe mental defects, seizures, and severe growth retardation. They are microcephalic with sloping foreheads, small eyes, and cutaneous scalp defects. Often they have polydactyly and characteristic clenching of the fists and either unilateral or bilateral clefting of the lips. Approximately 80% of children with Patau's syndrome have congenital heart defects.

*Cri-du-Chat Syndrome (5p−).* As the name implies, this syndrome is characterized by a shrill cat-like cry during infancy, which disappears within a few weeks of birth. These infants are characterized by low birth weight and slow growth. The affected infant's face is characterized by microcephaly, a round face, hypertelorism, downward-slanting eyes, strabismus, a broad nasal base, and low-set ears. There is usually severe mental retardation, hypotonia, congenital heart defects, and a single palmar crease.

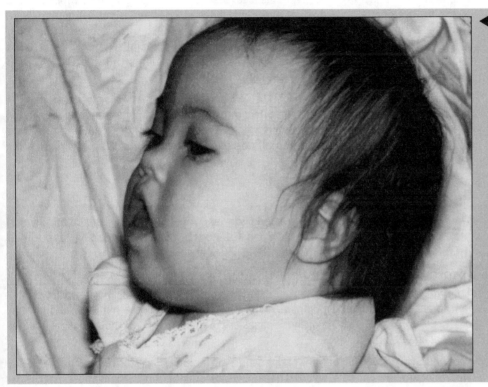

◀ **FIGURE 12-8**
**Profile of an Infant with Down's Syndrome.** *This infant, in addition to being quite hypotonic, has the facial features of a depressed nasal bridge, a very flat facial profile, and an open mouth with a prominent tongue. Almost all infants with Down's syndrome are mentally retarded to some extent.*

## X-CHROMOSOME SYNDROMES

*Turner's Syndrome (45,X).* Most fetuses (98%) with a 45,X karyotype spontaneously abort. Individuals born with Turner's syndrome usually can be recognized at birth because they have marked edema of the distal extremities, particularly the dorsal aspect of the hands and feet. They frequently have webbing of the neck, a shield-like appearance of the chest with widely spaced nipples, coarctation of the aorta, and multiple pigmented nevi. At adolescence, girls with Turner's syndrome exhibit sexual infantilism, primary amenorrhea, and sterility.

*Klinefelter's Syndrome (47,XXY).* Males with Klinefelter's syndrome are usually not diagnosed at birth. Frequently Klinefelter's syndrome is diagnosed during adolescence, when features of the secondary sex characteristics are noticed at the time of puberty. These patients are taller than normal males and often have abnormal body proportions, (e.g., legs relatively long in proportion to the trunk and arms). Other abnormalities are gynecomastia, unusually firm testes, and pubic hair with female distribution.

## MICRODELETION SYNDROMES

*Prader-Willi Syndrome [del(15q11−q13)].* Prader-Willi syndrome is marked by hypotonia in infancy, obesity, polyphagia, small hands and feet, short stature, hypogonadism, and mental retardation. This microdeletion is found in approximately 70% of children with Prader-Willi syndrome and is present in the number 15 chromosome inherited from the father. A number of patients with Prader-Willi syndrome have maternal uniparental disomy (UPD) for chromosome 15, meaning that these patients inherited two copies of chromosome 15 from their mother and no copy from their father [10, 11].

*Angelman's Syndrome [del(15q11−q13)].* This syndrome is frequently referred to as the "happy puppet syndrome" and is manifested by severe mental retardation, hyperactivity, seizures, ataxia, hand flapping, absence of speech, and inappropriate laughter. In Angelman's syndrome, cases are attributed to the deletion of 15q11–q13 in the number 15 chromosome inherited from the mother. Some cases are caused by paternal UPD for chromosome 15.

**DiGeorge Syndrome [del(22q11.2)].** This syndrome is manifested by hypertelorism, short philtrum, downward-slanting palpebral fissures, aortic arch anomalies, hypoplasia to aplasia of the thymus, and hypoplasia to absence of the parathyroid, which causes severe hypocalcemia and seizures in early infancy [12]. When other syndromes were evaluated by using fluorescence in situ hybridization (FISH), it was discovered that other syndromes had the same 22q11.2 deletion. The velocardiofacial syndrome (VCFS), the conotruncal anomaly face syndrome (CFAT), and some patients with the Opitz G/BBB syndrome have the same 22q11.2 deletion [13, 14].

**Miller-Dieker Syndrome [del(17p13.3)].** The most marked feature of this syndrome is incomplete development of the brain, often with a smooth surface (lissencephaly). Facial features include microcephaly with bitemporal narrowing, high forehead, and furrowing in the central forehead. There is a small nose with anteverted nostrils, upward-slanting palpebral fissures, and a small jaw.

## Mendelian Syndromes

Syndromes and congenital malformations caused by mendelian inheritance are divided into autosomal dominant and recessive and X-linked dominant and recessive, as outlined in Table 12-3.

### AUTOSOMAL DOMINANT

**Achondroplasia.** This disorder is manifested by a large head, low nasal bridge with prominent forehead, midfacial hypoplasia, short limbs (particularly the humeri), and lumbar lordosis. Approximately 80% of these cases are thought to be new mutations, with older paternal age being a factor.

**Noonan's Syndrome.** The features of this syndrome include short stature, epicanthal folds, hypertelorism, low-set ears, webbed neck, low-posterior hairline, shield chest, pectus excavatum, pulmonic stenosis, small penis, cryptoorichidism, and mental retardation.

> *Ectrodactyly-ectodermal dysplasia-clefting (EEC) syndrome includes defects in the midportion of the hands and feet varying from syndactyly to claw hand, cleft lip and cleft palate, and atresia of the lateral duct system.*

**Ectrodactyly-Ectodermal Dysplasia-Clefting (EEC) Syndrome.** This syndrome includes defects in the midportion of the hands and feet, varying from syndactyly to claw hand, cleft lip or cleft palate, and atresia of the lacteral duct system. This syndrome is noted for its variable expressivity, and in the same pedigree, some patients have clefting of only the hands, others have clefting of only the feet, and still others have clefting of the lip. This syndrome is mentioned elsewhere in this chapter under examples of negative eugenics in the United States.

**Apert's Syndrome.** Patients with this syndrome have a high forehead and occiput, irregular craniosynostoses, flat facies, shallow orbits, hypertelorism, downward-slanting palpebral fissures, small nose, narrow palate, occasional cleft palate, syndactyly, and broad distal phalanx of the thumbs and toes. Older paternal age is a factor, and most cases represent new mutations.

### AUTOSOMAL RECESSIVE

**Ataxia-Telangiectasia Syndrome.** Ataxia, telangiectasia first evident in ocular conjunctiva and later in other areas, frequent sinopulmonary infections, decreased-to-absent serum immunoglobulin A (IgA), and immunodeficiency are common.

**Meckel-Gruber Syndrome.** Features include prenatal growth deficiency, posterior or dorsal encephalocele, microcephaly with sloping forehead, microphthalmia, cleft palate, micrognathia, short neck, dysplasia of kidneys with varying degrees of cyst formation, polydactyly, and cryptoorchidism.

**Rhizomelic Chondrodysplasia Punctata Syndrome.** There is symmetric proximal shortening of the humeri and femora, punctuate epiphyseal mineralization, coronal cleft in vertebrae, mental deficiency with or without spasticity, microcephaly, low nasal bridge and flat face, occasional upward-slanting palpebral fissures, and cataracts.

| Disorders Caused by Mendelian Inheritance | Associated Birth Defects |
|---|---|
| **Autosomal dominant inheritance** | |
| Achondroplasia | Large head, low nasal bridge, prominent forehead, midfacial hypoplasia, short limbs (particularly the humeri), lumbar lordosis, short stature, and hydrocephalus. |
| Noonan's syndrome | Short stature, epicanthal folds, hypertelorism, low-set ears, webbed neck, low posterior hairline, shield chest, pectus excavatum, cubitus valgus, pulmonic stenosis, small penis, cryptoorchidism, and mental retardation. |
| Ectrodactyly-ectodermal dysplasia-clefting (EEC) syndrome | Defects in midportion of hands and feet varying from syndactyly to claw hand, cleft lip or palate, atresia of lacrimal duct system, and hypoplastic nipples. This syndrome is noted for its variable expressivity in succeeding generations. One individual can have clefting of feet and normal hands, whereas a member of the next generation can have clefting of hands and normal feet. |
| Apert's syndrome | High forehead and occiput, irregular craniosynostoses, flat facies, shallow orbits, hypertelorism, downward-slanting palpebral fissures, small nose, narrow palate, occasional cleft palate, syndactyly, and broad distal phalanx of thumbs and toes. |
| **Autosomal recessive inheritance** | |
| Ataxia-telangiectasia syndrome | Ataxia, telangiectasia (which may first be evident in the ocular conjunctiva and later in other areas), frequent sinopulmonary infections, decreased-to-absent serum immunoglobulin A (IgA), and immunodeficiency. |
| Meckel-Gruber syndrome | Prenatal growth deficiency, posterior or dorsal encephalocele, microcephaly with sloping forehead, microphthalmia, cleft palate, micrognathia, short neck, dysplasia of kidneys with varying degrees of cyst formation, polydactyly, and cryptoorchidism. |
| Rhizomelic chondrodysplasia punctata syndrome | Symmetric proximal shortening of the humeri and femora, punctate epiphyseal mineralization, coronal cleft in vertebrae, mental deficiency with or without spasticity, microcephaly, low nasal bridge and flat face, occasional upward-slanting palpebral fissures and cataracts. |
| Radial aplasia thrombocytopenia syndrome (TAR) | Most severe in early infancy because of thrombocytopenia associated with absence or hypoplasia of megakaryocytes, bilateral absence or hypoplasia of radius, abnormalities of ulna (including hypoplasia or bilateral or unilateral absence). Frequent defects of hands, legs, or feet; however, thumbs are present. Thirty percent have a congenital heart defect, approximately 40% of infants die during early infancy, usually as result of hemorrhage. |
| **X-linked dominant inheritance** | |
| Fragile X syndrome | X-linked dominant condition with 80% penetrance in males and only 30% penetrance in females. Affected males have long, prominent ears; a long face; hyperextensible finger joints; and after puberty, macroorchidism, as well as hyperactivity and poor eye contact. Mental retardation ranges from mild to severe. Female carriers are usually not dysmorphic, but approximately one-third have significant mental retardation, whereas other heterozygous females are cognitively unaffected. |

**TABLE 12-3** ▶
Continued

| Disorders Caused by Mendelian Inheritance | Associated Birth Defects |
|---|---|
| Incontinentia pigmenti, type I | Irregular pigmented skin lesions, patchy alopecia, conical or missing teeth, strabismus, retinal dysplasia, and blue sclera. The disease is seen only in affected females; hemizygous affected males do not survive to term. |
| *X-linked recessive inheritance*<br>Menkes' syndrome | Sparse, stubby, lightly pigmented hair that is twisted and appears kinky. Prenatal growth deficiency, lack of expressive movement, and thick and dry skin. Severe degenerative process in the cerebral cortex can result in profound CNS defects by 2 months of age. Death usually occurs by the age of 3 or 4 years. Responsible gene is Xq13. |
| Hunter's syndrome (mucopolysaccharidosis, type II; gargoylism) | Growth deficiency, clear corneas, coarse face, stiff joints, mental retardation, death before age 15 years in severely affected. Both mild and severe forms are caused by a deficiency of iduronate sulfatase. Responsible gene is Xq27–q28. |

*Radial Aplasia Thrombocytopenia Syndrome (TAR).* This syndrome is most severe in early infancy because of the thrombocytopenia with absence or hypoplasia of the mega-karyocytes. There is usually bilateral absence or hypoplasia of the radius, which is often associated with defects of the hands, legs, or feet. However, the thumbs are present. Thirty percent of patients have a congenital heart defect, and approximately 40% of the patients die during early infancy, usually as a result of hemorrhage.

## X-LINKED DOMINANT

*Fragile X Syndrome.* This is an X-linked dominant condition with 80% penetrance in males and only 30% penetrance in females. Fragile X syndrome is the most common cause of mental retardation from a single gene defect. Affected males have long, promi-nent ears; a long face; hyperextensible finger joints; and after puberty, macroorchidism, as well as behavior such as hyperactivity and poor eye contact. Mental retardation ranges from mild to severe. Female carriers of fragile X syndrome exhibit a wide variation, with some heterozygotes being cognitively unaffected and others having learning disabilities. Approximately one-third have significant mental impairment and intelligence quotient (IQ) scores lower than 85. The etiology of the fragile X syndrome is a mutation involving expansion of a CGG trinucleotide repeat segment in the fragile X mental retardation gene (FMR) on the long arm of the X chromosome at Xq27.3.

*Incontinentia Pigmenti, Type I.* This disorder is characterized by abnormal skin pigmen-tation, conical or missing teeth, strabismus, retinal dysplasia, and blue sclera. This disease is seen only in females, and it is thought that hemizygous males with this gene are so severely affected that they do not survive to term.

## X-LINKED RECESSIVE

*Menkes' Syndrome.* The outstanding feature of this syndrome is sparse, stubby, lightly pigmented hair that is twisted and appears kinky. Prenatal growth deficiency, lack of expressive movement, thick and dry skin, and severe degenerative process in the cerebral cortex that result in profound CNS defects by 2 months of age are also seen. Death is usually by the age of 3 or 4 years. The responsible gene is located at Xq13.

*Hunter's Syndrome.* Mucopolysaccharidosis type II, originally called gargoylism, is mani-fest by growth deficiency, clear corneas, coarse face, stiff joints, mental retardation, and, in the severe form, death usually before the age of 15 years. The differentiation between the mild and severe forms is based on the age of onset, degree of CNS involvement, and rapidity of deterioration. The primary defect is a deficiency of iduronate sulfatase. The gene for Hunter's syndrome has been mapped to Xq27–q28.

> *Fragile X syndrome is the most common cause of mental retardation from a single gene defect.*

# Multifactorial Inheritance

Multifactorial simply indicates that there are many factors, both environmental and genetic (polygenic), that are involved in the etiology of these birth defects and syndromes. Often, open neural tube defects, pyloric stenosis, and some forms of cleft lip or cleft palate are inherited in a multifactorial way (Table 12-4). Many common conditions are also multifactorially inherited, including coronary artery disease and some forms of hypertension.

◀ **TABLE 12-4**
*Examples of Multifactorial Inheritance as Causes of Birth Defects*

| Disorders Attributed to Multifactorial Inheritance | Associated Birth Defects |
|---|---|
| Open neural tube defects (ONTDs) | Defects vary depending on which portion of the neural groove fails to close. |
| Anencephaly | Cephalic end of neural groove fails to close, resulting in absence of bones of the vault of the skull, absent meninges, and defects in forebrain. Many infants with anencephaly are stillborn; those born live survive for a few hours or a few days at most. |
| Spina bifida cystica (meningocele, myelomeningocele) | Defects depend on location and extent of the defect and whether it is covered with skin or is open. Defects in closure of cervical and upper thoracic region can cause retroflexion of the upper spine with short neck and trunk, defects of the thoracic cage, anterior spina bifida, diaphragmatic defects with or without hernia, and hypoplasia of the lung or heart. Defects in closure at the lower thoracic or lumbar area can result in clubfeet, paralysis of lower extremities, and a lack of control of bladder and sphincter. |
| Pyloric stenosis | Infant seems normal for the first 2–6 weeks after birth, then forceful, projectile vomiting begins. A firm nodule can be palpated in the region of the pylorus. |
| Cleft lip and cleft palate | Cleft lip may be unilateral or bilateral and may occur with or without cleft palate. More than 100 syndromes include cleft lip or cleft palate and represent all modes of mendelian inheritance, environmental exposure, and chromosome abnormalities. However, most patients with cleft lip or cleft palate are affected because of multifactorial inheritance. |

*Open Neural Tube Defects (ONTDs).* Anencephaly and spina bifida cystica (i.e., meningocele, myelomeningocele, open spine) are the common forms of ONTDs that frequently occur together in families and are considered to have a common pathogenesis. In anencephaly, the forebrain, overlying meninges, vault of the skull, and skin are all absent. Many infants with anencephaly are stillborn; those born alive survive for a few hours or, at most, a few days. In spina bifida cystica, there is failure of fusion of the arches of the vertebrae, most typically in the lumbar regions (see Clinical Case). The incidences of both anencephaly and spina bifida are a little higher in females than in males. The ONTDs are a major cause of stillbirth, death in early infancy, and a handicap in surviving children. There has been much suggestive evidence that at least some cases of ONTDs are caused by a deficiency of enzymes in the folic acid pathway. Recently, the U.S. Public Health Service has recommended that all women of childbearing age take a daily dose of 0.4 mg of folic acid starting at least 2 months prior to conception to reduce their risk of having a child with a neural tube defect [15–17].

*In **anencephaly**, the forebrain, overlying meninges, vault of the skull, and skin are all absent.*

*In **spina bifida cystica**, there is failure of fusion of the arches of the vertebrae, most typically in the lumbar regions.*

**Pyloric Stenosis.** Pyloric stenosis is caused by hypertrophy and hyperplasia of the smooth muscle surrounding the distal aperture of the stomach so that it causes obstruction of the stomach, which causes the landmark feature of very forceful projectile vomiting. The obstruction can be relieved surgically. This condition is five times more common in males than in females (approximately 5/1000 male births compared to 1/1000 female births).

**Cleft Lip and Cleft Palate.** There are more than 100 syndromes that include cleft lip and cleft palate, representing all modes of mendelian inheritance, environmental exposure, and chromosomal abnormalities. However, a number of patients with cleft lip and cleft palate have it on the basis of multifactorial inheritance. This again demonstrates the complexity of counseling in certain forms of birth defects and syndromes and the need for an accurate diagnosis when evaluating the etiology of congenital malformations.

**Sporadic Cases.** Fifty years ago, most syndromes and congenital malformations had unknown causes; however, today, more than 75% of these conditions have a known, well-defined etiology. Two sporadic syndromes, the VATER association and Sotos' syndrome, are discussed below.

*VATER association* is an acronym to designate the features that include vertebral defects, anal atresia, tracheoesophageal fistula with esophageal atresia, and radial and renal dysplasia. Cardiac defects, a single umbilical artery, as well as prenatal growth deficiency are also nonrandom features. This pattern of malformations has been a sporadic occurrence in otherwise normal families. Although in infancy many of these children are quite ill, once they survive into childhood, they usually have good brain function and can have a productive life.

*Sotos' syndrome*, otherwise known as cerebral gigantism syndrome, is characterized by variable mental deficiency, prenatal onset of excessive size with large hands and feet, advanced osseous maturation, macrocephaly with downslanting palpebral fissures, hypertelorism, high narrow palate with prominent lateral palatine ridges, and coarse-looking facies. The cause is unknown, and most cases have been sporadic.

> There are more than 100 syndromes that include cleft lip or cleft palate, representing all modes of inheritance, environmental exposure, and chromosomal abnormalities. However, a number of patients with cleft lip or cleft palate have it on the basis of multifactorial inheritance.

# EUGENICS

> *Positive eugenics* is a systematic effort to maximize the transmission of genes that are considered desirable.

> *Negative eugenics* is a systematic effort to minimize the transmission of genes that are considered deleterious.

The term eugenics was first used in 1883 by Sir Francis Galton, who described it as "the study of the agencies under social control that may improve or impair the racial qualities of future generations, either physically or mentally." A more recent definition of eugenics is the science that deals with all influences that improve the inborn quality of the human race, particularly through the control of hereditary factors. *Positive eugenics* is a systematic effort to maximize the transmission of genes that are considered desirable. *Negative eugenics* is a systematic effort to minimize the transmission of genes that are considered deleterious. This might seem like a worthwhile effort, until the techniques used in negative eugenics, including involuntary sterilization, involuntary euthanasia, and genetic discrimination, are considered [18].

In the latter part of the 19th century, many eugenicists and social workers became concerned about what they considered to be a dilution of the "Anglo-Saxon superiority" in the United States. They lectured widely about the evils of continued unfettered immigration, and after 1910, when the Eugenics Record Office was established, they used this base to lobby Congress for an immigration restriction act. Because of the lobbying of the eugenicists, particularly Dr. Harry H. Laughlin, the Johnson Immigration Restriction Act was passed in 1924 by Congress and signed into law by President Calvin Coolidge. At the suggestion of the eugenicists, future immigration quotas were established based on the 1890 census because this favored immigration of the so-called Nordic or Anglo-Saxon stock (namely people from Northwestern Europe and Great Britain) and decreased the immigration of those from southern and eastern Europe and from Asia, and Jewish and black immigrants [18]. In 1924, a well-known German revolutionary wrote a book in which he praised the United States' 1924 Immigration Restriction Law and suggested that Germany should use this concept to control its "undesirable" ethnic groups. The author's name was Adolf Hitler, and the book was *Mein Kampf*. The

1924 Immigration Restriction Law, which discriminated on the basis of racial or ethnic groups, was further misused in the late 1930s and early 1940s by the State Department under Breckenridge Long. This law was used as a basis for refusing entry into this country of hundreds of thousands of Jewish refugees who were eventually killed in the Nazi concentration camps [19].

Another misuse of medicine was the passage of the eugenics sterilization laws. The first involuntary sterilization law in the United States was passed by the State of Indiana in 1907. In 1937, Georgia became the last state to enact an eugenics sterilization law [20]. It is important to realize that in the 32 states with sterilization laws, the laws were written and passed by the state representatives and senators and signed into law by the 32 governors. It is known that at least 60,000 people were involuntarily sterilized in the United States, and perhaps more than 100,000 were sterilized during the period of the active utilization of these laws. Most states included mental retardation, insanity, and criminality as reasons for involuntary sterilization. However, many states included conditions such as chronic alcoholism, epilepsy, and Huntington's disease. Orphans, derelicts, paupers, and prostitutes were sterilized as well [19, 20].

There is still a pro-eugenics sentiment in the United States regarding the treatment of individuals because of their race, ethnicity, social class, the presence of genetic disease or birth defects and, indeed, just because of poor health. This has been evidenced by many factors, including the publication in 1991 of a documentary video named "Blood in the Face." One of the characters who was interviewed frequently during this video was David Duke, who, at the time, was one of the leaders of the Ku Klux Klan, and who more recently was a gubernatorial candidate in the state of Louisiana. Duke stated that the poor are a drain on the existing resources of society. Others interviewed claimed that Hitler was a humanitarian and that in the next "revolution" all of the Jews and blacks would be eliminated from America. Another example of the pro-eugenics sentiment was the negative reaction from thousands of people when a bright, articulate, and successful television anchorwoman, Bree Walker, became pregnant. Ms. Walker has the EEC syndrome, which is manifested by clefting in the hands, feet, and sometimes of the lip (see autosomal dominant syndromes in this chapter). Ms. Walker had only clefting of her hands, and her infant had the same condition. There was an outpouring of anger on radio and television talk shows from thousands of our fellow citizens who claimed that she did not have the right to bring an infant with this syndrome into the world, although Ms. Walker is a highly productive member of society.

Utilitarian reasoning in health care was the basis of the early Nazi eugenics policy [21–25]. In the United States, pressure for utilitarian cost control is coming from health maintenance organizations (HMOs) and health insurance companies who want to keep the cost of health care to a minimum. One of the ways of doing this is to deny health care at the two ends of the lifespan, namely to newborns who have birth defects and genetic diseases and to the elderly. They have also advocated that for newborns to be covered in instances where there was genetic disease in the family, the mother should have prenatal diagnosis. The HMOs have stated that they will pay for the cost of the amniocentesis and the abortion, if necessary, but would not pay for the care of the child if the child is born with defects. This is an extreme example of negative eugenics, which, in many cases, is against the moral and ethical principles of the parents. The 21st century may be filled with many problems because of this utilitarian approach to the implementation of health care.

> *Utilitarian reasoning in health care* was the basis of the early Nazi eugenics policy. In the United States, pressure for utilitarian cost control is coming from HMOs and health insurance companies, which want to keep the cost of health care to a minimum.

# RESOLUTION OF CLINICAL CASE

Mary and Edmund's obstetrician referred the couple to a clinical geneticist who reviewed their medical and family histories. The obstetrician explained that ONTDs are usually inherited multifactorially, meaning that there were many genes in both Mary and Edmund that dictated a slower-than-usual closure of the neural groove and that perhaps there were environmental factors that precipitated the birth defect in their unborn son. The medical team also discussed with the parents the necessity to monitor carefully the spina bifida cystica as to its size and whether it remained skin covered. They indicated that if the cyst remained relatively small and skin covered, they could recommend a vaginal delivery. However, if the cyst ruptured (i.e., if it did not remain skin covered), and the spinal cord was exposed to the pressure of the contracting uterus, it would be safer to do a cesarean section before Mary went into labor.

Mary and Edmund asked their medical geneticist and the neurosurgeon about the future of their unborn son. They replied that, to some extent, this would depend on the extent of the lesion and whether the meninges and skin covered the spina bifida cystica. If the cyst opened, exposing the spinal cord to infection, there would be a much greater chance that the child would have some disturbances of function of the nervous system, particularly in the lower back and legs. Many of these children also have lack of bladder and bowel control because of damage during development and delivery to the nerves that supply these organs.

The parents were also told that, unfortunately, most infants with spina bifida cystica (approximately 80%) developed hydrocephalus. In most neurosurgical spina bifida clinics, this is followed very closely so that at the first indication of hydrocephalus, it is very quickly corrected. The philosophy in treating spina bifida cystica today is much different than it was in the 1940s and 1950s, in that it is realized that many of the children with spina bifida cystica, if treated early and promptly, can lead very productive lives and take part in most of the normal functions in society.

The medical geneticist counseled Mary and Edmund concerning some new findings in the prevention of ONTDs. He quoted the extensive research that has been done during the past 40 years, first in animals and later in children, involving folic acid, one of the B vitamins. Research has shown that starting this vitamin 2 months prior to conception can reduce the number of cases of ONTDs. Recently, to reduce the frequency of ONTDs and their resulting disability, the U.S. Public Health Service recommends that all women of childbearing age in the United States who are capable of becoming pregnant should consume 0.4 mg of folic acid per day to reduce their risk of delivering an infant affected with spina bifida or other ONTDs [15]. Because the effects of higher intakes are not well known but include complicating the diagnosis of vitamin $B_{12}$ deficiency, care should be taken to keep total folate consumption at lower than 1 mg/d, except under the supervision of a physician. Women who have had a prior pregnancy affected with a neural tube defect are at a higher risk of having a subsequent affected pregnancy. When these women plan a pregnancy, they should consult their physicians for advice. The medical geneticist indicated to the couple that because they now have a child with spina bifida cystica, they have an increased risk of having a child with anencephaly or spina bifida cystica as compared to the general population. Ordinarily, they would have approximately a 4% chance of having a second child with either of the ONTDs. However, by starting the maintenance dose of 0.4 mg of folic acid daily beginning 2 months prior to conception, this risk could be considerably reduced. The physician also suggested that Edmund and Mary change their eating habits and include folate-rich foods such as green leafy vegetables, fortified flour, and others in their daily diet.

Edmund and Mary were mystified as to why this birth defect occurred in their child. The medical geneticist quoted some very new research, including work by Mills and colleagues, which demonstrated that an abnormality in homocysteine metabolism, apparently related to methionine synthase, is present in many women who give birth to children with neural tube defects [16]. Overcoming this abnormality is likely the mechanism by which folic acid prevents neural tube defects. These findings suggest that the most effective periconceptional prophylaxis to prevent neural tube defects may require vitamin $B_{12}$ as well as folic acid. Other research is now evaluating the DNA in the gene that might be involved. Van der Put and colleagues studied the frequency of the 677C→T mutation in the 5,10-methylenetetrahydrofolate reductase (MTHFR) gene in 55 patients with spina bifida and the parents of such patients (70 mothers, 60 fathers). They found that 5% of 207 controls were homozygous for the 677C→T mutation compared with 16% of mothers, 10% of fathers, and 13% of patients and that the mutation was associated with decreased MTHFR activity, low plasma folate, high plasma homocysteine, and red-cell folate concentrations. They stated that the 677C→T mutation should be regarded as a genetic risk factor for spina bifida [17].

The value of finding the specific mutation was explained to the parents and several of their collateral relatives, who also attended the counseling session. It was mentioned that, in the future, by doing a simple DNA test, it might be possible to give exact figures for the risk of occurrence or risk of recurrence of ONTDs.

Edmund and Mary's son, Johnny, has already learned to walk. He has not yet shown any signs of hydrocephalus and has good control of his bowel and bladder sphincters. He socializes well with the team at the spina bifida clinic and also with the neighborhood children. He is bright, speaks well in sentences, understands, and can point to figures and objects. Mary and Edmund have decided to plan a third pregnancy.

# REVIEW QUESTIONS

**Directions:** For each of the following questions, choose the **one best** answer.

1. Spina bifida cystica (meningocele or myelomeningocele) and anencephaly are major malformations with an occurrence rate of
    (A) 1/10,000 pregnancies
    (B) 1/1000 pregnancies
    (C) 1/100 pregnancies
    (D) 1/10 pregnancies

2. Prospective genetic counseling is best defined as counseling that occurs
    (A) after an individual develops a genetic disease
    (B) after a couple has a child with a birth defect or genetic disease
    (C) when a couple is told their fetus will be born with a genetic disease
    (D) when a couple is identified as at risk for developing a genetic disease

3. An example of a minor malformation (minor anomaly) includes which one of the following?
    (A) Bilateral cleft lip and palate
    (B) Transposition of the great vessels
    (C) Polycystic kidney disease
    (D) Syndactyly

**Directions:** The group of questions below consists of lettered choices followed by several numbered items. For each numbered item, select the appropriate lettered option with which it is most closely associated. Each lettered option may be used once, more than once, or not at all.

### Questions 4–6

Match each of the following definitions to the term that it best defines.
    (A) Eugenics
    (B) Positive eugenics
    (C) Negative eugenics
    (D) Neutral eugenics

4. A systematic effort to maximize the transmission of genes that are considered desirable

5. A science that deals with all influences that improve the inborn quality of the human race, particularly through the control of hereditary factors

6. A systematic effort to minimize the transmission of genes that are considered deleterious

# ANSWERS AND EXPLANATIONS

**1. The answer is B.** Spina bifida and anencephaly are classified as major congenital malformations because they are life-threatening and have major surgical implications. Some of these serious malformations can be prevented if the mother ingests folic acid 2 months prior to becoming pregnant as well as during her pregnancy. These open neural tube defects occur in approximately 1 of every 1000 pregnancies in the United States.

**2. The answer is D.** Prospective genetic counseling is anticipatory in that an individual or couple is identified as being at increased risk for developing a genetic disease or birth defect or of having a child with a similar problem before its occurrence. Candidates for prospective genetic counseling include couples in which either the father or mother is older or in which one member of the couple is at risk for having a particular genetic disease that is more prevalent in their ethnic or racial group. For example, whites are at increased risk for cystic fibrosis, blacks are at increased risk for sickle cell anemia, Ashkenazi Jews are at increased risk for Tay-Sachs disease, people of Mediterranean heritage are at increased risk for β-thalassemia, and Asians are at increased risk for α-thalassemia. Retrospective counseling is given after an individual develops a genetic disorder or after a couple has a child with a particular birth defect or genetic disease. Until recently, most genetic counseling was retrospective. Currently, more at-risk situations can be identified before conception, and the individual or couple can be given prospective counseling.

**3. The answer is D.** A minor malformation (minor anomaly) is a minor defect that occurs in fewer than 4% of patients. Examples of minor malformations include single palmar crease, hypoplastic nails, inner epicanthic folds, low-set ears, and syndactyly (webbing between fingers or toes). A syndrome is a pattern of multiple anomalies thought to be pathogenetically related and not known to represent a single sequence or polytropic field defect. Malformations are morphologic defects of an organ, part of an organ, or a larger region of the body resulting from an intrinsically abnormal development process. Congenital malformations are anatomic or structural abnormalities that are present at birth, although they may not be diagnosed until later. A major congenital malformation is one that may be lethal or of serious surgical, medical, or cosmetic importance to the patient. Examples of major malformations are bilateral cleft lip and palate, anencephaly, spina bifida cystica, and transposition of the great vessels.

**4–6. The answers are: 4-B, 5-A, 6-C.** Eugenics is a science that deals with all influences that improve the inborn quality of the human race, particularly through the control of hereditary factors. Positive eugenics is a systematic effort to maximize the transmission of genes that are considered desirable. Negative eugenics is a systematic effort to minimize the transmission of genes that are considered deleterious. Negative eugenics might seem worthwhile until the techniques used to achieve the goal are considered. These techniques included involuntary sterilization, involuntary euthanasia, and genetic discrimination.

# REFERENCES

1. Warkany J: Teratology of the past. In *Congenital Malformations*. Chicago, IL: Year Book, 1971, pp 6–20.
2. Brent RL: Comments on the history of teratology. *Teratology* (2):199–200, 1979.
3. Greenblatt RB: *Search the Scriptures*, 1st ed. Philadelphia, PA: J. B. Lippincott, 1963.
4. Jones KL: *Smith's Recognizable Patterns of Human Malformation*, 5th ed. Philadelphia, PA: W. B. Saunders, 1996.

5.  Garver KL, Marchese SG: *Genetic Counseling for Clinicians*. New York, NY: Year Book, 1986.
6.  Graham JM: *Smith's Recognizable Patterns of Human Deformation*, 2nd ed. Philadelphia, PA: W. B. Saunders, 1988.
7.  Tjio JH, Levan A: The chromosome number in man. *Hereditas* 42:1–6, 1956.
8.  Ad Hoc Committee on Genetic Counseling. *Am J Hum Genet* 27(2):240–242, 1975.
9.  Garver KL: Genetic counseling in the delivery of health care. *PA Med* 80:40–43, 1977.
10. Saitoh S, Buiting K, Rogans PK, et al: Minimal definition of the imprinting center and fixation of a chromosome 15q11–p13 epigenotype by imprinting mutations. *Proc Natl Acad Sci* 93:7811–7815, 1996.
11. White LM, Rogan PK, Nicholls, RD, et al: Allele-specific replication of 15q11–q13 loci: a diagnostic test for detection of uniparental disomy. *Am J Hum Genet* 59:423–430, 1996.
12. Driscoll DA, Budarf ML, Emanuel BS: A genetic etiology for DiGeorge syndrome: consistent deletions and microdeletions of 22q11. *Am J Hum Genet* 50:924–933, 1992.
13. Wilson DI, Burns J, Goodship J: DiGeorge syndrome: part of CATCH 22. *J Med Genet* 30:852–856, 1993.
14. Morrow B, Goldberg R, Carlson C, et al: Molecular definition of the 22q11 deletions in velocardiofacial syndrome. *Am J Hum Genet* 56:1391–1403, 1995.
15. Centers for Disease Control and Prevention: Recommendations for use of folic acid to reduce number of spina bifida cases and other neural tube defects. *JAMA* 269(10):1233–1238, 1993.
16. Mills JL, McPartin JM, Kirke PN, et al: Homocysteine metabolism in pregnancies complicated by neural-tube defects. *Lancet* 345:149–151, 1995.
17. Van der Put NMJ, Steegers-Theunissen RPM, Frosst P, et al: Mutated methylenetetrahydrofolate reductase as a risk factor for spina bifida. *Lancet* 346:1070–1072, 1995.
18. Garver KL, Garver BL: Eugenics, euthanasia and genocide. *Linacre Q* 59(3):24–51, 1992.
19. Garver KL, Garver BL: The human genome project and eugenic concerns. *Am J Hum Genet* 53:148–158, 1994.
20. Reilly P: *The Surgical Solution: A History of Involuntary Sterilization in the United States*. Baltimore, MD: Johns Hopkins University Press, 1991.
21. La Chat MR: Utilitarian reasoning in Nazi medical policy: some preliminary investigations. *Linacre Q* 42(4):14–37, 1975.
22. Garver KL, Garver BL: Eugenics: past, present, and the future. *Am J Hum Genet* 49:1109–1118, 1991.
23. Franzblau MJ: Ethical values in health care in 1995: lessons from the Nazi period. *J Med Assoc Georgia* April:161–164, 1995.
24. Lerner BH, Rothman DJ: Medicine and the holocaust: learning more of the lessons. *Ann Intern Med* 122(10):793–794, 1995.
25. Ernst E: Killing in the name of healing: the active role of the German medical profession during the Third Reich. *Am J Med* 100:579–581, 1996.

# 13 CLINICAL ASPECTS OF MEDICAL GENETICS

W. Allen Hogge, M.D.

---

## CHAPTER OUTLINE

---

## INTRODUCTION OF CLINICAL CASE

A 38-year-old woman presents to her physician at 8 weeks from her last menstrual period for routine prenatal care. This is her first pregnancy, following many years of infertility. Her family history is negative for any factor that would place the pregnancy at risk for genetic disorders. Past medical history reveals no medical conditions in the patient. Because of the patient's age, the obstetrician refers her to a genetic counselor for a discussion regarding prenatal diagnosis.

---

## INTRODUCTION

Recent advances in prenatal and neonatal care have decreased perinatal morbidity and mortality rates. However, there has been little change in the incidence of birth defects, which now account for a significant proportion of both morbidity and mortality. During the last 2 decades, the ability to obtain information about the fetus has increased dramatically, and the prenatal diagnosis of birth defects has developed into one of the most powerful tools of the medical geneticist. These techniques make it possible for parents to alter the genetic risks to which their offspring are exposed.

In addition to prenatal monitoring of pregnancies at risk, there are three other approaches to the prevention of genetic diseases. These approaches are:

1. Screen newborns for disorders in which early treatment prevents the development of severe clinical abnormalities.

2. Detect carriers of a genetic disorder and counsel couples at risk prior to their first pregnancy.
3. Diagnose affected individuals early, then counsel parents and other relatives at risk.

# PRENATAL DIAGNOSIS

## Indications

The object of prenatal diagnosis is to determine whether a fetus who is at risk for a genetic disease is or is not affected with that disease. "Fetal diagnosis," which implies a broader scope of both finding and defining genetic disorders in the unborn, should result in attempts at finding appropriate methods of fetal treatment.

Because it is not possible to assess every pregnancy with a prenatal diagnostic procedure, attention must be focused on those pregnancies at a high risk for genetic disorders. Unfortunately, the methods for determining at-risk pregnancies are limited, and couples are often identified by the previous birth of an affected child. The most common indications for prenatal diagnosis are:

*The most common indication for prenatal diagnosis is advanced maternal age.*

1. The mother is of advanced age (i.e., 35 years or older at the time of delivery).
2. The couple has a previous child with a chromosome trisomy (e.g., trisomy 13, 18, 21).
3. The pregnancy is at risk for a neural tube defect (NTD), such as spina bifida or anencephaly. Risk factors include a previous child with a NTD or either parent affected with spina bifida.
4. The pregnancy is at risk for a biochemical or metabolic disorder. Most commonly, this risk is ascertained because of the previous birth of an affected child. However, some cases involve couples found to be carriers of genetic disorders detected by preconception screening. Currently, this form of screening is limited to high-risk groups, such as Tay-Sachs disease in individuals of Jewish ancestry or thalassemia in individuals with Mediterranean or Asian ancestry.
5. Either parent is a carrier of a chromosomal rearrangement, such as a translocation.
6. Abnormal maternal serum biochemical markers exist.
7. The pregnancy is at risk for a diagnosable genetic condition. It is imperative that the physician providing prenatal diagnosis remain abreast of current developments in both molecular biology and medical imaging to be aware of all the diagnostic options for fetal assessment.

## Techniques

### AMNIOCENTESIS

The mainstay in the armamentarium for prenatal diagnosis is amniocentesis, which is the aspiration of amniotic fluid. It is traditionally performed at 16–18 weeks gestation (defined as weeks from the last normal menstrual period). Between 16 and 18 weeks gestation, the amniotic fluid volume is approximately 200 ml, and 20–30 ml can be removed safely.

Prior to an amniocentesis, an ultrasound is performed to assess fetal age, viability, number, and anatomy, and to determine placental location. Continuous ultrasound guidance is used to place a small-bore (22-gauge) needle into the amniotic cavity, and the fluid is aspirated (Figure 13-1).

Maternal risks from amniocentesis are negligible, with infection being the only significant, albeit rare, complication. With concurrent use of ultrasound, fetal injury should not be a significant risk. However, there is an increased incidence of spontaneous abortion (miscarriage) in women undergoing amniocentesis. That risk is approximately 0.5% (1 in every 200 procedures). The cause for this increased risk is unknown [1].

*Risk of miscarriage following amniocentesis is 0.5%.*

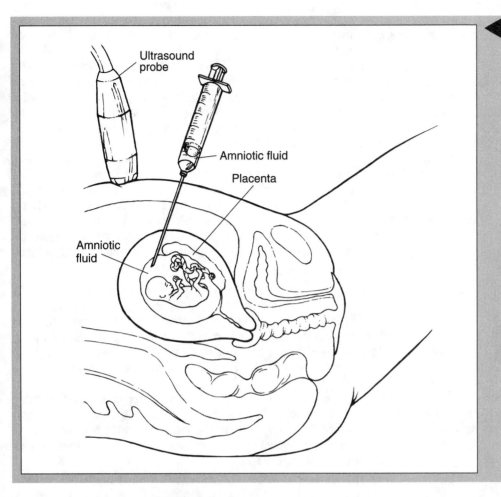

**FIGURE 13-1**
*Prenatal Diagnosis by Amniocentesis at 16 Weeks of Pregnancy.*

Labels in figure: Ultrasound probe; Amniotic fluid; Placenta; Amniotic fluid

For most genetic testing, cultured amniotic fluid cells are required, and therefore, there is usually a 2–3-week delay from the time of the procedure until results are available. For some disorders, such as NTDs, direct assays of amniotic fluid proteins (e.g., alpha-fetoprotein) can be used for diagnostic purposes. Molecular techniques, such as polymerase chain reaction, also provide a rapid diagnosis because uncultured cells can be used.

## EARLY AMNIOCENTESIS

One of the most significant disadvantages of amniocentesis is the relatively late time in pregnancy that it is performed. If the procedure is performed at 16 weeks gestation, and the results are not available for 2–3 weeks, the patient is almost halfway through her pregnancy before the status of her fetus is known. Because of the significant anxiety this causes, research is underway to assess the feasibility of doing amniocentesis at earlier times. A number of small studies have been completed that intended to determine the safety of amniocentesis between 11 and 14 weeks gestation (i.e., early amniocentesis). Despite the fact that most of the study procedures were performed at 13 and 14 weeks gestation, the preliminary results appear to show a higher risk for miscarriage, failed samplings, and vaginal leakage of amniotic fluid than is seen following traditional amniocentesis [2, 3]. Currently, early amniocentesis remains an investigational procedure.

***Case 1.*** A 25-year-old woman is referred to a genetics center for amniocentesis, which her obstetrician recommended because a maternal nephew is mentally retarded.

Amniocentesis is a powerful tool for the prenatal diagnosis of a large number of conditions associated with mental retardation. However, it is not possible to screen for all diagnosable conditions that cause mental retardation, and a large number of disorders with mental retardation are not diagnosable by amniocentesis. To be diagnosable, there must be an abnormality that can be detected in cultured cells (e.g., chromosome abnormality, enzymatic defect), or there must be an abnormal metabolic product that can be assayed in amniotic fluid. Alternatively, if the disorder is diagnosable by direct molecular testing or by linkage analysis, amniotic fluid cells (cultured or uncultured) can be used for diagnostic purposes.

In this case, the first step is to obtain the medical records of the nephew to determine the precise cause for his mental retardation. If the nephew has a nongenetic condition, the patient can be reassured that her risk for a similar problem in her child is no greater than that for anyone else in the general population. On the other hand, if the nephew has fragile X syndrome, evaluation of the patient can be undertaken to determine if she is a carrier of the fragile X gene. If she is a carrier, amniocentesis is an appropriate diagnostic tool. Another possibility is that the nephew has an X-linked genetic disease, for which there are no diagnostic or carrier tests. In that circumstance, the patient could choose amniocentesis to determine fetal sex and elect to continue only a female fetus.

This case illustrates that amniocentesis is a diagnostic test, not a screening test. It is useful only when a precise diagnosis is known. Likewise, it cannot guarantee a "normal" child. There is a 2%–3% chance for a birth defect in any pregnancy. Amniocentesis excludes only the precise condition for which testing is undertaken, not all birth defects.

### CHORIONIC VILLUS SAMPLING (CVS)

Another approach to early prenatal diagnosis is the technique of CVS, which is essentially a biopsy of the placenta. In the first trimester of pregnancy, the chorionic villi are loosely connected and may be easily aspirated, using a small catheter placed through the cervix (transcervical CVS) or using a needle placed through the maternal abdomen (trans-abdominal CVS). Because the villi have the same origin as the fetus, assessment of the chromosomal or biochemical status of the villi provides information identical to the information obtained by amniocentesis.

*CVS is performed between the tenth and twelfth week of pregnancy.*

As with amniocentesis, an ultrasound examination should precede CVS to determine fetal viability, age, and number, as well as placental location. Continuous ultrasound visualization is used to place the catheter or needle into the chorion frondosum (i.e., developing placenta). CVS is commonly performed between the tenth and twelfth week from the last menstrual period (Figure 13-2).

**FIGURE 13-2** ▶
*Transcervical Chorionic Villus Sampling at 10 Weeks of Pregnancy.*

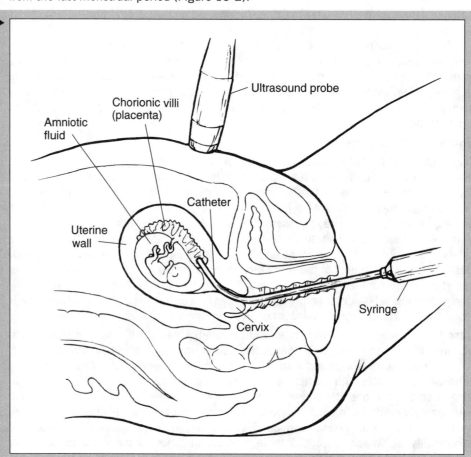

The indications for prenatal diagnosis using CVS are identical to those for amniocentesis with one exception. If a pregnancy is at risk for a NTD, amniocentesis—not CVS—should be performed. Alpha-fetoprotein (AFP) can be measured in amniotic fluid but not on a sample of villi.

> CVS is not indicated for a couple whose pregnancy is at risk for NTDs.

CVS does not place the mother at any risk except for the rare occurrences of infection. As with amniocentesis, there is an increased risk of spontaneous abortion after CVS. In experienced centers the risk of spontaneous abortion from CVS appears to be comparable to that of amniocentesis (0.5%) [4].

There are several other differences between CVS and amniocentesis. Failure to obtain a fluid sample by amniocentesis is quite rare, but failed CVS occurs in approximately 1% of cases [5]. In approximately 1% of CVS procedures, the chromosome analysis indicates a chromosome mosaicism (i.e., one normal cell line, one abnormal cell line) [6]. Approximately 20% of these results represent a true fetal mosaicism, but the remainder are confined to placental tissue only (i.e., confined placental mosaicism). Although some of these cases represent false-positive results, many indicate a high risk for obstetric complications and other nonchromosomal genetic disorders.

Despite these limitations, CVS has two major advantages over amniocentesis. First, CVS is performed much earlier in pregnancy; therefore, the period of anxiety for the family is reduced. Second, because the amount of material obtained by CVS is much larger than that obtained by amniocentesis, many assays can be performed directly on uncultured villi, shortening the time necessary for a final diagnosis.

*Case 2.* A 30-year-old woman requested CVS because she has a brother with Down's syndrome. The first step in the counseling of this individual is to determine the type of chromosomal abnormality that caused Down's syndrome in her brother. If chromosome studies indicate a nondisjunctional event (i.e., "free" trisomy 21) or mosaic trisomy 21, then this woman would have no increased risk of having a child with Down's syndrome. However, if her brother has an unbalanced translocation, then the woman is at risk for being a carrier of a balanced translocation. In this circumstance, or if the affected relative is no longer living and did not undergo chromosome studies, peripheral blood chromosome studies should be performed on the patient to rule out this possibility.

If this woman is found to have a balanced 14;21 translocation, her risk for having a child with Down's syndrome would be 10%, and she would be an appropriate candidate for CVS. However, if the brother did not have the unbalanced translocation form of Down's syndrome or the patient is not a carrier of a translocation, her risk for having a child with Down's syndrome would be no different than any other 30-year-old woman (approximately 1 in 1000). Therefore, the risk of miscarriage from the procedure (1 in 200) would be significantly higher than the risk of an affected child. In this circumstance, nondirective counseling is essential. For some individuals, the increased risk of miscarriage is of more concern than the risk of Down's syndrome. They may choose a screening test, such as multiple-marker screening (see below), rather than take this risk. For others, the concern about having a child with Down's syndrome may be paramount, and they may choose diagnostic testing such as CVS despite the risk of miscarriage. The final decision to undergo prenatal testing should always rest with the patient, not with the physician or other health care providers.

## PERCUTANEOUS UMBILICAL BLOOD SAMPLING (PUBS)

Also called cordocentesis, PUBS is a technique for obtaining a small sample of fetal blood. Most commonly performed after the eighteenth week of gestation, it involves continuous ultrasound guidance of a small needle into the umbilical vein, either at the placental or fetal end of the umbilical cord (Figure 13-3). PUBS can be used when a rapid chromosome study is needed, because results are available in 48–72 hours. It can be used to assess fetal hematologic status (e.g., thalassemia, coagulation disorders, platelet disorders) or to detect fetal infections. PUBS also is a tool for fetal treatment. It can be used to transfuse blood components, to provide direct intravascular delivery of medications, and, in some circumstances, to transplant fetal bone marrow.

> PUBS can provide a chromosome result in 48–72 hours.

As with other invasive procedures, the major risk of PUBS is miscarriage or fetal death when performed late in pregnancy. In centers that have extensive experience with this procedure, the risk for miscarriage or fetal death is approximately 1% [7]. Therefore,

**FIGURE 13-3** ▶
*Percutaneous Umbilical Blood Sampling.*
*This test is performed by inserting a needle into the umbilical vein at the placental insertion of the umbilical cord.*

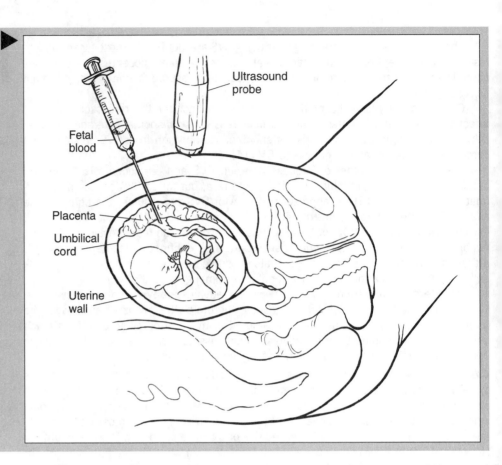

PUBS should be used only when the needed diagnostic information cannot be obtained by a less risky procedure or when a rapid diagnosis is essential to pregnancy management.

*Case 3.* Mrs. W. is referred to a medical center because an ultrasound performed in the eighteenth week of her pregnancy revealed fluid in the pleural cavity (pleural effusion), abdomen (ascites), and skin (edema) of the fetus. This condition, known as fetal hydrops, can result from certain chromosomal abnormalities, fetal infection, or a severe fetal anemia.

Although amniocentesis can provide information regarding chromosome status and can be used for the diagnosis of certain viral infections, it cannot assess fetal hematologic parameters. PUBS can be used for rapid karyotyping, for rapid assessment of blood counts, and for diagnosis of certain infections. In this case, a careful history from the mother reveals that her 4-year-old child was recently diagnosed with fifth disease (characterized by a fever and a rash that is most prominent on the cheeks). This mild condition in children, which is caused by a parvovirus, rarely causes symptoms in affected adults. However, if a pregnant woman becomes infected, and the virus is passed to the fetus, fetal infection can result in a severe anemia and often fetal death.

Mrs. W. immediately underwent PUBS, and the fetal hematocrit was 8% (normal is 30% at this gestational age). An immediate transfusion of the fetus was performed. During the following week the hydrops resolved. The infant was born at term and was normal.

This case illustrates the value of taking a careful history and the necessity of knowing how fetal pathophysiology differs from that of children and adults. A mild or inconsequential disease in children can be a devastating one in the fetus. Likewise, choosing the correct diagnostic tool allows a precise diagnosis, and in this case an appropriate therapy.

## FETAL TISSUE SAMPLING

In some genetic disorders, the gene in question is not expressed in amniotic fluid cells, chorionic villi, or fetal blood. Therefore, a number of techniques for fetal biopsy have been developed, including fetal skin, liver, and muscle biopsy [8]. Because of the rarity of the conditions that require these forms of invasive testing, there is insufficient information from which to ascertain the risks of each procedure. Because the biopsy needles are larger than those used for amniocentesis, and the length of the procedures is longer, the risk of fetal biopsy is expected to be greater than the risk of amniocentesis.

The second major limitation to fetal tissue biopsy is the limited knowledge about when a genetic disease manifests in the fetus or about when an enzymatic pathway is "turned on" in a normal fetus. For example, dermatologic diseases involving defective keratinization do not manifest until the time at which fetal skin is normally keratinized (approximately 20 weeks gestation). Fetal skin biopsy performed at an earlier time would appear normal, even in an affected fetus. For several metabolic disorders, enzyme activity is expressed only in hepatocytes. Fetal liver biopsy can be performed to measure enzyme activity, and the absence of activity would indicate an affected fetus. However, if the enzyme is not normally expressed until later in pregnancy, a fetal liver biopsy performed prior to normal expression would indicate an apparent deficiency, even in an unaffected fetus. It is essential that physicians performing these highly specialized procedures have a thorough understanding of normal fetal embryology and physiology.

*Case 4.* An amniocentesis performed for advanced maternal age on a 35-year-old woman reveals an apparently balanced translocation between an X chromosome and chromosome 20 in a female fetus. Evaluation of the translocation chromosome indicates that the breakage is at Xp21. Because the process of X inactivation involves only the normal X chromosome (to prevent inactivation of the chromosome 20 material on the translocation chromosome), the child would be expected to have Duchenne's muscular dystrophy (DMD) if the gene has been disrupted during the translocation.

The usual ways to diagnose DMD prenatally are linkage analysis and direct detection of molecular deletions in those cases in which affected male relatives are known to have significant deletions of the DMD gene as the cause of their disease. In the present case, there is no previously affected individual to provide the molecular information.

It is possible to take advantage of the knowledge that the deficient protein in DMD is dystrophin. Dystrophin is detectable in normal skeletal muscle by immunofluorescence, and it is absent in the muscle of individuals with DMD. Previous studies have found that dystrophin is present in muscle even in fetal life. Therefore, an option for the patient is to undergo fetal muscle biopsy to determine if normal dystrophin staining is seen in the muscle biopsy.

At 20 weeks of pregnancy, the patient underwent fetal muscle biopsy without complication. Immunostaining of the muscle tissue with an antibody for dystrophin indicated absence of dystrophin, which is consistent with the diagnosis of DMD.

This case not only illustrates the benefit of specialized prenatal procedures but also points out the significant concerns raised when a de novo (sporadic) translocation is found during amniocentesis. With this type of break and rearrangement of chromosomes, there is a risk that an important gene will be disrupted and rendered nonfunctional. Only in very rare circumstances, such as the one illustrated in this case, is a specific gene known to reside in the area of breakage (or, more importantly, is a diagnostic test available to determine if the gene has been disrupted). Fortunately, in more than 90% of the cases the translocation does not result in any clinically significant finding, and the child is normal.

Table 13-1 compares the various techniques of prenatal diagnosis discussed above.

**TABLE 13-1**
*Techniques for Prenatal Diagnosis*

|  | CVS | Amniocentesis | PUBS | Fetal Tissue Sampling |
|---|---|---|---|---|
| Gestational age performed | 10–12 weeks | 16–18 weeks | Greater than 18 weeks | Greater than 20 weeks |
| Risk of miscarriage | 0.5% | 0.5% | 1% | 1%–5% |
| Benefits of screening | Early diagnosis | Relative safety | Provides blood sample for hematologic studies | Allows direct assays on specific tissue type |

*Note.* CVS = chorionic villus sampling; PUBS = percutaneous umbilical blood sampling.

# SCREENING FOR GENETIC DISORDERS IN PREGNANCY

Invasive diagnostic tests are offered only to couples whose pregnancy is known to be at an increased risk for a specific genetic disorder. Because most children with genetic disorders are born to couples without any family history of birth defects, this approach detects only a very small number of affected fetuses. A more effective approach would be a screening test to determine pregnancies that are at a higher risk for a genetic disorder. Currently, an acceptable screening test is available for only two genetic conditions—NTDs and Down's syndrome.

> *Screening tests in pregnancy are only available for NTDs and Down's syndrome.*

## Neural Tube Defects (NTDs)

As a group, NTDs are among the most common congenital malformations, occurring with a frequency of 1 to 2 per 1000 live births. Only 10% of children with NTDs are born to couples known to be at risk because of a family history of NTDs. Therefore, screening all pregnant women would be necessary to detect a significant number of NTDs prior to birth.

In the fetus, the major plasma protein is AFP, which is analogous to albumin in adults. AFP is produced in the fetal liver and excreted into fetal urine. Because the major component of amniotic fluid is fetal urine, AFP is present in significant amounts in amniotic fluid and gains access to the maternal circulation via the placenta and across fetal membranes. The concentration of AFP in the maternal serum increases with increasing gestational age, reaching a peak at 32 weeks of pregnancy.

In pregnancies in which the fetus has an NTD, large amounts of AFP leak across exposed vessels, resulting in increased amounts of AFP in amniotic fluid and maternal serum. Using an absolute cutoff value of 2.5 multiples of the median (MoM), approximately 80% of fetuses with an NTD would be detected. A patient's MoM value is calculated by dividing her AFP level by the normal median level of AFP for the gestational week when the patient is tested. Maximum accuracy requires that the serum be drawn between 16 and 18 weeks gestation.

There are a number of other reasons for maternal serum AFP (MSAFP) levels greater than 2.5 MoM, including underestimation of gestational age, twin or triplet pregnancies, fetal death, or other fetal malformations. Therefore, an algorithm similar to that in Figure 13-4 must be used to detect pregnancies with a NTD effectively without placing a significant number of normal pregnancies at risk for invasive diagnostic procedures. As a

**FIGURE 13-4** ▶
*Algorithm for Maternal Serum Alpha-Fetoprotein (MSAFP) Screening.* MoM = multiple of the median.

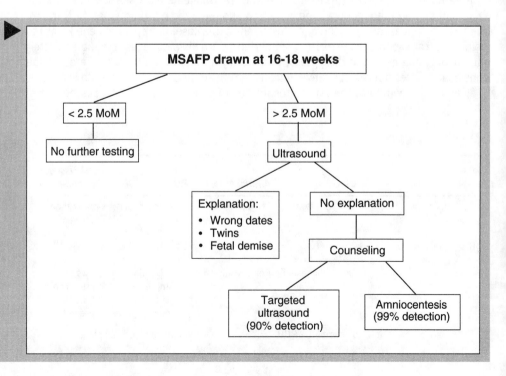

test for NTD, measuring the level of MSAFP meets the criteria of a "good" screening tool with a low false-positive rate (5%) and a relatively low false-negative rate (20%).

**Case 5.** A 25-year-old woman presents to her primary care physician at 16 weeks gestation requesting the "AFP test." Her history is remarkable for the birth of a child 2 years ago with a myelomeningocele.

A couple who has a previous child affected with a NTD has a 2%–3% chance of a recurrence in subsequent pregnancies. Therefore, a diagnostic test (e.g., amniocentesis to measure amniotic fluid AFP) should be performed. Assessment of amniotic fluid AFP detects 95% of open (not skin-covered) NTDs. If an assay for acetylcholinesterase (a neural-specific protein) also is performed on the amniotic fluid, the detection rate is greater than 99%. Measuring MSAFP detects only 80% of open NTDs and should be used for pregnancies in which there is not an increased risk above that of the general population (1 to 2 per 1000) for having a child with a NTD.

The marked increase in the resolution of ultrasound equipment and the increased expertise of sonologists in detecting NTDs by ultrasound allows a second option for parents with a previous child having a NTD. In centers with experienced sonographers, ultrasound detects 90%–95% of NTDs, which is a detection rate not greatly different from that of amniocentesis. However, the added benefit is that there is no increased risk of miscarriage associated with sonography.

## Down's Syndrome

In 1984, Merkatz and coworkers reported that MSAFP levels were lower in pregnancies with Down's syndrome [9]. Subsequently, other maternal serum components have been analyzed, and two additional components have been found to be useful in screening for Down's syndrome [10]. Similar to MSAFP, maternal serum estriol ($uE_3$) is decreased in pregnancies with Down's syndrome. The third marker, human chorionic gonadotropin (HCG), is elevated in pregnancies in which the fetus has Down's syndrome.

The most significant factor that determines the risk for having a child with Down's syndrome is the mother's age. Therefore, any screening test must take into account the risk that is specific to the patient's age. Using a combination of the three serum markers (i.e., AFP, $uE_3$, HCG) and the patient's age, a risk calculation is performed to determine the likelihood that the fetus has Down's syndrome. For example, a patient at age 31 years has a risk of 1 in 650 for having a fetus with Down's syndrome (for purposes of screening, risk figures are those for the likelihood of an affected fetus at midpregnancy, when the testing is performed). However, serum values for AFP of 0.75 MoM, $uE_3$ of 0.60 MoM, and HCG of 2.1 MoM would modify that risk to 1 in 200, a risk comparable to that of a 37-year-old woman. The cutoff value for calling a test "positive" is a risk of 1 in 270 or greater, the risk for a 35-year-old woman based on age alone. Using this approach, a risk of less than 1 in 270 is called "normal," regardless of the original age-related risk.

One of the most important parameters in this form of screening is gestational age. Because the serum components change significantly with increasing age of the pregnancy, an error in dating the pregnancy of as little as 14 days can modify the risk figures tenfold. If the 31-year-old patient above had her testing done at what she believed to be 18 weeks from her last menstrual period, but a subsequent ultrasound found her to be only 16 weeks at the time of testing, a recalculation of her risk would be necessary. Based on an earlier gestational age, her new value for AFP, $uE_3$, and HCG would now give her a risk of approximately 1 in 2000, a "normal" test result.

The algorithm for the evaluation of a patient with an increased risk for Down's syndrome based on multiple marker screening is outlined in Figure 13-5. Using this approach to screening in women younger than age 35 years, approximately 60% of Down's syndrome pregnancies would be detected with a false-positive rate of 4%; that is, 4% of women carrying fetuses who do not have Down's syndrome would have a risk of 1 in 270 or greater based on multiple marker screening.

**Case 6.** A 36-year-old woman requests that her obstetrician perform multiple marker screening to determine if her fetus has Down's syndrome. At age 36 years, this woman has a risk of 1 in 100 for having a fetus affected with either a trisomy (i.e., 21, 18, 13)

**FIGURE 13-5** ▶

*Algorithm for Multiple Marker Screening for Down's Syndrome.* LMP = *last menstrual period.*

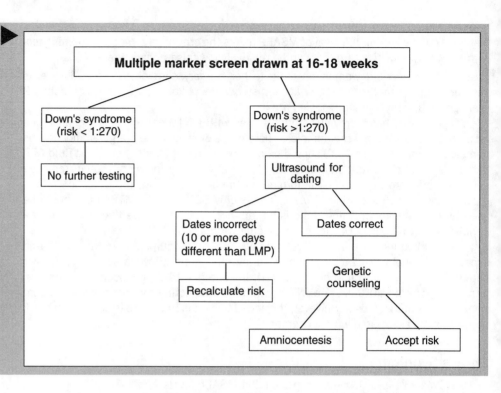

or one of the sex chromosome abnormalities (i.e., 47,XXY or 47,XXX). Amniocentesis detects all of these abnormalities with essentially 100% accuracy. However, there is a risk of miscarriage of 1 in 200.

There has been limited testing of the effectiveness of multiple marker screening in women older than 35 years. A study by Haddow and colleagues found that at age 36 years, the detection rate for Down's syndrome was approximately 70% [11]. Only a limited number of the other chromosomal abnormalities were detected. This patient should be counseled that multiple marker testing is a screening test not a diagnostic test. If, after being fully informed about the benefits and limitations of both amniocentesis and multiple marker screening, the patient decides to have serum testing only, it should be performed. Documentation that amniocentesis was offered to the patient must be made in the patient's record to protect the physician against subsequent liability if a child with Down's syndrome is born.

In this case, the patient chose multiple marker screening and her risk was found to be 1 in 20. After genetic counseling regarding the implication of this finding, the couple chose to have amniocentesis performed. The karyotype of the fetus was 46,XY. It is important to remember that a risk of 1 in 20 means that there is a 95% chance that the fetus does *not* have Down's syndrome.

# NEWBORN SCREENING

## Screening Criteria

Newborn screening tests are performed on blood samples from more than 4 million infants annually in the United States, making it the most common type of genetic testing. There are certain criteria by which a disease process is selected for a screening program.

1. The disease should be serious, or potentially serious, if not treated early.
2. The condition should be relatively common.
3. It must be possible to differentiate affected from unaffected individuals (most importantly affected individuals must be differentiated from carriers of that disorder).

4. It must be possible to reverse, slow down, or ameliorate the disease. In some cases, even if the disease is untreatable, detection of the index case allows the family to make informed decisions regarding future childbearing.

5. There must be an advantage to early treatment over treatment begun at the usual time of diagnosis.

6. The screening time (i.e., time from test to time of initial treatment) must be adequate to provide a benefit.

7. There must be adequate health care resources to diagnose and treat all individuals with a positive finding on a screening test.

8. The cost of screening, diagnosis, and treatment in the asymptomatic stage should be less than the expenditures if the problem is not discovered until the symptomatic stage.

Screening for a disease also requires an acceptable, reliable, and valid test. A valid screening test is one with a high sensitivity (i.e., accuracy of the test in correctly identifying individuals with the disease) and a high specificity (i.e., ability of the test to identify all unaffected individuals as such). When screening for serious or life-threatening illnesses, it is preferable to have a test with high specificity because it will have few false negatives (i.e., affected individuals with a "normal" test result). In most cases this type of test results in a lower specificity and a higher rate of false positives (i.e., unaffected individuals with an "abnormal" test result). In other words, it is preferable to detect as many affected individuals as possible, even if a reasonably large proportion of unaffected individuals are required to undergo further testing to exclude the disease in question.

*Sensitivity describes the accuracy of a test in correctly identifying affected individuals. Specificity describes the ability of a test to identify accurately unaffected individuals.*

## PRESENT STATUS

Currently, there are six genetic conditions for which at least 20 states screen newborns: phenylketonuria (PKU), congenital hypothyroidism, hemoglobinopathies (e.g., sickle cell disease, thalassemia), galactosemia, maple syrup urine disease, and homocystinuria [12]. These diseases fit the basic principles that have been established to govern newborn screening, which include the following.

1. There is a clear indication of benefit to the newborn.
2. A system is in place to confirm the diagnosis.
3. Treatment and follow-up are available for affected newborns.

The first population-based screening program was developed for the purpose of presymptomatic treatment of infants with PKU, an inborn error of metabolism caused by a deficiency of the enzyme phenylalanine hydroxylase. If PKU is not diagnosed and corrected early in life, the resulting high blood levels of phenylalanine can lead to severe, progressive mental retardation. Dietary restriction of phenylalanine, started in early infancy, is highly effective in preventing mental retardation. Because of the significant benefit of early diagnosis and treatment, screening for PKU is mandatory in many states (parents may refuse testing only on religious grounds).

A second disease for which early diagnosis and treatment are essential for normal physical and mental development is congenital hypothyroidism. As with PKU, newborn screening for congenital hypothyroidism is mandatory in many states.

For the other diseases that are included in state newborn screening programs, the benefit of early detection is less clear. Many screening tests have been added to state programs without careful evaluation of their benefits in pilot programs. When the principles outlined above are not followed, unnecessary health care costs result, and parental anxiety increases, without any clear benefit of early diagnosis and treatment. For optimal results from a newborn screening program, the primary reason for performing a test should be the identification of a treatable disease for which presymptomatic treatment offers the most benefit.

*The primary purpose of a newborn screening test should be to identify a disease for which presymptomatic treatment offers the most benefit.*

## Laboratory Approaches to Newborn Screening

Newborn screening for PKU is the prototype for newborn screening of genetic diseases. In the Guthrie test, blood from a heel stick on a newborn is collected on filter paper, and the dried blood samples are then mailed to a reference laboratory for testing. In the laboratory, small disks are punched out of the blood-impregnated filter paper. Therefore, a large number of samples can be analyzed simultaneously, so the cost per test performed is quite low. By using dried blood spots, sample acquisition and transport is also simple and inexpensive.

Many of the newborn tests are done by Guthrie-type bacterial inhibitor tests (Figure 13-6). The principle of the Guthrie test is that the inhibition of bacterial growth by a toxic compound can be reversed in a competitive manner by the presence of a structurally similar physiologic compound. To test for PKU, a strain of bacteria is used that is sensitive to beta-2-thienylalanine. However, the growth inhibition produced by this compound can be reversed by phenylalanine. In the newborn screening test, the bacteria and the toxic compound are mixed with agar and poured onto a plate. The filter paper disks are placed on the agar and then incubated. The amount of bacterial growth is directly proportional to the amount of phenylalanine present in the blood. The actual concentration can be estimated by comparison with a series of standards that are placed in the center of the plate.

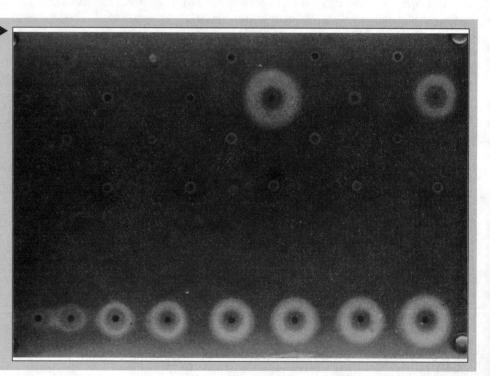

**FIGURE 13-6** ▶
*Guthrie Test for Newborn Screening. The controls on the* bottom row *are used to quantitate the levels in patients with abnormal results* (row 2).

Recently, the introduction of automated equipment that assays for a large number of blood compounds has raised the possibility of screening for as many as 40 genetic diseases from a single blood spot. Although this approach meets the criteria of being a simple, inexpensive test, it fails the major criteria because it detects many uncommon disorders, most of which have no effective treatment. Thus, early detection has no benefit to the newborn. In the future, as effective treatments are found for these rare, metabolic diseases, this multiplex screening approach may offer significant benefit.

*Case 7.* Mrs. J. is notified by her pediatrician that the newborn screening test performed on her infant indicates the possibility of cystic fibrosis (CF). The pediatrician orders a sweat test, which is normal (i.e., chloride is not elevated) and does cheek brushings for molecular studies. Those results indicate that the child has a single CF mutation (ΔF508) and is a carrier for CF. Additional molecular studies on both Mr. and Mrs. J. reveal that Mrs. J. also has a single CF mutation and is a carrier like her child.

The couple requires extensive counseling that emphasizes the following points: (1) being a carrier for a CF mutation has no clinical or health-related implications, and (2) false-positive results can occur with newborn screening tests. The child then is referred to his pediatrician for routine care.

This case illustrates the significant anxiety and concern that is raised by an initial positive screening test. A system must be in place to provide counseling for the family and to perform appropriate diagnostic tests as quickly as possible. Once the result is confirmed to be a false positive, it is essential that the parents understand that their child does not have the disease and that nothing will arise later.

Although CF is common, with an incidence of 1 in 2000 live births, newborn screening is controversial. There is a reasonably high false-positive rate with the test used for screening, and little scientific information exists to suggest that early diagnosis provides any benefit over a diagnosis made at the time of initial clinical symptoms. Currently, only a few states offer newborn screening for CF.

# ETHICAL IMPLICATIONS OF GENETIC TESTING

Each new genetic test that is developed raises questions about the circumstances under which the test should be administered, how testing should be implemented, and how results are to be used. Other questions include the following: (1) Should testing be voluntary or mandatory? (2) Should the individual be able to control who has access to the results of his or her tests? (3) If test results are released to an insurer, what protection must be in place to ensure that individuals are not discriminated against because of their genotype?

## Ethical Principles

There are four important principles that help provide answers to these questions: autonomy, confidentiality, privacy, and equity [12]. In the context of genetic screening, respect for *autonomy* refers to the right of persons to make informed, independent judgments about whether they wish to be tested and whether they wish to know the results of that testing. Once individuals have undergone genetic testing, *privacy* includes the right to make an informed decision about whether others (e.g., insurers, employers, spouses, other family members) may know details of their genotype. *Confidentiality* implies that any information regarding the patient or individual cannot be released to others without the consent of the individual, unless required by law (e.g., certain communicable diseases). In the context of genetic testing, *equity* implies equal access to health care, employment, and insurance without regard to one's genotype.

## Voluntary Testing

Voluntariness should be the cornerstone of any genetic testing program. There is no justification for any mandatory public health program involving genetic testing. Obtaining informed consent is the method of ensuring that the patient undergoes testing voluntarily. Informed consent involves giving information about the risks, benefits, efficacy, and alternatives to testing; information about the severity, potential variability, and treatability of the disorder being tested; and information about subsequent decisions that will be likely if the test is positive.

## Implications

### DISCRIMINATION

Genetic testing has implications beyond the scope of an individual's health care. Results of genetic tests could be used to preclude a person from life or health insurance coverage or from certain employment opportunities. Although several states have passed legislation protecting against this form of discrimination, the key factor is maintaining confidentiality at all times.

## NONPATERNITY

The results of genetic testing on one individual may have implications for the health of other family members or for their reproductive decisions. Patients must be encouraged to share appropriate information with other family members, but rules of confidentiality do not allow health care providers to contact family members, unless the information being withheld is potentially life-threatening. Also, the results of testing children for diagnostic purposes may reveal cases of nonpaternity. Current recommendations are that paternity information be shared with the mother but not volunteered to her partner.

## REPRODUCTIVE OPTIONS

Finally, in cases of prenatal diagnosis, couples should be informed about all reproductive options in a nondirective fashion, regardless of the religious or ethical views of the health care provider. Likewise, intent to abort an affected fetus should never be a prerequisite to offering prenatal diagnosis.

*Case 8.* A 37-year-old woman underwent amniocentesis because of advanced maternal age. The chromosome results were 46,XX with a variant of chromosome 15. This apparent duplication of a portion of chromosome 15 could represent noncoding material and have no clinical significance, or it could represent expressed genes and result in significant mental impairment or birth defects in her fetus. The patient was contacted and asked to come in with her husband to have blood drawn for chromosome analysis to determine if one of them had a similar chromosome 15. When the results indicating that neither had the variant chromosome were communicated to the mother, she volunteered that the father of her fetus may not be her husband. Subsequent chromosome analysis on the woman's boyfriend revealed the variant chromosome 15, and the mother was reassured that there would be no clinical implications to her child.

This case illustrates that inadvertent information may be obtained even in the course of routine cytogenetic studies for advanced maternal age. It is always important to consider nonpaternity in cases with apparent discrepant results. If nonpaternity is indeed the explanation, that information should be communicated only to the mother.

Advances in molecular biology have markedly broadened the ability to diagnose genetic conditions in fetuses, in children, and in adults. Good medical care requires that any testing that is contemplated be undertaken only with the fully informed consent of the patient or parents in the case of a minor. Any testing should have a clear benefit to the person's health or future reproductive decisions.

# RESOLUTION OF CLINICAL CASE

A 38-year-old woman has an approximately 1 in 75 risk of having a pregnancy affected with a chromosomal abnormality. During the genetic counseling session, the various diagnostic options (e.g., traditional amniocentesis, early amniocentesis, CVS) and screening options (e.g., multiple marker testing) were discussed in a nondirective fashion. The risk and benefits of both approaches were presented. Because the patient desired precise diagnostic information and wished to have that information as early in the pregnancy as possible, she chose CVS, which was performed at 10 weeks gestation. The results were normal.

Because CVS does not test for NTDs, and family and medical history had indicated no factor to place the patient at increased risk, she elected to have screening for NTD by MSAFP testing. This result, obtained at 16 weeks gestation, was 0.9 MoM, reducing her risk for a child with an NTD to approximately 1 in 10,000.

Upon completion of all genetic testing, the patient was reminded that not all birth defects are detected by these diagnostic and screening tests. However, based on these normal results, her risk for any birth defect is now lower than the overall population risk for birth defects, which is 2%–3%. In other words, prenatal diagnosis does not guarantee a normal child.

*Remember: prenatal diagnosis does not guarantee a normal child.*

# REVIEW QUESTIONS

**Directions:** For each of the following questions, choose the **one best** answer.

1. Chorionic villus sampling is indicated in which of the following situations?

    **(A)** The couple had a previous child with spina bifida

    **(B)** Fetal omphalocele was detected by ultrasound at 20 weeks gestation

    **(C)** The father is a carrier of a 14;21 translocation

    **(D)** The patient's brother has Down's syndrome

2. Mrs. Z. is a 40-year-old attorney with a long history of infertility who is now 10 weeks pregnant. The diagnostic tests recommended for Mrs. Z. should be

    **(A)** maternal serum multiple marker screening

    **(B)** amniocentesis

    **(C)** percutaneous umbilical sampling

    **(D)** detailed ultrasonography

3. As the head of the state's newborn screening program, a physician is asked to evaluate a new test for a biochemical disorder that is fatal in childhood. The decision to add this new test to the newborn screening program is based on which of the following factors?

    **(A)** The test is both highly sensitive and specific

    **(B)** The test is simple and inexpensive

    **(C)** The information can be useful to the parents in planning future pregnancies

    **(D)** There is an effective treatment for the disease if it is diagnosed early

**Directions:** The group of questions below consists of lettered choices followed by several numbered items. For each numbered item, select the appropriate lettered option with which it is most closely associated. Each lettered option may be used once, more than once, or not at all.

**Questions 4–7**

Match each of the following prenatal diagnostic methods with the optimal time in pregnancy for their performance.

    **(A)** 6–8 weeks

    **(B)** 10–12 weeks

    **(C)** 12–14 weeks

    **(D)** 16–18 weeks

    **(E)** 18–20 weeks

4. Chorionic villus sampling

5. Percutaneous umbilical cord sampling

6. Traditional amniocentesis

7. Multiple marker screening

# ANSWERS AND EXPLANATIONS

**1. The answer is C.** Chorionic villus sampling (CVS) provides a sample that can be cultured for chromosome studies. A father carrying a 14;21 translocation has a 3% chance of having a child with Down's syndrome. A pregnancy at risk for a neural tube defect is evaluated by measurement of the level of alpha-fetoprotein, which is a substance present only in amniotic fluid and maternal serum, not in chorionic villi. CVS is normally performed between the tenth and twelfth week of pregnancy and would not be indicated as a method for chromosome studies as late as 20 weeks gestation. A family member with Down's syndrome is not an indication for prenatal diagnosis in general, unless the individual has a translocation form of Down's syndrome. Prior to offering diagnostic testing, blood chromosomes should be done on the patient to determine if she is a carrier of a translocation, if the chromosomes of her brother are not known.

**2. The answer is B.** Mrs. Z. is at an increased risk for chromosome abnormalities caused by nondisjunction (e.g. trisomies 13, 18, 21; 47,XXY; 47,XXX). Amniocentesis provides precise diagnostic information for all of these abnormalities. Maternal serum markers are a screening test for Down's syndrome only, and they have only a 70% detection rate for Down's syndrome. Percutaneous umbilical blood sampling is a precise diagnostic tool, but has a higher risk of causing miscarriage than amniocentesis. Ultrasonography is a screening tool with a detection rate for Down's syndrome of approximately 20%. It has a slightly higher detection rate for trisomy 18 and 13 but would not detect sex chromosome abnormalities.

**3. The answer is D.** Tests that are highly sensitive, highly specific, simple, and inexpensive are important in designing a screening program, but the overriding principle should always be that early detection allows the institution of an effective therapy to prevent the effects of the disease. A screening test should never be added just because a disease can be diagnosed early and inexpensively. Although some screening programs have provided information to parents planning future pregnancies, this is not considered an appropriate reason for instituting newborn screening of a disorder for which there is no effective therapy.

**4–7. The answers are: 4-B, 5-E, 6-D, 7-D.** Chorionic villus sampling is most commonly performed between 10 and 12 weeks of pregnancy. Percutaneous umbilical cord sampling can be performed at any time from 18 weeks to the end of pregnancy but most commonly is done between 18 and 20 weeks. The optimal time for traditional amniocentesis *and* multiple marker screening is 16–18 weeks of pregnancy.

# REFERENCES

1. Tabor A, Madsen M, Obel EB, et al: Randomized controlled trial of genetic amniocentesis in 4606 low risk women. *Lancet* 1:1287–1293, 1986.
2. Henry GP, Miller WA: Early amniocentesis. *J Reprod Med* 37:396–402, 1992.
3. Byrne D, Marks K, Azar G, et al: Randomized study of early amniocentesis versus chorionic villus sampling: a technical and cytogenetic comparison of 650 patients. *Ultrasound Obstet Gynecol* 1:235–240, 1991.
4. Ferguson JE, Vick DJ, Hogge JS, et al: Transcervical chorionic villus sampling and amniocentesis: a comparison of reliability, culture findings and fetal outcome. *Am J Obstet Gynecol* 163:926–931, 1990.
5. Rhoads GR, Jackson LG, Schlesselman SE, et al: The safety and efficacy of chorionic villus sampling for early prenatal diagnosis of cytogenetic abnormalities. *N Engl J Med* 320:609–617, 1989.

6. Ledbetter DH, Martin AO, Verlinsky Y, et al: Cytogenetic results of chorionic villus sampling: high success rate and diagnostic accuracy in the United States collaborative study. *Am J Obstet Gynecol* 162:495–501, 1990.

7. Ghidini A, Sepulveda W, Lockwood CJ, et al: Complications of fetal blood sampling. *Am J Obstet Gynecol* 168:139–144, 1993.

8. Simpson JL, Golbus MS: *Genetics in Obstetrics and Gynecology*, 2nd ed. Philadelphia, PA: W. B. Saunders, 1992, pp 216–218.

9. Merkatz I, Nitowsky HM, Macri JN, et al: An association between low maternal serum alpha-fetoprotein and fetal chromosome abnormalities. *Am J Obstet Gynecol* 148:886–894, 1984.

10. Haddow JE, Palomaki GE, Knight GJ, et al: Prenatal screening for Down's syndrome with use of maternal serum markers. *N Engl J Med* 327:588–593, 1992.

11. Haddow JE, Palomaki GE, Knight GJ, et al: Reducing the need for amniocentesis in women 35 years of age or older with serum markers for screening. *N Engl J Med* 330:1114–1118, 1994.

12. Andrews LB, Fullarton JE, Holtzman NA, et al: *Assessing Genetic Risks: Implications for Health and Social Policy*. Washington, DC: National Academy Press, 1994, pp 65–70, 247–254.

# A LOOK TO THE FUTURE

# POTENTIALS OF GENE THERAPY

Gene therapy in the broadest sense is the treatment or prevention of human disease by genetic manipulation. This definition includes not only the correction of a clinical phenotype in a patient by the introduction of genetic material but also the ability to use genetic techniques to produce large amounts of therapeutic products and genetically engineered vaccines. Although most of the current excitement surrounding the use of gene therapy relates to genetic modification of human cells, the other uses of gene therapy should be appreciated.

The availability of many commonly used therapeutic agents depends on the genetic manipulation of cloned genes. For example, one of the first successes in this area was the production of genetically engineered insulin. The use of artificial insulin by patients with diabetes prevents problems of an immune response to the previously used pig insulin product. Other therapeutic products that have been produced by genetic engineering include human growth hormone to treat patients with growth hormone deficiency, factor IX for treating patients with hemophilia B, and interleukin-2 (IL-2) for treating patients who have renal carcinomas.

## Use of Gene Therapy to Treat Inherited and Acquired Diseases

Gene therapy potentially can be used to treat inherited disorders in which a single gene product is missing or abnormal, as well as to treat acquired diseases such as cancers or infectious diseases. In all cases, a cloned gene must be transferred into the patient's cells; once inside, it must be properly expressed in sufficient quantities to alleviate the clinical symptoms of the disease. It is also crucial to be able to target the cloned gene

*Clinical trials using gene therapy are under way in attempts to treat both inherited disorders and acquired diseases.*

to the tissue or organ that expresses the defect. For example, genes can be targeted to liver cells for diseases of hepatic origin, to muscle cells for muscles diseases such as Duchenne's muscular dystrophy (DMD), or to stem cells for diseases of hematopoietic origin.

Two major approaches are used in gene therapy protocols to introduce cloned genes into cells: *ex vivo methods* and *in vivo methods* (Figure 14-1) [1]. In the ex vivo approach, cells are removed from the patient, grown in culture, genetically altered in vitro, and then returned to the patient. The in vivo approach involves the introduction of a gene directly into the patient's cells. The in vivo approach is preferred because it is theoretically more efficient and less expensive. Unfortunately, this approach remains limited because of the inefficiency in targeting genes to specific tissues and the inability to control which cells take up the transferred genes. The ex vivo method is more difficult and more expensive, but it is easier to control. Because the genetic modification takes place in vitro, the types of cells that are genetically altered can be selected and specifically reintroduced to the patient in the gene therapy treatment.

**FIGURE 14-1**

**In Vivo and Ex Vivo Approaches to Gene Therapy.** *In vivo gene delivery involves the injection of the vector directly into the patient in an attempt to modify the cells in situ. In an ex vivo gene-delivery approach, cells are removed from the patient, modified in vitro, and then returned to the patient.*

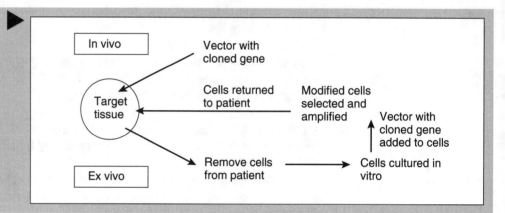

## Gene Delivery Systems

*Both ex vivo and in vivo approaches to gene therapy require an efficient method to transfer a gene into cells and to express that gene at a level sufficient to relieve the clinical phenotype.*

Both in vivo and ex vivo gene therapy require an efficient method to introduce the gene into cells as well as a need for a suitable level of expression of the introduced gene. Numerous ways have been developed to achieve the transfer of genes to mammalian cells. All of the methods have advantages and disadvantages, and no method is ideal. The methods used to transfer genes into cells can be classified into two major groups: *chemical and physical methods* and *biologic methods*, as listed in Table 14-1 [1, 2].

Once a gene or DNA enters a cell, it can have several fates. In some cases, portions of the added DNA are degraded in lysosomes, in which case the DNA never reaches the nucleus and is not expressed in the cell. Alternatively, the DNA can enter the nucleus undegraded, undergo recombination with the chromosomes of the cell, and become stably integrated into the genome. When integration of the exogenous DNA occurs, it most often occurs by a process called nonhomologous recombination, which results in a random insertion of the DNA into the chromosome at various sites. Only rarely does the exogenous DNA integrate into the chromosome at the site of its homologous gene. Even if the added DNA is not integrated at its normal site, it is associated with a chromosome and thus is stably expressed and transmitted to all daughter cells at mitosis. In some methods of gene transfer, the exogenous gene is not integrated into a chromosome but remains extrachromosomal in an episomal state. Because the episomal DNA does not contain a centromere, stable transmission to daughter cells does not occur. In the absence of any selective pressure, the genetic information is only transiently expressed because the episomal DNA is lost after several cell divisions.

*When exogenous DNA integrates into the human genome, it most often does so randomly at various sites within the chromosomes.*

### NONVIRAL VECTORS

***Calcium Phosphate Precipitation.*** The transfer of DNA relies on an element called a vector that helps deliver the gene of interest to the cells. Chemical and physical elements used in transferring genes into cells are also referred to as nonviral vectors.

TABLE 14-1
Vector Systems Used for Gene Transfer into Mammalian Cells

| Vector System | Advantages | Disadvantages |
|---|---|---|
| **Chemical and physical vectors** | | |
| Calcium phosphate | Easy to use | Random integration; inefficient DNA transfer |
| Electroporation | Easy to use; effective with many cell types; effective with embryonic stem cells | Random integration unless targeted; inefficient DNA transfer |
| Cationic liposomes | Noninfectious; nonimmunogenic; effective for in vivo gene transfer; can carry large DNA fragments | Unstable; remain episomal; poor gene expression |
| Polylysine-DNA conjugates | Targeted delivery; can carry large fragments of DNA | Unstable; remain episomal; poor gene expression |
| **Biologic (viral) vectors** | | |
| Retrovirus | Stable expression; broad host range; efficient transduction | Random integration; infects only dividing cells; shows variable expression; limit to size of DNA transferred |
| Adenovirus | Receptor-mediated uptake; infects nondividing cells; broad host range | Transient expression; remains episomal; elicits immune response; limit to size of DNA transferred |
| Adeno-associated virus | Potential site specific integration; noninfectious; infects nondividing cells; broad host range | Difficult to produce; limit to size of DNA transferred; in development stages |
| **Biologic (nonviral) vectors** | | |
| Human artificial chromosomes | Stable, noninfectious, can carry large fragments of DNA; nonimmunogenic; no integration into the genome | Still in developmental stages |

Calcium phosphate precipitation was the first method developed to introduce DNA into mammalian cells. This method involves purifying DNA or cloning a desired gene, mixing the DNA with calcium and phosphate to produce a flocculating precipitate, then allowing the mixture to precipitate on the surface of the cells. The cells take up the DNA by phagocytosis and deliver it into lysosomes. When the DNA is released from the lysosomes into the cytoplasm, it transverses the cytoplasm to the nucleus, where it crosses the nuclear membrane. Once inside the nucleus, the DNA can remain episomal to give transient expression, or a small portion may integrate into a chromosome to give stable expression.

In early experiments using calcium phosphate precipitation to transfer genes to cells, the frequency of obtaining a cell in which the exogenously introduced DNA was integrated and stably expressed was extremely low (i.e., 1 in 1 million cells treated). Additionally, experiments indicated that, contrary to expectations, the introduced DNA was not present at the site of the corresponding endogenous gene. Instead, the introduced DNA was randomly inserted into the cell's genome, with every cell having multiple copies of the introduced gene integrated and each cell having the DNA present on different chromosomes. The random integration of exogenously added DNA has since been found to be the most frequent event that occurs when DNA is introduced into mammalian cells and is a major drawback to many of the methods of gene transfer currently in use for gene therapy. Attempts to correct this problem have required the development of very sophisticated methods to direct the incoming DNA to a target site within the genome (i.e., targeting). Calcium phosphate precipitation is generally not used for gene therapy studies; it is used mainly in vitro for preparing certain viral vectors.

*Electroporation.* The technique of electroporation involves exposing cells in the presence of DNA or a cloned gene to an electric field. The intensity of the electric charge causes localized pore formation in the plasma membrane, which allows the DNA to be taken up by the cells into endosomes. When the DNA exits the endosomes, it transverses the cytoplasm and enters the nucleus, where it is expressed. Electroporation prevents DNA from entering the lysosomes, where it can undergo degradation. However, this method of gene transfer produces genetically altered cells at a very low efficiency and generally results in random integration of the DNA into the chromosomes. The technique of electroporation is used in the production of transgenic mice and in some ex vivo gene therapy protocols.

***Cationic Liposomes.*** A cationic liposome is made up of a cationic lipophile and a neutral lipid. When mixed together these two chemicals form a small unilamellar liposome that has the capacity to complex with DNA and create a lipid bilayer with the DNA sandwiched in between the bilayers (Figure 14-2) [3]. The positive charge of the DNA-liposome complex facilitates the binding of the complex to the negatively charged surface of mammalian cells, allowing it to be taken up by endocytosis into endosomes. Once released from the endosomes into the cytoplasm, the DNA enters the nucleus, where it remains extrachromosomal. In this episomal state, the DNA is transiently expressed for only 7–10 days before it is lost from the cells.

*The positive charges on DNA-liposome complexes facilitate binding of the complex to the negatively charged surface on human cells.*

**FIGURE 14-2**
*A Cationic Liposome-DNA Complex. Positively charged cationic liposomes form bilayers, with the DNA molecule sandwiched between the layers. The positive charge of the complex facilitates its binding to cell surface membranes. (Source: Adapted with permission from Gao X, Huang L: Cationic liposome-mediated gene transfer. Gene Ther 2:710–722, 1995.)*

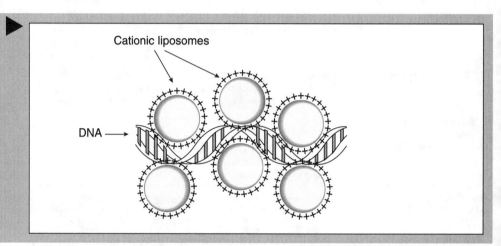

Cationic liposomes

DNA →

Cationic liposomes have several advantages as gene therapy vectors. They are capable of transferring large pieces of DNA, they are not infectious, and they do not elicit an immune response in the patient. The disadvantages of using cationic liposomes for gene transfer are their inefficient delivery to cells, the transient expression of the transferred genes, and the inability of the complexed DNA to be expressed at high levels. Currently, attempts are being made to increase the efficiency by which cationic liposomes transfer DNA into cells and to develop ways to promote their integration into chromosomal DNA so that the stability and rate of expression of the transferred genes increase. Despite their limitations, cationic liposomes are potentially very useful as nonviral vectors for gene therapy and currently are being used in human clinical trials.

***Polylysine-DNA Conjugates.*** Complexes made up of DNA coupled to polylysine offer another chemical method of gene transfer that is potentially useful for gene therapy protocols [4]. This method is particularly attractive because it allows the exogenous DNA to be targeted in vivo to specific tissues or organs. In this procedure, the DNA is linked to a specific antibody by a polylysine chain (Figure 14-3). The presence of the antibody facilitates the interaction of the DNA complex with other proteins, such as receptor ligands or viral proteins.

The most recent version of this DNA-ligand complex uses an antibody that recognizes the adenovirus protein coat [4]. This antibody facilitates the formation of a complex that has the DNA hooked through the polylysine chain to the antibody that is bound to the virus shell. When this adenovirus-polylysine-DNA complex is mixed with mammalian cells, the complex binds to specific receptors on the surface of the cell and is internalized

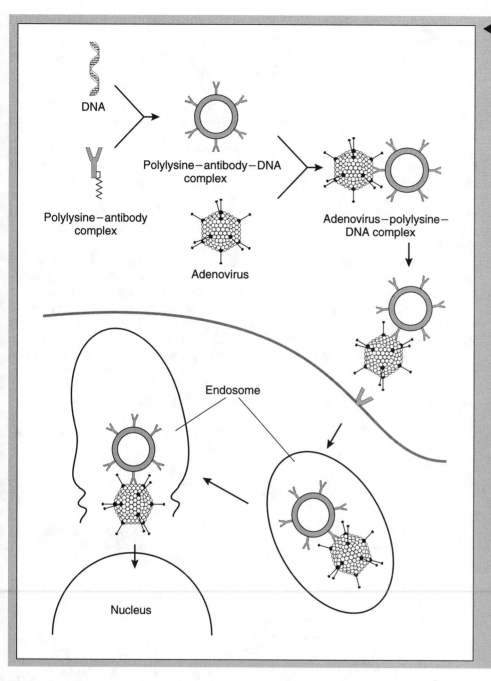

**FIGURE 14-3**
**Gene Transfer by Receptor-Mediated Endocytosis Using Adenovirus-Polylysine-DNA Complexes.** *An antibody specific for the adenovirus coat protein and attached to a polylysine chain interacts with a DNA molecule. The polylysine-antibody-DNA molecule forms a complex with adenoviral proteins. This complex binds to adenoviral-specific receptors on human cells and is taken into the endosomes of the cell by receptor-mediated endocytosis. Release of the complex from the endosomes is facilitated by adenoviral-directed disruption of the endosomes. (Source: Reprinted with permission from Curiel D: Adenovirus facilitation of molecular conjugate-mediated gene transfer. Prog Med Virol 40:1–18, 1993.)*

by receptor-mediated endocytosis directly into endosomes (see Figure 14-3). The adenovirus proteins serve two functions. One function is to bind to the cell surface receptors and to facilitate receptor-mediated endocytosis. The second function is to use the capacity of the proteins to stimulate disruption of the endosomes, thus releasing the DNA more efficiently into the cytoplasm. A more efficient release of DNA from the endosomes—a limiting step in several methods of gene transfer—results in an increased amount of the DNA reaching the nucleus, thereby resulting in an increased expression of the transferred gene.

Additionally, the complex can include proteins that recognize cell surface receptors other than the virus receptors. This modification helps target the complex to specific cells, although these cells may not have the adenovirus receptor. One protein that has been used in the complex is the asialoglycoprotein, a protein that is recognized by receptors on liver cells and acts to target the complex specifically to liver cells. Using these polylysine-DNA complexes for gene therapy is a promising approach, but currently this method of gene transfer has had limited success because of the transient retention of the DNA in the nucleus and the poor expression of the transferred gene.

*When adenoviral proteins are complexed to polylysine-DNA conjugates, uptake of DNA into endosomes is facilitated, and the DNA is more efficiently released from the endosomes into the cytoplasm.*

## VIRAL VECTORS

The major biologic vectors currently being used in gene therapy protocols are derived from mammalian viruses, which have evolved with the capacity to enter cells efficiently and then express their genes. Viral vectors are viruses that are genetically altered in such a way that they can still enter cells and express genes, but they can no longer replicate and become infectious. Both RNA- and DNA-containing viruses are currently being used to develop vectors for gene therapy.

The most frequently used and currently the best-developed viral vectors for gene transfer studies are the retroviral vectors. Retroviruses are RNA-containing viruses that efficiently infect many types of growing cells and randomly insert a copy of their genome into the host chromosomes, producing a cell with a stably integrated viral genome. Most of the available retroviral vectors in use today are derived from Moloney murine leukemia virus, a virus that is well studied at the molecular level and is easily manipulated in vitro [5].

*Life Cycle of a Retrovirus.* The life cycle of a retrovirus is shown in Figure 14-4. The genome of a retrovirus is made up of RNA and contains information for three viral genes known as *gag, pol*, and *env*. The *gag* gene codes for the core protein of the virus, the *pol* gene codes for a reverse transcriptase needed to form a DNA intermediate from RNA, and the *env* gene codes for the envelope proteins of the virus. These genes, which code for trans-acting proteins, are required for the replication and production of the virus. In addition to the three genes that code for trans-acting proteins, the virus genome contains several important nucleotide sequences that are cis-acting and do not code for proteins. These include the 5' and 3' long terminal repeats (LTRs) that contain the promoter, polyadenlyation, and integration signals and a packaging sequence known as *psi*.

> ***RNA-containing retroviruses**, used to develop viral vectors for gene therapy studies, randomly integrate into human chromosomes to provide stable gene expression.*

### FIGURE 14-4 ▶

***Life Cycle of a Retrovirus.*** *The RNA-containing retrovirus attaches to specific receptors on the cellular membrane and is internalized. The protein coat of the virus is removed, releasing the single-stranded RNA, which is then converted by reverse transcriptase to a double-stranded DNA intermediate. The double-stranded DNA molecule moves to the nucleus, where it randomly integrates into the cellular genome. Once integrated, the viral genome codes for RNA molecules that contain information for the Gag, Env, and Pol proteins. RNA molecules containing the* psi *sequence are assembled into new infectious viral particles, which are released by budding from the host membrane. LTR = long terminal repeat; gag = the gene coding for the viral core protein; pol = the gene coding for reverse transcriptase; env = the gene coding for the coat protein.*

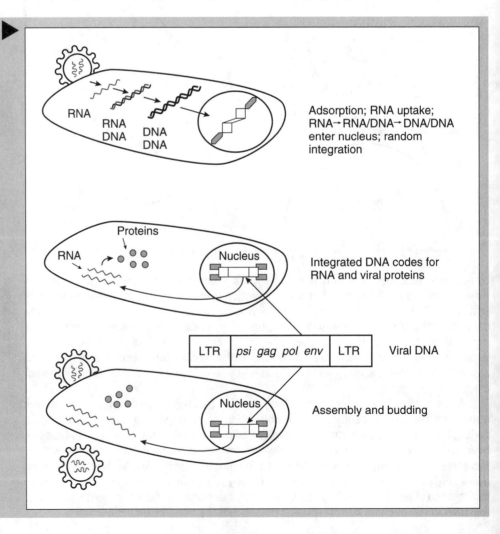

Adsorption; RNA uptake; RNA→RNA/DNA→DNA/DNA enter nucleus; random integration

Integrated DNA codes for RNA and viral proteins

| LTR | *psi gag pol env* | LTR |   Viral DNA

Assembly and budding

The virus infects growing cells by binding to a specific receptor on the surface of cells that interacts with the viral envelope protein and results in a fusion of the virus and the cell membrane. Following this fusion, the virus undergoes endocytosis into endosomes. Release from the endosomes is accomplished by a viral protein-directed action. Once released from the endosome, the RNA genome is reversibly transcribed into DNA by the reverse transcriptase, resulting in a double-stranded DNA molecule that enters the nucleus and integrates randomly into the chromosomes of the cell. Entry into the nucleus requires breakdown of the nuclear membrane, a process that occurs only in mitosis, thus limiting retrovirus infection to cells that are undergoing cell division and active growth.

Once the DNA of the virus is integrated into the chromosomal DNA, it remains stable and is transcribed into RNA molecules that contain the entire viral gene sequences. Using the *psi* sequence, this RNA molecule is then packaged into new viral particles, which are released from the cell by a budding process. The presence of the viral DNA in the cell's chromosome does not affect the growth or metabolism of the cell, which allows stable cells that continually produce virus particles to survive and proliferate.

*Genetic Modification of Retroviruses to Produce Vectors for Gene Therapy.* One important feature of retrovirus replication and packaging is that the proteins coded for by the *gag, pol,* and *env* genes can be provided in trans; that is, the genes do not need to be present on the virus genome for replication and packaging of the RNA to occur. Instead, cell lines, called *packaging cell lines*, can be created that maintain these genes integrated into their chromosomes and provide the function of these genes to defective incoming retroviruses. However, the ability to package an RNA retroviral genome into a viral coat to produce a virus particle requires that the *psi* sequence be physically present on that RNA molecule. Thus, in creating a retroviral vector, the *gag, pol,* and *env* genes can be removed from the viral RNA without affecting the packaging of the RNA into new virions, providing that the viral LTRs, the packaging sequence, and a few other sequences important for reverse transcription are left intact. The trans-acting functions of the Gag, Pol, and Env proteins are provided by the packaging cell line.

In preparing a retroviral vector for gene therapy protocols (Figure 14-5), a DNA copy of the viral RNA genome is genetically altered in vitro by first removing the *gag, pol,* and *env* gene sequences. These DNA sequences are replaced by the gene that is to be transferred into the host cells along with a second gene, which serves as a selective marker gene to allow the detection of cells that carry the modified viral DNA. It is important when creating the vector to have the newly inserted genes placed under the control of a promoter site to obtain proper gene expression. This promoter function can be performed by the viral LTR sequences or alternatively other promoter sequences can be added to the genetically modified vector [5].

*Preparing a retroviral vector for gene therapy protocols involves replacing the viral genes* gag, pol, *and* env *with the genes to be transferred.*

For increased safety, additional modifications of the viral genome are performed to eliminate the possibility that recombinational events might take place inside the cell and result in a virus that can carry out a productive infection. These modifications include making various deletions and rearrangements of sequences within the vector DNA as well as in the DNA present in the packaging cell lines that provides the trans-acting proteins.

*Packaging Cell Lines.* When preparing a defective retroviral vector, a packaging cell line is used to provide the Gag, Pol, and Env proteins in trans. Packaging cell lines are prepared by transfecting cells with viral DNA that has the *psi* packaging signal sequence deleted but contains genetic information for the Gag, Pol, and Env functions. These cell lines have the viral genes stably integrated into their chromosomes and produce RNA transcripts that code for the Gag, Pol, and Env proteins. These RNA molecules, however, cannot be packaged into infectious virions because of the absence of the packaging signal, *psi.*

*Packaging cell lines are used to provide the viral proteins Gag, Pol, and Env in trans for packaging of retroviral vectors.*

The procedure to produce a retroviral vector using these packaging cell lines is illustrated in Figure 14-5. In this example, a modified viral DNA is created in which the *gag, pol,* and *env* gene sequences are replaced with the adenosine deaminase (*ada*) gene and the gene called *neo*, which is a selective marker gene that makes a cell resistant to an antibiotic analogue of neomycin called G418. This vector DNA molecule retains the LTRs and the *psi* sequence. The modified viral DNA is transfected into a packaging cell line using calcium phosphate precipitation or electroporation, and the cells are plated in a

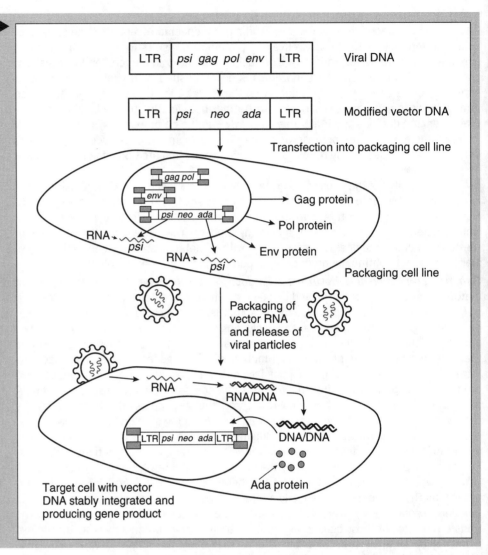

**FIGURE 14-5** ▶

*Producing Retroviral Vectors for Gene Therapy. Using a DNA copy of the retroviral genome, the viral genes gag, pol, and env are deleted and replaced by the genes neo and ada (adenosine deaminase). The modified vector is transfected into a packaging cell line by electroporation or calcium phosphate precipitation, and the DNA is randomly integrated into the host genome. The packaging cell line contains genes that code for the Gag, Pol, and Env proteins, which act in trans to produce the viral proteins. The RNA that is transcribed from the genetically modified vector carries, in addition to the neo and ada genes, a packaging signal, psi, which is necessary for the assembly of the RNA molecule into mature virus particles. The viral particles carrying the modified genome are released from the cell by budding and are capable of infecting a target cell only once. The double-stranded DNA produced from the modified vector genome integrates into the host genome of the target cell and produces a cell that stably expresses the adenosine deaminase protein.*

medium containing G418. Cells that have integrated the defective vector DNA now contain the *neo* gene and thus are stably resistant to G418. The cell lines with the modified viral vector stably integrated into a chromosome are the producer cell lines, which have the capacity to produce retroviral vectors that carry the *ada* gene.

The cell lines contain the viral genes that provide the Gag, Env, and Pol proteins in trans, but the RNA molecules produced that code for these functions do not carry a *psi* sequence. The defective viral vector DNA produces an RNA molecule containing the LTR, the *psi* sequence, the *ada* gene, and the *neo* gene. In the presence of the trans-acting proteins, only the RNA molecules containing the *psi* sequence, and thus the *ada* gene and the *neo* gene, are encapsulated and packaged into virus particles, which now bud from the cell into the medium.

The viral particles released from the cell can be directly added to a target cell or first purified and then added to target cells. These mature virus particles, although defective, can still infect a target cell by the normal procedure, but only once. Because they carry a functional reverse transcriptase within the viral particle, the defective viral vector makes a DNA copy of its RNA and stably integrates the DNA into the chromosomes of the target cell. By doing so, the target cell contains a stably integrated and functional gene such as the *ada* gene. The absence of any viral genes in the target cell and on the incoming defective viral vector genome prevents any further production of virus particles from the target cell.

Although retroviruses are currently being used in gene therapy clinical trials using ex vivo transfer, they still retain some disadvantages. Retroviruses are known to cause cancer in mammals. Because a potential, although slight, possibility remains that replication-proficient retroviruses might be generated sometime during the process, the vector preparations must be thoroughly tested before they can be used in humans. The viral vectors are capable of carrying only 8–9 kb of DNA, thus limiting the transfer process to smaller genes. Additionally, retroviruses infect only actively growing cells, making them unsuitable for transferring genes to nongrowing or differentiated cells. Finally, the random integration of the viral DNA into the host chromosomes poses a small chance for insertional mutagenesis if the integration occurs within a crucial cellular gene.

Despite their limitations, retroviral vectors offer certain clear advantages over other methods of gene transfer. First and foremost is their capacity to integrate a single copy of their genome into the host chromosomes. Although the integration is random, it provides a means for creating a cell that stably maintains the transferred gene, allowing for long-term expression of the gene. The virus has a broad host range and gives efficient transduction with many kinds of growing cells, making it useful for many types of ex vivo gene therapy approaches. For these reasons, a large number of current clinical trials are using retroviral vectors to transfer exogenous genes into human cells.

*Adenoviruses as Vectors for Gene Therapy.* Adenoviruses are DNA-containing viruses that normally produce respiratory infections in humans. The adenoviral genome is 38 kb in size and is modified for gene therapy protocols by deleting two early genes known as *E1* and *E3*, leaving approximately 7–8 kb of space for insertion of the gene of interest. Adenoviruses have a broad host range and infect both dividing and nondividing cells. They enter the cell by receptor-mediated endocytosis and have an effective mechanism to disrupt endosomes, making them potentially ideal virus vectors for gene therapy. On the negative side, once the adenoviral DNA is inside the nucleus, it does not integrate into the chromosomes but remains episomal, making stability and long-term expression of transferred genes a serious problem.

> *DNA-containing adenoviral vectors, genetically modified to carry human DNA, remain episomal and provide only transient gene expression.*

Results from initial experiments that used the E1-, E3-deleted adenoviral vectors for gene therapy have been disappointing because of inflammation and neutralizing antibodies against the viral proteins developing in the patients [10]. This adverse host reaction limits repeated administration of the virus, which would be necessary because of the lack of stable long-term gene expression seen with the episomal adenoviral DNA.

Recently, two new types of adenoviral vectors have been developed. The first, described previously (the adenovirus-polylysine-DNA complexes), utilizes only the adenoviral coat proteins, which are coupled to DNA via antibodies and polylysine. This new form of adenovirus vector may help overcome the problem of evoking the nonspecific inflammation and cellular immunity associated with previous adenovirus vectors, yet it still retains the important properties of receptor-mediated endocytosis and the ability to facilitate release of the transferred genes from the endosomes.

The second type of adenoviral vector being developed is one in which all of the protein-coding viral genes have been deleted. This allows the vector to accommodate larger DNA inserts (up to 28 kb) and to avoid the production of any viral proteins that might elicit an immune response. This new adenoviral vector retains only the LTR sequences and the packaging sequences of the virus and requires a helper virus for its preparation. It is currently being developed for use in gene therapy for DMD.

*Adeno-Associated Viruses as Vectors for Gene Therapy.* Adeno-associated viruses are single-stranded DNA viruses that require the help of another virus, such as adenovirus, to carry out a productive infection. When a helper virus is not available, the adeno-associated viral DNA integrates into the chromosome. The integration process is highly specific, occurring on the long (q) arm of chromosome 19 at 19q13 and results in stable, long-term persistence of the viral genome without any obvious harm to the host cells. The virus is small and can accommodate only 4.5 kb of DNA, limiting its usefulness for transferring larger genes. For smaller genes however, adeno-associated virus remains an attractive candidate for gene therapy and is capable of transducing genes to a wide range of cell types, at least in vitro. At present the virus is difficult to produce in large quantities and is still in the developmental stages as a gene therapy viral vector.

## MAMMALIAN ARTIFICIAL CHROMOSOMES AS VECTORS FOR GENE THERAPY

From the previous discussions it should be clear that no single ideal vector for gene therapy has yet been developed. Each currently used vector has both advantages and disadvantages as listed in Table 14-1, and the vector of choice depends on the gene being transduced, the cells being targeted, and the approach being taken in the gene therapy protocol. As an alternative to viral vectors, a new biologic nonviral vector—the human artificial chromosome (HAC)—recently has been developed (Figure 14-6).

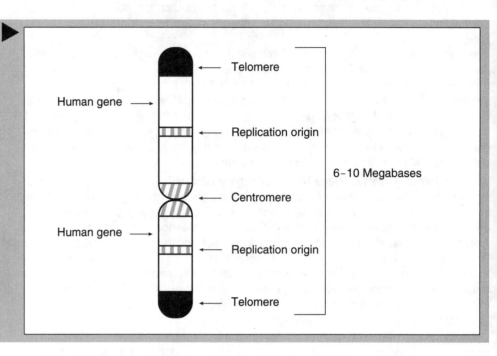

**FIGURE 14-6**

**Elements of a Human Artificial Chromosome (HAC).** *HACs that remain episomal and stable in human cells are created by combining telomere DNA sequences, centromere DNA sequences, and DNA fragments containing DNA replication origins with cationic liposomes. The DNA-liposome complex enters human cells, with the DNA sequences combining in vivo to produce a stable microchromosome that is capable of undergoing controlled DNA replication and proper segregation to daughter cells.*

*Human gene* → 
*Telomere* ←
*Replication origin* ←
*Centromere* ←
*Human gene* →
*Replication origin* ←
*Telomere* ←
6–10 Megabases

**HACs** *can carry large fragments of DNA, are stable in a cell during cell division, and replicate in a normal fashion, which may make them the ideal vector for gene therapy protocols.*

The creation of a mini-HAC vector containing human DNA sequences, a centromere sequence, telomere sequences, and DNA replication origins offers a number of advantages for gene therapy that many other vectors lack. First, the HAC can accommodate large human DNA inserts, which removes the limitation on the size of the gene to be transferred. Second, the presence of a centromere sequence within the chromosome assures its stability and its equal distribution to daughter cells at mitosis. Third, the addition of telomere sequences to the chromosome ends provides stability for the chromosome structure and prevents its degradation and its fusion with other normal chromosomes present in the nucleus. Finally, the presence of DNA replication origins on the HAC maintains the normal copy number from one cell cycle to the next by regulating the replication of the HAC during the normal S phase of the cell cycle.

The production of HACs has come closer to reality with the cloning of human telomere sequences and the identification of repetitive elements that make up human centromeric DNA. One approach has involved creating the HAC in vitro and then transferring it to cells by liposomes or receptor-mediated uptake as discussed above. Alternatively, the HAC can be created in vivo, a feat that has recently been demonstrated [6]. In this experimental approach, human telomere DNA sequences, alpha satellite DNA sequences that serve as centromeres, a selectable marker gene, and genomic DNA fragments were transfected into human cells using cationic liposomes. Stable clones were isolated and analyzed by fluorescent in situ hybridization (FISH) and shown to contain mini chromosomes that remained stable in dividing cells over a period of 6 months. In one case, the mini chromosome appeared to be formed de novo inside the cells by the association of the added DNA fragments. If this novel system is reproducible, it offers a new and unique approach for developing HACs that can serve as important gene therapy vectors.

## Targeting Exogenous Genes

Currently, most of the methods of gene transfer that produce stable gene expression result in random insertion of the exogenous DNA into the chromosomal DNA. The exceptions to this are adeno-associated virus vectors that are still under development and HACs that remain stable in cells without being integrated into the genome. Although it is certainly advantageous for long-term expression of transferred genes that the DNA be integrated into the chromosomal DNA, obvious disadvantages are associated with random integration. In particular, the probability that the integration might activate a proto-oncogene or inactivate a tumor suppressor gene, both of which could lead to the production of a malignant cell, is of serious concern. Thus, the ideal situations are either to have extrachromosomal DNA stably maintained in the cells (e.g., with HACs) or to be able to target the incoming DNA to a specific site within the chromosomal DNA, resulting in stable integration without disruption of any endogenous genes.

Targeting genes to specific DNA sites requires that the added exogenous gene sequences undergo recombination with the homologous gene sequences present in the chromosomal DNA. For example, the *ada* gene would be targeted to integrate into one of the chromosomal alleles coding for adenosine deaminase. In the ideal situation, the integrated gene would be functional, would be under the same controls as the normal cellular gene, and would be expressed at the appropriate time.

Unfortunately, gene targeting by homologous recombination currently is only possible in vitro using very specific vector designs and selection procedures. The frequency of integration of exogenous DNA into cellular DNA can be very low, perhaps only 1 in 1000 cells. Within 1000 such cells, moreover, there may be only one that is the result of homologous recombination of the incoming DNA with the DNA sequence at the targeted site. It is important then to develop methods of identifying that one cell in 1000 that has the DNA targeted to the correct site within the chromosomal DNA.

In an in vitro system, where the cells can be cultured under different conditions, it is possible to use various selective marker genes to carry out a positive selection for all of the cells that have integrated the exogenous DNA followed by a negative selection that eliminates all of the cells that have integrated the DNA by random insertion. This positive–negative selection procedure allows only those cells that have undergone homologous recombination to be preferentially selected, grown into large populations, and then used in subsequent studies [7, 8]. Such a method has been very important in the creation of transgenic mice. Unfortunately, the method requires more refinement before it can be used for gene therapy studies in humans.

> *Targeting genes to specific sites* within a chromosome requires that the exogenous gene sequences undergo homologous recombination with the homologous endogenous gene sequences.

## Current Applications of Gene Therapy for Treating Human Diseases

Currently, there are a number of different strategies available for using gene therapy to treat both inherited and acquired human diseases. When dealing with inherited human diseases that are caused by the loss of function of a single gene, the approach most often used is to introduce into the appropriate cell copies of a functional gene, the so-called *gene addition approach*. In these experiments, it is difficult to control the level of expression within the proper cell type, and it is impossible to control the site of integration of the added gene when using retroviral vectors. Despite these limitations, if a low level of enzyme is sufficient to reverse the disease phenotype, then a gene addition procedure theoretically can be successful for treating diseases such as those listed in Table 14-2.

*Gene Therapy for Adenosine Deaminase (ADA) Deficiency.* The best known and currently the only successful example of using a gene addition approach to correct a disease phenotype, is the treatment of a form of severe combined immunodeficiency disease (SCID) that results from the loss of ADA activity. Absence of ADA activity leads to the loss of both B- and T-cell function and is usually lethal, with the patients dying of infections or cancer within the first few years of life. Treatments available for the disease include bone marrow transplantation, if a suitable donor is available, and enzyme replacement therapy using polyethylene glycol (PEG)–conjugated ADA protein. In 1990, the first ADA-deficient patient was treated by gene therapy using an ex vivo method and a retroviral vector system [9]. The normal *ada* gene of 1.5 kb was cloned into a retroviral vector, as illustrated in Figure 14-5. The genetically modified retroviral vector was transfected into T lymphocytes that had been removed from the patient and cultured in vitro. The modified T cells, which contained a retroviral vector carrying a normal *ada* gene integrated into their chromosomal DNA, were then reinfused into the patient.

The patient was 4 years old at the time and had been undergoing treatment with

**TABLE 14-2** ▶
*Examples of Inherited and Acquired Human Diseases Amenable to Gene Therapy*

| Disease | Cells Treated | Approach Used |
|---|---|---|
| **Inherited diseases** | | |
| Adenosine deaminase (ADA) deficiency | Stem cells | Ex vivo *ada* gene addition |
| Gaucher's disease | Stem cells | Ex vivo glucocerebrosidase gene addition |
| Cystic fibrosis | Respiratory epithelium | In vivo CFTR gene addition |
| Familial hypercholesterolemia | Liver cells | Ex vivo low-density lipoprotein receptor gene addition |
| Duchenne's muscular dystrophy | Muscle cells | In vivo Dystrophin gene addition |
| **Acquired diseases** | | |
| Malignant melanoma | Fibroblasts | Ex vivo interleukin-2 gene addition |
| Brain tumors | Tumor cells | In vivo addition of herpes simplex virus thymidine kinase gene |
| Ovarian cancer | Tumor cells | In vivo addition of p53 gene |
| Neck cancer | Tumor cells | In vivo infection with EB1 defective adenovirus |

*Note.* CFTR = cystic fibrosis transmembrane conductance regulator.

PEG-ADA protein with some limited success. After 5–6 months of gene therapy treatment, which involved repeated infusions of the retroviral-transfected T cells, the peripheral blood T-cell count of the patient had increased and stabilized, and her ADA activity had increased from nondetectable levels to 50% of the level seen in an individual heterozygous for the *ada* gene. After 2 years of treatment, the level of T cells and the levels of ADA activity remained stable, even when the gene therapy treatment was discontinued. As of October 1995, the patient had fewer symptoms of the disease although she was receiving decreased levels of PEG-ADA protein. She attends public school, has fewer infections, and is normal with respect to her weight and height. Her T-cell count and her level of ADA activity remain stable, although no more than 50% of her T cells contain the normal *ada* gene.

There seems little doubt that this patient has benefited from the gene therapy treatment that has been an important factor in her recovery. However, the fact that she is still receiving enzyme replacement therapy makes it difficult to reach any definite conclusions concerning how much of her recovery is due to the gene therapy treatment and how much is due to the continued enzyme replacement treatment. Another 9-year-old patient has undergone similar gene therapy treatment. However, in this case, fewer than 1% of the patient's cells express the normal gene, leaving many questions still to be answered about the important factors involved in this type of gene addition therapy when dealing with different patients.

The apparent success of the *ada* gene therapy clinical trials launched a whole series of additional clinical trials that attempted to use the gene addition approach to correct a variety of different inherited human diseases such as cystic fibrosis (CF), familial hypercholesterolemia (FH), and DMD.

***Gene Therapy for Cystic Fibrosis (CF).*** CF is an autosomal recessive disorder that has an incidence in the white population of 1 in 2500 live births (see Chapter 11). The disease is caused by a defect in the chloride ion ($Cl^-$) transport system, which results in the lungs of affected individuals accumulating excessive mucus, leading to chronic bacterial infections. The gene coding for the protein altered in CF patients—the cystic fibrosis transmembrane conductance regulator (CFTR)—was transferred to the epithelial cells of the lungs of 12 CF patients using an in vivo approach [10]. The vector of choice was an

*The reported success of the* ada *gene therapy clinical trials launched a series of additional clinical trials designed to treat single gene disorders and cancer.*

adenovirus-derived vector because the adenovirus naturally infects respiratory epithelial cells and could be administered in vivo directly to the target cells.

Analysis of nasal lavage fluids from the patients showed that the virus persisted for up to 8 days, but the percentage of cells transfected by the virus vector was less than 1%, and the clinical phenotype of the patients was not corrected. Additionally, the adenoviral vector proteins that were produced triggered a severe immune response as well as a local inflammatory response in the patients. With this adverse reaction elicited in the patients, the dose of the virus that could be administered was limited, and the efficiency of gene transfer remained low and thus was ineffective in treating the disease.

The recent development of new adenoviral vectors that are deleted from all viral genes may help eliminate the immunogenic response of the host and thus be more effective in future gene therapy trials [11]. Another gene therapy approach being tested that may avoid the immune reaction resulting from the presence of the adenovirus-vector proteins is the utilization of liposome-CFTR gene complexes administered in vivo to CF patients.

**Gene Therapy for Diseases of the Liver.** A large number of metabolic diseases, including FH, phenylketonuria (PKU), disorders of lipid metabolism, and cancers, are manifested in the liver, making this organ an important target for gene therapy. Because hepatocytes can be cultured in vitro and are susceptible to retroviral infection, an ex vivo gene therapy approach to treat diseases of the liver is theoretically possible. FH is a dominantly inherited disease caused by a deficiency of the low-density lipoprotein (LDL) receptor protein, with the disease phenotype being expressed in both the homozygous and heterozygous states (see Chapter 11).

In an initial clinical trial designed to use gene therapy to treat patients with FH, hepatocytes from five patients homozygous for the disease were obtained by surgically removing a lobe of each patient's liver and culturing the cells in vitro [12]. The cultured hepatocytes were transfected with a retroviral vector carrying the gene for the LDL receptor protein and then placed back into the patients by way of a catheter to the portal vein.

The five patients differed with respect to the type of receptor defect treated. Two of the patients had a receptor-negative (or null) form of the disease, whereas the other three patients expressed the receptor protein but at reduced levels. Significant variability was observed in response to the gene therapy treatment, with three of the five patients having little if any improvement and the other two patients showing only a small decrease in LDL catabolism. These variable results and the low level of correction of the clinical phenotype were attributed to the inconsistent engraftment of the replaced hepatocytes, which was considered to be less than 5%. The inefficiency of this gene therapy procedure precludes the treatment of liver diseases by this method until the level of gene transfer can be substantially improved.

**Gene Therapy for Muscle Diseases.** Another attractive candidate disease for gene therapy is DMD. DMD affects 1 in 3500 males, has no known cure, and is a particularly devastating disease that begins at 4–6 years of age and becomes progressively worse, with death occurring at an early age. Initial approaches to treat DMD involved the transplantation of normal human myoblasts to the muscles of affected boys in hopes of providing a functioning dystrophin gene. The goal was to achieve fusion of the normal donor myoblasts with host cells and thus provide a functioning protein that could repair the damaged muscle. To date, these experiments have not demonstrated any significant clinical change in the patients, necessitating the development of other approaches.

Retroviral vectors are not suitable vectors to treat DMD because of their inability to infect the nongrowing muscle cells. Additionally, the unusually large size of the dystrophin gene compounds the problem of finding a suitable vector that can transfer this gene to muscle cells. Recently, adenoviral vectors have been constructed that have all of their viral genes deleted and replaced by a mini dystrophin gene [11]. The mini gene is a complementary DNA (cDNA) for the dystrophin gene with all of the intron sequences deleted to make it small enough to fit into the deleted adenovirus-vector genome, which is capable of carrying 28 kb of foreign DNA. The adenoviral vector carrying the dystrophin gene is capable of transfecting skeletal muscle cells in vitro and expressing the modified dystrophin protein at high levels in vivo in dystrophin-defective mice (i.e., *mdx* mice).

This new adenovirus vector is potentially capable of being used to treat patients with DMD, but it requires additional modification and testing before it is used successfully in humans.

***Gene Therapy for Cancer.*** Most of the gene therapy clinical trials in progress are designed to treat human cancer, which is an acquired disease. Treating acquired diseases by gene therapy can involve several strategies. In one approach, tumor cells or fibroblasts are genetically modified to produce compounds that stimulate the normal host immune system. In this way, the enhanced immune system attacks and kills cancer cells. In a second approach, the cancer cells are directly killed by the production of toxic compounds that specifically attack the cancer cells. A third approach utilizes compounds such as anti-sense RNAs, which act by blocking the expression of oncogenes present in cancer cells. Finally, if the patient has a known defective tumor suppressor gene, such as the *p53* or the retinoblastoma (Rb) gene, theoretically one can add that gene back into the patient's cells by the methods discussed above. This approach is similar to those being used to correct diseases that result from the loss of a functioning gene.

Most of the clinical trials involved in treating cancers aim to insert into tumor cells ex vivo a gene coding for a cytokine, which is a compound that stimulates the natural immune system so that it recognizes tumor cells and either limits their growth or kills them. In a similar approach, fibroblasts can be removed from the patient then grown and genetically modified in culture with retroviral vectors so that they produce a cytokine (e.g., IL-2, a T-cell growth factor). The modified cells are irradiated and then injected directly into the tumor in the patient. An immune response is elicited at the site of the injection from the augmentation of natural killer cells that attack the cancer cells.

***Direct Killing of Tumor Cells.*** Two different approaches are being used to directly induce death of the cancer cells by the presence of specific toxic molecules. One approach uses a gene cloned from herpes simplex virus (HSV) that codes for thymidine kinase. The HSV thymidine kinase protein has the capacity to phosphorylate an analogue of thymidine known as ganciclovir. Human cells also express a thymidine kinase, but the human enzyme does not use ganciclovir as a substrate. Because human cells cannot convert ganciclovir to the inhibitory phosphorylated form, they are not sensitive to its killing action. However, when a human cell is transfected with a vector containing the HSV thymidine kinase gene and is then exposed to ganciclovir, that cell dies. Death results from the phosphorylation of ganciclovir by HSV thymidine kinase to a toxic molecule, ganciclovir phosphate, which blocks DNA replication and kills the cells.

The approach is to transfect actively growing tumor cells with retroviral vectors containing the HSV thymidine kinase gene and then administer ganciclovir to the tumor cells. This therapy is feasible for solid brain tumors because nearby normal brain cells are not actively growing and are not infected by the retroviral vectors, which can infect only the growing cancer cells. Because the normal brain cells are not infected by the retroviral vectors, they do not express the HSV thymidine kinase, and they remain viable in the presence of ganciclovir. However, the retroviral-infected tumor cells contain the HSV thymidine kinase, resulting in phosphorylation of ganciclovir and death of the cancer cells. Because the infection of the tumor cells within a solid tumor is not 100%, some of the cancer cells within the tumor are killed by a phenomenon called the "bystander effect," in which the inhibitory phosphorylated form of ganciclovir diffuses from the retroviral-infected tumor cells to the surrounding uninfected tumor cells.

***Use of Defective Adenovirus to Target and Kill Cancer Cells.*** A new approach being tested to kill tumor cells directly relies on the use of defective adenoviruses that cannot replicate in cells expressing the tumor suppressor protein (p53) [13]. Normal adenoviruses contain a gene that codes for a protein called E1B. During the infection process, the E1B protein binds to the p53 protein and stimulates the cells to enter the S phase of the cell cycle and support DNA replication of the virus. The replication and release of the virus kills the cells (Figure 14-7A). Fortunately, the immune system is capable of eliminating the virus within a few days.

The function of the p53 protein is to block cells in the G1 phase of the cell cycle in response to DNA damage or to the presence of foreign DNA (e.g., viral DNA). Thus, viruses have evolved proteins like E1B to block the function of p53 and stimulate cells to

*Gene therapy strategies to treat cancer aim to kill tumor cells either directly by the presence of toxic compounds or indirectly by stimulating the patient's immune response.*

bypass the G1 block and enter the S phase of the cell cycle. In theory, an adenovirus mutant that lacks the E1B protein should be unable to replicate in cells that express the p53 protein. In the presence of a functioning p53, the virally infected cells are blocked in the G1 phase of the cell cycle and will not enter into the S phase; thus, the virus is unable to replicate its DNA, and the infection is aborted. However, a cell lacking the p53 protein should serve as a receptive host for the defective adenovirus because the cell is no longer blocked in G1 phase in response to the viral DNA; thus, it progresses to the S phase, and it allows the defective virus to replicate and kill the host cells (see Figure 14-7B).

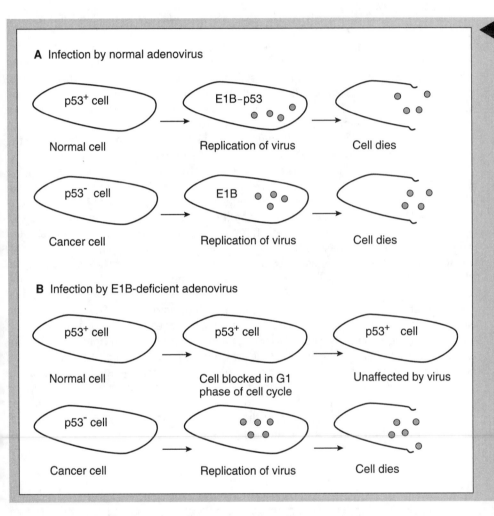

**A** Infection by normal adenovirus

p53⁺ cell → E1B–p53 → Cell dies

Normal cell   Replication of virus   Cell dies

p53⁻ cell → E1B → Cell dies

Cancer cell   Replication of virus   Cell dies

**B** Infection by E1B-deficient adenovirus

p53⁺ cell → p53⁺ cell → p53⁺ cell

Normal cell   Cell blocked in G1 phase of cell cycle   Unaffected by virus

p53⁻ cell → → Cell dies

Cancer cell   Replication of virus   Cell dies

**FIGURE 14-7**
*Use of a Mutant Adenovirus to Kill p53-Deficient Cancer Cells.* (A) Normal adenovirus infects and kills both normal and cancer cells. The E1B gene product of the virus blocks the function of p53 in normal cells, allowing the infected cells to enter the S phase of the cell cycle and support viral replication. Cancer cells deficient in p53 also enter S phase and actively support replication of the virus. (B) Adenovirus mutant for the E1B protein cannot kill normal cells because the functioning p53 protein blocks the infected cells in G1 phase, and the virus is unable to replicate. Cancer cells deficient in p53 are not blocked in G1 phase upon infection with the defective virus, and they enter S phase, allowing replication of the virus and specific killing of the p53-deficient cancer cells.

Because more than 50% of human cancer cells are defective in p53, a mutant adenovirus that does not express E1B could potentially be used to target these cancer cells and specifically kill them. Any normal cells present should use their p53 protein to prevent the replication of the defective virus and should remain viable.

Recently, this theoretical approach to selectively kill cancer cells defective in p53 has been demonstrated to be feasible [13]. A mutant adenovirus defective in the E1B protein was shown to lyse p53-deficient human cells but not kill cells that expressed a normal p53 protein. Furthermore, when the mutant virus was injected into human cervical carcinomas present in nude mice, there was significant regression of the tumor size and complete regression in 60% of the injected tumors. Currently, clinical trials are under way to test this therapy in individuals with head or neck cancers who have not responded to other therapies. Whether this therapy will work in humans is yet to be determined. The major obstacle in humans is the presence of an immune system that potentially can prevent the spread of the virus in the tumor. The therapy currently is limited to tumors that can receive direct injections, but even if it works on a few types of cancer, it will be an important and beneficial treatment.

*A mutant adenovirus defective in the E1B protein can be used to target cancer cells that are deficient in the p53 tumor suppressor protein.*

## Current Status of Human Gene Therapy

The inherited human diseases that have been chosen as targets for gene therapy have common features. They are diseases associated with single gene defects that have been identified at the molecular level and have a well-characterized pathology. All of these diseases are fatal and usually affect children, and most conventional therapies have proven inadequate in their treatment. At least in animal models, some success has been achieved when a normal gene is provided to cells and correction of the clinical phenotype is observed.

Human clinical trials have been less successful, and it is unclear whether any firmly documented clinical benefits have occurred. Attempts to treat CF with the CFTR gene using adenoviral vectors failed to alleviate the ion transport deficiency in the disease. Failure was attributed to the low efficiency of gene transfer to cells in vivo and to the inflammatory and immune response caused by the viral vector. The use of myoblasts to treat DMD has also been unsuccessful, although more recent approaches using completely defective adenoviral vectors carrying a mini dystrophin gene offer promise. The attempts to treat patients with FH by partial hepatectomy, transfection of hepatocytes with the gene for the LDL receptor in culture, and reinfusion of the modified cells into the portal vein of patients has likewise met with less-than-impressive results.

Although the initial trials treating SCID with a retroviral vector carrying the *ada* gene met with some success, it is still unclear how much of the improved clinical phenotype in the single patient who has shown improvement is due to the presence of the normal gene in the T cells and how much is due to the continued use of enzyme replacement therapy.

Despite the negative results with the current clinical trials, most scientists and clinical investigators remain optimistic that gene therapy will become a form of treatment for many inherited and acquired diseases. Clearly much more work is necessary to develop the ideal vectors to deliver the genes to the proper cells, to find ways to target the gene to the correct site in the genome, and to understand the biologic interactions between different human cells and foreign genes that enter them. The field of human gene therapy is still very young, and as some have suggested, it may have promised unrealistic expectations too soon [14]. Still, much has been learned from the less-than-successful trials concerning methods of transferring genes into human cells, how to achieve continued and efficient expression of the transferred genes in vivo, and the identification of the problems that must be met when dealing with a complicated organism like the human body rather than cells in culture. All of these factors will be addressed in the near future, offering the promise of eventual success in using this approach to treat different forms of human disease.

# ANIMAL MODELS FOR THE STUDY OF HUMAN DISEASE

## Transgenic Mice

*Transgenic mice with specific gene modifications serve as important animal models to study human disease.*

During the past few years, techniques have been developed for creating transgenic mouse models to study human disease. A transgenic mouse has a modified gene stably integrated into the genome of its germ cells, allowing that gene to be passed on from one generation to the next. The transgenic mice that are used as model systems today are mice in which a specific gene is altered or "knocked out." These defective mice are used to explore various effects that mutations have on development, immunity, and other metabolic processes. One approach that can be used to create a transgenic mouse is to inject DNA directly into the pronucleus of a fertilized egg. The injected DNA forms large concatemers and then randomly integrates as multiple copies into the genomic DNA. This random integration is impossible to control and results in extensive variability both in copy number and in the level of gene expression, requiring a large number of animals to be screened to find the animal of choice.

***Targeted Transgenic Mice Using Embryonic Stem (ES) Cells.*** A more recent and more reliable method developed to create transgenic mice involves inserting a single copy of a gene into a targeted site of the mouse genome by homologous recombination. The specificity of this process allows the creation of transgenic mice that harbor a corresponding mutation for any human disease gene that has been cloned. This approach became possible with the development of a method to culture pluripotent ES cells in vitro and maintain their pluripotency even after multiple cell divisions.

In this procedure, mouse ES cells are taken from the inner mass of a blastocyst of a congenic 129/SvJ strain of mice that has an agouti (brown) coat color. The ES cells are grown in culture and then transfected with a vector designed to promote recombination with homologous gene sequences in the mouse genome (Figure 14-8). The vector carries DNA sequences of the gene to be targeted along with a positive gene marker and a negative gene marker. The markers can be used in a positive–negative selection procedure, allowing the isolation of ES cells that have integrated the vector DNA into a specific site (Figure 14-9).

Any cell that incorporates the vector DNA, either by a random integration event or by homologous recombination at the targeted site, integrates the *neo* gene (the positive marker) and becomes resistant to the antibiotic G418. The G418-resistant cells are then

*ES cells can be cultured and genetically modified in vitro and then used to produce "knockout" mice that are mutant for a specific gene.*

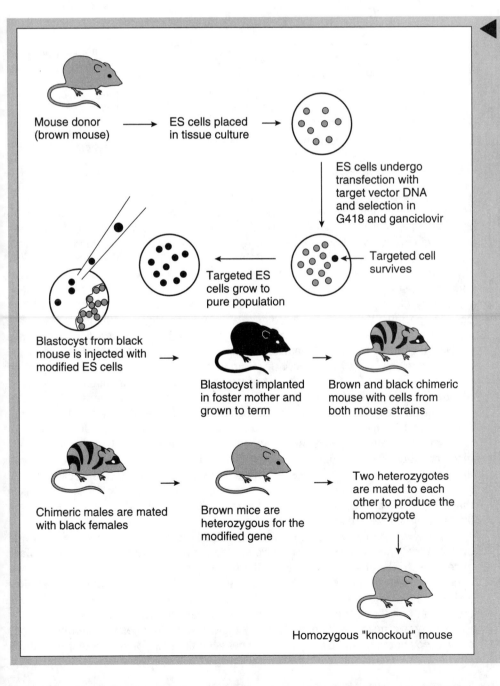

### FIGURE 14-8
**Generation of a Transgenic Mouse from Genetically Altered Embryonic Stem (ES) Cells.** *ES cells from a brown mouse are cultured and genetically modified in vitro using a targeting procedure that inactivates a specific gene. The modified ES cells are injected into a blastocyst of a mouse with a black coat color, and the blastocyst is implanted in a foster mother and grown to term. Chimeric mice with black and brown mixed coats represent mice that have developed from both cell types. Breeding experiments identify the male chimeric mice that have the genetically altered gene in the germ line because they will produce brown mice when mated with a black female. Subsequent breeding of two heterozygous mice results in a transgenic (knockout) mouse that is homozygous for the defective gene.*

Mouse donor (brown mouse) → ES cells placed in tissue culture →

ES cells undergo transfection with target vector DNA and selection in G418 and ganciclovir

Targeted cell survives

Targeted ES cells grow to pure population

Blastocyst from black mouse is injected with modified ES cells →

Blastocyst implanted in foster mother and grown to term →

Brown and black chimeric mouse with cells from both mouse strains

Chimeric males are mated with black females →

Brown mice are heterozygous for the modified gene →

Two heterozygotes are mated to each other to produce the homozygote

Homozygous "knockout" mouse

**FIGURE 14-9** ▶

*Gene Targeting Using the Positive–Negative Selection System. (A) A targeting vector contains the positive selective marker—the neo gene—located between DNA sequences of the gene to be targeted (the Y gene) and the negative marker, the gene from the herpes simplex virus (HSV) that codes for thymidine kinase, located at the ends. During the process of integration of the vector into the targeted Y gene by homologous recombination, the ends of the vector DNA are lost, and the resulting cell is defective in Y (Y−), positive for the neo gene (neo+, resistant to G418), and negative for the TK gene (TK−, resistant to ganciclovir). (B) During the process of random insertion of the vector DNA into the host chromosomes by nonhomologous recombination, the vector DNA inserts by its ends. The resulting cell is active for Y (Y+), positive for the neo gene (neo+, resistant to G418), and positive for the TK gene (TK+, sensitive for ganciclovir).*

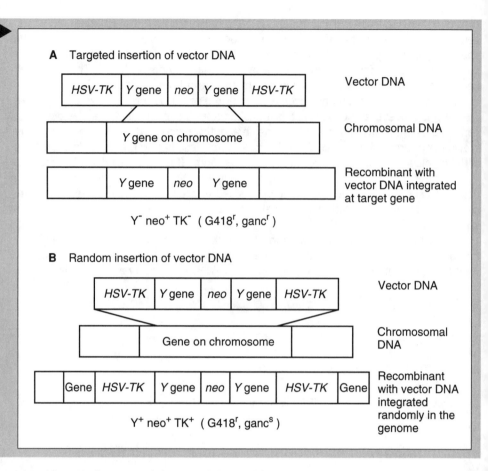

plated in a medium containing ganciclovir, which kills any cells that have integrated the HSV thymidine kinase gene (the negative marker) by random integration. Random integration, as shown in Figure 14-9B, involves a different form of recombination than homologous recombination (Figure 14-9A). When the vector DNA undergoes random integration, it inserts into the genomic DNA by the ends. The negative marker, the HSV thymidine kinase gene, is located at the ends of the vector DNA. During the process of random integration, it integrates and becomes a stable part of the genomic DNA. Conversely, during the process of homologous recombination, the ends of the vector DNA containing the HSV thymidine kinase gene are not integrated into the genome and are lost. Thus, a cell that undergoes homologous recombination at the target site is negative for HSV thymidine kinase and ganciclovir-resistant, whereas a cell that has undergone random insertion is positive for HSV thymidine kinase and sensitive to the killing action of ganciclovir. The use of the positive–negative selection procedure allows the isolation of the one cell in thousands that has integrated a specific gene sequence at its target site.

Once an appropriate ES cell is isolated by the positive–negative selection procedure and is shown to have a specific gene inactivated, the cell is amplified in culture to produce a population of genetically modified ES cells (see Figure 14-8). The genetically modified ES cells are microinjected into a blastocyst derived from a mouse line that has a different coat color (e.g., the C57BL/6J mouse or black coat mouse). The black coat color gene is recessive to the agouti coat color gene, allowing one to visualize chimeric mice that arise. The blastocyst is then implanted in a foster mother with a black coat, and the offspring are allowed to grow to term.

Any newborn mice with a chimeric or mixed coat color of brown and black provide evidence that the modified ES cells have survived and proliferated in the mouse. Black mice indicate the absence of the ES cells. It still is necessary to identify the chimeric mice that have the targeted mutation in the germ cells. This identification requires mating male chimeric mice to female black mice and looking for brown mice that are produced when the sperm is derived from the original genetically modified ES cells.

The brown mice are then analyzed by molecular techniques to determine if they carry a copy of the mutated gene at the proper genomic site. For genes other than those located on the X chromosome, only one allele of a gene is inactivated by the recombinational event that occurs between the vector DNA and the host chromosome. Selection of mice that have the second allele inactivated occurs during the subsequent breeding process. Matings carried out between mice that are heterozygous for the mutated gene produce offspring that are homozygous for the mutated gene at a 25% frequency. These homozygous "knockout mice" now serve as model systems to study the phenotypic consequences of a mutation in a specific gene. This technique, which may require approximately 1 year to complete, has been used to produce hundreds of transgenic mice with defects in many different corresponding human disease genes.

*Important Examples of Transgenic Mouse Models.* Many of the inherited human diseases discussed in this book have "knockout mice" available to study as model systems. Some of these are listed in Table 14-3 [15]. Mouse models are used to study the function of a disease gene in an otherwise normal genetic background, to understand the pathophysiology of the disease associated with a specific gene mutation and to provide models to develop gene therapy protocols for alleviating the clinical phenotype of a disease. There are, however, problems associated with some of the model systems because a mouse, after all, is not a human.

Mice may have alternative biochemical and developmental pathways that sometimes influence the expression of the transgene. Furthermore, the targeting procedure used to develop the transgenic animals provides a mouse model in which the mutation produced is a loss-of-function mutation resulting from the endogenous gene function being destroyed by the transgene integration. Such models exclude the study of diseases that result from gain-of-function mutations (e.g., the human diseases that result from trinucleotide expansion within a gene) like myotonic dystrophy (a trinucleotide expansion in the 3' untranslated region of the gene) or Huntington's disease (a trinucleotide expansion in the 5' translated region of the disease (see Chapter 11). With current

**TABLE 14-3**
Examples of Mouse Models for the Study of Human Diseases

| Human Disease | Protein Altered | Mouse Gene |
|---|---|---|
| Osteogenesis imperfecta | Collagen, alpha 1 | *Cola 1* |
| Achondroplasia | Fibroblast growth factor receptor 3 | *Fgfr3* |
| Alzheimer's disease | Amyloid beta precursor | *App* |
| Ataxia telangiectasia | Phosphoinositol kinase | *Atm* |
| Myotonic dystrophy | DM kinase | *DM15* |
| Duchenne's muscular dystrophy | Dystrophin | *Dma (mdx)* |
| Fragile X syndrome | FMR1 | *Fmr 1* |
| Huntington's disease | Huntingtin | *Hdh* |
| Acute intermittent porphyria | Porphobilinogen deaminase | *Pbgd* |
| Adenomatous polyposis coli | APC | *Apc* |
| Breast cancer, type 1 | BRCA1 | *Brca 1* |
| Malignant melanoma | Cyclin-dependent kinase inhibitor p16 | *Cdkn2A* |
| Hereditary nonpolyposis colorectal cancer | hMSH2, hMLH1, hPMS2 | *Msh2, Mlh1, Pms2* |
| Neurofibromatosis 1 | Guanosine triphosphatase | *Nf1* |
| Retinoblastoma | Rb protein | *Rb1* |
| Li-Fraumeni syndrome | p53 | *Trp53* |
| Severe combined immunodeficiency syndrome | Adenosine deaminase | *Ada* |
| α-Thalassemia | Hemoglobin alpha | *Hba* |
| β-Thalassemia | Hemoglobin beta | *Hbb* |
| Cystic fibrosis | CFTR | *Cftr* |
| Gaucher's disease, type I | Glucocerebrosidase | *Gba* |
| Tay-Sachs disease | Hexokinase A | *Hexa* |
| Lesch-Nyhan syndrome | Hypoxanthine-guanine phosphoribosyltransferase | *Hgprt* |
| Familial hypercholesterolemia | Low-density lipoprotein receptor | *Ldlr* |
| Phenylketonuria | Phenylalanine hydroxylase | *Pah* |

*Source:* Bedell MA, Largaespada DA, Jenkins NA, et al: Mouse models of human disease. *Genes Dev* 11:11–43, 1997.

technology, these mutations cannot be duplicated in a simple knockout model. Mouse models are available, however, that have the DM kinase gene (dystrophy myotonia) or the huntingtin gene (Huntington's disease) inactivated.

In addition to using knockout mice as important tools to study diseases in which the pathophysiology is well understood in humans (e.g., the thalassemias), knockout mice also offer a unique approach to identify unknown biochemical functions of genes. This is a particularly attractive approach to identify the molecular basis of protein actions coded for by tumor suppressor genes (e.g., the gene for p53), the breast cancer genes (i.e., BRCA1 and BRCA2), and genes involved in regulating the cell cycle (e.g., the cyclin-dependent kinase inhibitor, p27). The availability of both heterozygous and homozygous knockout mouse strains helps the investigator establish the function of these tumor suppressor proteins in cellular metabolism and to determine their role in preventing cellular proliferation and cancer.

Recently, knockout mice defective in the breast cancer gene BRCA1 have been created [16]. In contrast to humans homozygous for a BRCA1 mutation, mice that are homozygous for a mutation in BRCA1 die early in embryogenesis. The heterozygous mice do not die, but they also do not develop breast cancer. Despite these differences between the human phenotype and the mouse phenotype, the mouse model has been valuable in relating the function of the BRCA1 protein to the action of p53 and p21, which are protein regulators of the cell cycle. Data from the mice also support a role for the BRCA1 gene product as a transcription regulator and as a factor involved in DNA repair processes.

Mice with a disruption in the gene that codes for p27 have been very informative in elucidating the role of p27 in controlling cellular proliferation (see Chapter 5). Mice lacking the p27 protein show an enhanced cellular proliferation, increased body size, multiple organ hyperplasia, and pituitary tumors. The levels of CDK2—the cyclin-dependent kinase normally inhibited by the p27 protein in humans—are elevated approximately tenfold in the p27 knockout mice, which is consistent with the increased cellular proliferation and loss of cell-cycle control seen in the absence of a functioning p27 protein.

Other very important mouse models that appear to mimic human diseases are the ΔF508 CF mouse [17]; the *mdx* mouse, which is a model for DMD [18]; and the *Atm*-deficient mouse, which is a model for the study of ataxia telangiectasia [19].

The lethality of some of the knockout mice in the homozygous state, such as seen with the BRCA1 mouse, often limits the investigation of certain mutations in the pathophysiology of a disease or in development. New approaches aim to control the expression of the knockout gene, which would allow the investigator to turn on or off the expression of the transgene or to limit the knockout of the gene to certain tissues to avoid the lethality. Although more precise methods of targeting genes to allow exact modifications of DNA sequences are necessary to extend the mouse models to gain-of-function mutations, the knockout mouse will continue to be a valuable tool for the study of human disease.

## Cloning of Mammals from Adult Somatic Cells

A very different type of approach to create animal models has recently been reported; that is, the ability to clone a mammal from adult DNA [20]. This surprising accomplishment has sparked the interest of everyone not only for its scientific achievement but also for the many ethical questions it raises. Although species such as frogs, mice, cattle, and monkeys can be cloned by using the nuclei from embryonic cells (Figure 14-8), no one had been able to clone a mammal with DNA taken from adult cells. The problem in using adult DNA to clone an animal appears to be related to the irreversible changes that occur in DNA during development. These changes result in a DNA that is no longer capable of supporting the differentiation of different cell types.

In an attempt to overcome this irreversibility, Wilmut and colleagues removed cells from mammary tissue and blocked the adult cells in the G1 phase of the cell cycle (Figure 14-10). The block causes many genes to be turned off and many proteins to be removed from the DNA. When the blocked mammary cells were fused with an enucleated egg cell, the adult DNA received a signal to bypass the G1/S border and enter into the S phase of the cell cycle, where the DNA begins to replicate using the cytoplasmic proteins provided by the egg cell rather than the old proteins previously attached to the adult DNA. The early lack of gene expression when the cells are blocked in G1 is believed to reprogram

*Adult DNA can be used to clone a mammal, providing that the DNA is reprogrammed to undergo accurate differentiation.*

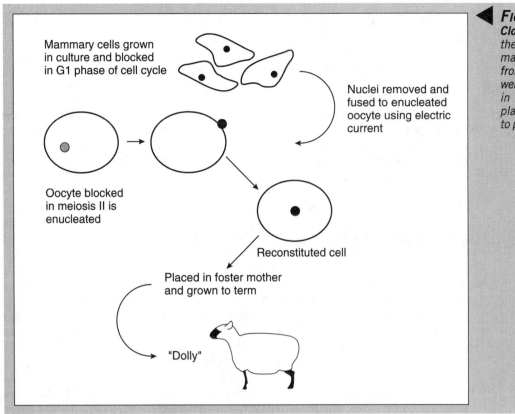

**Cloning of a Mammal from Adult DNA.** *In the first successful experiment to clone a mammal from adult DNA, nuclei derived from mammary cells of a 6-year-old ewe were fused to an enucleated oocyte blocked in meiosis II. The reconstituted cell was placed in a foster mother and grown to term to produce a viable lamb named Dolly [20].*

the DNA in some unknown manner so that it is capable of undergoing a normal development from an embryo to an adult (see Figure 14-10).

The efficiency of this cloning process was very low, with only one reported success (i.e., the sheep Dolly) in 277 attempts. The lack of understanding of the basic processes of developmental programming—or in this case deprogramming—will undoubtedly continue to contribute to the low success of this technique. Still, the promise of success offers unique possibilities for creating identical animal models without variation in genetic background to study human diseases, as well providing a method for creating farm animals that can produce more meat, wool, or milk.

As might be expected, the successful experiment received excessive media coverage and has raised serious concerns about the prospects of cloning human beings from adult DNA. Even politicians have been stimulated to develop legislation and ban the cloning of human beings. Scientists continue to inform the public that the success of this one experiment does not mean it is possible to create carbon copies of humans. Cloning humans is highly unlikely because of the strong influence of other factors besides our genetic makeup (e.g., emotional and cultural factors) that are involved in creating people. Most scientists suggest, however, there is no a priori reason why the technique should not work with humans; however, is there any justification for cloning humans? At the current time, the low rate of success probably precludes using this procedure to clone a human being. However, in the future, with improved technology, there is no doubt that the ethics of human cloning will be debated at all levels.

# GERM-LINE GENE MODIFICATION FOR THE PREVENTION OF INHERITED HUMAN DISEASES

All of the current studies on gene therapy as a means of treating human diseases involve the modification of somatic cells, which represent all cells in the body with the exception of the germ cells (i.e., the sperm, the egg). Somatic gene therapy does not differ in principle from other conventional therapies, such as organ transplantation, and only the patient who has consented to this therapy is directly affected. Currently, only patients

*Current technology is not applicable to germ-line modification of humans.*

who have life-threatening diseases and who do not respond to other conventional treatments are candidates for human gene therapy trials.

Modification of human germ-line cells adds an entirely new dimension to gene therapy. Currently, this is not practiced and probably is not likely to occur in the near future. With current technology, germ-line gene modification is not generally feasible because it requires replacing a defective gene with a functioning gene rather than simply adding the functioning gene, which is the approach now taken in somatic-cell gene therapy. Gene addition in the germ line is not acceptable because of the unknown effects that extra DNA could have on the germ cell during development. Therefore, a much more sophisticated and more controlled way of transferring, targeting, and regulating exogenous genes is required before human germ-line modification becomes a reality.

What types of clinical situations would justify germ-line gene modification? One case would be the unusual situation in which two individuals who are homozygous for the same genetic disorder wish to have a normal child. This situation is not the norm because the homozygous state for any inherited disease is rare and, in many cases, the affected individuals do not live long enough to reproduce. For this example, though, the assumption is made that somatic gene therapy becomes a standard treatment, and these individuals live long enough to have children. Although some of their somatic cells have been cured of the disease phenotype, their germ cells still carry the genetic defect, and all of their children would probably be affected with the disorder. Of course, their children could also be treated by somatic cell gene therapy, but, sooner or later, germ-line gene modification would become a possible option.

A second example involves two individuals who are heterozygous for the same genetic disorder. In most cases of single gene disorders, only 25% of the children are affected with the disorder. With 75% of the children predicted to be normal, there are alternative approaches available to identify the homozygous conception, making it less necessary to have the very expensive and more risky germ-line gene therapy to guarantee a normal child.

These scenarios and others like them continue to raise scientific debate on the medical and ethical issues surrounding germ-line modification [21]. Some of the arguments for and against germ-line modification are listed in Table 14-4. As the field of gene therapy continues to advance, more arguments on both sides will most likely develop.

**TABLE 14-4** ▶

*Ethical Arguments in Favor and against Germ-Line Gene Therapy*

**Arguments in Favor**
- The health profession has a moral obligation to use the best available method of treatment.
- Parents should have access to available technologies for purposes of having a healthy child.
- Germ-line modification would be more efficient than continued somatic cell gene therapy for inherited diseases.
- There should be freedom of scientific inquiry and the pursuit of new knowledge.

**Arguments against**
- The procedure is an expensive intervention with limited applicability.
- Alternative methods are available for preventing genetic diseases.
- There are unavoidable risks and irreversible mistakes.
- Pressures to use germ-line modification for enhancement will increase.

*Source:* Wivel NA, Walters L: Germ-line gene modification and disease prevention: some medical and ethical perspectives. *Science* 262:533–538, 1993.

# HUMAN GENOME PROJECT

*The Human Genome Project is designed to determine the entire base sequence of human DNA as well as that of a number of nonhuman organisms.*

The Human Genome Project is an international project designed to develop detailed genetic and physical maps of the human genome and to determine the entire nucleotide sequence of human DNA. The availability of the complete base sequence of human DNA will aid in identifying and studying the estimated 60,000 to 80,000 human genes.

Additionally, the project aims to determine the entire nucleotide sequence of the DNA of nonhuman organisms, including prokaryotes, yeast, worms, flies, the mouse, the rat, and plants.

## Types of Physical Maps in the Human Genome Project

Several different approaches are being used to construct the physical map of the human genome. Early physical maps used restriction enzymes and the natural polymorphisms within the human DNA to construct a human genetic map based on restriction fragment length polymorphisms (RFLPs). With the advent of chromosomal banding techniques, the studies on somatic cell hybrids, and FISH, a cytogenetic map was established, providing the initial framework for the identification and placement of individual genes on the 24 human chromosomes (see Chapter 7).

A major advance in human gene mapping came with the development of the polymerase chain reaction (PCR) [see Chapter 6]. This technique, which uses two short synthetic primers, allows the amplification and the isolation of a particular segment of DNA, starting from very small amounts of DNA. PCR has been instrumental in developing specific landmarks along chromosomes called sequence-tagged sites (STSs). STSs are short unique sequences of DNA that can be amplified by the PCR technique. These sequences can include microsatellite repeats or known sequences of cloned genes. Multiple STSs are placed on the gene map with respect to one another to form a STS content map. This is done by determining if two STSs are present or absent on the same yeast artificial chromosome (YAC) clone or the same DNA fragment present in a radiation hybrid (see Chapter 7). The first human map of the human genome based on STSs was completed in 1995 and had 15,086 sites placed on the 24 chromosomes, with an average spacing of 199 kb between them [22].

*Expressed-Sequence Tag (EST) or Transcript Map of the Human Genome.* The human genome is made up of $3 \times 10^9$ bp of DNA, only 3% of which contain coding information for RNA and protein. Some investigators question the wisdom of randomly sequencing the entire genome and suggest it would be more beneficial to focus on the 3% of the DNA that makes up the coding sequences. Thus, an approach has been developed to create cDNA libraries that represent all of the expressed genes in human DNA. These ESTs are used to develop a transcript map of the human genome. ESTs are similar to STSs except they represent a DNA sequence that is transcribed into messenger RNA (mRNA) rather than any base sequence present in the DNA.

EST DNA libraries are made by isolating total mRNA from individual tissues and using reverse transcriptase to prepare cDNA sequences complementary to the mRNA. The cDNA fragments are amplified by PCR and then sequenced by large-scale automated sequencing. The EST sequence fragments are clustered into groups representing distinct genes because a single gene may be represented by multiple ESTs. Using the 3' untranslated region of genes has simplified the ability to obtain unique fragments for each gene because these DNA sequences do not contain introns, and they tend to be less conserved than translated exon sequences. The sequence of each EST is compared to the known sequences of genes present in various data banks and then assigned to a specific gene. The ordered mapping of ESTs is carried out in a similar manner as described above for STSs by determining the togetherness of two sequences on the same YAC clone or radiation hybrid fragment. In October 1996, the gene map of the human genome, which represented the mapping of 20,104 STS and EST fragments, was published [23].

The Institute for Genomic Research headed by Craig Venter has been instrumental in developing EST cDNA libraries. As of May 1997, this institute has generated 300 cDNA libraries constructed from 37 distinct organs and tissues. From these libraries, nearly 175,000 ESTs, which represent 52 million nucleotides of human DNA sequence and approximately 80% of all of the human genes, have been generated.

*Approximately 80% of all human genes are represented in EST cDNA libraries.*

*Genome Maps of Nonhuman Organisms.* In addition to sequencing the human genome, the DNA genomes of a number of other model organisms are being or have been sequenced, some of which are listed in Table 14-5. Already completely sequenced are the genomes of *Mycoplasma genitalium, Methanococcus jannaschii, Haemophilus influ-*

**TABLE 14-5** ▶

*Genome Size and Approximate Number of Genes in Organisms Expected to Have Their Genome Sequenced*

| Organism | Genome Size (megabases) | Predicted Number of Genes |
|---|---|---|
| *Mycoplasma genitalium* | 0.58 | 470 |
| *Methanococcus jannaschii* | 1.66 | 1739 |
| *Haemophilus influenzae* | 1.83 | 1727 |
| *Escherichia coli* | 4.6 | 4100 |
| *Saccharomyces cerevisiae* | 12.1 | 5885 |
| *Caenorhabditis elegans* | 100 | 14,000 |
| *Drosophilia melanogaster* | 165 | 12,000 |
| *Fugu rubripes* | 400 | 70,000 |
| *Mus musculus* | 3000 | 70,000 |
| *Rattus novegicus* | 3000 | 70,000 |
| *Homo sapiens* | 3000 | 70,000 |

enzae, *Escherichia coli*, and *Saccharomyces cerevisiae*. Information from the DNA sequences of these organisms will be useful in understanding both the evolutionary relationships and the functions of many genes. The DNA sequence of the bacterium *H. influenzae* already has provided some interesting new insights into the pathogenicity of this organism with the identification of new virulence genes.

Even with a complete knowledge of the human DNA base sequence, the amino acid structure and the function of human gene products coded for by the DNA cannot be directly deduced. However, conservation of important pathways is seen between yeast and humans, as well as between flies and humans. Because flies and yeast offer excellent genetic systems to analyze the role of gene products in biochemical pathways, the job of placing the 70,000 human genes in some biochemical order should be simplified once the complete DNA sequence of nonhuman organisms is known [24].

Additionally, yeast, mice, and flies are useful organisms to carry out gene transfer experiments. The knowledge of the complete DNA base sequence of these organisms may offer new approaches to achieve accurate gene targeting and may provide model systems to study gene control and function. In any case, the information gained from knowing the DNA sequences of humans and other organisms promises to open up new avenues of treatment for human genetic diseases.

Molecular diagnosis of human diseases is expected to accelerate with the knowledge of the complete DNA base sequence in humans. Once disease genes are identified and sequenced, a large number of mutations associated with the clinical pathology of the disease will be identifiable. All of these mutations should be readily detected in a clinical setting with the use of PCR.

## Ethical, Legal, and Social Issues Related to the Human Genome Project

*Knowledge of the complete nucleotide sequence of human DNA will have many medical benefits but will also bring associated risks to patients, their families, and the medical profession.*

There are obvious potential benefits to knowing the complete DNA sequence of the human genome, as well as to knowing the DNA sequence of other model organisms. This knowledge, however, does not come without associated risks from unanticipated consequences of having such information. With the cloning of human genes and the identification of mutations associated with human disease, new diagnostic tests are becoming available to the public to identify those at risk for a particular disease. The availability of routine testing for predispositions to diseases certainly will have a strong impact on patients, their families, and their physicians.

How much does an individual want to know about his or her genetic makeup? Who should have access to information regarding a person's genetic heritage and his or her potential risk of disease? These important questions and others regarding the ethical, social, and legal issues surrounding the human genome project were recognized early in the development of the project. For this reason, The Ethical, Legal, and Social Implications (ELSI) group was formed as part of the Human Genome Project to study these issues. The

ELSI group consists of geneticists, ethicists, lawyers, sociologists, and theologians and has focused its attention in three main areas. The first is to look at the impact of the Human Genome Project on introducing new clinical testing into medical practice. The second is to determine who will have access to a person's genetic makeup, and the third is to define ways to educate both medical personnel and the public about genetic issues [25].

*Issues Involving Genetic Testing.* The questions surrounding clinical testing for human genetic diseases has become a very highly debated issue with the introduction of commercially available genetic tests for detecting mutations in the breast cancer genes *BRCA1* and *BRCA2* [26]. These mutations are implicated in 80% of cases of hereditary early-onset breast and ovarian cancer (see Chapter 11). In fact, these mutations account for less than 10% of the cases of sporadic breast cancer found in the general population. The debate is not whether these prepackaged tests can detect a mutation in these genes, but what the results mean and how they will affect the individual, her family, and her life. Because there is no guarantee that a negative result means a woman will not get breast cancer or that a positive result means she will develop breast cancer, should the test be available for everyone? The individual who plans on taking the test should be encouraged to seek genetic counseling. Genetic counselors are trained to explain both the benefits and the disadvantages associated with the test and to counsel the individual on how she should interpret the results and deal with them.

Another recent debate involving genetic testing surrounds a recommendation that all couples wishing to have children undergo screening for a CF gene mutation. The current screening test for CF can detect 90% of the potential CF mutations. Approximately 25,000 Americans have CF, and nearly 850 children are diagnosed with it each year. Treatments have improved to extend the life of an individual with the disease, but death still occurs, usually in the third decade of life. Routine screening of newborns is not encouraged because there is no therapeutic value to do so without an effective cure for the disease. The argument for offering couples a screening test is to provide the potential parents with some experimental data to allow them to make an informed consent about their decision to have children.

As more and more clinical tests for detecting human disease genes become available, decisions about whether or not to have the test will have an impact on family planning, family relationships, and confidentiality of a person's genetic composition, as well as raise social and legal questions. All of these matters are currently being addressed by the ELSI group, university committees, and biotechnology companies, and the debate will undoubtedly continue as the completion of the Human Genome Project approaches.

*Genetic testing will continue to raise important ethical, social, and legal issues regarding the confidentiality and the use of such information.*

Recently, the federal government issued some guidance dealing with genetic discrimination in the workplace [27]. As of 1995, the Americans with Disabilities Act (ADA) protects individuals subjected to discrimination on the basis of genetic information relating to illness, disease, or other disorders. In 1996, Congress passed the Health Insurance Portability and Accountability Act. This Act prohibits a group health insurance plan from using genetic information to prevent an individual from being eligible or from continuing to be eligible for insurance. It does not, however, deal with increasing the cost of insurance or making a blanket exclusion of coverage for a particular condition [27].

At the present time, there are some limited safeguards against discrimination for genetic disease, but there is no firm policy on who has access to genetic information, and additional protection is needed for everyone. Recent recommendations by the Hereditary Susceptibility Working Group of the National Action Plan on Breast Cancer (NAPBC) and the ELSI group to state and federal policymakers have addressed this issue in more detail in an effort to develop a comprehensive approach to the problem of genetic privacy and discrimination in the workplace [27].

# SUMMARY

The role of the physician and other health care professionals in genetic testing is becoming increasingly important. They must be able to interpret the new molecular-based tests, explain the results to patients, and discuss how the results will affect the patients and their families. Education of the public is needed so that there is an understanding of the advantages, the limitations, and the risks associated with genetic testing.

By the year 2005, or sooner, the Human Genome Project is expected to be completed, and society will need to be able to handle the new genetic information available to it. Nearly every human disease gene should be identified and isolated by that time. Continued studies and advancements in the field of somatic-cell gene therapy may provide new ways to treat and even cure some of the diseases caused by mutations in these genes. The era in which the study of human genetic diseases will be greatly accelerated and medical molecular genetics will become a fundamental base in the treatment of human disease is about to be entered.

# REVIEW QUESTIONS

**Directions:** For each of the following questions, choose the **one best** answer.

1. Which one of the following vectors used in gene therapy studies stably integrates into the host genome?

    **(A)** Adenoviral vectors

    **(B)** Cationic liposomes

    **(C)** Retroviral vectors

    **(D)** Adenovirus-polylysine-DNA complexes

    **(E)** Human artificial chromosomes

2. Which one of the following vectors used in gene therapy studies elicits an immune response in the host?

    **(A)** Adenoviral vectors

    **(B)** Cationic liposomes

    **(C)** Retroviral vectors

    **(D)** Adenovirus-polylysine-DNA complexes

    **(E)** Human artificial chromosomes

3. Which one of the following vectors can provide stable gene expression in a transfected cell although the vector remains episomal?

    **(A)** Adenoviral vectors

    **(B)** Cationic liposomes

    **(C)** Retroviral vectors

    **(D)** Adenovirus-polylysine-DNA complexes

    **(E)** Human artificial chromosomes

4. Transgenic mice models are associated with which one of the following characteristics?

    **(A)** They are excellent models to study gain-of-function mutations

    **(B)** They consistently reflect the phenotype seen with the corresponding human disease

    **(C)** They are excellent models to study loss-of-function mutations

    **(D)** They are created by using adult DNA

    **(E)** They cannot be used to study a gene with an unknown metabolic function

5. Targeted insertion into the human $Y$ gene by homologous recombination with a vector DNA molecule that carries the *neo* gene in the middle of the $Y$ gene sequences and the HSV thymidine kinase (*TK*) gene at the ends results in a cell with which one of the following genotypes?

    **(A)** $Y+$, *neo*−, *TK*−

    **(B)** $Y+$, *neo*+, *TK*+

    **(C)** $Y−$, *neo*+, *TK*−

    **(D)** $Y+$, *neo*−, *TK*+

# ANSWERS AND EXPLANATIONS

**1. The answer is C.** Retroviruses make a DNA copy of their RNA genome and integrate randomly into the host genome, resulting in stable gene expression. Adenoviral vectors, cationic liposomes, adenovirus-polylysine-DNA complexes, and human artificial chromosomes remain episomal.

**2. The answer is A.** In studies using adenoviral vectors to treat cystic fibrosis, the patients receiving the adenoviral vector carrying the CFTR gene had an immune response to the viral proteins. Additionally, a local inflammatory reaction occurred. This adverse reaction in the patients makes it impossible to use repeated administrations of the virus to correct the genetic defect. Cationic liposomes, retroviral vectors, adenovirus-polylysine-DNA complexes, and human artificial chromosomes have not elicited an adverse immune response in patients.

**3. The answer is E.** Human artificial chromosomes have centromere and telomere sequences and DNA replication origins, which allows them to remain stable in cells although they are episomal and not integrated into one of the normal cellular chromosomes. Adenoviral vectors, cationic liposomes, and adenovirus-polylysine-DNA are episomal but are only transiently retained and expressed in a cell. Retroviral vectors give stable gene expression but are integrated into the host genome.

**4. The answer is C.** The general procedure used to create a transgenic mouse is to modify embryonic stem cells in vitro and inactivate a particular gene, resulting in a loss-of-function mutation. Gain-of-function mutations have not as yet been reproduced in transgenic mice models. Reflection of the phenotype of a human disease often is seen, but differences in the mouse can result in a varied phenotype in the presence of a specific mutation. Transgenic mice are created from embryonic cells, not adult DNA. Transgenic mouse models are a very good approach to determine an unknown function of a gene, such as with tumor suppressor genes.

**5. The answer is C.** Targeted integration inserts the vector DNA into the *Y* gene, making it *Y*−. If insertion is by homologous recombination, the ends of the vector are not integrated, and the cell will be *TK*− and *neo*+. Any cell that is *TK*+ (options B and D) would be the result of random insertion because the ends of the vector carry the *TK* gene and are inserted only by nonhomologous recombination. Option A is incorrect because the genotype would indicate that the vector did not integrate into the genome of the cell.

# REFERENCES

1. Morgan RA, Anderson WF: Human gene therapy. *Ann Rev Biochem* 62:191–217, 1993.
2. Crystal RG: Transfer of genes to humans: early lessons and obstacles to success. *Science* 270:404–410, 1995.
3. Gao X, Huang L: Cationic liposome-mediated gene transfer. *Gene Ther* 2:710–722, 1995.
4. Curiel D: Adenovirus facilitation of molecular conjugate-mediated gene transfer. *Prog Med Virol* 40:1–18, 1993.
5. Salmons B, Gunzburg W: Targeting of retroviral vectors for gene therapy. *Hum Gene Ther* 4:129–141, 1993.
6. Harrington J, Van Bokkelen G, Mays R, et al: Formation of de novo centromeres and construction of first-generation human artificial chromosomes. *Nat Genet* 15:345–355, 1997.
7. Capecchi M: Targeted gene replacement. *Sci Am* 271:52–59, 1994.

8. Capecchi M: Altering the genome by homologous recombination. *Science* 244: 1288–1292, 1989.

9. Blaese R, Culver K, Miller A, et al: T lymphocyte-directed gene therapy for ADA-SCID. Initial trial results after 4 years. *Science* 270:475–480, 1995.

10. Knowles M, et al: A controlled study of adenoviral-mediated gene transfer in the nasal epithelium of patients with cystic fibrosis. *N Eng J Med* 333:823–831, 1995.

11. Haecker S, Stedman H, Balice-Gordon R, et al: In vivo expression of full-length human dystrophin from adenoviral vectors deleted of all viral genes. *Hum Gene Ther* 7:1907–1914, 1996.

12. Grossman M, et al: A pilot study of ex vivo gene therapy for homozygous familial hypercholesterolemia. *Nat Med* 1:1148–1154, 1995.

13. Bischoff J, Kirn D, Williams A, et al: An adenovirus mutant that replicates selectively in p53-deficient cells. *Science* 274:373–376, 1996.

14. Friedman T: Human gene therapy—an immature genie, but certainly out of the bottle. *Nat Med* 2:144–147, 1996.

15. Bedell M, Largaespada D, Jenkins N, et al: Mouse models of human disease Part II: Recent progress and future directions. *Gene Dev* 11:11–43, 1997.

16. Hakem R, de la Pompa J, Sirard C, et al: The tumor suppressor gene *Brcal* is required for embryonic cellular proliferation in the mouse. *Cell* 85:1009–1023, 1996.

17. Colledge W, Abella B, Southern D, et al: Generation and characterization of a delta F508 cystic fibrosis mouse model. *Nat Genet* 10:445–448, 1995.

18. Barlow C, Hirotsune S, Paylor R, et al: Atm-deficient mice: a paradigm of ataxia telangiectasia. *Cell* 86:159–171, 1996.

19. Fero M, Rivkin M, Tasch M, et al: A syndrome of multiorgan hyperplasia with features of gigantism, tumorigenesis, and female sterility in p27 (kip1)-deficient mice. *Cell* 85:733–744, 1996.

20. Wilmut I, Schnieke A, McWhir J, et al: Viable offspring derived from fetal and adult mammalian cells. *Nature* 385:810–813, 1997.

21. Wivel N, Walters L: Germ-line gene modification and disease prevention: some medical and ethical perspectives. *Science* 262:533–538, 1993.

22. Hudson T, Stein L, Gerety S, et al: An STS-based map of the human genome. *Science* 270:1945–1954, 1995.

23. Schuler G, Boguski E, Steware L, et al: A gene map of the human genome. *Science* 274:540–546, 1996.

24. Miklos G, Rubin G: The role of the genome project in determining gene function: insights from model organisms. *Cell* 86:521–529, 1996.

25. Guyer M, Collins F: How is the human genome project doing, and what have we learned so far? *Proc Natl Acad Sci* 92:10841–10848, 1995.

26. Brower V: Testing, testing, testing? *Nat Med* 3:131–132, 1997.

27. Rothenberg K, Fuller B, Rothstein M, et al: Genetic information and the workplace: legislative approaches and policy changes. *Science* 275:1755–1757, 1997.

# INDEX

NOTE: An f after a page number denotes a figure; a t after a page number denotes a table; an s after a page number denotes a sidebar.